GEOLOGY OF THE ARCTIC

Volume II

GEOLOGY OF THE ARCTIC

Proceedings of the First International Symposium on Arctic Geology

HELD IN CALGARY, ALBERTA
JANUARY 11–13, 1960
UNDER THE AUSPICES OF THE
ALBERTA SOCIETY OF
PETROLEUM GEOLOGISTS

Volume II
Editor: Gilbert O. Raasch

UNIVERSITY OF TORONTO PRESS

University of Toronto Press

Diamond Anniversary 1961

CONTENTS OF VOLUME II

Section Two: Glaciology, Permafrost, Climatology, Geomorphology, etc.

Section Three: Logistics and Exploration

NOTE

Figures too large to be accommodated in the text page may be found in the separate container which accompanies these volumes. References to these figures in the text are marked with an asterisk.

Section Two

GLACIOLOGY, PERMAFROST,

CLIMATOLOGY, GEOMORPHOLOGY, ETC.

Danish Glaciological Investigations in Greenland

BÖRGE FRISTRUP

ABSTRACT

From the Geographical Department, University of Copenhagen, glaciological investigations were carried out in Greenland in the years 1956-8 as part of the Danish programme for the IGY. Four specially selected glaciers in North, South, West, and East Greenland were studied in relation to glacial morphology, glacial meteorology, and oscillations in relation to change of climate.

A very broad geographical variation was found especially concerning the temperature in the ice. The glaciers in northern Greenland are real polar, while the glaciers in most parts of southern and central Greenland are temperate, according to the definition by Hans Ahlmann, with temperatures very near zero during most of the year. Therefore the glaciers in southern Greenland will also be very sensitive to increasing temperature, while the glaciers in north Greenland react only very slowly to rise in temperature. In accordance with this it has been found that the glaciers in Thule district started to retreat later than in south Greenland. Of special importance is the great area of superimposed ice.

AS PART OF THE DANISH CONTRIBUTION to the IGY, glaciological investigations were carried out in Greenland by an expedition organized by the Geographical Institute at the University of Copenhagen. Four specially selected glaciers were investigated. The ratio of ablation/accumulation in relation to climate and microclimate was studied. Special studies were made at the glacier front to determine the oscillations in relation to change of climate. Also, the glacier fronts were surveyed and fix points established which could be resurveyed in future years. All the investigated glaciers are local glaciers, outside the Greenland ice-cap, as it was assumed that they would be more sensitive to change of climate. Careful planning and reconnaissance ensured that the glaciers would be representative of the different geographical regions of Greenland with respect to position and exposure to the prevailing wind, and that they all belonged to a morphological glacier type which was characteristic for the region. The four glaciers investigated were: Hurlbut Gletscher in Thule district, 77°23′30″ lat. N, 67°57′ long.; Sermikavsak Gletscher in the Umanak district, 71°11′30″ lat. N, 53°03′ long. W; Sermikavsak suaq Gletscher on Sermersoq Ø in the Julianehåb district, 60°18′ lat. N, 45°13′ long. W, and Mitdluagkat Gletscher, in the Angmagssalik district, 65°40′40″ lat. N, 37°54′ long. W (Figure 1).

THE MORPHOLOGICAL CHARACTERISTIC OF THE GLACIERS

Napassorssuaq Gletscher is situated on Semersoq Island in southern Greenland. The physiography of the region here is alpine, with numerous high mountains, and is strongly glaciated. The valley glacier is the dominant glacier form and most of

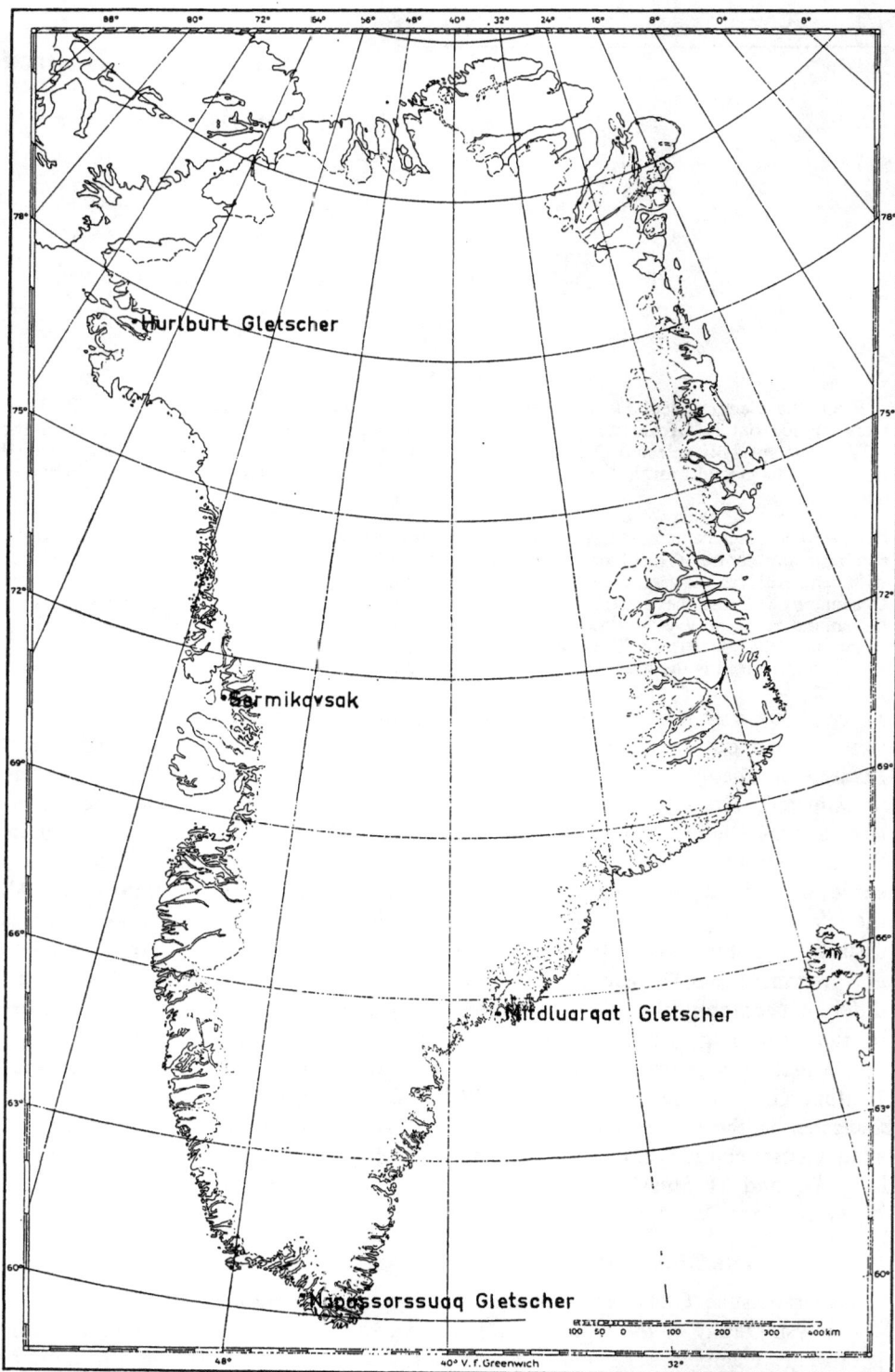

FIGURE 1. The investigated glaciers.

the glaciers are rather small, but frequently have very steep tongues and ice falls. Napassorssuaq Gletscher is 3 km long and 500 to 900 m wide; the highest point of the glacier is 1,125 m above sea-level, and from here the glacier descends to a lake 495 m above sea-level. The glacier front is nearly vertical, 12-16 m above the water surface; it is not floating, and there is no production of icebergs. The total area of the glacier is 2.1 sq. km. According to Ahlmann's classification, based on the height intervals, the glacier is a transition between the valley type and the piedmont type (see Figure 2). It is evident from the general morphology that Napassorssuaq is not a normal valley glacier, the reason being that the lateral parts of the glacier front and lower part of the glacier are formed of dead ice partly covered by debris and moraines — proof of the present retreat of the glacier. Since it was first visited in 1894 the front has receded 350 m; the retreat from 1951 to 1957 has been 150 m, or approximately 20 m per year. The type here described is very common in Greenland, and the name valley glacier of the Greenland type is therefore proposed. The glacier régime is a characteristic product of the south Greenland climate with its subarctic temperatures, and rather heavy precipitation in summer, when rain is frequent and typical radiation weather not common.

Sermikavsak on Upernivik Ø in Umanak Bay is quite another type of valley glacier. This glacier descends from a rather well-defined firn area, with a glacier

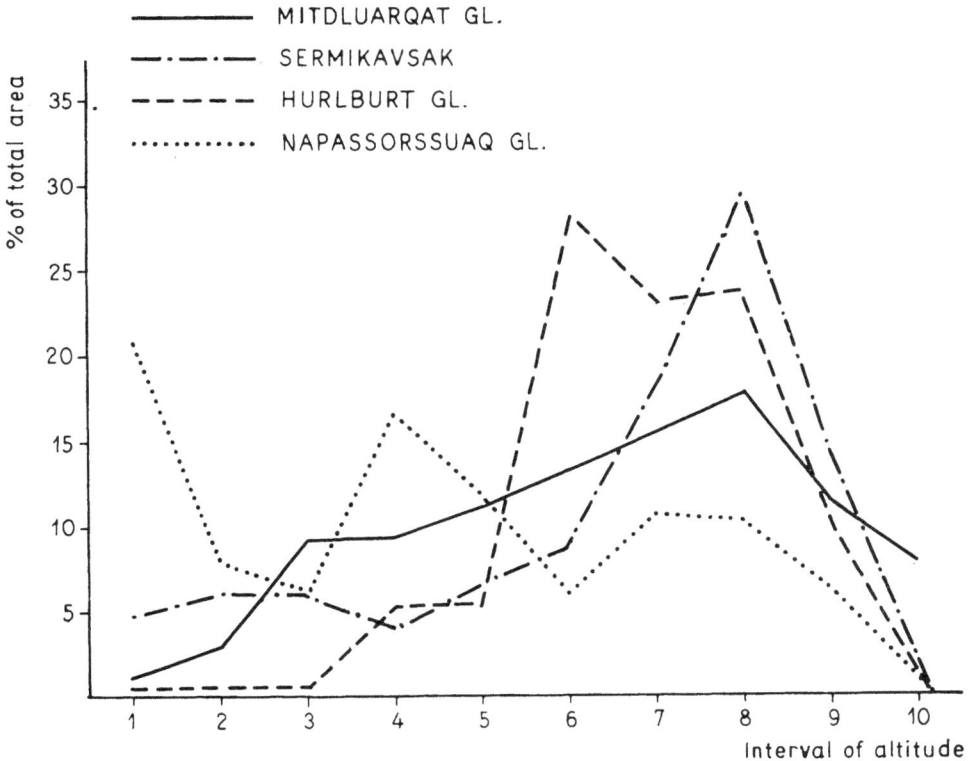

FIGURE 2.   Diagrams of the height intervals for the investigated glaciers.

tongue 15 km long and 1 km wide, to a flat outwash plain. The front is at present nearly one km from the sea. The glacier descends with three well-defined ice falls which are related to the differences in resistance of the substratum. The through-formed glacier valley is just at the contact zone between the West Greenland Cretaceous sandstone and the northern Archaean gneiss regions. The glacier is surrounded by steep mountains, 2000 m high. In winter the mountains collect the snow and the glacier is, therefore, partly nourished by avalanches which create a very complicated crevasse system. Nourishment by avalanches is characteristic of many glaciers in the Umanak region, but is not common in most other parts of Greenland and has never been observed in northernmost Greenland. Several peculiar glacier forms result from these avalanches from the steep mountains and also from the fact that, as many glaciers are so steep avalanches are sliding along the glacier from its higher parts to the glacier snout. Because of variations in climate, the central steep part of the glacier is

FIGURE 3.   The Napassorssuaq Gletscher.

now very thin in many places, and in some glaciers has even disappeared. Consequently the glacier has been divided in two parts; an upper part consisting of the firn area with a glacier tongue falling down very steeply over the mountain side; and below that nourished only by avalanches), the former glacier tongue, mostly stagnant and in the form of a very steep cone.

The total area of Sermikavsak is 21.6 sq. km, and studies of the different height intervals show it to be a typical valley glacier of type II (Ahlmann's definition) (see Figure 2). Photos from 1934, airphotos from 1953, and photographic records from the expeditions of 1956 and 1957 show a distinct retreat (with the exception of the southern moraine-covered ice lobe). From 1934 to 1953 the retreat was 600-700 m, and from 1953 to 1957 150 m, which gives us a withdrawal per year for the first period of 34 m, and for the last period of 38 m.

Special studies were made here of the glacier wind and glacial meteorology using a steel mast 12 m high with four contact anemometers and thermo-electric temperature readings. The results obtained have been published by Hans Kuhlman (1959). The results of the survey have been published by J. T. Møller (1959a, 1959b) who also made a special study of the geomorphology of the

FIGURE 4.   The glacier front of Sermikavsak.

periglacial phenomena, particularly the moraines, which here, as in most parts of Greenland, consist only of a thin cover of debris and stone over an ice ridge. According to Kuhlman the gravity wind (glacier wind or katabatic wind) was found in 78 per cent of the observations on the glacier, and 80 per cent of the ablation was caused by radiation. The total meteorological observations for the summer revealed radiation weather, with clear almost cloudless sky, 61 per cent of the time; overcast skies 17.7 per cent of the time; condensation weather 11.9 per cent of the time; and foehn conditions 6.9 per cent of the time.

In contrast to Sermikavsak and Napassorssuaq, the Hurlbut Gletscher is not a valley glacier, but a glacier cap. This glacier form is characteristic of the sandstone plateau areas of northern Greenland. The total area of the glacier is 188.1 sq. km. From the glacier cap proper some glacier tongues descend towards the coast especially to Inglefield Fjord north of the glacier (Figure 6). In ravines and narrow valleys we found several glacier tongues which are dead or nearly dead. Only one glacier tongue really reaches the coast, and the front of this is a nearly vertical ice-cliff standing on the beach above the sea-level even at low tide. No icebergs of importance are produced. Towards Olrik Fjord two glacier tongues occur. Exceptionally well-developed trimlines are found along both, indicating a

FIGURE 5.   Sermikavsak, the higher part.

greater extension previously. On the most westerly of these tongues a 15-25-m deep canyon has been formed in the ice by meltwater. Canyons of such dimensions can only be formed in stagnant ice. The thickness of the ice-cap proper is only 100-200 m and the firn area is not considerable. Superimposed ice was found by ice drillings down to at least 15 m and the glacier is partly nourished by a similar process to that described by P. D. Baird (1952) with respect to Barnes ice-cap on Baffin Island, that is, by the refreezing of melt-water and water-percolated snow and ice. Special investigations were made on the glacier tongue descending to the Inglefield Fjord. The movement was measured by surveying using bamboo poles placed in the ice; several crevasses were found, especially at the transition from the plateau to the valley. The lower part of the glacier consisted of very hummocky ice. While only small and insignificant moraines were found along the edge of the ice-cap proper, well-developed lateral moraines were found along the lower part of the glacier tongue. Most of the material here was sharp-edged and was transported on the surface of the ice, coming down on the glacier by landslides from the mountain sides. There are no terminal moraines, as the glacier at present terminates in a small bay formed

FIGURE 6. Hurlbut Gletscher, the glacier tongue descending towards Inglefield Fjord.

between the two lateral moraines that indicate the former greater extension of the glacier. Observations from the glacier front show that the total amount of transported material in the ice is very small, that the material can be transported to different heights above the glacier bottom, and that no true bottom moraines have been developed. The glacier is retreating very slowly at present. Observations from photos taken in 1939 by the Danish physicist, Aage Gilberg, and the present photographs show a very slight retreat of the central part of the glacier front, and a greater retreat of the lateral parts of the glacier snout, especially at the western side of the glacier. The total retreat in twenty years is probably less than 100 m, which gives less than 5 m withdrawal per year.

The glacier régime is dominated by the high arctic climate with low precipitation and very low winter temperatures; in summer, the ablation season, the diurnal temperature variation is insignificant, and the ablation is mostly caused by radiation.

Mitdluagkat Gletscher is a typical product of the climate of eastern Greenland which is humid with a great snowfall and frequently overcast skies. According to Ahlmann the glacier is a transection glacier (see Figure 2). The glacier is surrounded to the east and south by mountain chains culminating in the 1,084 m

FIGURE 7.   Hurlbut Gletscher, the glacier cap.

high Vegas Fjeld. To the north and west the glacier is more or less a highland glacier. The total area is 36.4 sq. km and the glacier drains towards the Sermilik Fjord, with a broad but rather short tongue which goes down a flat-bottomed valley nearly 800 m long and 200 m wide. The glacier front here is only 4.1 m above sea-level. The ice surface on the glacier tongue is rather hummocky, and several very deep, narrow glacier mills were found; on the glacier snout there is a very well-developed calderon. Besides the routine investigations a special study was made of the run-off from this glacier tongue. The discharge was studied by means of a self-recording river-gauge. The preliminary results have been published by Hans Valeur Larsen (1959), who found the discharge was primarily deter-mined by the radiation. A typical diurnal variation was found with the maximum at 17-1800 h and low water at 9-1000 h; the corresponding discharge values were about 4 cu m/sec and 2 cu m/sec. Along the southern and western margin of the glacier several ice-dammed lakes were found, and tappings from these lakes were observed in September. From the river-gauge it was found that the curve of discharge was relatively gentle, increasing to maximum value and then dropping

FIGURE 8.   Glacier tongue descending from Mitdluágkat Gletscher towards Sermilik Fjord. The nunataks and the sunny rocks on the left near the front were covered by the ice in 1933.

off suddenly. The total quantity of water tapped off from the ice-dammed lakes has been computed to 400-450,000 cu m. The water from the lakes drained through channels in or below the ice. The water from the tappings did not appear in the front of the glacier, but in the day came high up in the lateral drainage channel along the glacier tongue, the water coloured yellow from the suspended material. At one of the tapped lakes the drainage was so complete that it was possible to follow the channel in the ice for more than two hundred m. The tapping had taken place through a channel in the ice, not along the bottom or in a subglacial valley. The channel was 5-10 m high and 10-20 m broad; the roof over the tunnel was partly supported by thick ice columns formed by the erosion of the water; the floor of the tunnel was ice covered only by a thin clay layer, and at some places thin crevasses were found to a deeper level. Great ice blocks from the collapsed glacier front, which had been taken away with the water into the tunnel by the catastrophic water stream, had nearly filled the tunnel making the advance of water difficult to impossible. Just inside the glacier front the tunnel was bigger, forming a large cavern in the ice; the collapse of such a cavern in the ice might be responsible for the formation of caldera such as those found on the glacier tongue towards the Sermilik Fjord. The temporary tapping of the glacier-dammed lakes was most probably caused by hydrostatic pressure lifting the ice sufficiently for the discharge. The re-establishment of the lake was confirmed by a flight over the Mitdluagkat Gletscher in late August, 1959, one year after the investigations on the glacier took place, and the lake was again formed and filled with water.

The firn area of the glacier is very small and superimposed ice is of great importance. The front of the glacier tongue towards Sermilik Fjord was visited by K. Milthers in 1933, and exposures with a phototheodolite were made. Comparison of those photos with the present stage of the glacier shows a very distinct retreat. In twenty-five years the front has retreated approximately 400 m and a nunatak has formed dividing the glacier front into two emerging tongues. A new nunatak higher up on the glacier is developing and has lowered the ice surface. The final calculations have not been finished, but most probably the lowering averages 50 m. Similar retreat of other glacier fronts were found for most of the glaciers visited in the district.

TEMPERATURES IN THE GLACIER ICE

Due to the great climatic variation between south and north Greenland it was presumed that the physical conditions in the ice should vary from south to north. Special studies of the temperature of the ice were, therefore, conducted. Drilling with a SIPRE auger down to 15 m was planned for all glaciers, but technical difficulties prevented this in some cases. Electrical resistance thermometers were placed at different depths and a reading was taken at regular intervals during the whole observation period. The yearly temperature variation continues down to 10 m below the surface, but at greater depth the temperature is nearly constant. The results of the measurements are shown on Figure 9. All drillings figured here are from the ablation zone, and the depth in metres is the depth at the time when the thermometers were put in place. The upper part of the Mitdluagkat and the

Napassorssuaq glaciers have temperatures close to the melting point, while the temperature is low on the Hurlbut Gletscher in North Greenland and the Sermikavsak is a transition form. According to Ahlmann's definition the two southern glaciers may be regarded as temperate glaciers and this seems to be the general rule for most of the local glaciers in southern Greenland. The glaciers in northern Greenland are polar glaciers, according to Ahlmann's definition, but are not high polar glaciers, which are characterized by temperatures in the accumulation areas so low that even in summer there is little or no melting. Based on the observations from Peary Land by Fristrup (1951) it is most probable that only the Greenland ice-cap and a few of the greatest glacier caps in North Greenland can be considered as truly high polar. The rest of the Greenland glaciers may according to the definition be polar (or even temperate).

As the yearly temperature variations affect only the upper ten metres of the glacier it will be seen that the cold North Greenland glaciers will react only

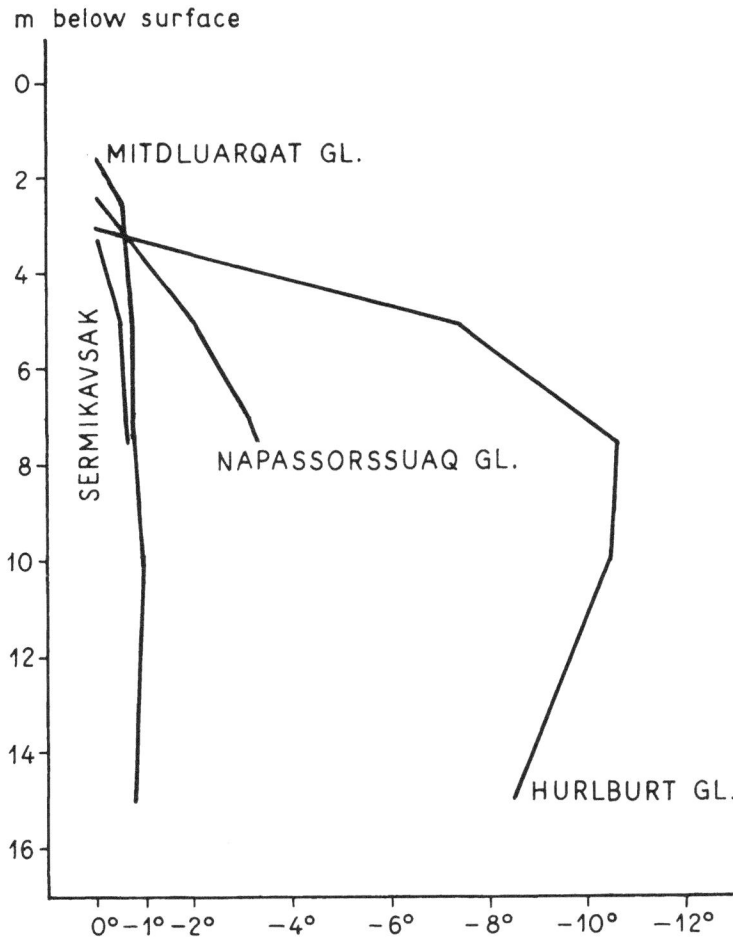

FIGURE 9.   Temperature in the ice in ablation zone, August.

very slowly to a change of climate, while the South Greenland glaciers with temperature near the melting point will be more sensitive to climatic fluctuations. This may be an explanation of why the glaciers in North Greenland, according to the observations by Lauge Koch (1928), seem to have started to retreat later than those in south Greenland. According to available observations it seems that the glaciers in North Greenland (with the exception of some of the outlet glaciers from the ice-cap) do not withdraw as fast as those in South and central Greenland.

REFERENCES

AHLMANN, HANS W:SON.   1948.   Glaciological research on the north Atlantic coasts; R.G.S. Research Series, no. 1.

BAIRD, P. D.   1952.   The glaciological studies of the Baffin Island expedition, 1950; J. Glacio., vol. 2, no. 11.

FRISTRUP, B.   1951.   Climate and glaciology of Peary Land; U.G.G.I. Assemblée genérale de Bruxelles.

———— 1960.   Studies of four glaciers in Greenland; Geogr. Tidsskr. vol. 60.

KOCH, LAUGE.   1928.   Contributions to the glaciology of North Greenland; Medd. om Grønland, vol. 65, no. 2.

KUHLMAN, HANS.   1959.   Weather and ablation observations at Sermikavsak in Umanak district; Medd. om Grønland, vol. 158, no. 5.

LARSEN, HANS VALEUR.   1959.   Run-off studies from the Mitdluagkat Gletscher in SE Greenland during the late summer 1958; Geogr. Tidsskr., vol. 59.

MØLLER, J. T.   1959a.   A West Greenland glacier front: A survey of Sermikavsak near Umanak in 1957; Medd. om Grønland, vol. 158. no. 5.

———— 1956b.   Glaciers in Uperrivik Ø, with special reference to the periglacial phenomena: Geograf. Tidsskr., vol. 59.

Late Pleistocene Glaciation
in Northeast Greenland

DANIEL B. KRINSLEY

ABSTRACT

Early Pleistocene glaciations of Greenland extended far into areas now covered by sea; only later Pleistocene glaciations are recorded by deposits on present land areas.

Ice-free land in northeast Greenland between J. P. Koch Fjord and Dijmphna Sound is mantled with glacial drift which is truncated near the sea by marine terraces now more than 200 m above sea-level. Aerial reconnaissance, photo interpretation, and recorded ground observations permit delineation of surficial deposits.

Modified moraine cut locally by marine terraces abuts less-modified moraine near Deichmans Oer in Hagens Fjord. Moraines near the entrance of Independence and Danmarks fjords are comparable to the less-modified moraine in topographic expression, degree of drainage integration, and relation to the present ice fronts. Less extensive moraines, modified locally by marine erosion, occupy fjords, valleys, and adjacent highlands.

Fresh moraines adjacent to the ice-cap front indicate recent advances. Currently the main ice-cap and isolated ice-caps are receding.

OBSERVATIONS MADE BY KOCH (1928), Troelsen (1952), and Davies and Stoertz (1957) have extended the information concerning the nature and distribution of glacial deposits in Peary Land. The distribution of the emerged Pleistocene marine deposits of Peary Land was summarized by Laursen (1954). It is now clear that most of southern Peary Land probably was covered at least once by ice moving north and northeast from a centre of accumulation in an area now occupied by the present Greenland ice-cap. Later withdrawal of the ice-sheet was followed by a marine invasion which either altered or removed the previous glacial deposits and left extensive marine materials. Subsequent isostatic adjustments have raised these marine deposits to more than 200 m above sea-level (Koch, 1928b). Locally, as at Brønlunds Fjord (Troelsen, 1952; Davies and Stoertz, 1957), there is stratigraphic evidence for a later but less extensive glaciation in Peary Land. This has been dated with C-14 by the United States Geological Survey (Rubin, written communication) as having occurred less than 5,370 ± 200 years ago.

As a result of aerial reconnaissance made during the summers of 1953 and 1959, and of a recent examination of aerial photographs and some ground photographs together with a study of recorded ground observations, the positions of numerous morainal systems and their related marginal channels were plotted on a 1:250,000 topographic base map of Danmarks Fjord and vicinity (Figure 1).

Examination of the positions of the ice front as inferred from the morainal systems and their related marginal channels indicates that the principal source of the ice was the Greenland ice-cap and to a lesser extent the ice-field north of Ingolfs Fjord. In most localities glacial deposits and features directly traceable to the expanded Greenland ice-cap have not been modified by the advance of glaciers from adjacent small ice-caps. Vandredal (Figure 2), for example, contains several distinct morainal loops convex towards the north. These moraines probably have been partially covered by lacustrine and marine deposits but they have not been destroyed by the advance of ice from adjacent glaciers to the east. Similarly, moraines north of the head of Ingolfs Fjord have not been disturbed by glaciers now only one km to the east. This evidence would seem to indicate that most of the small ice caps are remnants of the once expanded Greenland ice-cap.

Moraines and marginal channels at the mouth of Danmarks Fjord exhibit the best integration of drainage and are the most highly modified of any yet recognized in the area. Southward towards the Greenland ice-cap the glacial deposits and related features are progressively better preserved. Prominent recessional moraines, composed of cobble and boulder gravel with ridge lines 15 m above the flat upland surface, extend for several km in unbroken lines roughly parallel to the Greenland ice-cap front (Davies and Stoertz, 1957). In the absence of stratigraphic evidence, multiple glaciation of the Danmarks Fjord area is difficult to establish. It can only be assumed, from the position of the oldest, highest, and most northerly recognized morainal remnants and associated marginal channels, that the Danmarks Fjord area was completely glaciated at least once during the Pleistocene.

Northern Gluckstadt Land is partially mantled with recessional moraines which were deposited by ice lobes moving eastward from Hagens and Independence fjords and which merge with moraines deposited by ice moving north from Danmarks Fjord. Ten km northwest of Cape Kronborg, marginal drainage channels from the Hagens Fjord lobe intersect marginal channels from the Danmarks

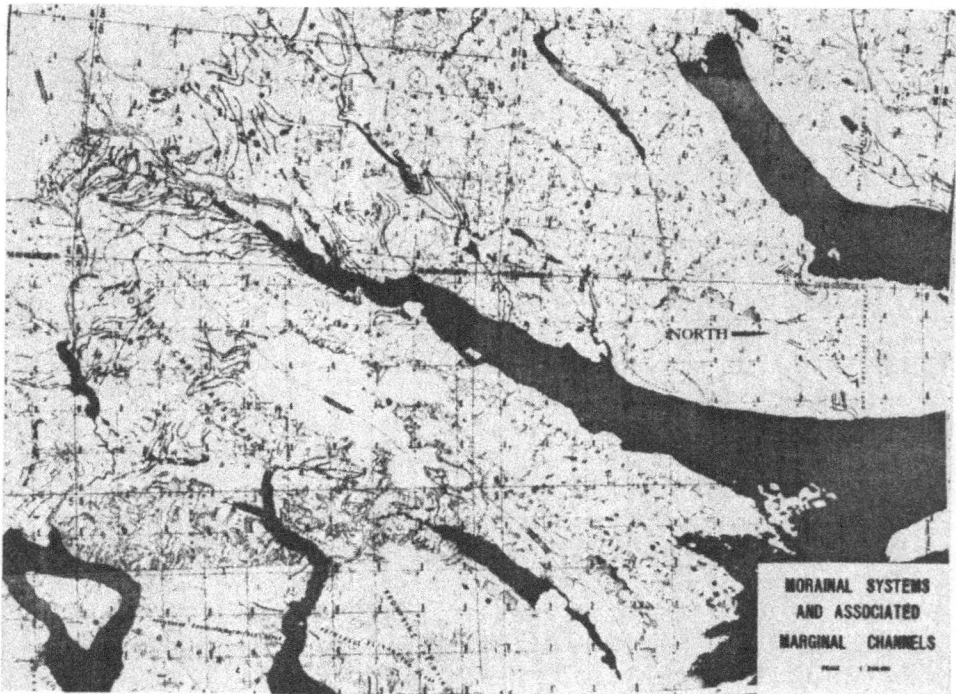

MORAINAL SYSTEMS
AND ASSOCIATED
MARGINAL CHANNELS

FIGURE 1.

Fjord lobe at an altitude of 200 m. At the time these marginal channels were cut, ice from Independence, Hagens, and Danmarks fjords probably coalesced to form a single broad ice front which extended northward beyond the coast of Gluckstadt Land and filled the trough at the mouth of Danmarks Fjord. North from the interlobate area the moraines separate, indicating that the ice from Danmarks Fjord was no longer in contact with the combined ice of Hagens and Independence fjords. At the same time, the ice of Hagens and Independence fjords began to assume individual lobe identity at its terminus as suggested by a slight bifurcation in the recessional moraine 5 km northwest of the interlobate area. Further withdrawal of the Hagens Fjord lobe is recorded by moraines near Deichmanns Øer and their equivalents along the east coast of J. C. Christensens Land.

Southward, 20 km upvalley from the mouth of Zig-Zag Dal a large moraine delta is graded to Zig-Zag Dal and adjacent valleys. The delta is characterized by a flat surface 65 m above sea-level which is pitted with kettles and punctuated by a steep northeast slope towards Danmarks Fjord. Freuchen (1915) reported terraces consisting of raised sea bottom from the same area. The topographic evidence suggests that the delta was formed in standing water in close proximity to glacier ice from Danmarks Fjord. This ice may have been floating in sea water in much the same way as the snout of Hagens Glacier is floating at present. The position and orientation of the moraine delta indicates that deglaciation of the western tributary valleys preceded the retreat of the Danmarks Glacier towards the head of Danmarks Fjord.

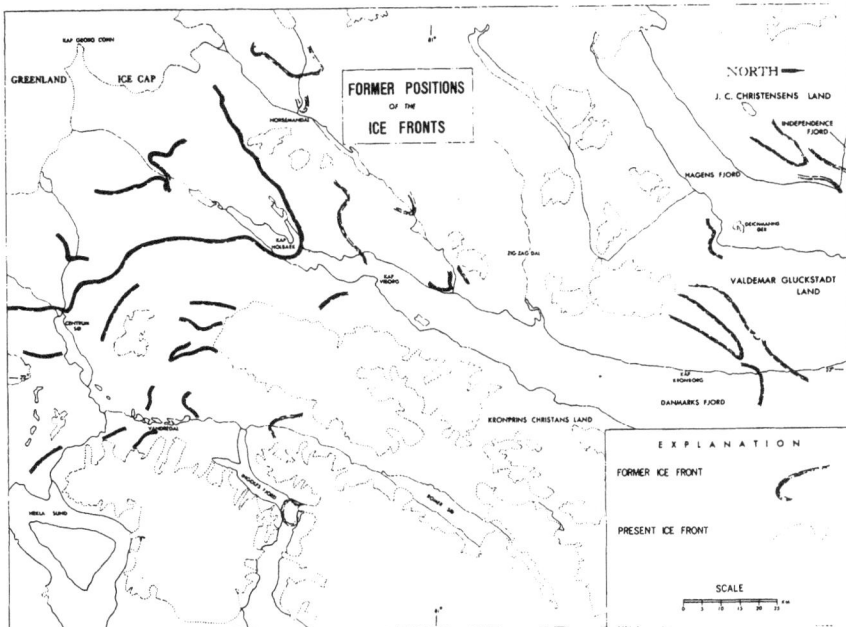

FIGURE 2.

Five km south of Cape Viborg, a marginal drainage channel system can be traced westward towards Norsemandal, and southeastward towards the adjacent ice cap. This same system probably is continuous to the southeast, across Vandredal.

When the Danmarks lobe had retreated to Cape Holbaek, Norsemandal was free of ice as was most of the highland between the head of Danmarks Fjord and Vandredal. The younger, better-preserved moraine southwest of Cape Holbaek shows a gradual diminution of the Danmarks Fjord lobe towards Cape George Cohn which is currently a re-entrant in the front of the Greenland ice-cap.

A prominent series of marginal channels and associated morainal remnants can be traced from the head of Norsemandal to Cape Holbaek and then across Amdrups Hojland to the west shore of Centrum Sø (Figure 3). A series of marginal channels near the mouth of Graesrig Elv along the east wall of the valley are similar to the marginal channels along the west shore of Centrum Sø in gradient, degree of modification, and relation to terrace deposits. These related marginal channels in Graesrig Elv continue southward to the Greenland ice-cap. A conspicuous terrace at an altitude of 240 m along the north and south slopes of Centrum Sø valley terminates upvalley at these prominent marginal channel systems. Downvalley from these points of termination, the terrace crosses numerous intervening marginal channels oriented normal to Centrum Sø valley with consequent diversion of the streams in these channels. Davies (1957)

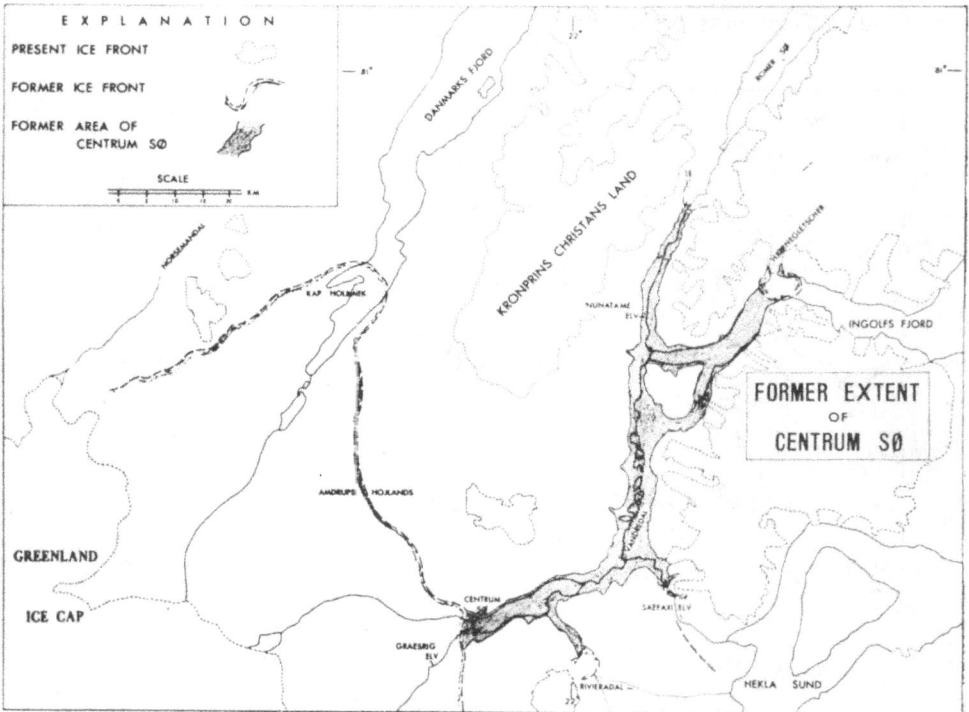

FIGURE 3.

visited this terrace and reports that it is composed of pea gravel and sand; no evidence was found for a marine origin. Photographs and aerial observation indicate that Vandredal is floored with fine materials, but in the absence of evidence for a marine origin the materials can also be explained by a lacustrine environment.

Vandredal, an extensive interior trough in Kronprins Christians Land, drains to Ingolfs Fjord and Hekka Sound. Some glaciers calve into Ingolfs Fjord, and several small glaciers lie on the slopes of the two tributary valleys which connect Vandredal with Hekla Sound. During the time that ice blocked the west end of Centrum Sø, the expanded glaciers in Nunatame Elv and Ingolfs Fjord could have provided an effective dam to impound drainage to a height of 240 m above sea-level. Moraines in Nunatame Elv and marginal channels at 900 m above the snout of the Hjornegletscher suggest this explanation. Saefaxi Elv probably was dammed by the glacier which is presently only 1.5 km from the 240-m contour line on the west slope of the valley. Riviera Dal probably was dammed by a tongue of the small ice-cap to the southwest, the front of which is 4.5 km from the east valley wall at the 240-m contour line. It is likely that Vandredal was subjected to both lacustrine and marine deposition as ice dams alternately formed and were destroyed. Radiocarbon dating of these deposits would provide a key to an absolute time-scale for the deglaciation of Kronprins Christians Land.

REFERENCES

DAVIES, W. E., and STOERTZ, G. E.   1957.   Contributions to the geomorphology of northeast Greenland; unpublished manuscript.
FREUCHEN, P.   1915.   General observation as to natural conditions in the country traversed by the expedition; Medd. om Grønland, bd. 51, nr. 9.
KOCH, L.   1928a.   Contributions to the glaciology of north Greenland; Medd. om Grønland, bd. 65 (2).
——— 1928b.   The physiography of north Greenland; Greenland, vol. 1, pp. 491-518.
LAURSEN, D.   1954.   Emerged pleistocene marine deposits of Peary Land (North Greenland); Medd. om Grønland, bd. 127, nr. 5.
TROELSEN, J. C.   1952.   Notes on the pleistocene geology of Peary Land, north Greenland; Medd. fra Dansk Geologisk Forening, bd. 12.

Surface Features of the Ice-Cap Margin, Northwestern Greenland[1]

LAURENCE H. NOBLES

ABSTRACT

The gross configuration of the Greenland ice-cap in marginal regions is determined by the relation between local accumulation and ablation, by the configuration of the underlying bedrock topography, by the rate of ice movement, and to a certain extent by the past history of the ice-cap. Local features of the ice surface develop within the framework of this gross configuration and are controlled by the thermal regimen of the ice, the underlying ice structure, and the nature of the accumulation and ablation régime.

The ice margin is formed by large outlet glaciers, small valley glaciers, ice cliffs, steep ramps, and gentle ramps. The melt zone is characterized by surface run-off, forming such features as slush avalanches, algal pits, and an integrated drainage pattern.

The surface of the wet firn zone is characterized in summer by very soft and sticky granular firn. Crevasses are particularly dangerous in this zone as they are always wholly or partially bridged by firn.

The surface of the dry snow zone, which forms the interior of the ice-cap, is marked by sastrugi and other features of wind-drifted snow.

MARGINAL TYPES

In northwestern Greenland the outer forty to sixty miles of the ice-cap is a highland ice-cap composed of a relatively thin ice carapace resting on a very rugged sub-ice topography. The nature of the margin of the ice-cap in this region is determined largely by the nature of the underlying topography. Differing configurations of this topography in the immediate vicinity of the ice margin have produced at least five different types of marginal features. Each of these marginal features presents unique problems, both from the standpoint of fundamental glaciological research and from the view of applied military investigations.

Large Outlet Glaciers

Most spectacular of the marginal features are the large outlet glaciers. The Sermerssuaq, or Moltke, Glacier (Figure 1; Koch, 1928; Wright, 1939) which calves into Wolstenholme Fjord near Thule is a typical example. These outlet glaciers are up to several miles in width. They drain large portions of the interior of the ice-cap, are heavily crevassed, and flow at rates as high as six to eight feet per day. As a result of the rapid outflow of ice through a large outlet glacier a "cone of draw-down" is produced in the interior of the ice-cap that is not dissimilar to the cone of draw-down in the water table surrounding a well. These draw-down areas are well back from the margin of the ice-cap and above the accumulation limit. They have relatively steep surface gradients and moderately

[1]Work for this paper was done under contract with the Cold Regions Research and Engineering Laboratory, Corps of Engineers, United States Army.

FIGURE 1.   Large outlet glacier. The Sermerssuaq or Moltke Glacier.

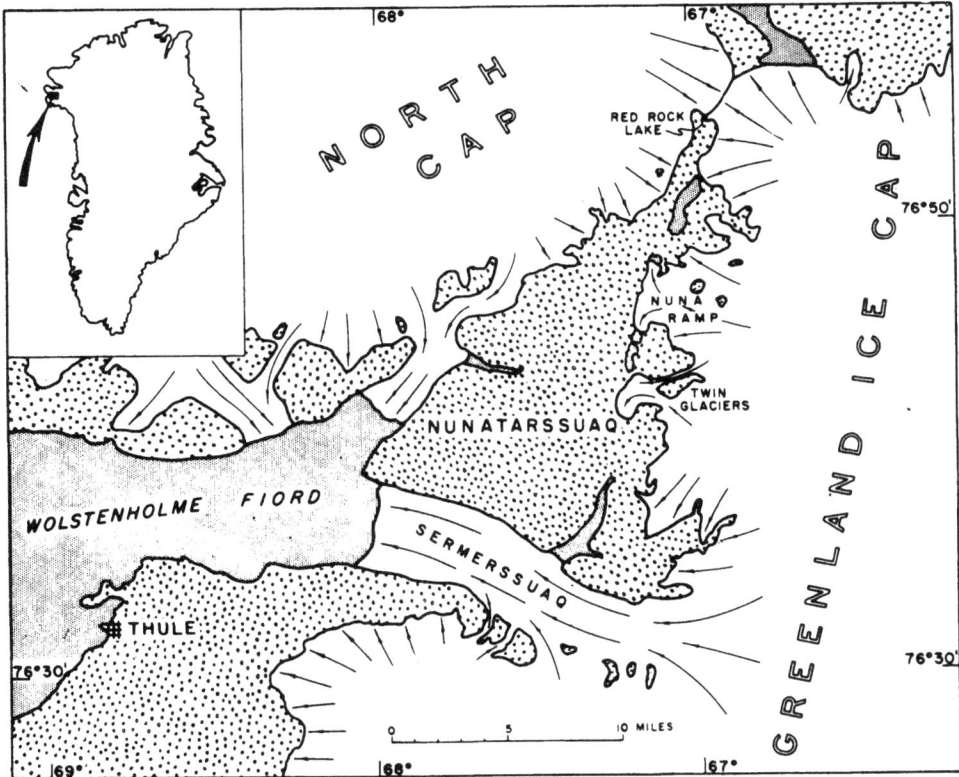

FIGURE 2.   Index map, Nunatarssuaq region.

rapid rates of flow. As a result large crevasses partially or wholly bridged with snow are common, making these areas the most dangerous part of the ice-cap in which to travel.

A large portion of the accumulation in the centre of the Greenland ice-cap is ultimately dissipated through these large outlet glaciers. A conservative estimate indicates an ablation of 7.0×10^8 cubic m of ice per year on the lower portion of the Moltke Glacier, at least three-fourths of which is by calving from the terminus of the glacier.

Small Valley Glaciers

Another type of marginal feature is represented by small valley glaciers such as the Twin Glaciers (Figures 2 and 3; White, 1956). These small valley glaciers drain relatively restricted accumulation basins at the margin of the ice-cap. Many of them appear to be deteriorating rapidly. Their rate of flow ranges from one to four inches per day (White, 1956, p. 21), and, in general, they have steep surface gradients and are moderately crevassed. They do not play a significant role in the over-all accumulation and ablation budget of the Greenland ice-cap.

Ice Cliffs

Ice cliffs are another important feature of the ice margin. In the Nunatarssuaq region of northwestern Greenland they range from 50 to 150 feet in height and are prevalent on the margins of the North ice-cap (Figures 2 and 4). A few occur at scattered localities on the margin of the main Greenland ice-cap. These cliffs have been intensively studied by Goldthwait *et al.* (1956, 1957) in the vicinity of Red Rock Lake (Figure 2), and their existence appears to be the result of a combination of topographic and climatologic factors. The ice immediately above these ice cliffs shows flow rates of about 0.5 inches per day (Goldwait *et al.*,

FIGURE 3.   Small valley glacier. South Twin Glacier, with Nuna ramp in the distance.

1956, p. 38; 1957, p. 74) and crevasses are uncommon in the areas above the cliffs. These ice cliffs, which form abrupt topographic discordances between the ice surface and the adjacent land, create major obstacles to transportation in parts of northwestern Greenland. In addition, dry calving of ice from the cliffs makes operation in their vicinity dangerous during certain portions of the year.

Steep Ramps

Marginal features having a height of 50 to 150 feet and a slope of 20° to 45° occur locally at the margin of the ice-cap and have been called steep ramps (Figure 5). Work now in progress in the Red Rock Lake area suggests that steep ramps may be a transitional stage in the decay of an ice cliff to form a gentle ramp.

Gentle Ramps

Most important from the standpoint of access to the interior of the ice-cap are marginal areas with relatively gentle slopes (generally less than 5°) which are called gentle ramps (Nobles, 1954a, 1954b, 1960; Schytt, 1955). A typical example is the Nunartarssuaq ice ramp (Figures 2 and 6) which averages two miles in width and has a total area (below regional accumulation limit) of twelve square miles, of which 8.5 square miles is exposed glacier ice and the remainder is superimposed ice and local snowfields. The average slope of the Nuna ramp is 300 feet per mile or 3° (6 per cent), and the slope ranges from 80 feet per mile to a maximum of 8°.

These gentle ramps are relatively slow moving (up to three inches per day), and are even stagnant in part. As a rule they are almost completely free of crevasses. They appear to owe their existence to a combination of favourable topographic circumstances. In most cases ice from the accumulation area is

FIGURE 4.   Ice cliff. Cliff bordering North Cap near Red Rock Lake.

being delivered to the ramp over a bedrock threshold that limits the amount of ice received on the lower ramp. In addition, ramps usually form where the ice flows out into a bedrock valley from the valley side and spreads out both up and down the valley, forming a broad topographic divide on the lower part of the ice ramp. The evidence seems to indicate that most of these ramps are at present deteriorating rather rapidly (Nobles, 1960, pp. 17-18).

SURFACE FEATURES

For purposes of studying the regimen of a glacier system it is most advantageous to consider the glacier in terms of a zone of net accumulation (surplus area) and a zone of net ablation (deficit area). The break between these two zones has been called the firn limit by Ahlmann (1946) and the accumulation limit by Schytt (1955). On temperate glaciers this boundary may be easily

FIGURE 5.   Steep ramp. The steep ramp northeast of Red Rock Lake.

FIGURE 6.   Gentle ramp. The Nunatarssuaq ice ramp.

recognized as it corresponds to the lower limit of firn on the glacier surface at the end of the ablation season. It is more difficult to recognize on subpolar glaciers (Nobles, 1960).

In winter the entire Greenland ice-cap is a zone of accumulation, with dry snow on the surface. On the other hand, during the summer ablation season the surface of the ice-cap may be divided into three zones, each possessing distinctive surface features.

Dry Snow Zone

On 80 to 90 per cent of the area of the ice-cap in northern Greenland melting is exceedingly rare and the surface is marked by features of drifting snow throughout the year. Sastrugi with a relief of as much as three feet are common in some areas (Figure 7). These features are normally quite asymmetrical in shape and as they are formed of hard-packed snow they do not break down readily under the weight of men or vehicles. The frequency of snow-moving winds is high in this region and the sastrugi form and re-form very rapidly when snow is moving.

Wet Firn Zone

The portion of the ice-cap that lies above the accumulation limit, but in which some melting takes place during the summer ablation season, is called the wet firn zone. In this zone crevasses, if present, are either wholly or partially bridged by snow and may form a major obstacle to transport. In addition the surface is formed by very soft and sticky granular firn that makes travel across the zone difficult. The boundaries of the wet firn zone are governed very largely by altitude, and the width of the zone therefore depends on the slope of the glacier surface.

FIGURE 7.   Sastrugi in the snow of the Greenland Ice Cap, latitude 76° N, March 1955.

Melt Zone

The melt zone lies below the accumulation limit and is characterized by bare ice at the surface. In this zone crevasses, if present, are either open or filled with water. The surface features that develop in the melt zone during the ablation season are controlled by the negative temperature, except in the upper few feet, of the underlying ice. This situation, which is analogous to that found in permafrost areas on land, causes all melt-water to flow off at, or near to, the ice surface, and this flow of melt-water across or immediately beneath the surface is the prime factor governing development of features on the ice surface.

Slush. At times early in the melt season, portions of the marginal ramps are covered by a thick layer of slush on top of the ice. This slush forms as a result of penetration of melt-water into the winter snow pack and its severity therefore depends partly on the thickness of this snow pack. The slush owes its existence in part to the fact that the ice beneath the snow pack is at negative temperature and all run-off must be through the snow pack itself.

The amount of water delivered to any particular segment of the snow pack (or firn) is a function of the rate of melting in the region. The rate of flow through this same firn segment is governed by the permeability of the firn and by the slope of the underlying ice surface. As long as the rate of melting is low, melt-water percolates vertically through the firn to the firn-ice interface and then flows along the surface at or near to this interface.

Each successive downstream segment of firn must pass a larger volume of melt-water than the one directly above it. Therefore, when the rate of melting increases to the point that the run-off cannot be carried by trickling along the firn-ice interface, a water-table forms in the firn. This water-table slopes downhill in the same direction as the surface slope but with a slightly lower gradient. At this stage flow is in response to a hydraulic gradient along this water-table. As the slope of the water-table is less than that of either the firn surface or the ice surface, the water

FIGURE 8.   Slush avalanche on the southern edge of Nuna ramp, June 26, 1954.

table must intersect the firn surface in some downstream area. Below this point of intersection the firn is completely saturated with water and normally appears blue in colour at the surface. As a result of this intersection of the water-table with the surface, a row of "springs" develops. These "springs" are not dissimilar to the springs that develop on the land surface at points of intersection with the underlying water-table. However, if the saturated firn pack, now called slush, is on a slight slope it is very unstable so that it normally fails rather quickly after the "springs" develop, and a large mass of the firn flows downhill as a slush avalanche or slusher (Figure 8). As the first point of intersection of the water-table with the surface is in a downstream area the slush avalanches normally develop on the lower parts of the gentle ramps and grow headward as the season progresses.

The size and severity of these slush avalanches depends upon the thickness of the winter snow cover, on the slope, and on the rate of melting. They have never posed a serious problem in northwestern Greenland but reports from elsewhere in Greenland indicate that under certain conditions they may be large enough to bury men and vehicles.

Stream Channels. Immediately after the disappearance of the winter snow cover from the ice — either by melting and run-off or by flow as slush avalanches — the 0°C isotherm is normally at, or only a foot or so beneath, the ice surface. Thus, as melting continues, all run-off of melt-water must be at or near to the surface. In the few hundred yards directly downstream from the "instantaneous firn limit" this run-off is by sheetwash or direct overland (over-ice) flow. However, in a short distance this overland flow is gathered into specific threads and these soon form distinct channels across the ice surface. In most areas the ice behaves as a completely homogeneous mass with respect to stream erosion, giving rise to a sub-parallel dendritic drainage pattern (Figure 9). In a few places a structural control

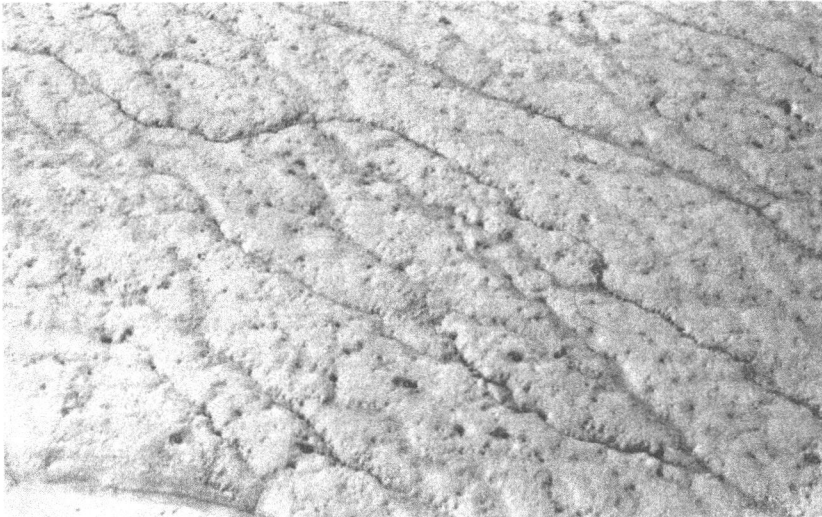

FIGURE 9.   Aerial view of a portion of the lower Nuna ramp showing stream pattern and frequency of algal pits, July 15, 1954.

of the drainage pattern may be seen in that the streams follow pronounced blue ice-bands for considerable distances, then cut abruptly across the white-ice areas between blue bands, thus forming a roughly rectangular drainage pattern.

The stream channels range from six inches to two feet in depth over most of the ramp and average fifteen to eighteen inches deep. The water depth in the streams (at peak flow in mid-afternoon) ranges from a few inches to a foot or more. The diurnal variation in flow is considerable and for some streams ranges from zero during the night hours to several second feet at mid-afternoon.

On the lower portions of the glaciers the main streams are carrying the water of many tributaries and have been incised to depths of ten feet and more. In general, even where deeply incised, the streams tend to meander, with the width of the meander belt being two to three times the width of the stream channel.

Dynamically these streams are unusual in that they either carry no load at all or their load consists of ice crystals, which are of lower density than the water that is transporting them. Therefore, the entire load is a floating load and is not readily subject to the classical analysis of stream dynamics. The lack of load carried by these streams means that they are incapable of erosion by mechanical corrasion. Hence the channels cut by these streams must result from greater melting at the site of the stream channel itself than in the area between the channels. An understanding of the development and sustenance of these stream channels can best be obtained by a consideration of the yearly cycle of events to which they are subjected.

At the close of any melt season (middle to late August) the marginal areas will be crossed by an integrated drainage pattern of well-developed stream channels, for the most part twelve to eighteen inches deep. Snowfall during the winter completely fills the stream channels and on some parts of the ramps buries the interstream areas to depths of up to a foot or two. However, on much of these marginal ramps the interstream areas are blown free of snow throughout the winter and only the channels are filled. In any circumstances, at the beginning of the melt season the snow depth over the whole of the area is less in the interstream areas than it is in the channels themselves.

Therefore melting of ice begins first in the interstream areas and the resulting melt-water flows into the channels forming slush from the snow present in them. When the slush in the channels becomes fluid enough it flows out, normally as a small slush avalanche. Most of these slush avalanches start in the lower courses of the stream channels and work their way gradually headward. However, in some cases slushers start well upstream while the lower course of the channel is still badly clogged with firn and slush. In these cases the stream is often forced out of the old channel and a new channel develops. The only changes in drainage pattern that take place from year to year are due to development of new stream courses early in the melt season when older channels are clogged with winter snow. Otherwise the channels that form in any given melt season are all inherited from the previous melt season and the drainage pattern, particularly of the larger and more deeply incised streams, remains the same for many years.

The inauguration of melting of the ice on the interstream areas at a time when the channels are still filled with the winter's snow means that when the old channels are finally opened their depth has been reduced to an average of six to nine inches. After the channels are cleared of slush, and bare ice is exposed over both the areas of the channels and the interstream areas, run-off is entirely through the old channels and these channels are then deepened with respect to the interstream areas until they reach an average depth of eighteen inches to two feet near the end of the melt season. As these channels cannot be deepened by corrasion, the deepening must be the result of melting. The source of heat for this melting must be the frictional heat generated by turbulence in the water. (The ultimate source of the energy is, of course, the potential energy loss of the falling water.) A simple calculation, made to estimate the amount of heat energy available from this source, shows that for a stream flowing in a rectangular channel one foot wide and one foot deep on a 6 per cent slope (average for the gentle ramps), an average water velocity of half a mile per hour would be sufficient to melt 1 cm of ice in twenty-four hours if all of the heat energy gained could be expended on the *bottom* of the channel. It is difficult to evaluate the exact manner in which the heat energy generated by the running water is finally dissipated but in any event this heat energy appears to be of sufficient magnitude to account for the deepening of the melt-water streams during the course of the melt season.

FIGURE 10.   Small algal colonies in the bottom of a
stream channel, lower Nuna ramp, July 12, 1954.

Algal Pits. Cryoconite holes, which are formed by more rapid melting in the vicinity of surface pebbles or small cobbles because their albedo is lower than that of the surrounding ice, are common features on the surface of glaciers in many areas. These cryoconite holes are usually roughly circular in plan view, have vertical walls, are ten to twenty inches deep, are filled nearly to the brim with melt-water, and contain a rock fragment, or group of fragments, at the bottom of each hole. Their formation has been studied in some detail by many authors (Brant, 1932; Phillipp, 1912; Steinback, 1936; and others).

In northwestern Greenland many cryoconite holes have a somewhat different form in that their formation is controlled not by fragments of mineral debris (cryoconite holes of this type are present but are not common) but instead they are related to the presence of masses of blue-green algae that are living on, or only slightly below, the glacier surface.

Algae living on glacier ice have been studied by Nordenskjold (1885, pp. 217-24), Wittrock (1885, pp. 63-124), Drygalski (1897, pp. 430-6), Peterson (1924), Benninghoff and Nobles (1957), and Gerdel and Drouet (1959), and the following species have been identified: *Ancylonema Nordenskjoldii, Pleurococcus vulgaris, Scytonema gracilis, Pinnularia lata*, and *Sphærella nivalis* (all by Wittrock); *Calothrix parietina* and *Anacystis limnetica* (by Drouet).

These algae, which are thought to have been originally blown onto the ice while in a quiescent state, have assumed a number of different forms. They were found by Schytt (1955, p. 84) on the Thule ramp southeast of Thule in colonies, in small cryoconite holes, and also in widely disseminated very small colonies. On the Nuna ramp thirty miles northeast of Thule they are also found in colonies in small cryoconite holes, particularly along the bottoms of melt streams (Figure 10), but their most common form is as larger colonies in large cryoconite holes or algal pits.

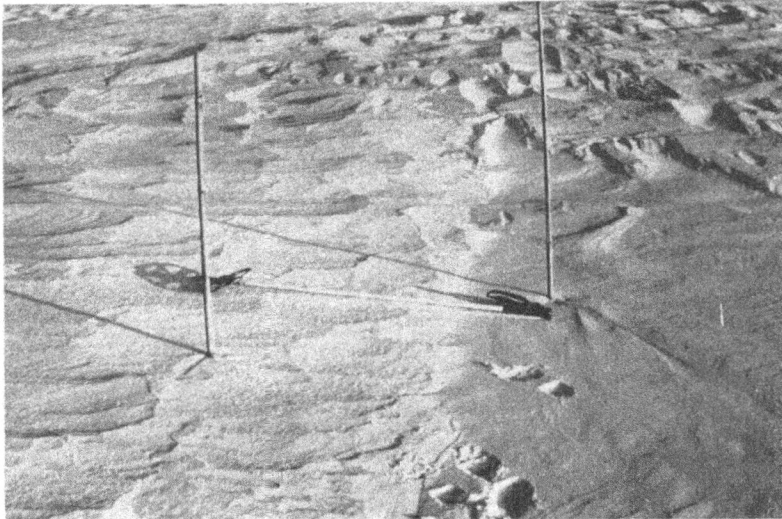

FIGURE 11. Snow-covered algal pit in mid-winter, March 21, 1954.

These pits are very prevalent on the lower mile to mile-and-a-half of the Nuna ramp. In midsummer they are open, are a maximum of nineteen to twenty-one inches deep, and usually contain fifteen to seventeen inches of water. The algae cover the bottom of the pit as a loose mass one-half to one inch thick. The pits are roughly circular in plan, ranging from a few inches to two or three feet in diameter. (The largest found in the area was eight feet long, four feet wide, and nineteen inches deep.) Their frequency may best be seen on an aerial view of the lower portion of the Nuna ramp (Figure 9).

Excavations in March 1954 and March 1955 indicate that the algae spend the winter frozen into the ice in a quiescent state at a depth of nineteen to twenty-one inches, and at a temperature that at times may be as low as —30°C (Figures 11, 12, and 16). Since the maximum depth of penetration of radiation into the ice in this region is of the order of nine inches, the algae remain dormant during the early part of the ablation season or until the upper ten to twelve inches of ice have melted away. Then insolation combined with the lower albedo of the dark algal mass causes local melting in the vicinity of the colonies and the algae resume growth (Figure 13). This melting soon results in the formation of open, water-filled pits about nine inches deep (Figure 14).

After first opening with a depth of nine inches, the pits increase in depth during the summer to a maximum of nineteen to twenty-one inches. This consistent maximum depth must represent the maximum depth of penetration of significant solar radiation through the water in the pits. As some of the pits are twenty-one inches

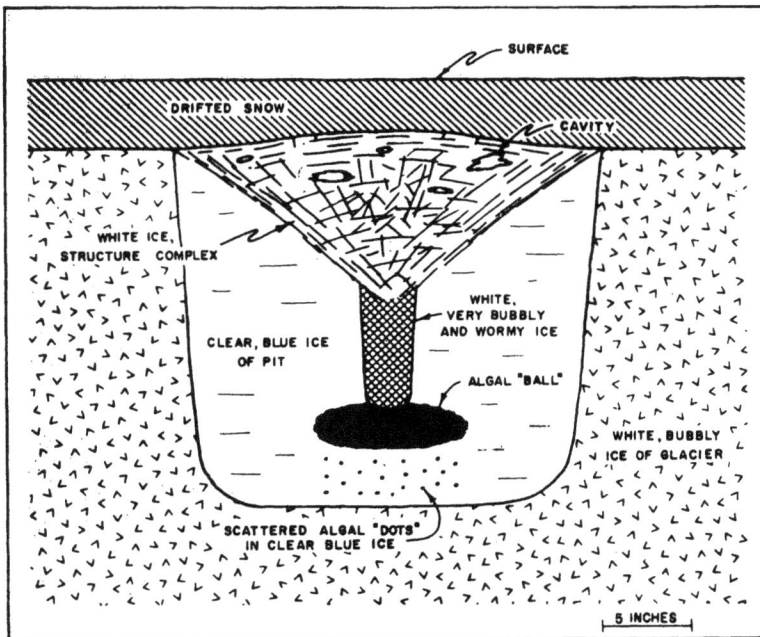

FIGURE 12.   Cross-section of algal pit in mid-winter. Compare with Figure 16.

deep and less than two inches in diameter, the penetration of radiation to the algal colonies must also require reflection off the pit walls.

The condition of the pits during the summer is governed by current weather conditions. In warm, sunny weather the pits will be completely open and water-filled. The depth of water at any given time is a function of the local "ground water" régime and is governed by the balance between the rate of melting and the rate of run-off in streams and through intercrystal openings in the upper foot or so of ice. In cloudy, cool weather the water in the pits may be covered by a thin

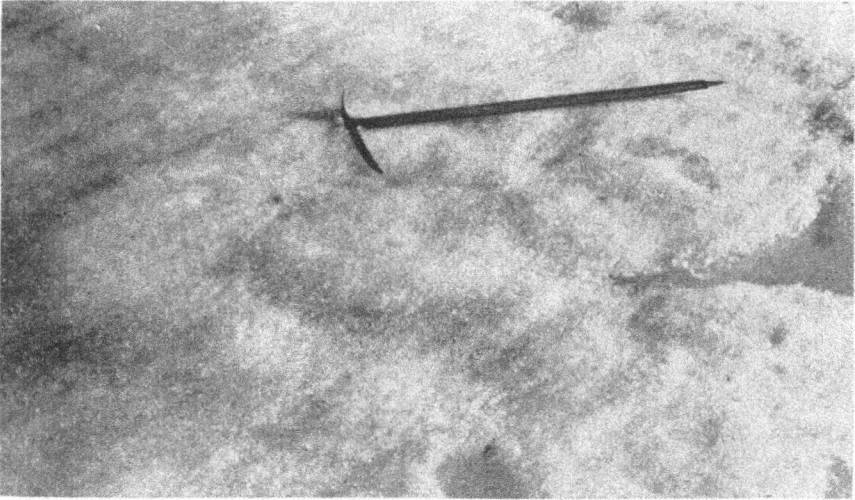

FIGURE 13.   Algal pit beginning to open early in the melt season, July 3, 1954.

FIGURE 14.   Open algal pit on the lower Nuna ramp—mid-ablation season. Note kidney shape with low dividing rim, probably resulting from formation of the pit by coalescence of two smaller pits, July 29, 1954.

film of ice (Figure 15). After midsummer snows this thin film of ice may in turn be covered by an inch or two of snow, making the pits very difficult to detect at the surface. Late in the ablation season, when the angle of the sun's rays is very low, the pits are covered almost continuously with a thin film of ice. With the advent of the lower temperatures of winter this film of surface ice becomes thicker and finally the pits freeze up completely, entrapping the algae in the ice at a depth of nineteen to twenty-one inches.

FIGURE 15. Algal pit beginning to freeze over—late ablation season, August 29, 1953.

FIGURE 16. Cross-section of algal pit in mid-winter, March 22, 1955. See Figure 12 for description.

The first freezing of the pits is by formation of a two to three inch thick crust on the upper surface of the water. Then freezing begins around the whole of the perimeter of the pit forming clear, bubble-free ice that may easily be distinguished from the adjacent bubbly glacial ice (Figures 12 and 16). As the bottom of the pit freezes most of the algae move upward a few inches leaving behind scattered algal "dots" frozen into clear, blue ice. As freezing continues the main mass of algae is compressed into a ball-like mass a few inches above the original bottom of the pit. The oxygen in the algal colony and that dissolved in the water of the pit is trapped in a cone-shaped mass of bubbly ice in the centre of the original pit. These bubbles of oxygen are placed under considerable pressure by the expansion during freezing of the water in the pit and they cause this bubbly ice to "explode" when tapped with an ice axe. Expansion during freezing also warps the early-formed surface crust into a gentle, anticlinal fold and in some cases forces material out of the pit along gentle thrust faults with a displacement of an inch or so (see Figures 12 and 16).

The factors controlling the growth and persistence of the pits are not completely understood. It seems clear from the size and distribution of the pits on the Nuna ramp that individual small colonies must grow in size to form pits at least two to three feet in diameter. Most of the pits of larger size are kidney-shaped to irregular in outline, and appear to have been formed by merging of two or more smaller pits. Often the remnants of the wall that separated the two small pits may still be seen as an ice ridge four to six inches high on parts of the bottom of the new large pit (Figure 14). The rate of ice movement on the lower Nuna ramp is approximately one mile per hundred years. It is not impossible, then, that some of the larger pits on the lowermost part of the ramp may have formed as small colonies a mile or two out on the ramp one to two hundred years ago. The slightly longer growing season for the algae on the lower portion of the ramp may also be partially responsible for the larger average size of pits on the lower ramp.

REFERENCES

AHLMANN, H. W., and ERIKSSON, B. E.   1946.   Deposition of fluid water in firn and on ice surface; Geog. Annal., vol. 28, pp. 227-57.

BENNINGHOFF, W. S., and NOBLES, L. H.   1957.   Algal pits on marginal areas of the ice cap, northwestern Greenland (Abstr.); Bull. Ecol. Soc. Amer., vol. 38, p. 79.

BRANDT, B.   1932.   Beobachtungen und Versuche uber die Entwicklung der Kryokonitformer; Zeit. f. Gletscherk., vol. 20, pp. 84-93.

DRYGALSKI, E.   1897.   Grönland—Expedition 1897, Band I; pp. 430-6.

GERDEL, R. W., and DROUET, FRANCIS.   1959.   The cryoconite of the Thule area; Snow Ice and Permafrost Res. Est., Res. Rept. 50.

GOLDTHWAIT, R. P.   1956.   Study of ice cliff in Nunatarssuaq, Greenland; Ann. Rept. Ohio State Univ. Res. Found. to Snow Ice and Permafrost Res. Est., Proj. 636, Rept. 11, 150 p.

———   1957.   Study of ice cliff in Nunatarssuaq, Greenland; Ann. Rept. Ohio State Univ. Res. Found. to Snow Ice and Permafrost Res. Est., Proj. 636, Rept. 17, 166 p.

KOCH, L.   1928.   Contributions to the glaciology of north Greenland, Medd. om Grønland, vol. 65, pp. 181-464.

NOBLES, L. H.   1954a.   Glaciology of the Nuna ice ramp, in Operation ice cap, 1953; Trans. Res. and Dev. Com., pp. 53-72.

———   1954b.   Characteristics of high polar type glaciers in northwestern Greenland (Abstr.); Bull. Geol. Soc. Amer., vol. 65, p. 1290.

———   1960.   Glaciological investigations on the Nunatarssuaq ice ramp, northwestern Greenland; Snow Ice and Permafrost Res. Est., Tech. Rept. 66, 57 pp.

NORDENSKJOLD, A. E.   1885.   Report on 2 Dicksonske Expedition to Greenland; Stockholm, pp. 217-24.

PETERSON, J. B.   1924.   Freshwater algae from the north coast of Greenland; Medd. om Grønland, vol. 64, no. 13.

PHILLIPP, H.   1912.   Ueber die Beziehungen der Kryokonitlocher zu den Schmelzschalen und ihren Einfluss auf die ablations-verhältnisse arktischer Gletscher; Zeit. Deut. Geol. Gesell., vol. 64, pp. 489-505.

SCHYTT, V.   1955.   Glaciological investigations in the Thule ramp area; Snow Ice and Permafrost Res. Est., Rept. 28, pp. 88.

STEINBOCK, O.   1936.   Ueber Kryokonitlocher und ihre biologische Bedeutung; Zeit. f. Gletscherk., vol. 24, pp. 1-21.

WHITE, S. E.   1956.   Glaciological studies of two outlet glaciers, northwest Greenland, 1953; Medd. om Grønland, vol. 137, pp. 1-30.

WITTROCK, V. B.   1885.   Om snons och isens flora, in A. E. NORDENSKJOLD, Studier och forskninger; pp. 63-124.

WRIGHT, J. W.   1939.   Contributions to the glaciology of northwest Greenland; Medd. om Grønland, vol. 125, no. 3, 42 pp.

Periglacial Phenomena in Canada[1]

FRANK A. COOK

ABSTRACT

The study of periglacial geomorphology has made great advances throughout the world in recent years. However, in Canada periglacial research is still in an early stage, having emerged as a field of study only in the past decade. Since 1956, however, Canada has been a member of the Commission on Periglacial Geomorphology of the International Geographical Union, and recently a Canadian National Committee of the Commission was formed to co-ordinate research in Canada in periglacial geomorphology.

The opening of the Canadian north since World War II has given geologists and geographers a vast new area in which to conduct field surveys. This same vastness accounts, perhaps, for the paucity of detail on the distribution of periglacial phenomena in Canada, for most field surveys are in the nature of a general reconnaissance. In addition, there are very few scientists actually working on periglacial problems. Large areas still remain unreported. Research is further complicated by the small scale of the map coverage. For most northern areas adequate base maps have been published only during recent years.

In Canada the emphasis on the type of periglacial phenomena studied has differed from that in other countries. Alpine and fossil forms have received little attention; the main interest has been such active forms as patterned ground and permafrost.

The present paper attempts to document references in the literature to periglacial phenomena in Canada. Many well-known features are absent because as yet they are undocumented.

PERIGLACIAL GEOMORPHOLOGY is a comparatively recent branch of study relating to certain soil and landform features produced under very cold climatic conditions. Lozinski (1909, 1910, 1911), in his now famous papers, was one of the first to recognize traces of this special climatic environment. He introduced the term "periglacial" to describe it, and although there has been continuous controversy over the use of the term, it is now almost universally accepted, and by extension is applied to all features, either active or fossil, produced by very cold climates. Troll (1944) considered the alpine zones of high mountains in temperate and tropical regions to be periglacial areas although frost forms were miniature in contrast to the larger forms of the arctic regions.

The memorable excursion of the Stockholm Geological Congress to Spitsbergen in 1910 produced a flood of papers on the more obvious periglacial forms such as polygonal structure. Unfortunately, enthusiasm was short lived, and during the next three decades there was only occasional reference to periglacial phenomena in the literature. Since the end of World War II, however, there has been renewed and very active study in most parts of the world of the problems of periglacial geomorphology. Many workers now consider that the geological and morphological effects of periglacial processes are perhaps more important than those produced by

[1]Published by permission of the Director, Geographical Branch, Department of Mines and Technical Surveys, Ottawa. Her Majesty the Queen in Right of Canada, reserves the right to reprint this article.

glaciers. This new attention is revealed in the publication of hundreds of papers on periglacial problems in many international journals. In 1949 a Commission on Periglacial Geomorphology was set up within the International Geographical Union, and since that time it has been very active in all phases of periglacial research. Also, the appearance in 1954, in Poland, of *Biuletyn Peryglacjalny*, the first bulletin devoted exclusively to periglacial geomorphology, did much to encourage research in this field of study.

DEVELOPMENT OF PERIGLACIAL STUDY IN CANADA

The only publications on periglacial phenomena in Canada in the early years of this century were those of the late J. B. Tyrrell (1904, 1910). In his first paper he suggested the name "crystophene" for ground ice or ice wedges existing below the surface to replace the term "glacier" which was being used locally; in the second paper he described rock glaciers in the Dawson area, Yukon Territory, and attributed them to frost-thaw action. Other papers on periglacial phenomena appearing before World War II were the studies of Nichols (1931, 1936), Paterson (1940), and Polunin (1934) on solifluction and polygonal structure; and of Porsild (1938), who published the first description of pingos in Canada.

The opening of the Canadian north following World War II, with the resulting improvement in transportation and communication facilities, gave natural scientists an opportunity to conduct reconnaissance studies over a vast new area where periglacial processes are still active. The establishment of the northern weather stations, described by Thomson (1948) and Rae (1951), provided new bases of operation. Fraser (1957), has shown that in the last ten years field surveys have been carried out in the Arctic and sub-Arctic from the northern coast of Ellesmere Island to the Hudson Bay coastal plain in Ontario, and from the Alaska boundary to Labrador.

In 1956 Canada became a member of the Commission on Periglacial Geomorphology of the International Geographical Union. The paper by Brochu (1956) is the first report of the status of periglacial research in Canada. Hamelin (1957a, 1959a, 1959b) has since outlined plans for future study, and a national committee of the periglacial commission has recently been formed to co-ordinate research. The Geographical Branch of the Department of Mines and Technical Surveys is developing a long-term plan for the study of periglacial phenomena and processes in Canada (Cook, 1960). Among projects already completed in this programme are (*a*) the reduction of all published data to punch cards; (*b*) the publication of annotated bibliographies on Canadian permafrost and periglacial phenomena (Cook, 1958b, 1959c); and (*c*) a review of the current status of periglacial research (Cook, 1956b). A preliminary map of the known distribution of the numerous forms was prepared by Hamelin for presentation to the 19th Congress of the International Geographical Union held in Stockholm in 1960.

Knowledge of the distribution of the various periglacial forms in Canada is limited because of the vast areas involved and the reconnaissance nature of most surveys to date, and because very few scientists are actually working on periglacial problems. At best there is only scattered reference to isolated phenomena at specific locations, and extensive areas remain unreported. The problem is further complicated by the absence both of an accepted classification for periglacial phenomena and also of a universally accepted nomenclature. The problem of mapping the distribution has been hindered until recently by the lack of base maps on an adequate scale.

In Canada, there has been a different emphasis on the type of phenomena studied from that in most countries. Little attention has been given to alpine forms, whereas elsewhere, in the Alps, the Andes, and other high mountain areas, these have received considerable attention. Fossil forms produced during or immediately after Pleistocene glaciation have recently received a great deal of attention in Europe and in the United States, but have been almost completely neglected in Canada. These forms — involutions, ice wedges, and loess — have been widely used in the study of Pleistocene chronology, or as past climatic indicators. The major interest in Canada has been the active forms such as patterned ground and solifluction. Washburn's important papers (1950, 1956), coming as they did when Canadian geographers and geologists were moving into areas rich in all types of patterned ground, have strengthened interest in this particular phenomenon.

This paper attempts to bring together the scattered references to periglacial phenomena in the Canadian literature. Many well-known forms, however, are absent because as yet they have not been documented.

PERIGLACIAL PHENOMENA IN CANADA

Permafrost

Permafrost is a periglacial phenomenon of considerable importance because of the engineering difficulties it presents. The first comprehensive article discussing the formation, growth, age, and distribution of permafrost in Canada was published by Jenness (1949). Since that time many references have been made to isolated observations of permafrost encountered by mining and engineering specialists. Selected references to permafrost in Canada have been published by Cook (1958b) as an annotated bibliography. The only detailed scientific work on permafrost in Canada has been confined to permafrost research stations at Resolute, in the eastern Arctic, and at Norman Wells and Aklavik in the Mackenzie River area.

The drilling operations at Resolute were carried out during the summers of 1950 to 1953 under a project sponsored jointly by the Dominion Observatory, Department of Mines and Technical Surveys; the Meteorological Division, Department of Transport; the United States Weather Bureau; and the Associate Committee on Soil and Snow Mechanics of the National Research Council of Canada. Many important papers have been published on this project. Thomson and Bremner (1952) provided preliminary observations from shallow-hole readings obtained during the first two years of drilling. Bremner (1955) discussed the problems of drilling in permafrost. Cook (1955, 1958c) analysed near-surface temperatures for the years 1950-3, and temperatures to depths of 650 feet for the period 1952-7. Misener (1955) studied heat flow in the permafrost, and concluded that the lower depth of permafrost at Resolute was 1,280 ± 10 feet. He also found that the temperatures near the surface were higher than would be expected from the gradient at depth and that the heat flow was several times greater than in the southern part of the Canadian shield. Misener and others (1956) discussed the heat-flow measurements and drilling programme at Resolute in general terms. Lachenbruch (1957) offered nearby bodies of water as an explanation for the abnormally large outward heat flow reported by Misener (1956).

The drilling operations at Norman Wells, carried out by the Imperial Oil Company Limited, in association with the National Research Council of Canada, have been summarized by Hemstock (1953). The permafrost investigations in the Mackenzie delta carried out by the National Research Council in connection with the relocation of Aklavik have been outlined by Brown (1956).

Much work remains to be done on the characteristics and distribution of permafrost in Canada. The programme of the Division of Building Research of the National Research Council of Canada has been discussed by Legget and others (1960). Recently, Brown (1959) described the difficulties of mapping permafrost.

Patterned Ground

Patterned ground has the widest distribution of all periglacial phenomena in Canada, and as previously stated, has been the form most intensely studied. Washburn (1956, p. 824) defines patterned ground as "a group term for the more or less symmetrical forms, such as circles, polygons, nets, steps and stripes that are characteristic of, but not necessarily confined to mantle subject to intensive frost action." As such, it includes a wide variety of forms. Although most observations in Canada have been chiefly descriptive, Washburn (1947), Mackay (1953), and Cook (1956, 1958a) have attempted some detailed work, principally on circles. A plan outlining necessary field observations of patterned ground was prepared by Mackay (1957), and recently Cook (1959a) published a photographic study of selected types of patterned ground in Canada. All known references to patterned ground in Canada have been compiled (Table I).

Pingos

Pingos are probably the most spectacular individual periglacial feature found in northern Canada and are easily distinguishable both on the ground and from aerial photographs. Porsild (1938) first described them in the Mackenzie delta–Kotzebue area. Later, in the same general area, Stager (1956) examined and mapped the distribution of 1,380 pingos from aerial photographs. Pihlainen and others (1956) examined one pingo in the Mackenzie delta area in some detail; Craig (1959) found contained organic material in a pingo in the Thelon River basin to have a radiocarbon date of 5,500 ± 250 years B.P.; Mackay (1958a, 1958b) has reported pingos from the Anderson River area; and J. G. Fyles (personal communication) has observed them on Victoria Island. The most comprehensive study of pingos in the western Canadian Arctic was made by Müller (1959). He found them concentrated in a belt coincident with specific climatic and permafrost conditions and extending roughly between latitudes 65° and 75° N, where permafrost is still continuous but where areas of talik, or unfrozen layers of soil, exist under lakes.

Ice Wedges

Ice wedges were first studied in Alaska by Leffingwell (1915, 1919) and have been widely reported elsewhere along the Arctic coast; little, if any, detailed work has been attempted on ice wedges in Canada, and fossil forms are unreported.

TABLE I

PATTERNED GROUND OBSERVATIONS

Type		Description	Location	Source
ACTIVE FORMS				
Polygons	1.	Incomplete polygons	Southampton Island	Bird, 1953
	2.	Non-sorted, average diameter less than 2 feet, depressed margin, 4- to 6-sided, maximum diameter 40 feet	Alert, Ellesmere Island	Blackadar, 1954
	3.	Not well-developed polygons	Admiralty Inlet, Baffin Island	Blackadar, 1956
	4.	Small desiccation 20 cm in diameter, medium 50 cm in diameter, tundra 50 m in diameter	Near Knob Lake	Derruau, 1956
	5.	Nets, average diameter less than 2 feet with high centres and depressed rims, tundra 150 feet in diameter with enclosed secondary polygons	Floeberg Bay, Ellesmere Island	Gadbois et Laverdière, 1954
	6.	Depressed centre 20 inches diameter with raised rims, tundra 50 to 100 feet in diameter	Darnley Bay	Mackay*
	7.	Tundra 50 to 100 feet in diameter	Valley of lower Anderson River	Mackay, 1958
	8.	Non-sorted	Coronation Gulf	Marsden*
	9.	Stone about 7 feet in diameter	Resolute, Northwest Territories	Nichols, 1953
	10.	Stone with maximum diameter of 5 feet	Mugford Tickle, Labrador	Odell, 1933
	11.	Stone 2.5 feet in diameter, giant tundra	Akpatok Island	Polunin, 1934
	12.	Tundra 25 to 75 feet in diameter with depressed border 2 feet wide	Mould Bay, Northwest Territories	Robitaille*
	13.	Tundra, both depressed and raised centre	Fosheim Peninsula	Sim*

TABLE I—*continued*

14.	Occur in peat swamps, approx. 30 feet in diameter with interpolygon area barren peat in V-shaped depressions 1 to 4 feet wide and 1 to 3 feet deep	Shethanei Lake, Manitoba	Taylor, 1958
15.	Mud polygons 1 to 2 feet in diameter, raised centre	Walker Bay, Victoria Island	Washburn, 1947
Stripes			
1.	Non-sorted	Alert, Ellesmere Island.	Blackadar, 1954
2.	Non-sorted	Darnley Bay, Northwest Territories	Mackay*
3.	Non-sorted	Coronation Gulf, Northwest Territories	Marsden*
4.	Non-sorted and sorted	Fosheim Peninsula, Northwest Territories	Sim*
Circles			
1.	(Mud) non-sorted	Resolute, Northwest Territories	Mackay, 1953 Cook, 1956
2.	(Stone) sorted	Resolute, Northwest Territories	Cook, 1958
3.	Non-sorted	Coronation Gulf, Northwest Territories	Marsden*
FOSSIL			
Polygons	Stone 6 to 10 feet in diameter	Southwest Yukon	Raup, 1951
Rings			
1.	Stone	Lower Margaree valley, Cape Breton Island	Raup, 1951
2.	Stone	Alaska Highway, Yukon	Denny, 1952
3.	Stone	Near Portobello Creek, New Brunswick	Lee, 1956
Nets	Stone	Wolf Creek area, Yukon	Sharp, 1942
Stripes	Stone	Northern British Columbia	Denny, 1952

*Personal communication.

Fissures

Fissures or furrows occur practically everywhere along the Arctic coast, including the coasts of many islands. They are especially common on raised beaches and are usually underlain by ice wedges. Washburn (1947) at Cambridge Bay, Victoria Island, and Mackay (1953) at Resolute Bay, Cornwallis Island, have made detailed observations on fissures.

Ground Ice

Ground ice is known to be very common in the Canadian Arctic, but there is little reference to it in the literature. Occurrences were observed by Denny (1952) along the Alaska Highway and by St-Onge (1960) in the Isachsen area, Ellef Ringnes Island.

Ice Mounds

Although small seasonal ice mounds have been reported from isolated points along the Arctic coast and in the Queen Elizabeth Islands, no detailed study has been made of them. Henderson (1952) described till mounds in the Watino area of Alberta as possible fossil forms built in a periglacial environment when the ground was permanently frozen. Gravenor (1955, 1956) has interpreted prairie mounds in the western prairies and in the Peace River area as originating from debris-filled pits which became mounds when ground ice eventually melted.

Lake Ramparts

Lake ramparts or ice-shoved ridges are common periglacial features in Canada, although, as is the case with most other periglacial phenomena, they are seldom reported. The writer has observed many of these forms following the inflections of smaller lakes in the Queen Elizabeth Islands, and has watched them being formed during periods of strong prevailing winds in the Resolute Bay area, Cornwallis Island. On one occasion during August, 1959, great masses of ice were seen over-riding the beach, pushing large amounts of rock and beach material before them. Malcolm (1912) reported fossil ramparts from along the shores of lakes in the Gold Field area of Nova Scotia. They were later observed by Jarvis (1928) paralleling Clear Lake in Renfrew County, Ontario, and by Denny (1952) along the Alaska Highway in northern British Columbia. Charlesworth (1957, plate facing p. 576 after Kupsch) gives an excellent illustration of a lake rampart on the shore of Sucker Lake, Shellbrook, Saskatchewan.

Blockfields

Blockfields, often referred to as felsemmeer or mountain-top detritus are undoubtedly widespread in Canada, both as active forms being formed at present in the Canadian Arctic, and as fossil forms to the south of the zone of intense frost action. Blockfields in the Labrador-Ungava area have received some attention. They were described in the Torngat Mountains by Odell (1933) and Tanner (1944) who believed they post-dated Wisconsin glaciation. Coleman (1921) had previously studied the Torngat blockfields and considered them pre-Wisconsin. Ives (1957, 1958a, 1958b) spent the summers of 1956 and 1957 in the area and concluded that they ante-dated the last glaciation, and that, following a short maximum stage of inundation, the higher summits of the Torngat projected as nunataks through the ice-sheet for a considerable period. Restricted occurrences of blockfields have been observed on Kingaite Peninsula in southernmost Baffin Island by Mercer (1954, 1956) but they could not be dated with certainty. Recently Bergeron (1959) has reported on blockfields on Ungava Peninsula.

Boulder Chains

Boulder chains are common features along the tidal flats of the St. Lawrence River. It is thought that they are numerous in the fjords of the Labrador coast and Baffin Island. At Pangnirtung, for example, the very sharply defined boulder chain limits ship unloading operations to approximately four-and-a-half hours each tide. If the boulders are removed by explosives in the summer to form a channel they tend to be replaced during the following winter. Brochu (1954, 1957) was the first to describe boulder chains in detail, and has attempted to attribute their origin to the movement of ice over tidal flats. He offered statistical evidence to show the ice pushed the blocks to the lower limit of the flats, and was thus the major factor in their development.

Rock Glaciers

Rock glaciers are a periglacial phenomenon found in high mountains throughout the world. They are widespread in the Cordilleran area, in the mountains of Baffin and Ellesmere islands, and in the Torngat but have been rarely reported. Tyrrell (1910) used the concept of icing in talus to explain rock glaciers on the outskirts of the city of Dawson. This locality is outside the limits of Pleistocene glaciation and several hundred feet below the theoretical firn line on northward-facing glaciers, yet in a region of climate favourable to permafrost.

Nivation Hollows

Large nivation hollows have been described by Henderson (1956, 1959) as common features in the highlands of the Labrador Lake plateau in the vicinity of Knob Lake, Quebec. Twidale (1956) has discussed features in the same general area and refers to them as "ravins de gélivation." However, these features have the form of a V-shaped valley and owe their origin more to the work of frost-shattering and solifluction than to nivation. Robitaille (1959, 1960) described and mapped nivation hollows both in the Mould Bay area and in southeastern Cornwallis Island in the high arctic.

Oriented Lakes

The origin of oriented lakes in the Liverpool Bay area, N.W.T., was attributed by Mackay (1956) to contemporary cross-winds rather than to Pleistocene winds as suggested for the origin of similar lakes on the Alaskan coastal plain. The work of Rex (1960) would appear to confirm this view.

Ventifacts

The study of ventifacts is well advanced in Europe but has received little attention in Canada, although Mackay (1958a, 1958b) observed classic examples in the Anderson River area of the western Canadian Arctic.

Sand Dunes

Studies of Pleistocene sand dunes in Alberta were made on aerial photographs by Odynsky (1958) in an attempt to show the probable effective wind direction responsible for their formation. It was concluded that the effective wind direction in the Grande Prairie area was from the southwest, in the central part of Alberta from the northwest, whereas in the southern and southwestern part of the province it was from the southwest. Stalker (1956) noted that wind action was widespread in the Beiseker area of Alberta following the disappearance of an ice-sheet and later a lake, and that dunes then developed over a wide area. Sand dunes in the Deloraine area of Manitoba were attributed to periglacial action by Elson (1956).

Loess

Loess is by far the most important periglacial accumulation, and its study in Europe is well advanced, particularly in Poland. Recently, considerable interest has been shown in the Pleistocene loess deposits that cover tens of thousands of square miles in the United States. In Canada, however, there has been a general neglect of the study of periglacial loess. Millette and Higbee (1958) compared the physical and mineral properties of the periglacial loess and alluvial deposits of the Susquehanna River valley, Pennsylvania, with the deposits in a section of the Laurentians, in Quebec. These properties were used as a means of correlating the field and laboratory observations of the morphological properties of loess and alluvium.

String Bogs

String bogs have been noted on aerial photographs from many northern areas, and have received considerable attention in the Labrador-Ungava region. However, some confusion exists over their formation, and the term has not been clearly defined. Hamelin (1957b) has described them in the Labrador-Ungava peninsula in the zone between 50° and 55° N latitude, and south of the line of continuous permafrost. Allington (1958, 1959) has studied string bogs in detail in the Knob Lake area of central Labrador and found them related to seasonally frozen ground, but not to permafrost. Patterned fens, a form of string bog, have been reported from the Hudson Bay lowland by Sjörs (1959). Although many writers consider string bogs to be restricted to an area outside the zone of continuous permafrost, Henoch (1960) has reported very similar forms from Adelaide Peninsula and King William Island in the western Arctic some 600 miles north of the tree-line, and in an area of continuous permafrost. Mackay (1959c) has discussed "vegetation arcs" in the western Arctic, specifically in the area to the north and west of Great Bear Lake, and on Cornwallis Island, District of Franklin. Williams (1959) suggested that the features observed on Cornwallis Island either were different from typical string bogs, or were localized occurrences as the area is outside the climatic region associated with string bogs.

Palsa Bogs

Palsa bogs are bogs with ice-heaved mounds of peaty soil rising a few feet above wetter ground, and have been observed in areas north of the occurrence of string bogs. Hare (1959), for example, states that in the wetter parts of the central and northern Labrador Trough area — which is underlain by permafrost — palsa bogs replace the familiar string bogs to the south. G. Falconer (1960, personal communication) observed palsa bogs in the Great Whale River area on the eastern coast of Hudson Bay.

Involutions

Involutions have been widely reported as fossil forms in Europe, and recently many references to them have appeared in the United States. The only observations of involutions in Canada, however, have been by Lee (1956) in the Fredericton area, New Brunswick, by Hamelin (1958a) along the banks of some streams in the Abitibi district in northeastern Quebec.

Dry Valleys

It is thought that dry valleys may develop by stream action over permafrost in a periglacial climate. In Canada, Brochu (1956) noted them in the Saguenay area.

Solifluction Forms

Solifluction forms such as slopes, fronts, lobes, and plains are widespread in Canada, and there are many casual references to them in the literature. Nichols (1932) discussed solifluction forms observed on aerial photographs. P. J. Williams (1960, personal communication) has recently undertaken quantitative studies of solifluction features in the Knob Lake area.

CONCLUSION

The principal conclusion that may be drawn from this discussion of periglacial phenomena in Canada is that the knowledge of the distribution is very incomplete. Nevertheless, many of the larger details of the distributional pattern have recently become known. The increasing attention being given periglacial geomorphology by natural scientists in Canada today, the co-ordination of research under the Commission on Periglacial Geomorphology, and the considerable increase in personnel actually working in the field, will do much in the next few years to establish periglacial geomorphology as a major field of research in Canada.

The writer gratefully acknowledges the continuing assistance given by Professor J. Ross Mackay in all phases of periglacial research. He is indebted to members of the northern Canada section of the Geographical Branch of the Canadian Department of Mines and Technical Surveys for stimulating discussion on the merits of works reviewed, and for bringing to his attention many references unknown to him. He wishes to thank Mr. Bernard Gutsell of the Manuscript Division of the Geographical Branch for editorial and technical assistance during the assembly of this paper.

REFERENCES

ALLINGTON, KATHLEEN R.   1958.   Bogs of central Labrador-Ungava; McGill University Sub-Arctic Laboratory Research Paper no. 4, pp. 88-92.

———   1959.   The bogs of central Labrador-Ungava, an examination of their characteristics; McGill University Sub-Arctic Research Paper no. 7, 89 pp.

BERGERON, R.   1959.   Champ de blocs dans le Québec arctique; unpublished manuscript, presented to Symposium of Canadian Committee of Commission on Periglacial Geomorphology of the IGU held Quebec City, December 9, 1959.

BIRD, J. B.   1953.   Southampton Island; Geog. Branch Canada Mem. 1, 84 pp.

BLACKADAR, R. G.   1954.   Geological reconnaissance north coast of Ellesmere Island, Arctic Archipelago, N.W.T.; Geol. Surv. Canada, Paper 53-10, 22 pp.

———   1956.   Geological reconnaissance of Admiralty Inlet, Arctic Archipelago, N.W.T.; Geol. Surv. Canada, Paper 55-6, 40 pp.

BREMNER, P. C.   1955.   Diamond drilling in permafrost at Resolute Bay, Northwest Territories; Publications of the Dominion Observatory, vol. 16, no. 12, pp. 267-390.

BROCHU, MICHEL.   1954.   Un problème des rives du St. Laurent: blocaux erratiques observés à la surface de terrasses marines; Rev. Géomorphologie Dynamique, no. 2, 5ième année, pp. 76-82.

———   1956.   Canada; Builetyn Peryglacjalny, no. 4, pp. 9-14.

———   1957.   Movement of boulders and other sediments by ice on the tidal flats of the St. Lawrence River; Defence Research Board, Directorate of Physical Research Report no. G-1, 8 pp.

BROWN, R. J. E.   1956.   Permafrost investigations in the Mackenzie delta; Canadian Geographer, no. 7, pp. 21-6.

———   1960.   Mapping permafrost in Canada; Arctic, vol. 13, no. 3, pp. 163-77.

CHARLESWORTH, J. K.   1957.   The Quaternary era with special reference to its glaciation, vol. 1, Edward Arnold Ltd., 591 pp.

COLEMAN, A. P.   1921.   Northeastern part of Labrador and new Quebec, Canada; Geol. Surv., Mem. 124, 68 pp.

COOK, FRANK A.   1955.   Near surface soil temperatures at Resolute Bay, Northwest Territories; Arctic, vol. 8, no. 4, pp. 237-49.

———   1956.   Additional notes on mud circles at Resolute Bay, Northwest Territories; Canadian Geographer, no. 8, pp. 9-17.

———   1958a.   Sorted circles at Resolute, N.W.T.; Geog. Bull. no. 11, pp. 78-81.

———   1958b.   Selected bibliography on Canadian permafrost; Geog. Branch Canada, Bibliographical Series no. 20, 23 pp.

———   1958c.   Temperatures in permafrost at Resolute, N.W.T.; Geog. Bull., no. 12, pp. 5-18.

———   1959a.   Some types of patterned ground types in Canada; Geog. Bull., no. 13, pp. 73-81.

———   1959b.   A review of the study of periglacial phenomena in Canada; Geog. Bull., no. 13, pp. 22-53.

———   1959c.   Selected bibliography on periglacial phenomena in Canada; Geog. Branch Canada, Bibliographical Series no. 24.

———   1960.   Geographical branch studies in periglacial geomorphology; Cahiers de géographie de Québec, no. 7.

CRAIG, B. G.   1959.   Pingo in the Thelon valley, Northwest Territories; radiocarbon age and historic significance of the contained organic material; Bull. Geol. Soc. Amer., vol. 70, no. 4, pp. 509-10.

DENNY, W. M.   1952.   Late Quaternary geology and frost phenomena along Alaska Highway, northern British Columbia and southeastern Yukon; Bull. Geol. Soc. Amer., vol. 63, pp. 883-921.

ELSON, J. A.   1956.   Surficial geology of Deloraine, Manitoba; Geol. Surv., Canada, Paper 55-19.

FRASER, J. KEITH.   1957.   Activities of the Geographical Branch in northern Canada, 1947-57; Arctic, vol. 10, no. 4, pp. 246-50.

GADBOIS, PIERRE, et LAVERDIÈRE, CAMILLE.   1954.   Esquisse géographique de la région Floeberg Beach, nord de l'Île Ellesmere; Geog. Bull., no. 6, pp. 17-44.

GRAVENOR, C. P.   1955.   The origin and significance of Prairie mounds; Amer. J. Sci., vol. 253, pp. 475-81.

———   1956.   Air photographs of the plains region of Alberta; Research Council, Alberta, Preliminary Report 56-5.

HAMELIN, LOUIS-EDMOND.   1957a.   Projet de coordination des recherches périglaciaires dans

l'Est Canadien: Notes et nouvelles; Cahiers de Géographie de Québec, no. 3, pp. 141-2.
———— 1957b.   Les tourbières reticulées du Québec-Labrador Subarctique—Interprétation morpho-climatique; Cahiers de Géographie de Québec, no. 3, pp. 87-106.
———— 1958a.   Les cours d'eau à berges festonnées; Canadian Geographer, no. 12, pp. 20-4.
———— 1958b.   Dallage de pierres au lac Lichen: Notes et nouvelles; Cahiers de Géographie de Québec, no. 14, pp. 250-1.
———— 1959a.   La Commission internationale de géomorphologie périglaciaire et le Canada; Canadian Geographer, no. 13, pp. 14-16.
———— 1959b.   La Commision de géomorphologie périglaciaire se réunit en pologne: Notes et nouvelles; Cahiers de Géographie de Québec, no. 5, pp. 146-7.
HARE, F. KENNETH.   1959.   A photo-reconnaissance study of Labrador-Ungava; Geog. Branch Canada, Mem. 6.
HEMSTOCK, R. A.   1953.   Permafrost at Norman Wells, N.W.T.; Imperial Oil Company Ltd., 100 pp.
HENDERSON, E. P.   1952.   Pleistocene geology of the Watino quadrangle, Alberta; Ph.D. thesis, Indiana University.
———— 1956.   Large nivation hollows near Knob Lake, Quebec; J. Geol., vol. 64, no. 6, pp. 607-16.
———— 1959.   A glacial study of central Quebec-Labrador; Geol. Surv., Bull. 50, 94 pp.
HENOCH, W. E. S.   1960.   String bogs in the Arctic 400 miles north of the tree-line; Geog. J. vol. CXXVI, pt. 3, pp. 335-9.
IVES, J. D.   1957.   Glaciation of the Torngat Mountains, Northern Labrador; Arctic, vol. 10, no. 2, pp. 67-87.
———— 1958a.   Mountain-top detritus and the extent of the last glaciation in northeastern Labrador-Ungava; Canadian Geographer, no. 12, pp. 25-31.
———— 1958b.   Glacial geomorphology of the Torngat Mountains, Northern Labrador; Geog. Bull., no. 12, pp. 47-75.
JARVIS GERALD.   1928.   Lacustrine littoral forms referable to ice pressure; Canadian Field Naturalist, vol. 42, no. 2, pp. 29-32.
JENNESS, JOHN L.   1949.   Permafrost in Canada; origin and distribution of permanently frozen ground with special reference to Canada; Arctic, vol. 2, no. 1, pp. 13-27.
JOHNSON JR., J. PETER.   1952.   Information collected about Dumbbell Bay and north Ellesmere Island regions of the Canadian Arctic Archipelago; Dartmouth College Paper 102, 100 pp.
LACHENBRUCH, ARTHUR H.   1957.   Thermal effects of oceans on permafrost; Bull. Geol. Soc. Amer., vol. 68, no. 11, pp. 1515-30.
LEE, HULBERT A.   1956.   Superficial geology of Fredericton, York and Sunbury counties, New Brunswick; Geol. Surv., Canada, Paper 56-2.
LEFFINGWELL, E. DE K.   1915.   Ground-ice wedges; the dominant form of ground-ice on the north coast of Alaska; J. Geol., vol. 23, pp. 635-54.
———— 1919.   The Canning River region, northern Alaska; U.S. Geol. Surv., Prof. Paper 109, 251 pp.
LEGGETT, R. F., DICKENS, H. B., and BROWN, R. J. E.   1960.   Permafrost investigations in Canada; The geology of the Arctic, Toronto, vol. II, pp. 956-69.
LOZINSKI, W.   1909.   Ueber die mechanische Verwitterung des Sandsteine im gemassigten Klima; Bull. Inst. Acad. Sci., Cracovie, Cl. Sci., Math. et Nat. ni. 1.
———— 1910.   O mechanicznym wietrzeniu piaskowcow w klimacie umiarkowanym; P.A.U.
———— 1911.   Die periglaziale Fazier der mechanischen Verwitterung; CR. XI, Intern. Geol. Congr. 1910, Stockholm.
MACKAY, J. ROSS.   1953.   Fissures and mud circles on Cornwallis Island, N.W.T.; Canadian Geographer, no. 3, pp. 31-7.
———— 1956.   Notes on oriented lakes of the Liverpool Bay area, Northwest Territories; Rev. Canadienne de Géographie, vol. X, no. 4, pp. 169-73.
———— 1957.   Field observations of patterned ground; Canadian Alpine J., vol. 40, pp. 91-6.
———— 1958a.   The valley of the lower Anderson, N.W.T.; Geog. Bull., no. 11, pp. 37-48.
———— 1958b.   Anderson River map area; Geog. Branch Canada, Mem. 5, 127 pp.
———— 1958c.   Arctic "vegetation arcs"; Geog. J., vol. 124, pt. 2, pp. 294-5.
MALCOLM, W.   1912.   Goldfields of Nova Scotia; Geol. Surv., Canada, Mem 20E, 331 pp.
MERCER, J. H.   1954.   The physiography and glaciology of southernmost Baffin Island; Ph.D. thesis, McGill University.
———— 1956.   Geomorphology and glacial history of southernmost Baffin Island; Bull. Geol. Soc. Amer., vol. 67, no. 5, pp. 553-70.
MILLETTE, J. F., and HIGBEE, HOWARD W.   1958.   Periglacial loess: I— Physical properties; Amer. J. Sci., vol. 256, pp. 284-93.

MISENER, A. D.   1955.   Heat flow and depth of permafrost at Resolute Bay, Cornwallis Is., N.W.T., Canada; Amer. Geophys. Union, Trans., vol. 36, pp. 1055-60.

MISENER, A. D. et al.   1956.   Heat flow measurements at Resolute Bay, N.W.T.; J. Roy. Astron. Soc. Canada, vol. 50, no. 1, pp. 14-24. Reprinted as Contributions of the Dominion Observatory, vol. 1, no. 19.

MÜLLER, FRITZ.   Beobachtungen ueber pingos; Medd. om Grønland, Bd. 153, no. 3, pp. 1-127.

NICHOLS, D. A.   1932.   Solifluction and other features in northern Canada shown by photographs from the air; Trans. Roy. Soc. Canada, ser. 3, sec. 4, vol. 26, pp. 267-75.

――――   1936.   Physiographic studies in the eastern Arctic; Canadian Surveyor, vol. 5, no. 10, pp. 2-7.

NICHOLS, R. L.   1953.   Geomorphological observations at Thule, Greenland and Resolute Bay, Cornwallis Island, N.W.T.; Amer. J. Sci., vol. 251, no. 4, pp. 268-75.

ODELL, N. E.   1933.   The mountains of northern Labrador; Geog. J., vol. 82, no. 3, pp. 193-210.

ODYNSKY, W.   1958.   U-shaped dunes and effective wind directions in Alberta; Can. J. Soil Sci., vol. 38, no. 1, pp. 56-62.

PATERSON, T. T.   1940.   The effects of frost action and solifluction around Baffin Bay and in the Cambridge district; Geol. Soc. London Quart. J., vol. 96, pp. 99-130.

PIHLAINEN, J. A. and others.   1956.   Pingo in the Mackenzie delta; Bull. Geol. Soc. Amer., vol. 67, pp. 1119-22.

POLUNIN, NICHOLAS.   1934.   The vegetation of Akpotak Island—Part I; J. Ecol., vol. 22, pp. 337-95.

PORSILD, A. E.   1938.   Earth mounds in unglaciated arctic, Northwestern America; Geog. Rev., vol. 28, no. 1, pp. 46-58.

RAE, R. W.   1951.   Joint Arctic weather project; Arctic, vol. 4, no. 1, pp. 18-26.

――――   1960.   Hydrodynamic analysis of circulation and orientation of lakes in northern Alaska; The Geology of the Arctic, vol. II.

ROBITAILLE, BENOÎT.   1959.   Présentation d'une carte géomorphologique de la région de Mould Bay, Île-du-Prince Patrick, Territoires du Nord-Ouest; Canadian Geographer (in press).

――――   1960.   Géomorphologie du sud-est de l'île Cornwallis territoires du Nord-Ouest; unpublished manuscript.

ST-ONGE, DENIS.   1960.   The ground ice in the Deer Bay area, Ellef Ringnes Island, N.W.T.; unpublished manuscript.

SJÖRS, HUGO.   1959.   Bogs and fens in the Hudson Bay Lowlands; Arctic, vol. 12, no. 1, pp. 3-20.

STAGER, JOHN K.   1956.   Progress report on the analysis of the characteristics and distribution of pingos east of the Mackenzie delta; Canadian Geographer, vol. 7, pp. 13-20.

STALKER, A. MacS.   1956.   Beiseker, Alberta: Surficial geology, map with marginal notes; Geol. Surv., Paper 55-7.

TANNER, V.   1944.   Outlines of the geography, life and customs of Newfoundland–Labrador (the eastern part of the Labrador Peninsula); Acta Geog. vol. 8, no. 1, p. 216.

TAYLOR, F. C.   1958.   Shethanei Lake, Manitoba; Geol. Surv., Canada, Paper 58-7, 11 pp.

THOMSON, ANDREW.   1948.   The growth of meteorological knowledge in the Canadian Arctic; Arctic, vol. 1, no. 1, pp. 34-43.

THOMSON, ANDREW, and BREMNER, P. C.   1952.   Permafrost drilling and soil temperature measurements at Resolute Bay, Cornwallis Island, Canada; Nature, vol. 170, no. 4330, pp. 705-6.

TROLL, CARL.   1944.   Strukturböden, Solifluktion und Frostklimate der Erde; Geol. Rundschau, bd. 34, pp. 545-694.

TWIDALE, C. R.   1956.   Vallons de gélivation dans le centre du Labrador; Rev. Geomorphologie Dynamique, no. 1-2, pp. 18-23.

TYRRELL, J. B.   1904.   Crystosphenes or buried sheets of ice in tundra of North America; J. Geol., vol. 12, no. 3, pp. 232-6.

――――   1910.   "Rock glaciers" or crystocrenes; J. Geol., vol. 28, pp. 549-53.

WASHBURN, A. L.   1947.   Reconnaissance geology of portions of Victoria Island and adjacent regions, Arctic Canada; Geol. Soc. Amer., Mem. 22, 142 pp.

――――   1950.   Patterned ground; Rev. Canadienne Géographie, vol. 4, no. 3-4, pp. 5-59.

――――   1956.   Classification of patterned ground and review of suggested origins; Bull. Geol. Soc. Amer., vol. 67, pp. 823-66.

WILLIAMS, M. Y.   1936.   Frost Circles; Trans. Roy. Soc. Canada, ser. 3, sec. 4, vol. 3, pp. 129-32.

WILLIAMS, P. J.   1959.   Arctic "Vegetation arcs"; Geog. J., vol. 125, pt. 1, pp. 144-5.

Comparison of Gravitational and Seismic Depth Determinations on the Gilman Glacier and Adjoining Ice-Cap in Northern Ellesmere Island

J. R. WEBER

ABSTRACT

During the summers of 1957 and 1958 twelve bedrock profiles on the Gilman Glacier and on the ice-cap between Gilman Glacier and Mount Oxford were determined from seismic reflections. During the second field season more than 200 gravity stations were established over the same general area. The regional Bouguer anomaly was calculated from the known ice thickness at a few selected locations along the seismic profiles, and was then extrapolated for the whole area. With this information and with assumed specific gravities of ice and bedrock of 0.9 and 2.71 respectively, the ice thickness was calculated from the gravity measurements. Agreement between the bedrock profiles as determined by the two methods was very close. It is concluded that a gravity survey, when supplemented by a few seismic soundings, can give a good indication of the shape of the bedrock and the ice thickness.

THE THICKNESS OF VALLEY GLACIERS and ice-caps is usually measured by the seismic method. The thickness can also be estimated from the measurement of gravity changes at the surface of the ice (Bull and Hardy, 1956; Littlewood, 1952; Thiel *et al.*, 1947). The advantage of the gravity over the seismic method lies in the simplicity and rapidity of the field work. However, the interpretation of gravity changes is not always simple and can be formidable task in the case of narrow valley glaciers between high mountains (Jacobs *et al.*, 1960), while the interpretation of the seismic data is straightforward, provided that good reflections are obtained.

In the present investigation some 200 gravity stations were established over an extended glacier region, where the bedrock elevations, in certain areas had previously been fairly well established from seismic reflections (Figure 1). (Sandstrom, 1959; Weber and Sandstrom, 1960).

The purpose of the work was to evaluate the merit of the gravity method by comparing the bedrock profiles obtained from gravity interpretations with the seismic reflection profiles. It was hoped that these studies would reveal to what degree gravity interpretations could replace or supplement seismic soundings under certain conditions.

SURVEY AND SEISMIC CONTROL

The region of seismic and gravity surveys extended from the snout of the Gilman Glacier (82°03′20″N; 70°18′42″W) to Mount Oxford (82°09′53.7″N;

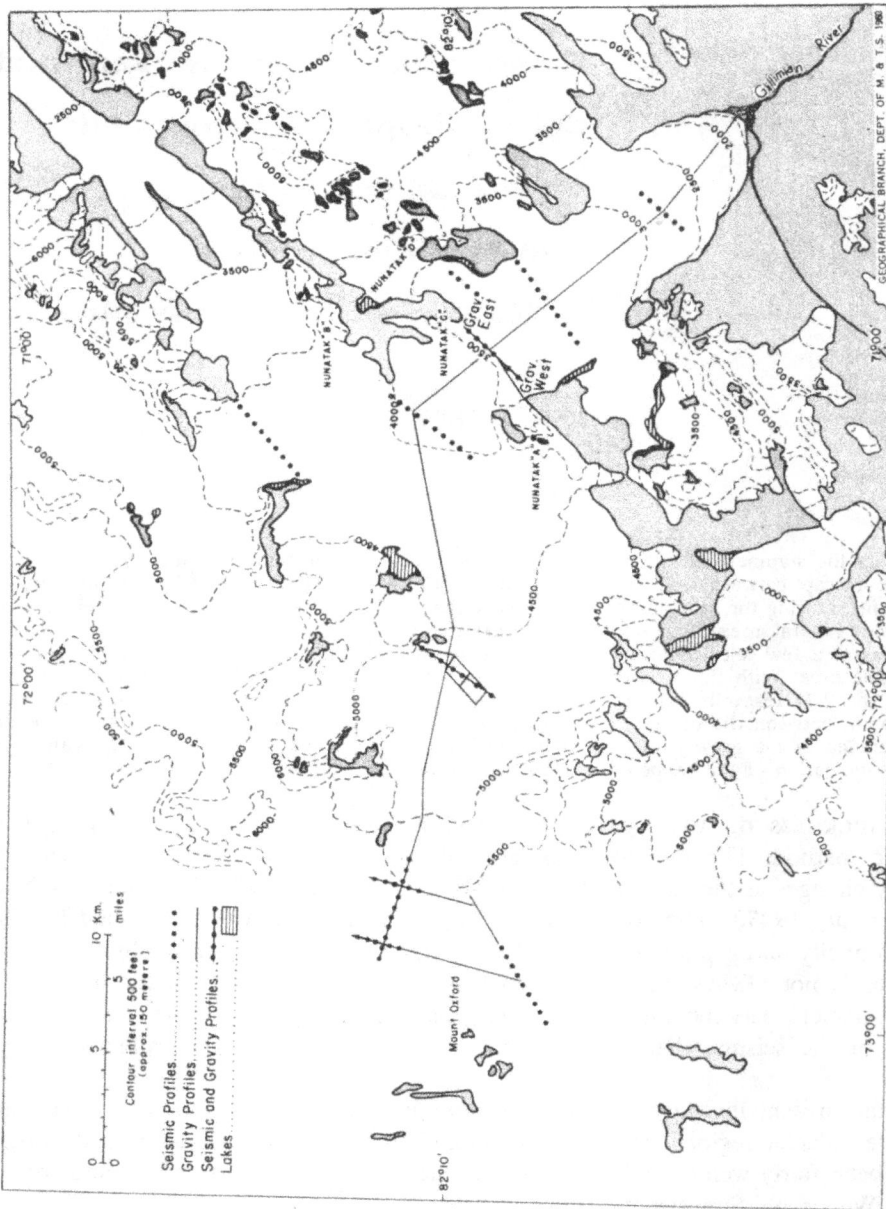

FIGURE 1. Gilman Glacier, northern Elesmere Island.

73°11'00.1"W). Some forty movement and ablation stakes, which were fixed by theodolite resection, were used as gravity stations. The remaining gravity stations were surveyed by theodolite resection and with steel tape and altimeter. It is believed that differences in elevation between stations are accurate to 0.3 m, whereas the absolute elevation above sea-level is accurate to about 1 m. Most of the survey work was carried out by K. C. Arnold and H. Sandstrom. In the summer of 1957 the seismic field work on the Gilman Glacier was carried out by F. S. Grant and H. Sandstrom; in May, 1958, a refraction profile on the Gilman Glacier and a few reflection profiles on the ice-cap near Mount Oxford were shot by Sandstrom and the writer. The locations of the seismic profiles are shown in Figure 2.

ESTABLISHMENT OF GRAVITY STATIONS

The "Operation Hazen" IGY programme included carrying out a gravity traverse from the head of Clements Markham Inlet over the United States Range *via* Gilman Glacier to the head of Chandler Fjord. For this purpose Dr. Grant tied the local gravity network to the gravity network of the Dominion Observatory by observations at Fort Churchill, Resolute, and Alert, and by establishing new, absolute stations at the Lake Hazen base camp, the Gilman Glacier camp, and on the east shore of Clements Markham Inlet, which were also stations in the local network. Values of gravity in the local network are, therefore, absolute values, although in fact relative gravity differences would have been adequate for ice thickness determinations. Dr. Grant used a Worden gravimeter (geodetic model, serial no. 44) lent by the Dominion Observatory, and established his stations by aircraft; the writer used a Worden gravimeter (standard model, serial no. 10) lent by the University of Toronto, and established the local stations by motor toboggan, by dog team, on skis, and on foot. The observations were made in loops so that corrections could be made for instrument drift, which was remarkably slight, probably due to the almost constant temperature during a twenty-four hour period.

FIGURE 2.   Location of the seismic profiles.

COMPUTATION OF ICE THICKNESS

The gravity traverse from Clements Markham Inlet to Chandler Fjord showed that the Bouguer anomaly decreased gradually from about zero at Clements Markham Inlet to a minimum of —80 mg near Mount Oxford, and increased again to about —10 mg at Chandler Fjord.

The regional anomaly, A, can be expressed by the equation

(1) $$A = g_0 - \gamma + C_B + C_T$$

where

$g_0 =$ observed gravity

$\gamma =$ theoretical gravity

$= 978{,}049 \ (1 + 0.0052884 \sin^2\theta - 0.0000059 \sin^2 2\theta)$

where $\theta =$ latitude

$C_B =$ Bouguer correction for free air, rock, and ice

$C_T =$ terrain correction

(All values expressed in milligals).

The equation

$$a = (0.3086 - 0.04185 \ \delta) \text{ milligals per metre}$$

gives the correction factor combining Bouguer's reduction and free-air reduction, where δ is the density of the material between the point of observation and sea-level. The surface rock for the whole Gilman Glacier and Mount Oxford region is quartzite, and density determination of rock samples showed an average density of 2.71 g/cm³. Assuming an ice density of 0.9 g/cm³, the correction factors become

$a_1 = 0.1952$ milligals per metre for rock,

$a_2 = 0.2709$ milligals per metre for the ice.

If it is assumed as an approximation that the rock and ice under the gravity station form a slab of infinite extent, then the Bouguer correction can be calculated from the equation

(2) $$C_B = a_1 h_1 + a_2 h_2$$

where

$h_1 =$ bedrock elevation in metres,

$h_2 =$ ice thickness in metres.

The ice thickness is obtained by substituting $h_1 = h - h_2$ in equation (2), where h is the surface elevation of the ice in metres.

Hence

$$h_2 = \frac{C_B - a_1 h}{a_2 - a_1} = 1.321 \ (C_B - 0.1952 \ h)$$

where

$$C_B = \gamma - g_0 + A - C_T \text{ (from (1) above).}$$

The values that have to be known for the reduction of each gravity station are, therefore, elevation, latitude, observed gravity, regional anomaly, and terrain correction.

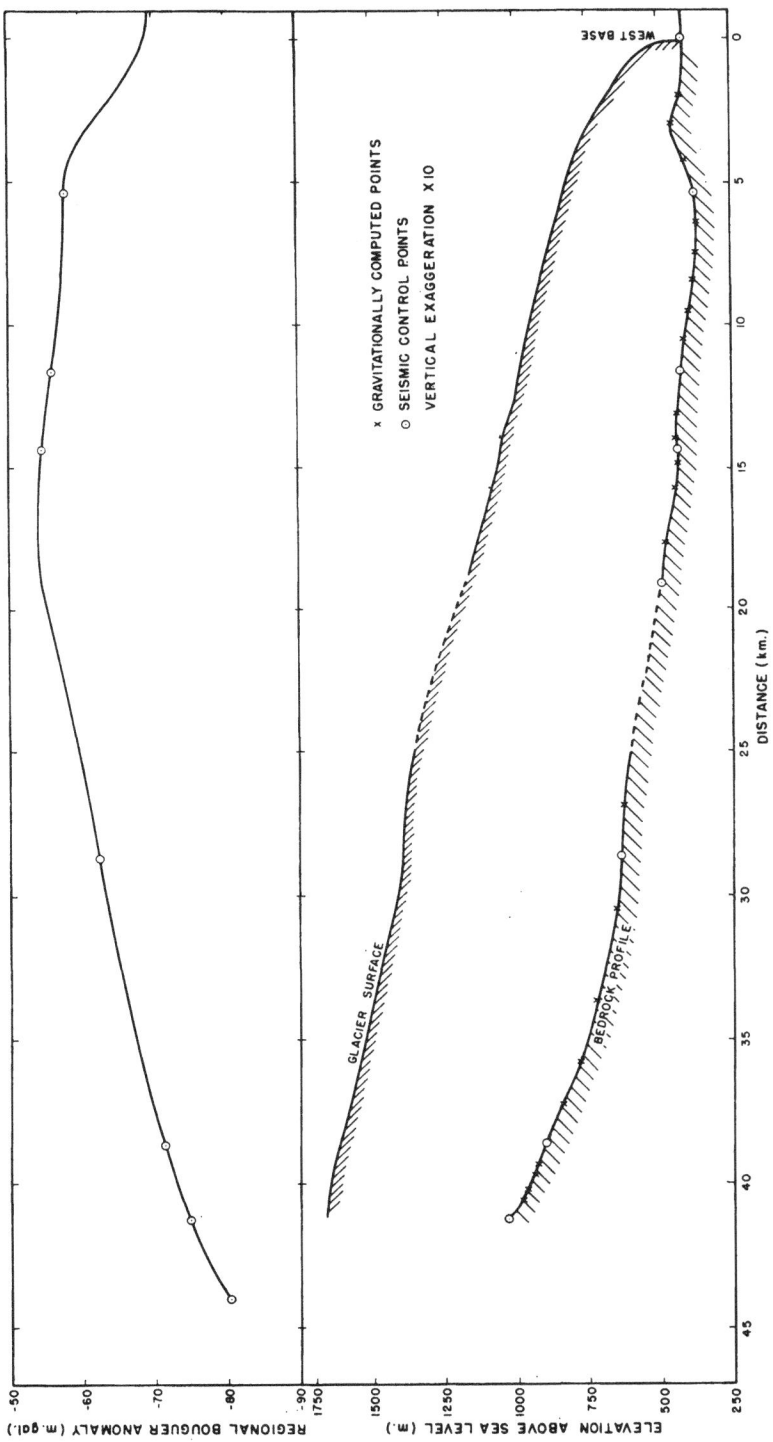

FIGURE 3 (top). Plot of regional anomaly as it changes from the snout of Gilman Glacier to Mount Oxford, some 45 km away.

FIGURE 4 (bottom). Corresponding bedrock profile computed from the curve.

The regional anomaly is the quantity most difficult to assess. It cannot be calculated from gravity observations on the bedrock of nunataks near the glacier because it is impossible even to estimate the terrain correction without very accurate contour maps. However, in the case of an arctic glacier, such as the Gilman Glacier, conditions are ideal for determining the regional anomaly at the centre of the glacier at points where the ice thickness is known from seismic reflections. The Gilman Glacier has an average width of 5 km; its bedrock is quite flat and gently sloping so that the terrain correction at the centre is negligible and the regional anomaly can be simply computed from equations (1) and (2). As the surface geology is fairly uniform, abrupt changes in regional anomaly are unlikely, and a few depth soundings along the centre of the glacier will give enough information for interpolating the regional anomaly as it changes relative to the mountain range. Figure 3 shows a plot of the regional anomaly as it changes from the snout of the Gilman Glacier to Mount Oxford, some 45 km away. Over that distance six seismic depth soundings are available for control. The corresponding bedrock profile computed from the regional anomaly curve is shown in Figure 4. It is interesting that certain bedrock features which accord with the topography of the surrounding mountains, but were not suspected from the seismic reflections, show up from the gravity interpretation.

Comparison between Seismic and Gravity Profiles

All the gravity stations on ice were reduced and the bedrock elevations plotted. It was found that the agreement between seismic and gravity profiles was excellent as long as the stations were not too close to rock outcrops. Even for profiles across the glacier the agreement was good, except at stations near the glacier edge. As an example Figure 5 shows the reduced gravity profile together with the computed reflection points of seismic profile 108.

In order to check the assumption that the terrain effect in the central area of the glacier can be neglected, a gravity profile of thirty stations was established between the off-glacier stations "Gravity E" and "Gravity W," where the glacier is relatively narrow (4200m). From a 500-foot (152 m) contour map (Figure 1) the terrain effect of the surrounding four nunataks was computed for the six points of the traverse Gravity E, M1, M2, M5, G, and Gravity W. The evaluation of the terrain effect was done by means of a mechanical integrator described by A. J. E. Siegert (1942), and the effect of each individual nunatak on each of the six profile points is tabulated in milligals in Table I.

TABLE I

Station	Nunatak "A"	Nunatak "B"	Nunatak "C"	Nunatak "D"	Total Contribution
Gravity E	0.1	0.6	(1.2)	0.7	(2.6)
M1	0.1	0.6	0.3	0.5	1.5
M2	0.2	0.4	0.1	0.3	1.0
M5	0.5	0.4	0	0.2	1.1
G	2.5	0.2	0	0.2	2.9
Gravity W	(7.2)	0.2	0	0.1	(7.5)

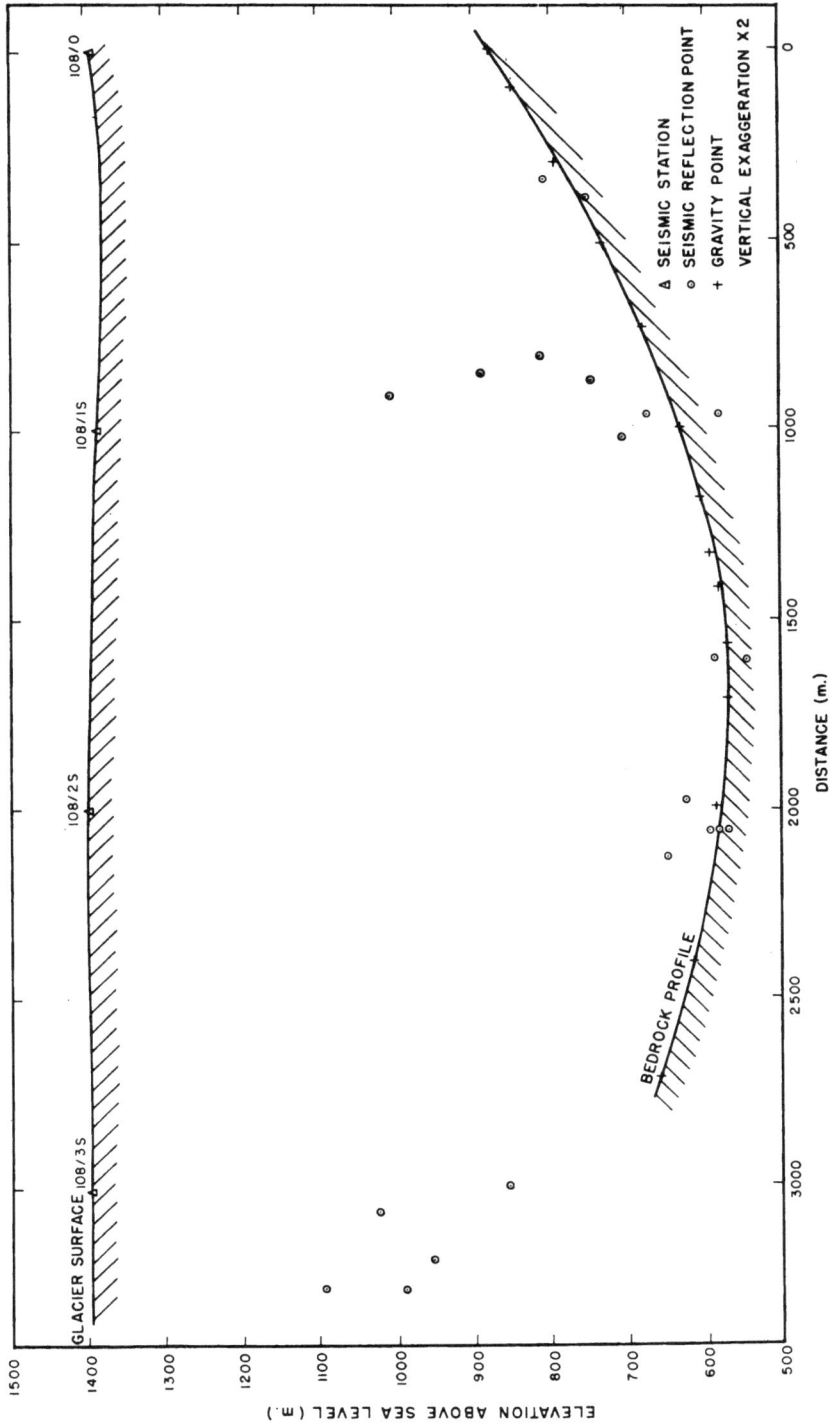

FIGURE 5. Reduced gravity profile together with the computed reflection points of seismic profile 108.

The results are plotted in Figure 6, in which (a) shows the Bouguei anomaly computed by assuming the ice to be of the same density as the surrounding rock, thus showing the combined effects of ice and terrain; (b) shows the terrain effect alone; (c) shows the bedrock profile computed from the gravity values with constant regional anomaly; (d) shows a curve similar to that in (c), except that here the terrain effect has been added to the regional anomaly; and (e) shows the bedrock of the seismic profile 101 nearby. For the determination of the regional anomaly, the assumption that the terrain effect at the centre of the glacier can be neglected is clearly justified, because the effect is less than 1 milligal at the centre of the narrow part of the glacier. The curves also show that the gravitationally-computed bedrock profile is most accurate where the terrain effect is small compared to the effect caused by the mass deficiency of the ice, and that the dis-

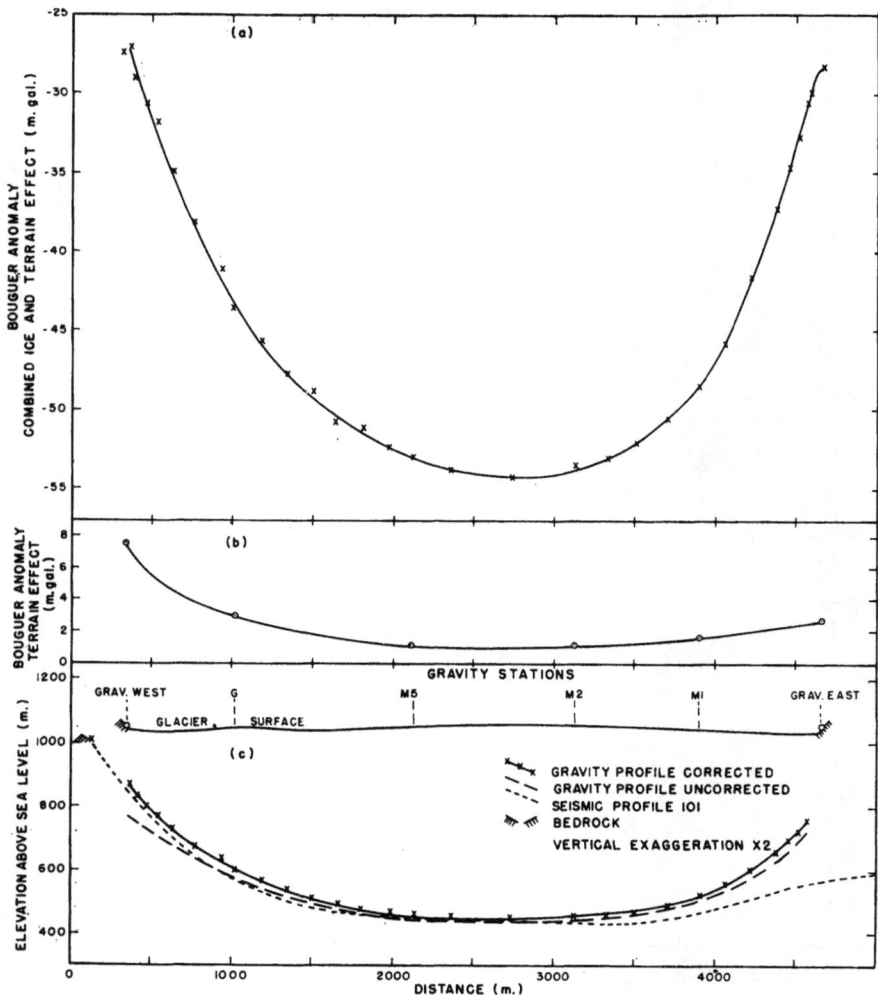

FIGURE 6

crepancy between the real and the computed profile increases towards the glacier edges. The terrain-corrected gravity profile does not extend to the off-glacier stations, because it is impossible to make an accurate assessment of the terrain effect of stations so close to the mountains. Some of the seismic profiles shot on the ice-cap near Mount Oxford, where firn and ice are some 800 m deep, gave very poor reflections, so that it was impossible to determine the shape of the bedrock from the seismic records. On the other hand, the reduction of the gravity observations taken over the same profiles revealed a very undulatory bedrock floor, which was belied by the flat snow surface and was the reason for the poor seismic reflections.

CONCLUSIONS

Gravity measurements alone, without seismic depth control, can give a good picture of the shape but no accurate elevations of the bedrock. When supplemented by a few seismic soundings, the gravity method provides an excellent yet simple means of determining the shape and elevation of the bedrock beneath glaciers and ice-caps. But, when no contour maps are available and terrain correction is impossible, its easy application to valley glaciers is limited to geologically uniform regions with relatively wide and flat glaciers. Although these conditions rarely occur in the glaciated regions of the temperate zones, they are common in the polar regions. Without elaborate terrain corrections the gravity method fails to give the exact shape of the walls near the edge of a valley glacier. It is the writer's experience, however, that seismic reflections near the glacier edge are usually so scattered as to give even less information than the gravity method.

ACKNOWLEDGMENTS

The writer wishes to express his thanks to the Defence Research Board of the Department of National Defence for sponsoring "Operation Hazen," and for making available funds for both the field work and the working-up of the results; to Mr. T. A. Harwood for his efficient planning of the operation; and to Dr. Hattersley-Smith for his excellent and inspiring leadership. The author is particularly indebted to Mr. H. Sandstrom for his co-operation and the sharing of his valuable experience in gravity and survey work, as well as for computing the results of the seismic programme; and to Mr. K. C. Arnold for his help in the field and for working out all the survey data. The advice and assistance of Dr. G. D. Garland of the University of Alberta in preparation of this paper are also gratefully acknowledged. The paper is published by permission of the Chairman, Defence Research Board.

REFERENCES

BULL, C., and HARDY, J. R. 1956. The determination of the thickness of a glacier from measurements of the value of gravity; J. Glac., vol. 2, pp. 755-62.

JACOBS, J. A., GRANT, F. S., and RUSSELL, R. D. 1960. Gravity measurements on the Salmon Glacier and adjoining snow field, British Columbia.

LITTLEWOOD, C. A. 1952. Gravity measurements on the Barnes icecap, Baffin Island; Arctic, vol. 5, no. 2, pp. 118-24.

SANDSTROM, H. 1959. Operation Hazen: Geophysical methods in glaciology, Part I; Defence Research Board. Ottawa, Report no. D, Phys. R (G) Hazen 3.

SIEGERT, A. J. F.   1942.   A mechanical integrator for the computation of gravity anomalies; Geophysics, vol. I, pp. 354-66.

THIEL, E., LaCHAPELLE, E., and BEHRENDT, J.   1947.   The thickness of Lemon Creek Glacier, Alaska, as determined by gravity measurements; Trans. Amer. Geophys. Un., vol. 38, no. 5, pp. 745-9.

WEBER, J. R., and SANDERSON, H.   1960.   Operation Hazen: Geophysical methods in glaciology, Part II; Defence Research Board, Ottawa, Report no. D, Phys. R (G) Hazen 6.

Some Glaciological Studies

in the Lake Hazen Region

of Northern Ellesmere Island

G. HATTERSLEY-SMITH

ABSTRACT

Glaciological research on the ice-cap and glaciers to the north of Lake Hazen in northern Ellesmere Island was one of the main objectives of the Canadian IGY expedition to the area in 1957-8. The work was concentrated on Gilman Glacier, but surveys extended over the ice-cap from Clements Markham Inlet in the northeast to Tanquary Fjord in the southwest. Studies of accumulation and ablation indicated a budget deficit of about 60 per cent for Gilman Glacier and its accumulation area for the year 1957-8; a budget deficit which was appreciably greater in 1956-7; and from preliminary data a smaller budget deficit in 1958-9. At the highest level of the ice-cap (1800 m) the mean annual accumulation over the last twenty years is estimated at 12.8 gm cm^{-2}; the mean height of the equilibrium line is about 1200 m. The main glaciers from the high ice-cap are thinning slightly in their lower reaches but their snouts are not receding; it may be expected that similar glaciers elsewhere in Ellesmere Island and in Axel Heiberg Island are reacting in the same way to the present climate. Marginal features in the ablation area of Gilman Glacier provide some evidence of recent thinning and of recession from the rock walls of the glacier. South of the main ice-cap, small local ice masses, which do not rise above about 1200 m, and associated glaciers are both thinning and receding at their edges year by year. It seems certain that thinning and recession is occurring on small, low-level ice-caps throughout the Queen Elizabeth Islands.

NORTHWEST OF THE LINE Tanquary Fjord – Lake Hazen – Alert, little more than half of northern Ellesmere Island is covered by ice at the present time. The ice cover of this region of high mountains, but of low precipitation, is relatively thin and limited in contrast to the great mass of the Greenland ice-cap centred many miles to the southeast.

In the north coastal region, which is intersected by fjords, collecting grounds are too restricted and the general elevation not great enough for an extensive ice cover with the present scanty snowfall; glaciers only occur in locally favoured areas. Thus, a strip along the north coast, 30 to 50 km wide, is predominantly ice free, although alpine glaciers occur where the height of the mountains exceeds about 1200 m and small corrie glaciers and masses of snowdrift ice occur at lower elevations with suitable exposure to snow-bearing winds. It is probable that the glaciers and local ice masses of the immediate coastal area are thinning at the present time. Between Cape Columbia and Cape Discovery the north coast is fringed for about 90 km by an ice-shelf; further west, as far as Cape Bourne, there are discontinuous ice-shelves in most of the fjords and bays (Koenig *et al.,* 1952). The genesis of the ice-shelves was due primarily to the ability of great thicknesses of sea ice to form off this coast and remain fast; ice from the land played only a minor part in their development (Hattersley-Smith *et al.*, 1955). The formation of sea ice is favoured by the great winter cold, low summer melt, and low precipitation of the region. Upward growth of the ice-shelf took place through freezing of sea water at the lower surface and through firn formation and refreezing of melt-water at the upper surface (Marshall, 1955). There is evidence that these ice-shelves have developed *de novo* as a result of the deterioration of climate which has taken place since the Climatic Optimum, 4000 to 6000 years ago (Crary *et al*, 1955). At the present time they are wasting through calving to form ice islands (Koenig *et al.*, 1952) and through a net surface ablation which may amount to 60 cm in a single summer (Hattersley-Smith *et al.*, 1955). Much of the present land ice cover near the north coast may be as recent a development as the ice-shelves, and advances of glaciers over raised beaches may signify a limited reglaciation since the Climatic Optimum (Hattersley-Smith *et al.*, 1955). In contrast the main ice-cap is a survival of the much greater, probably complete, ice cover over northern Ellesmere Island in earlier times. Much of the evidence for this former glaciation, such as moraines and striae, is doubtless beneath the sea or has been destroyed by frost action and river erosion. But erratics found at a height of

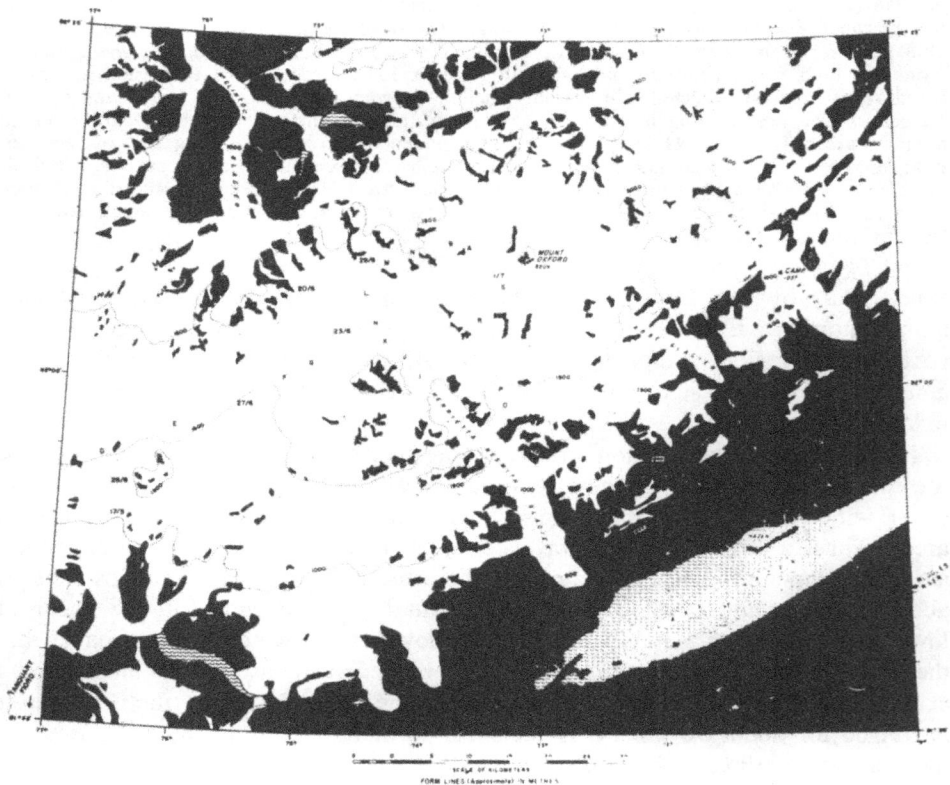

FIGURE 1. Map of part of the ice-cap of northern Ellesmere Island, showing glaciological stations. Cartography: Surveys and Mapping Branch, Department of Mines and Technical Surveys, Ottawa. Elevations and form lines from surveys by K. C. Arnold on "Operation Hazen."

over 300 m on Ward Hunt Island (Hattersley-Smith *et al.*, 1955), and at elevations up to 1000 m on the ice-free mountains between Lake Hazen and the main ice-cap (Smith, 1959) testify to the extent of the former ice cover.

The elevation of the central part of the ice-cap in the area shown on the map (Figure 1) varies between about 1500 and 2000 m; few nunataks rise above 2300 m. The highest mountains are situated on the flanks of the ice-cap, namely, between the heads of M'Clintock and Milne fjords and between the head of Tanquary Fjord and Henrietta Nesmith Glacier. A mountain of about 2515 m in the latter location is probably the highest in northern Ellesmere Island (Arnold, 1959). The ice moves out towards the periphery of the high land, whence some of it moves in great trunk glaciers down to the long fjords of the north coast, some escapes through gaps in the mountains to form piedmont glaciers in high plateau regions, and some spills out as glacier tongues into valleys at lower

FIGURE 2.   Central part of the ice-cap of northern Ellesmere Island from the east. Photo: R.C.A.F., June 24, 1950, from 20,000 feet.

elevations. On the north side of the ice-cap, the glaciers occupy valleys extending far back into the mountains, which were presumably carved by rivers in pre-glacial time; they were thus able to channel ice from the highest accumulation areas, so that their valleys underwent powerful glacial erosion, which may also have been favoured by rather higher precipitation on the northern than on the southern slopes of the ice-cap. The central part of the ice-cap (Figure 2) is by no means a plateau, for where not broken by nunataks it shows an undulating hill and dale topography with a local relief of about 120 m. Seismic soundings showed that within 8 km of Mount Oxford there are variations in thickness of the ice from 230 to 850 m, indicating buried mountains (Weber and Sandstrom, 1960). On the south side of the ice-cap, the glaciers flow away much less steeply and are probably slower moving than on the north side.

LAKE HAZEN

Before examining the results of glaciological work on the glaciers and ice-cap, it is worth considering briefly some of the climatic conditions at Lake Hazen,

FIGURE 3.   Base camp on north shore of Lake Hazen from the southwest, showing ice-free conditions on the lake. Photo: R.C.A.F., August 24, 1957, from 1500 feet.

situated at an elevation of 158 m above sea level. With an area of some 525 sq. km, Lake Hazen is by far the largest body of fresh water in the Queen Elizabeth Islands. Mean annual temperature, annual precipitation at the lake, and ice conditions on the lake are of interest in reference to climatic conditions on the ice-cap. A thorough study of the meteorology at the Lake Hazen base camp was made by Jackson from observations over one year (Jackson, 1959); the base camp was situated on the north shore of the lake, opposite the western end of Johns Island (Figures 3 and 4). For the year August 1957 to August 1958 the estimated mean annual temperature was —6°F (—21°C), or nearly 5°F colder than the mean annual temperature at a height of 1037 m on Gilman Glacier, where a steady temperature of approximately —1.2° F (—18.4° C) was observed at a depth of 24.5 m in the ice (Hattersley-Smith, 1960). Jackson concluded that the Lake Hazen trough, lying close below the ice-cap, acts as "a gigantic frost-hollow."

The difficulties of measuring precipitation in the Arctic are great, but with due allowances it seems unlikely that the mean annual precipitation at Lake Hazen

FIGURE 4.   Johns Island and Lake Hazen base camp from the south, showing ice-choked conditions on the lake. Photo: R.C.A F., August 28, 1959, from 10,000 feet.

exceeds a water equivalent of 6.5 cm (Jackson, 1959). The precipitation is appreciably greater on the ice-cap, where the snow also has a very different texture. On the lake between the base camp and Johns Island, in the first week of May 1957, there was a mean depth of 23 cm of soft, loose snow composed of platy crystals up to 1 cm in diameter, with a water equivalent of 5.1 cm, if a mean density of 0.22 gm cm^{-2} is assumed. In the same area on May 3, 1958, there was a mean snow depth of 31 cm, with a water equivalent of 7.7 cm. Half-way down the snow pack there was a conspicuous layer of wind-packed snow, formed by a mid-January blizzard. It seems certain that part of the wind-packed layer of snow, if not all, had been blown off the land to the lake. On May 21 and May 22, 1959, the mean snow depth was 28 cm, with a water equivalent of 6.6 cm; a layer of wind-packed snow, containing dust particles, had formed near the bottom of the snow pack, indicating a high wind in late fall or early winter. Typical profiles in 1958 and 1959 are shown in Figure 5. These profiles do not, of course, take account of precipitation during the melt season or of snow falling into the lake during the period of freeze-up in late August and the month of September. The snow conditions on the lake have a practical importance in the landing of aircraft. The shallow depths, which are normal, allow wheeled landings and take-offs by certain types of aircraft with little or no preparation of the surface; an absence of wind-packed snow, as in the spring of 1957, gives conditions of maximum safety. Snow profiles on the lake should be compared with those on the ice-cap and on Gilman Glacier (Figures 5 and 7). On the ice-cap in 1957 and 1958, the snow accumulation was two or three times greater than on the lake; on the glacier, although precipitation is higher than on the lake, the snow accumulation measured in the spring was similar owing to wind deflation.

In some years the ice on Lake Hazen has melted completely by the end of August, while in others it melts only round the margin to give a narrow shore lead, with large pools at the ends of the lake (Dunbar and Greenaway, 1956). In 1957, the ice thickness on the lake in the spring varied from 1.5 to 1.56 m and in 1958, from 1.37 to 1.98 m, according to measurements by Deane (1958, 1959); the wide variation in 1958 was due to the irregularities of the winter snow cover. On May 19, 1959, the ice was 1.86 m thick between the base camp and Johns Island. The ice thickness varies from year to year according to the number of degree-days of cold in the winter, to whether or not any ice remains from the previous winter, and to the depth of snow cover. The summer break-up on the lake is dependent to some extent on the ice thickness, but from early July onwards is particularly affected by the influx of rivers of melt water from the ice-cap to the north and the plateau to the south, and by the incidence of high winds after the ice has begun to candle. The latter factor is probably the most disruptive of all. In 1957, the lake was completely ice free by the end of the first week in August, but the mean ice thickness was less than in 1958 and 1959, the summer warmer, and there were high southwesterly winds in late July. In 1958, the lake was ice free to the east of Johns Island, but a concentration of four-tenths ice remained in the western part of the lake at the end of August, which Deane (1959) attributed primarily to the absence of strong winds during the summer. In 1959, a concentration of nine-tenths ice remained over the whole lake at the end

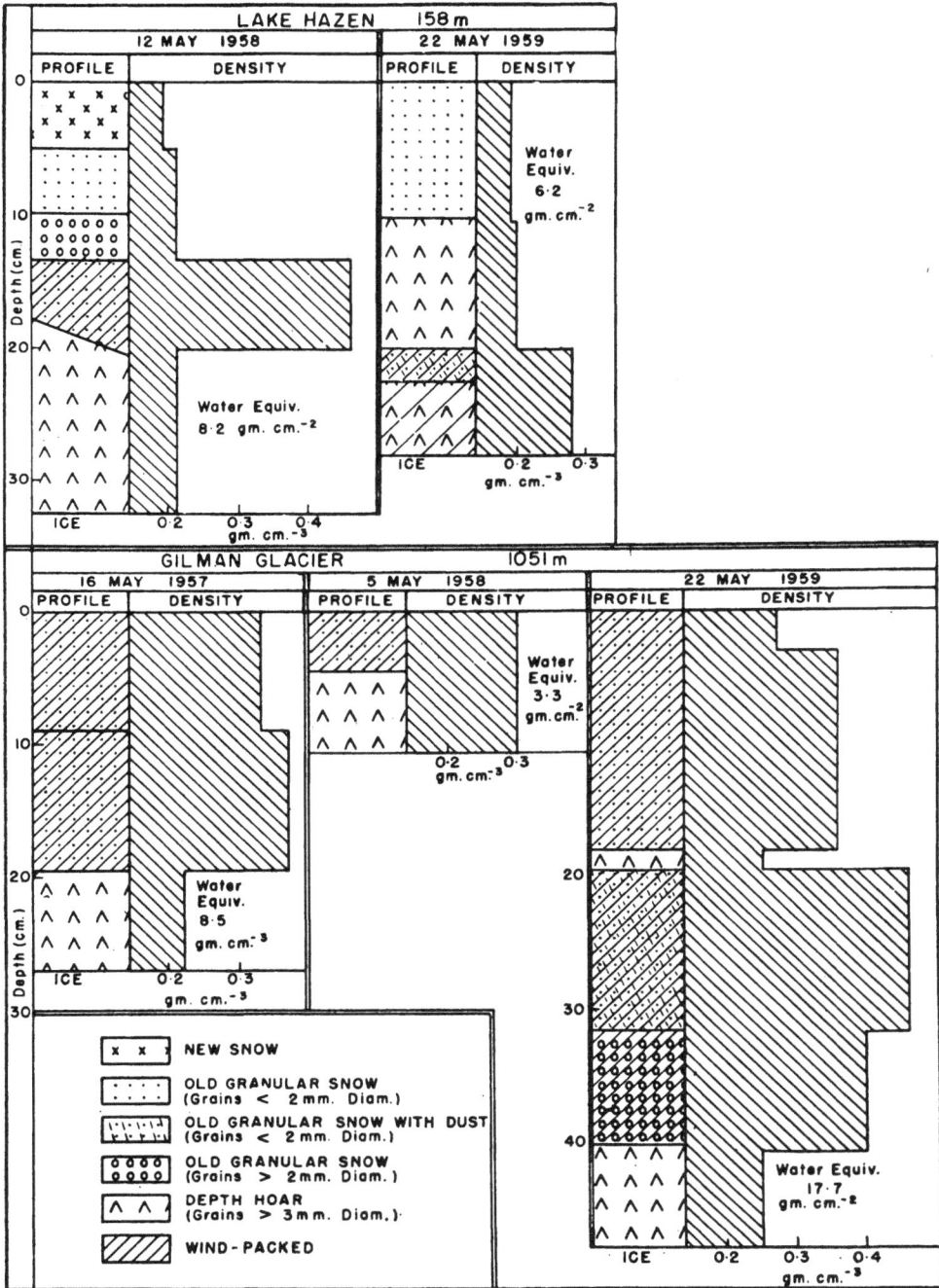

FIGURE 5.   Snow profiles on Lake Hazen and on Gilman Glacier, 1957-9.

of August. The 1959 summer was cooler than the two previous summers, when the mean ice thickness on the lake may have been less; however, the main cause of the heavy ice conditions was probably the absence of high winds. The contrasting ice conditions in 1957 and 1959 are well shown in two air photographs of the lake between Johns Island and the north shore, taken late in the month of August (Figures 3 and 4). A "Canso" flying-boat landed on this part of the lake on August 20, 1958, but a landing was out of the question in late August, 1959; instead, a landing was made on a wide shore lead near the outlet of Ruggles River (Figure 1). Because the factors which affect ice conditions on the lake are not the same as those controlling melting on the glaciers, it may be coincidental that the summers of 1957, 1958, and 1959 were seasons of progressively more ice concentration on the lake in late August and of progressively less melting on Gilman Glacier.

INVESTIGATIONS ON THE ICE-CAP AND GLACIERS

The glaciological work in 1957-8 included studies of accumulation and ablation, firn and ice stratigraphy, englacial temperatures, melt-water features, and moraines. A geophysical and survey team carried out measurements of ice thickness and movement. In 1957, glaciological work was confined to Gilman Glacier and its accumulation area between the glacier camp and Mount Oxford; measurements were made at some fifty stations. In 1958, measurements were made at nearly all the 1957 stations in the environs of Gilman Glacier, and at about fifty additional stations, some over areas not covered in 1957, namely, the lower 7 km of Gilman Glacier, the tributary glaciers east of the camp, and small local ice masses. Measurements were also made at some thirty stations in the area between Mount Oxford and a glacier at the head of Tanquary Fjord, about 90 km west of Gilman Glacier. In this paper some of the data obtained are used to form an estimate of the 1957-8 budget of Gilman Glacier and its accumulation area, and an assessment of the regimen of the main ice-cap and glaciers.

ACCUMULATION

The accumulation was studied in pit profiles at various elevations. The equilibrium line[2] lies at a mean elevation of about 1200 m. From the highest part of the ice-cap at 2000 m down to a level of about 1450 m, accumulation is by firn formation; from 1450 m down to 1280 m, surface increments show an interfingering of firn and superimposed ice; in a narrow belt below 1280 m down to the equilibrium line, the annual accumulation consists almost entirely of superimposed ice (Hattersley-Smith, 1960). There is a pronounced difference between the snow profiles above and below an elevation of about 1400 m. It is near this level that the major glaciers debouch from the ice-cap, and the ice cover becomes channelled into valleys. Below 1400 m the snow cover in spring is less deep and more wind-hardened than at higher levels owing to the action of strong katabatic winds blowing off the ice-cap. By deflation these winds promote surface loss from the glacier, and thus tend to raise the equilibrium line.

[2]The line below which there is no net surface increment either of firn or of superimposed ice (Baird, 1952).

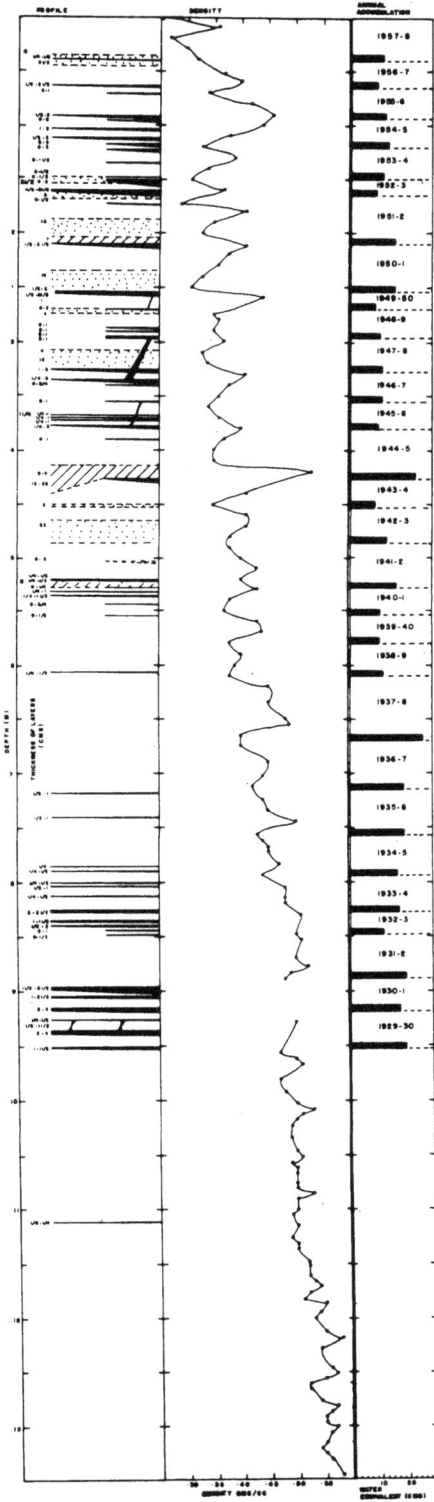

FIGURE 6. Firn profile on the ice-cap at 1805 m (station 1/7).

The following observations were made on traverses of the ice-cap between Gilman Glacier and a glacier above Tanquary Fjord; the line of traverse passed over the highest part of the ice-cap near Mount Oxford (Figures 1 and 2). It was interesting to find that the detail of snow, ice, and rock in a photograph taken northwestward from the summit of Mount Oxford (Hattersley-Smith, 1958) was almost identical with that shown in a photograph taken by Moore on the only previous visit to the ice-cap more than twenty years before (Moore, 1936). In 1958, a pit was dug at station 1/7, 4 km west of Mount Oxford; it was 6.5 m deep and situated at an elevation of 1805 m, near the centre of an extensive névé where the accumulation seemed to be fairly uniform over an area of about 80 sq. km at the head of the northeast tributary of Henrietta Nesmith Glacier. From the bottom of the pit a bore-hole was sunk for a further 7 m into the firn; density, annual accumulation, and temperature measurements were made in the pit and bore-hole, the results of which are described in detail elsewhere (Hattersley-Smith, (1960). The annual stratigraphy based on density measurements and occurrence of ice layers is shown in Figure 6. On July 1 the snow depth was 40 cm; the subsurface temperatures fell rapidly from near freezing point at the surface to a steady temperature of —24°C at a depth of 3 m, which was maintained to the bottom of the bore-hole at 13.5 m. During the summer at this altitude, the air temperature may rise to freezing point, or a little above, from about the last week in June to the third week in July. Below the surface, freezing temperatures are unlikely to penetrate below a depth of about 30 cm, or below the previous winter's snow pack. Percolating melt-water will therefore not form ice layers below this stratum, but may form them at any level within it, which leads to difficulties in interpretation of annual stratigraphy based solely on the presence of ice layers. However, density variations in the firn indicate which of the ice layers should be regarded as annual markers. Calculations for the years 1938 to 1958 give a mean annual accumulation of 12.8 gm cm^{-2}. Although ice layers up to 5 cm thick were observed in the upper part, the lower part of the section was almost free of ice layers, indicating little or no summer melting in the decade or so before 1930. Subsequent summers were marked by appreciable melting, particularly in the early nineteen-thirties and in the years since 1944. Although no attempt was made to separate annual strata in the lower part of the section, the bottom of the bore-hole probably corresponded to about the year 1916. The accumulation for the year 1957-8, up to early July, was 12.0 gm cm^{-2}.

The névé where station 1/7 was situated is separated by rock ridges from an undulating névé, which extends about 60 km to the westward and is the ice shed of glaciers flowing into Disraeli, M'Clintock, Tanquary, and Yelverton fjords on the north coast, to the Yelverton-Tanquary through-valley in the west, and to the Lake Hazen trough in the south. Snow and firn profiles were examined at eleven stations on this undulating névé; the stations were situated 5 to 7 km apart and at an elevation of 1500 to 1800 m, or above the level where the glaciers debouch from the ice-cap (Figure 1). Measurements were made between June 20 and July 1, 1958, during which period there was negligible precipitation. Snow depths varied from 36.5 to 53.5 cm, with a mean depth of 45.4 cm and a mean snow density of 0.30 gm cm^{-3}; the mean accumulation for 1957-8 was thus

13.6 gm cm^{-2}. Figure 7 shows the profile in snow and firn at Station 27/6 at an elevation of about 1480 m; the snow here was 53.5 cm deep, or rather deeper than the mean, probably because the station was situated in the depression which marks the upper end of the valley occupied by M'Clintock Glacier. Wind action is mainly responsible for variations in density and hardness of the snow, and for the variations in snow depth from station to station on the ice-cap. Comparisons of mean accumulation over several years at different stations, rather than comparison of snow depths, may serve to indicate areas of greater or less accumulation. Thus the mean accumulation at station 27/6 over the three years from 1955 to 1958 was 17.8 gm cm^{-2}, as compared with 11.6 gm cm^{-2} at station 1/7. The effect of wind action was well shown in the uppermost part of Henrietta

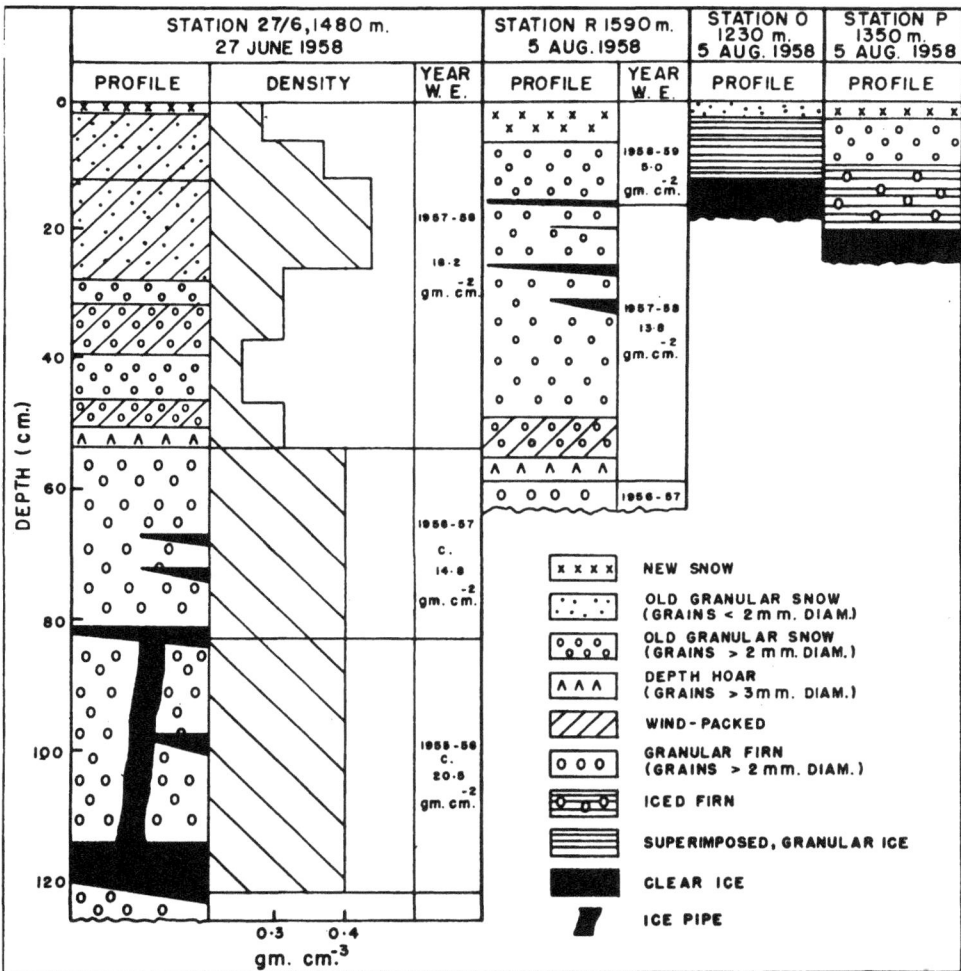

FIGURE 7.   Snow and firn profiles on the ice-cap at 1480 m (station 27/6) and on the northeast tributary of Henrietta Nesmith Glacier at 1590 m (station R), 1350 m (station P), and 1230 m (station O).

Nesmith Glacier between elevations of 1500 and 1600 m. At Station K at 1640 m, the snow depth was 44.5 cm; at Station J at 1600 m, 35.5 cm; and at Station I at 1540 m, 15.5 cm. Snow depths were less variable on the uppermost part of Disraeli Glacier where measurements at elevations between 1450 and 1780 m at five stations — 29/6, A, B, M, and N — gave a mean snow depth of 39 cm with a water equivalent of about 11.7 gm cm^{-2}; snow depths ranged from 33 to 42 cm.

For the part of the ice-cap to the west of Mount Oxford, pit data in the late summer of 1958 are available only from stations situated on the névé at the head of the northeast tributary of Henrietta Nesmith Glacier. Profiles, examined on August 5 at stations 1/7 (1805 m), S (1735 m), and R (1590 m), help to define the end of the ablation season at the higher levels on the ice-cap. Snow depths at the three stations were respectively 56, 62, and 58.5 cm, but these depths included snow deposited after the short period of ablation, which ended before July 21 when there was a severe blizzard. It was concluded from the position of thin ice layers and hard frozen layers and from the earlier measurements at station 1/7 (Figure 6) that approximately 22, 15.5, and 15.5 cm of snow had accumulated after the end of the ablation period. The profile at station R (Figure 7) is typical; it shows the accumulation for the budget year 1957-8, 13.8 gm cm^{-2}, and the early accumulation for 1958-9, 5.0 gm cm^{-2}. After the late June and July 1 pit measurements on the ice-cap, it is believed that there was negligible accumulation before the end of the ablation period above the 1500 m level. The observations indicate a mean accumulation of about 14 gm cm^{-2} in the budget year 1957-8, with values as high as 18 gm cm^{-2} at station 27/6 and as low as 10 gm cm^{-2} at station 23/6.

The névé which runs east from Mount Oxford forms the main accumulation zone of Gilman Glacier which is the only significant outlet of the névé (Figure 2). In 1958 the equilibrium line was situated at an elevation of about 1200 m (see below). In the part of the accumulation zone above 1400 m, data from snow pits indicated a mean accumulation not exceeding 15 gm cm^{-2}, with snow depths varying from 30 to 50 cm; between an elevation of 1400 m and the equilibrium line the mean accumulation was appreciably smaller, for less snow was deposited at these levels and loss through melting and run-off was considerable (Hattersley-Smith, 1960; Sagar, 1960). The area of accumulation of Gilman Glacier was approximately 362 sq. km. It can therefore be said that the total accumulation over the whole area, expressed as water equivalent, did not exceed $362 \times 15 \times 10^{-5}$ or 5430×10^{-5} km^3.

ABLATION

A network of ten stakes was established in two rows across Gilman Glacier, 1 to 3 km north of the camp, in May 1957 (Figure 1); the mean elevation of these stakes was 1049 m (Arnold, 1959). Table I shows the mean snow depth and water equivalent of the snow in spring, and the mean ablation across the glacier at these stakes during three seasons of observations, 1957-9; the water equivalent (W.E.) for ablated ice is given on the assumption that the mean density of the glacier ice is 0.9 gm cm^{-3}.

TABLE I

Year	Snow depth (cm)	W.E. of snow (gm cm^{-2})	Ice ablation (cm)	W.E. of ice (gm cm^{-2})
1957	23	6.5	78.2	70.4
1958	22.5	6.5	54.9	49.4
1959	21.5	7.3	38.0*	34.2*

*At four stakes only up to 15 August.

Although snow depths at the stakes in the spring showed considerable variations owing to wind action from no snow cover at all to a depth of 48 cm, the mean depths during the three years were similar. On an average the snow was 16 cm in 1957, 6 cm in 1958, and 12 cm in 1959 deeper at the western four stakes than at the eastern four. Variations in the spring snow profiles near the west side of the glacier at the same stake over the three years are shown in Figure 5. The 1959 profile was characterized by a dirty layer of exceptionally hard and dense wind-packed snow near the bottom, which probably corresponded to the dirty layer observed in the snow on Lake Hazen. Very strong late fall or early winter winds had blown dust on the glacier from the bare rock of nunataks at least 4 km away. From the profiles on the east side of the glacier the dirty layer was absent, indicating that the glacier was mainly stripped of snow in this area. It was evident that northerly and northeasterly winds in winter and spring tended to blow the snow off the eastern part of the glacier, where as a result ablation was rather higher than on the western part, particularly in the summer of 1957.

The stake network did not cover the margins of the glacier, but ended about 500 m from the east side and 1.5 km from the west side. Owing to a preponderance of wind-blown rock material and to the insolation effect at bare rock surfaces, ablation is much greater near the margins of the glacier than elsewhere. Ablation data for 1958 are available from three stakes set 50 to 100 m from the eastern margin of the glacier at an elevation of 950 m. The mean ablation at the three stakes was 192.5 cm of ice, with a water equivalent of 173 gm cm^{-2}, up to August 14, or more than three times the ablation measured during the same period at the stake network at an elevation of only 100 m greater. Excessive wastage at the margins is partly compensated for by lateral spreading of the glacier which was measured at 6 m over a width of 1800 m at the stake network between July 1957 and July 1958 (Arnold, 1959). Although the rate of ablation falls off rapidly with distance from the margins, the mean ablation at the stake network should be regarded as a minimum approximation for the ablation over the width of the glacier.

In 1958 ablation was measured above the main stake network at two stations at 1080 and 1140 m, where the ablation was 47 and 18.5 cm of ice respectively for the whole summer. A station at 1224 m showed a net accumulation of 14.5 cm of snow and superimposed ice, with a water equivalent of about 7 gm cm^{-2}. The equilibrium line can safely be placed at an elevation of approximately 1200 m for the summer of 1958. Ablation was also measured along four transverse profiles of three to six stakes at mean elevations of 830, 860, 985, and 1015 m, as well as at a number of single stakes at intermediate levels. A level survey (Arnold, 1959) showed that the gradient of the glacier between elevations of 830 and 1200 m over a distance of 14.2 km is nearly uniform at about 1 in 40. It has therefore seemed permissible to plot ablation against elevation for the upper part of the ablation zone in order to arrive at a mean value for the ablation over this part of the glacier. This has been done in Figure 8, which shows a roughly linear variation of ablation with elevation. At the mean elevation of the whole area, 1015 m, a theoretical value for the mean ablation is read off from the graph, and is found to be approximately 58 gm cm^{-2}. The total area of the ablation zone of Gilman Glacier above 830 m is 85 km^3. As the glacier maintains a roughly constant width, the area is multiplied by the mean ablation to give a total ablation loss above 830 m of $98 \times 58 \times 10^{-5}$ or 5684×10^{-5} km^3, expressed as water equivalent.

Below the 830 m level in a distance of 5.4 km the glacier falls through 415 m to the lowest part of its snout, which rests on the outwash gravels of the river that drains it; the slope is steeply convex in the lower 3 km (Arnold, 1959). Ablation

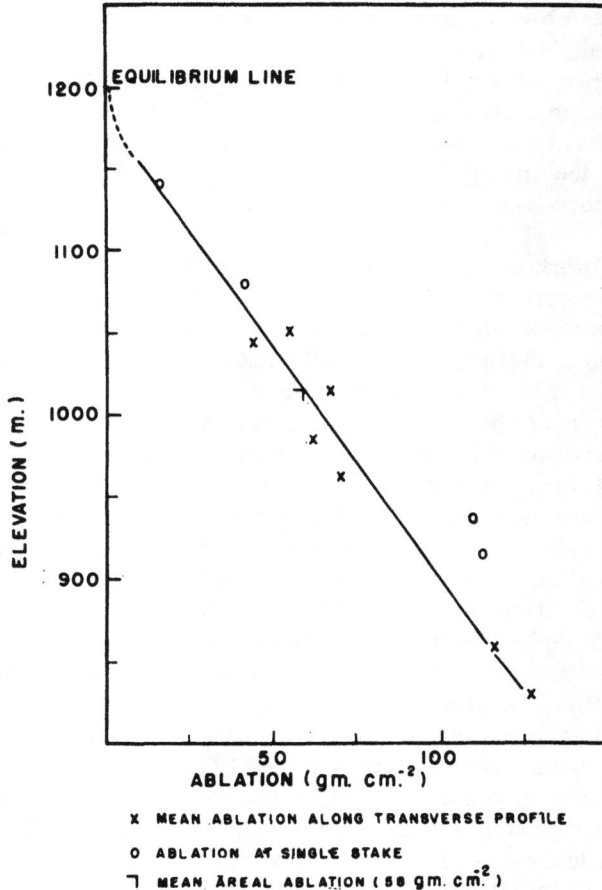

FIGURE 8.   Ablation of ice on Gilman Glacier, 1958.

data for the summer of 1958 are available from three stakes (M.22, 23, and 24) on this part of the glacier; the elevation and ablation at these stakes, and their distance from the lowest part of the snout are shown in Table II. In 1958, the ablation was not measured below the M.24 stake; in 1959, the ablation at stake M.25 (see Table II) was 144.5 cm of ice (Sagar, 1959), when the ablation at M.22, 23, and 24 was on average 29 per cent less than in 1958. Table II shows an estimate from these data of the 1958 ablation at M.25, but does not take into account the ablation closer to the snout. A few metres upglacier from the 20 m ice cliff the mean ablation in 1959 was 201.5 cm of ice (Sagar, 1959), and in 1958 is estimated to have been of the order of 285 cm of ice. The mean ablation at the four stations in the table was thus 161 cm of ice, with a water equivalent of 145 gm cm^{-2}. Since the stations were more or less evenly spaced and situated 1.5 km or more away from the margins of the glacier, this value is regarded as a minimum estimate for mean ablation over the part of the glacier below 830 m. The area of this part of the glacier was 21 sq. km, and its total ablation in 1958 was therefore $21 \times 145 \times 10^{-5}$ or 3045×10^{-5} km^3, expressed as water equivalent.

TABLE II

Stake	Distance from snout (km)	Elevation (m)	Ablation of ice (cm)	W.E. of ice (gm cm^{-2})
M.22	4.4	794	154	138.5
M.23	3.3	740	138	124
M.24	2.5	671	151	136
M.25	1.5	610 (est.)	203 (est.)	182 (est.)

Quantitative ablation data are available only for Gilman Glacier, but snow pit data in 1958 allow the elevation of the equilibrium line to be assessed with reasonable accuracy on two other southward-flowing glaciers. On a glacier above Tanquary Fjord in the middle of May, 19.5 cm of snow, with a water equivalent of 6.5 gm cm^{-2}, rested on glacier ice at station 17/5. This station at an elevation of about 1080 m was clearly below the equilibrium line. On the same glacier at the end of June, 22 cm of snow, with a water equivalent of 8.0 gm cm^{-2}, rested on alternating layers of coarse, iced firn and superimposed ice at station 26/6. This station at about 1280 m was above the equilibrium line. It is safe to say that the equilibrium line on this glacier lay at an elevation of approximately 1200 ± 30 m, with a margin of error due to the aneroid barometer in use. On the northeast tributary of Henrietta Nesmith Glacier, station 0 at about 1230 m was very close to the equilibrium line, for here on August 5, 2.5 cm of crusted snow rested on 10 cm of granular, superimposed ice which in turn rested on a surface of clear ice (Figure 7). Station P at 1350 m was certainly above the equilibrium line, as here on August 5 a 10-cm layer of iced firn was encountered below 10 cm of snow (Figure 7). It is concluded that the equilibrium line on this glacier lay at an elevation of 1230 ± 30 m. On both glaciers therefore the equilibrium line was at approximately the same level as on Gilman Glacier. Although no ground data are available, air photographs and observations from vantage points on the north side of the ice-cap indicate that the elevation of the equilibrium line is not widely different on the northward-flowing glaciers.

BALANCE OF GILMAN GLACIER IN 1957-8

The equation for the mass balance of Gilman Glacier in the budget year 1957-8 is as follows:

$$\left\{ \begin{array}{l} \text{Total areal accumulation} \\ \text{above equilibrium line} \end{array} \right\} \sim \left\{ \begin{array}{l} \text{Net areal ablation} \\ \text{above 830 m} \end{array} \right. + \left. \begin{array}{l} \text{Net areal ablation} \\ \text{below 830 m} \end{array} \right\} = \begin{array}{l} \text{Increment} \\ \text{or} \\ \text{Deficit} \end{array}$$

Substitution in this equation of the values found above shows a deficit of $(5684 + 3045 - 5430) \times 10^{-5}$ km^3, or 3299×10^{-5} km^3, which is 61 per cent of the total accumulation. The deficit thus derived is a low rather than a high estimate, because, as stated above, the values used for mean accumulation and ablation were respectively maximum and minimum approximations. In the 1956-7 budget year the deficit must have been considerably greater, for although the accumulation was similar the elevation of the equilibrium line was about 40 m higher, and the mean ablation near the Gilman Glacier camp about 40 per cent greater. In the 1958-9 budget year ablation figures only are available (Sagar, 1959); they suggest that the glacier was more nearly in a state of balance than in 1956-8.

Studies in the deep pit at station 1/7 indicated that summers had been rather warmer in the last twenty years than in the previous twenty. These studies and the inferred budget deficits in two years of observations on Gilman Glacier suggest that the glacier has been thinning in recent years. Evidence of thinning and slight recession from the rock walls comes from marginal lakes, from ice-cored ridges of moraine and debris cones near the sides of the glacier, and from fillings of old surface drainage channels. Nevertheless, the position of the snout appears to have remained the same, for at the snout there is a rough balance between the amount of movement during the year, about 3 cm per day (Arnold, 1959), and the ablation during the summer. The glacier, rising very steeply through 200 m in the first 1.5 km and attaining a thickness of 400 m 5.5 km from the snout (Sandstrom, 1959), is so massive that only the excess ablation of many years could be expected to have an appreciable effect on its contours.

CONCLUSIONS

It is concluded that the main glaciers flowing southwards from the high ice-cap are thinning in their lower reaches, and that the present wastage is due, at least in part, to the climatic trend of the last few decades towards higher summer temperatures, as in the North Atlantic area generally (Ahlmann, 1953). The present period of wastage is most strikingly portrayed to the south of the main ice-cap (Figure 9), where a number of local ice masses and corrie glaciers, situated almost entirely below the equilibrium line, are undoubtedly thinning and in many cases receding at their edges by up to 10 m or more in a summer; small patches of snowdrift ice below about 1220 m are disappearing from year to year. These conditions are matched near the north coast of Ellesmere Island by the deterioration of the ice-shelves, and the probable thinning of local ice cover on the land.

It may be expected that the main, high-level ice-caps and trunk glaciers else-where in Ellesmere Island and in Axel Heiberg Island are showing little or no change in areal extent from year to year, but that thinning is taking place in the lower reaches of the trunk glaciers, and both thinning and recession of small, low-level ice masses throughout the Queen Elizabeth Islands.

FUTURE WORK IN NORTHERN ELLESMERE ISLAND

The thickness and rate of movement of at least one main glacier flowing towards the north coast of Ellesmere Island and the ablation in its lower reaches should be investigated. Further data are required on accumulation on the north side of the ice-cap. Without doubt most valuable information would come from stratigraphic and thermal studies in a deep bore-hole on the highest part of the ice-cap. Because the present accumulation on this ice cap is one-third to one-quarter the accumulation on the ice-cap in northern Greenland, a deep core in northern Ellesmere Island might cover three or four times the time span of a core to the same depth in northern Greenland. A deep core from this far northern ice-cap

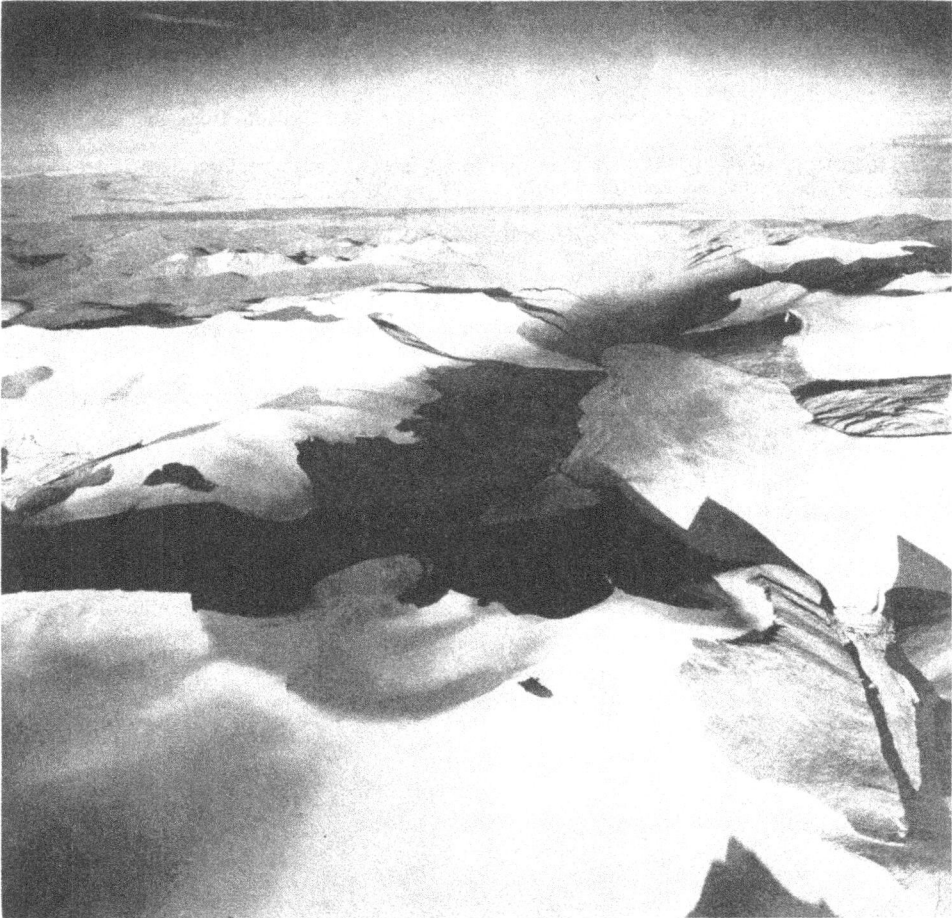

FIGURE 9.   Local ice cover between the main ice-cap and Lake Hazen. Photo: R.C.A.F., August 2, 1958, from 15,000 feet.

would help to elucidate both the glacial history of northern Ellesmere Island and the climatic history of the Polar Basin.

ACKNOWLEDGMENTS

Grateful acknowledgment is made to the following for their valuable help and advice both in the field and subsequently: Mr. K. C. Arnold, Dr. F. S. Grant, and Messrs. J. R. Lotz, R. B. Sagar, H. Sandstrom, and J. R. Weber. References to some of their work have been made above. This paper is published by permission of the Chairman, Defence Research Board of Canada.

AHLMANN, H. W.   1953.   Glacier variations and climatic fluctuations; Amer. Geog. Soc.

ARNOLD, K. C.   1959.   Operation Hazen: Survey 1957-58; Defence Research Board, Ottawa; Report no. D Phys. R (G), Hazen 5.

BAIRD, P. D.   1952.   The glaciological studies of the Baffin Island expedition, 1950. Part I: Method of nourishment of the Barnes Ice Cap; J. Glaciology, vol. 2, no. 11, pp. 2-9.

CRARY, A. P., KULP, J. L., and MARSHALL, E. W.   1955.   Evidences of climatic change from ice island studies; Science, vol. 122, pp. 1171-3.

DEANE, R. E.   1958.   Pleistocene geology and limnology in Operation Hazen: Narrative and preliminary reports for the 1957 season; Defence Research Board, Ottawa, pp. 19-23.

———   1959.   Pleistocene geology and limnology in Operation Hazen: Narrative and preliminary reports 1957-58; Defence Research Board, Ottawa; pp. 61-3.

DUNBAR, MOIRA, and GREENAWAY, K. R.   1956.   Arctic Canada from the air; Defence Research Board, Ottawa.

HATTERSLEY-SMITH, G.   1958   A note on Mount Oxford, northern Ellesmere Island; Geog. J., vol. 124, part 2, pp. 280-1.

———   1960.   Studies of englacial profiles in the Lake Hazen area of northern Ellesmere Island; J. Glaciology, vol. 3, no. 27, pp. 610-25.

HATTERSLEY-SMITH, G., CRARY, A. P., and CHRISTIE, R. L.   1955.   Northern Ellesmere Island, 1953 and 1954; Arctic, vol. 8, no. 1, pp. 3-36.

JACKSON, C. I.   1959.   The meteorology of Lake Hazen, N.W.T. Part I: Analysis of the observations; McGill University, Publication in Meteorology no. 15, pp. 1-194.

KOENIG, L. S., GREENAWAY, K. R., DUNBAR, MOIRA, and HATTERSLEY-SMITH, G.   1952. Arctic ice islands; Arctic, vol. 5, no. 2, pp. 67-103.

MARSHALL, E. W.   1955.   Structural and stratigraphic studies of the northern Ellesmere ice shelf; Arctic, vol. 8, no. 2, pp. 109-14.

MOORE, A. W.   1936.   Oxford University Ellesmere Land expedition. Part III: The sledge journey to Grant Land; Geog. J., vol. 87, no. 5, pp. 419-27.

SAGAR, R. B.   1959.   Personal communication.

———   1960.   Glacial-meteorological observations in northern Ellesmere Island . . . 1958; McGill University, Publication in Meteorology no. 29.

SANDSTROM, H.   1959.   Operation Hazen: Geophysical methods in glaciology; Defence Research Board, Ottawa; Report no. D, Phys. R (G), Hazen 3.

SMITH, D. I.   1959.   Geomorphology in Operation Hazen: Narrative and preliminary reports 1957-58; Defence Research Board, Ottawa; Report no. D Phys. R (G), Hazen 4, pp. 58-60.

WEBER, J. R., and SANDSTROM, H.   1960.   Geophysical methods in glaciology. Part II: Seismic measurements on Operation Hazen in 1958; Defence Research Board, Ottawa; Report no. D Phys. R (G), Hazen 6.

Induction and Galvanic Resistivity Studies on the Athabasca Glacier, Alberta, Canada

G. V. KELLER AND

F. C. FRISCHKNECHT

ABSTRACT

The United States Geological Survey has studied the use of electrical methods for measuring ice thickness on the Athabasca Glacier, Alberta, Canada. Two methods for measuring resistivity were tested: one, a conventional resistivity method in which current was fed galvanically into the glacier through electrodes; and the other, an electromagnetic method in which a wire loop laid on the ice was used to induce current flow. Both the galvanic and the inductive techniques for measuring resistivity were capable of measuring ice thicknesses up to a thousand feet, the maximum thickness of the Athabasca Glacier. The electromagnetic method appeared to be the better of the two for measuring thickness of ice and for measuring the resistivity of the rock beneath the ice, while the galvanic method was better for studying differences within the ice.

LARGE AREAS OF THE EARTH'S SURFACE in the polar regions are covered with a permanent ice-sheet. It is probable that these regions may be called upon to provide raw materials in the forseeable future. The exploration of ice-covered areas will require geophysical methods operative through several hundreds or thousands of feet of ice. Ice is known to be a very poor conductor, electrically. It should normally have a greater resistivity than any rock with which it is associated. This contrast in electrical properties makes it reasonable to assume that electrical methods would be especially applicable to the study of ice thickness.

An opportunity for studying the use of electrical methods over glacial ice came about during the summer of 1959, when the United States National Bureau of Standards planned a field study on the Athabasca Glacier in Alberta, Canada, and invited the United States Geological Survey to participate. The Athabasca Glacier was an excellent location for preliminary studies because of its accessibility and because of the extensive programme of glaciological work that is being carried on there by the Universities of Alberta and British Columbia.

ACKNOWLEDGMENTS

We are indebted to Dr. James Wait and Donald Watt, United States National Bureau of Standards, for suggesting the work on Athabasca Glacier; to Dr. George Garland of the University of Alberta, for his invitation to share the facilities of the glaciological party on Athabasca Glacier; and to the Department of Northern Affairs and National Parks of the Dominion of Canada for permission to work in Jasper National Park. We especially appreciate the assistance given by Mr. W. R. Ruddy of Snowmobile Tours Limited, in providing transportation on the glacier.

ATHABASCA GLACIER

The Columbia ice-field, which is the source of the Athabasca Glacier, lies astride the British Columbia–Alberta border, approximately 110 miles north of the town of Banff, Alberta. The Columbia ice-field is drained by three large valley glaciers: the Saskatchewan to the east, providing the headwaters of the Saskatchewan River; the Athabasca, flowing to the north and providing the headwaters of the Athabasca River; and the Columbia, flowing to the northwest and

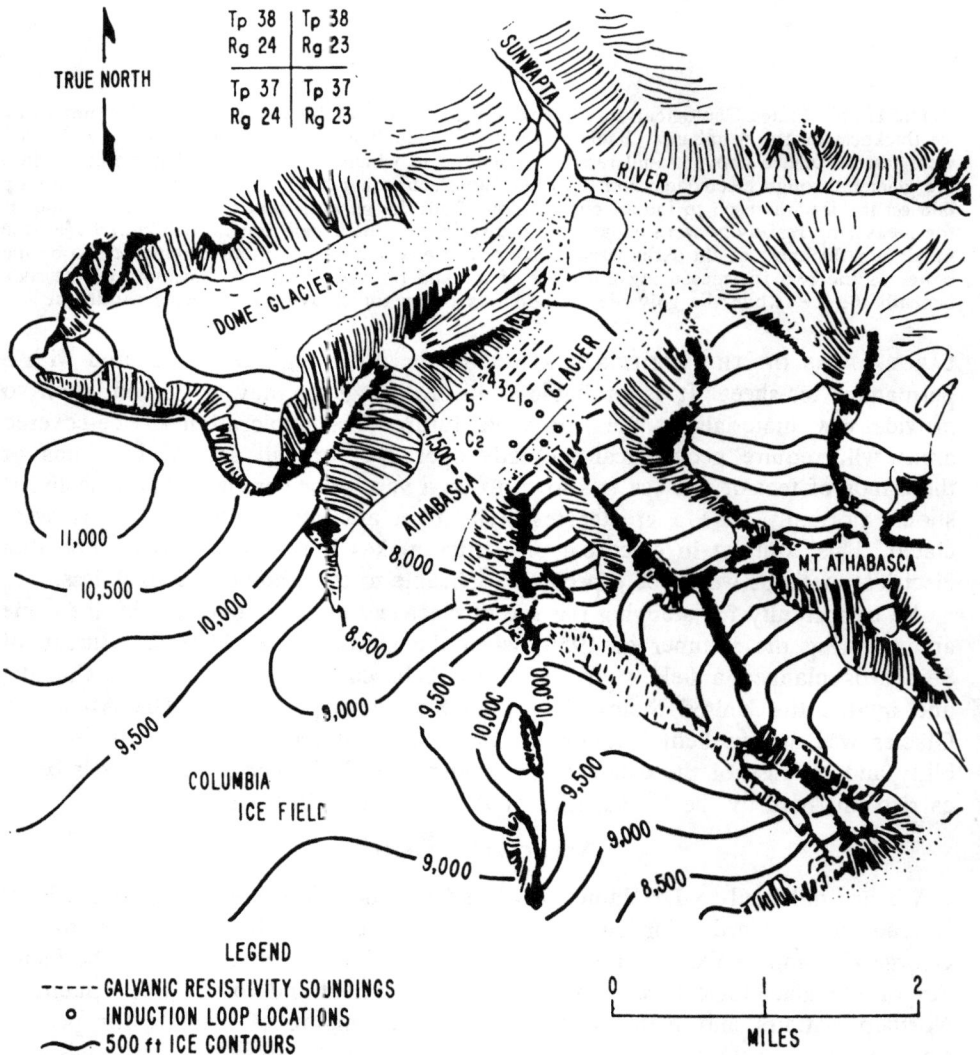

FIGURE 1.   Sketch map showing the Athabasca Glacier.

providing the headwaters of the Columbia River. The Athabasca Glacier is the smallest of the three but the most accessible. The Banff-Jasper highway runs within less than a mile of the glacier, and a side road goes to the toe of the glacier, from where snowmobile transportation is provided by a private company.

The Universities of British Columbia and Alberta are conducting a programme of glaciology on the Athabasca Glacier which includes micrometeorology, seismic depth determinations, gravity measurements, ice-movement studies, and drilling (oral communication, G. D. Garland, University of Alberta, 1959). The work done by these two universities provides a great deal of scientific information about the glacier.

In recent years, the glacier has been retreating at a rate of a hundred feet per year. Large, well-preserved, lateral moraines give strong evidence of the rapid recession of the glacier. As the glacier is receding so rapidly, one would expect that the ice would be at or only slightly below the freezing point, with abundant water percolating through it, especially during the summer melt season.

Athabasca Glacier (Figure 1) descends approximately two-and-one-half miles from the firn line at 8000 feet to the toe at an elevation of 6300 feet. There are two ice falls in the first three-quarters of a mile after the glacier leaves the ice-field, each with a drop of several hundred feet. In this area the ice is heavily crevassed, and in the upper ice fall many seracs are formed, making access with equipment difficult. The lowermost step of the glacier is relatively flat and smooth for a distance of more than a mile, before the ice surface drops off to the terminal lake. Figure 2 is a photograph of this lower step, taken from the centre of the second step.

The lower step is only slightly crevassed and provides an excellent working surface. During the melt season, parallel hummocks three to five feet high and

FIGURE 2.   Photograph of the lower step of Athabasca Glacier.

spaced about ten feet apart develop over most of the surface. Commonly, the valleys between these hummocks provide drainage for melt-water, with streams a few hundred to a thousand feet in length. These streams end in moulines, or melt holes, which are probably several hundred feet deep. The width of the lower step is approximately 3000 feet, including the moraine-covered edges. The glacier valley is steeply walled, and considerable detrital material is supplied to the edges of the glacier by landslides. This material provides a few inches to a few feet of cover, insulating the ice so that the margins of the glacier are twenty to thirty feet higher than the main surface.

All of the electrical studies were carried out on the lower step. Along the centre line of the lower step, seismic reflection data have given ice thicknesses of 900 to 1050 feet (oral communication, P. J. Savage).

FIELD WORK WITH ELECTRICAL METHODS

Two methods of measuring resistivity were used. One was a conventional resistivity method in which current was introduced galvanically into the ice through electrodes, and the other an electromagnetic method in which the mutual coupling between two wire loops laid on the ice was measured.

Galvanic Method

The resistivity work consisted of five depth soundings made at 300-foot intervals spaced along a line from the mid-point of the lower step of the glacier to the northwest margin, and a resistivity profile along this same line (see Figure 1).

A single-moving electrode array was used for the depth soundings. The positions of the two current electrodes and one of the pickup electrodes were fixed (see Figure 3). The current electrodes were separated a distance of 2,500 feet, and the distant pickup electrode was placed another 2,500 feet down the glacier. The moving pickup electrode was used to measure the electric fields at distances of five to sixteen hundred feet from one of the current electrodes.

The single-moving electrode array has several practical advantages over multiple-moving electrode arrays. It requires only two people for efficient operation and because none of the wiring is laid between the pickup spread and the current spread, the possibility of errors caused by leakage are minimized. The disadvantage

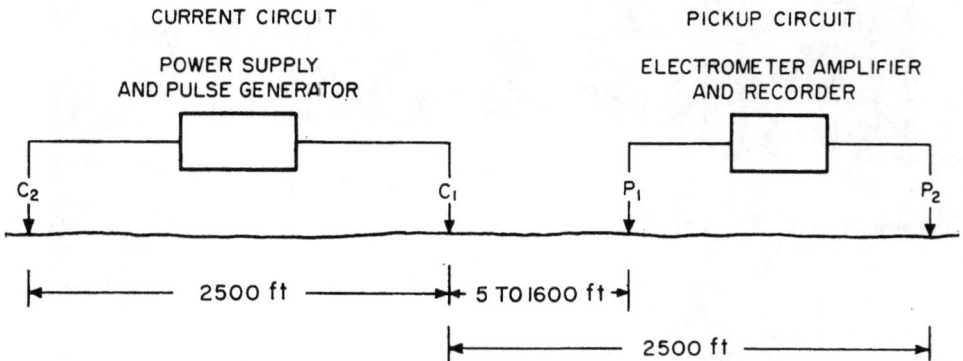

CURRENT CIRCUIT

POWER SUPPLY
AND PULSE GENERATOR

PICKUP CIRCUIT

ELECTROMETER AMPLIFIER
AND RECORDER

C_2 C_1 P_1 P_2

2500 ft   5 TO 1600 ft

2500 ft

FIGURE 3.   Block diagram of the single moving electrode array used in measuring ice resistivity.

20 VOLTS

ONE SECOND

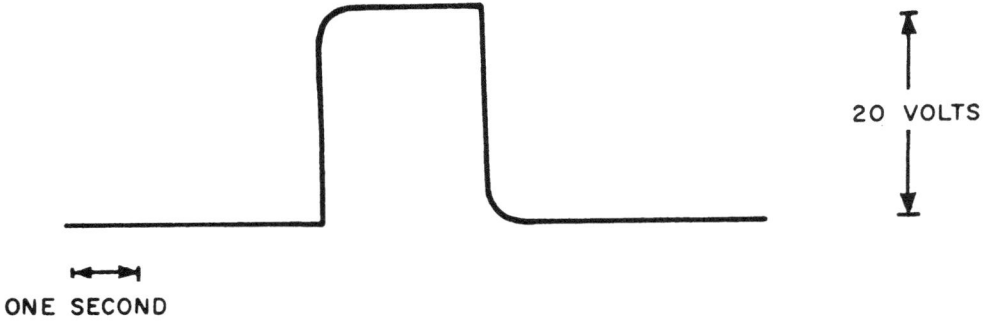

A. NORMAL RECORDING IN REGION OF HIGH FIELD STRENGTH

20 MILLIVOLTS

ONE SECOND

B. NORMAL RECORDING IN REGION OF LOW FIELD STRENGTH

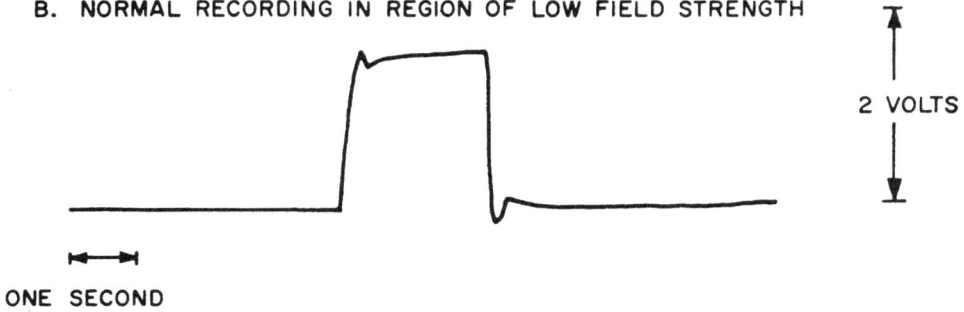

2 VOLTS

ONE SECOND

C. RECORDING SHOWING UNUSUALLY LARGE SWITCHING TRANSIENT

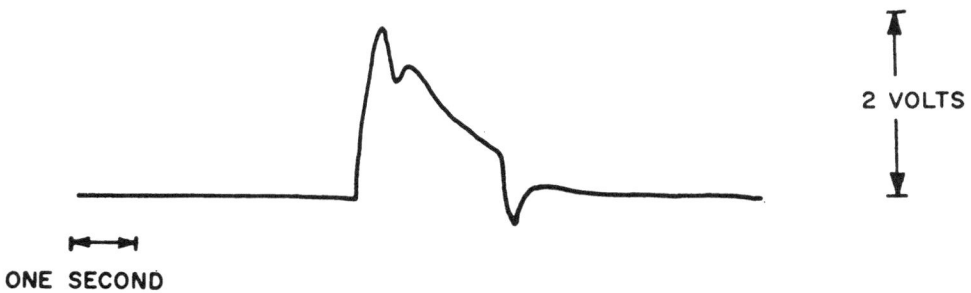

2 VOLTS

ONE SECOND

D. RECORDING SHOWING ANOMAIOUSLY LOW D.C. PICKUP VOLTAGE

FIGURE 4. Examples of voltages recorded at the pickup electrodes during current pulses.

of this array is that it is difficult to detect the effects of horizontal resistivity changes.

Copper tubes, one inch in diameter and fifteen inches long, were used as current electrodes. They were placed in stagnant melt ponds with a few ounces of salt. The ground circuit resistance between current electrodes was approximately one megohm. Pulsed direct current with a period of 0.1 to 3 seconds was used to energize the current electrodes. The current used was approximately one-half milliampere. Steel pins, an eighth of an inch in diameter and ten inches long, were used as pickup electrodes. These were laid in shallow melt ponds or driven several inches into the ice. The ground circuit resistance between pickup electrodes varied from 0.5 to 5 megohms, though most commonly it was about 5 megohms.

The voltage developed across the pickup electrodes during current pulsing was recorded on a hot-pen oscillograph having an input resistance of 60 megohms. Examples of some typical recordings are shown in Figure 4. The first example (A) shows ideal conditions; the pickup voltage is large compared to background noise and the transient rise and fall of the pickup voltage may be detected in spite of the switching transient. The second example (B) shows the voltage recorded for a large electrode separation; the pickup voltage is comparable in amplitude with the variations in telluric background. The third recording (C) shows a common occurrence; a very large capacitive surge from the current wire as the current is switched on and off caused by the relatively large capacity to ground of the current wire. The most serious difficulty experienced is illustrated by example (D) in Figure 4. The voltage is seen to rise normally during a capacitive surge but following this it

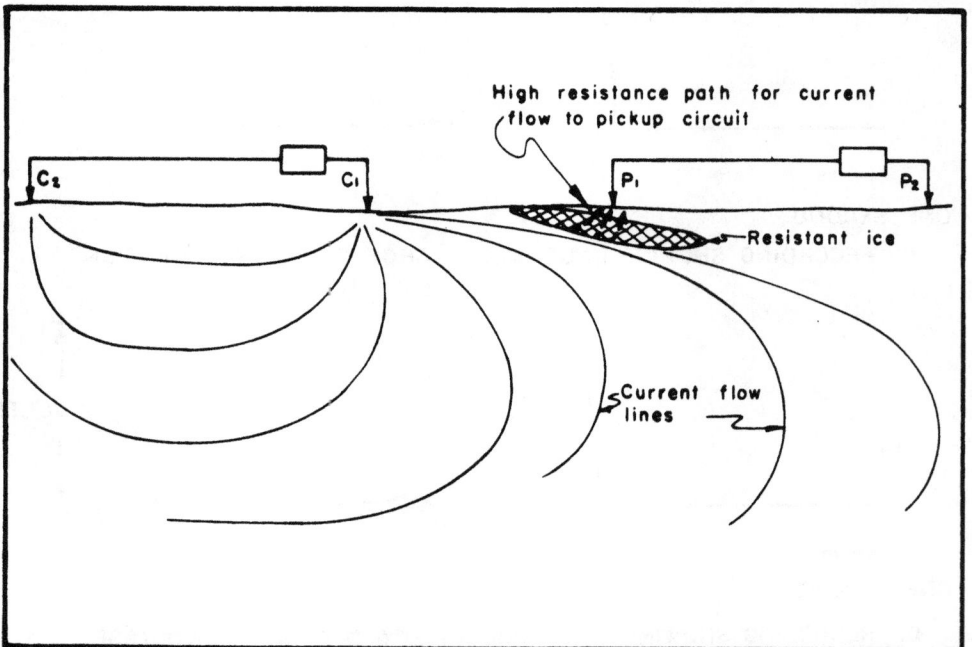

FIGURE 5.   Possible explanation of erroneously low values for the measured electric field.

falls to a very low level, indicating what appears to be an erroneously low resistivity. Many times, the pickup voltage would have a normally high value a few feet away from the point where this phenomenon was noted. A possible explanation is given by the situation shown in Figure 5. The pickup electrode is located over a lense of dense ice having virtually no galvanic conductivity, so that the electrode is insulated from the ice beneath where current is flowing. To be detected, the recording system must drain an infinitesimal amount of current through this insulating layer making it behave as though there were a large source resistance in series with the pickup circuit. The resistance of the volume through which this current is drained may be much higher than the ground circuit resistance through the pickup electrodes which can be measured with an ohmmeter. In ground circuit measurements, the current travels along the thin film of moisture covering the glacier.

With recordings as shown in Figure 4(A), it is possible to calculate both the apparent resistivity and the apparent dielectric constant of the ice. Resistivity is calculated directly from the voltage, current, and electrode geometry:

(1)
$$\rho_a = 2\pi \, \frac{E}{I} \left(\frac{1}{\dfrac{1}{d_1} - \dfrac{1}{d_2} - \dfrac{1}{d_3} + \dfrac{1}{d_4}} \right)$$

where

ρ_a is the apparent resistivity to direct current,

E is the plateau voltage recorded following the capacity surge and charging transient,

I is the current,

d_1 and d_2 are the distances between the moving electrode and the near and far current electrodes, respectively,

d_3 and d_4 are the distances from the fixed pickup electrode to the near and far current electrodes, respectively.

The dielectric constant at the frequencies contained in the current pulse may be calculated by a Fourier analysis of the pickup voltage and its transient. The Fourier analysis gives the phase shift for the sinusoidal harmonics comprising the square wave pulse transmitted to the current electrode. The dielectric constant is calculated from the expression:

(2)
$$\epsilon = \frac{\tan \delta}{\rho_a \, \omega \, \epsilon_0}$$

where

ϵ is the apparent dielectric constant,

δ is the phase shift determined by Fourier analysis,

ρ_a is the apparent resistivity calculated for the same electrode array,

ϵ_0 is the dielectric constant for free space, 8.854×10^{-12} farads per metre,

ω is the angular frequency for which the phase shift was determined.

Results of Galvanic Resistivity Measurements

Apparent D.C. resistivities measured on Athabasca Glacier are shown in Figure 6, plotted as a function of the current electrode-pickup electrode $(C_1 - P_1)$ separation. The most striking feature of the date is the large scatter of the resistivity values, far more than normally can be accepted in a resistivity sounding. This diffi-culty was partially overcome by making a large number of field measurements on each sounding so that the shape of the field curve would be apparent in spite of the scatter.

All of the soundings have the same general shape: a surface layer, a second layer with higher resistivity, and a bottom layer with low resistivity. Resistivity departure curves were prepared from tables given by Mooney and Wetzel (1956) in order that a more exact interpretation could be made. The family of curves most

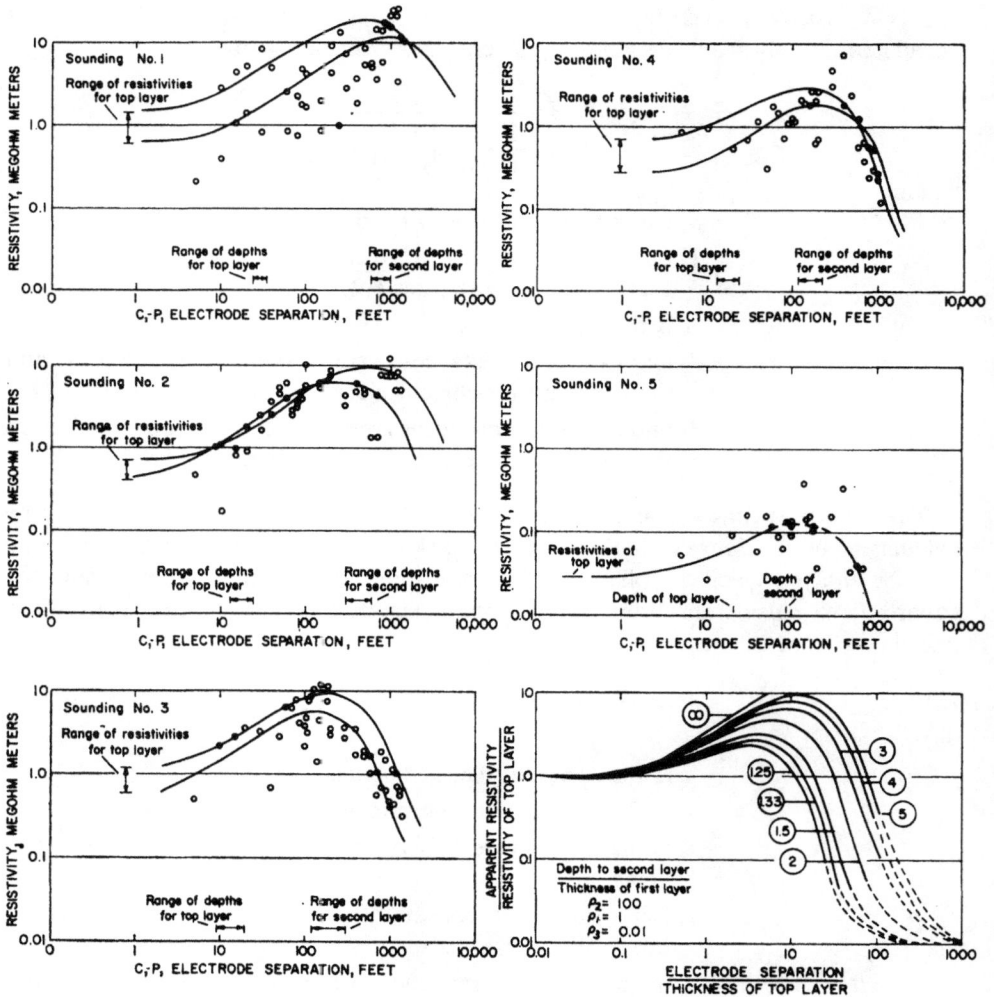

FIGURE 6.   Resistivity depth soundings.

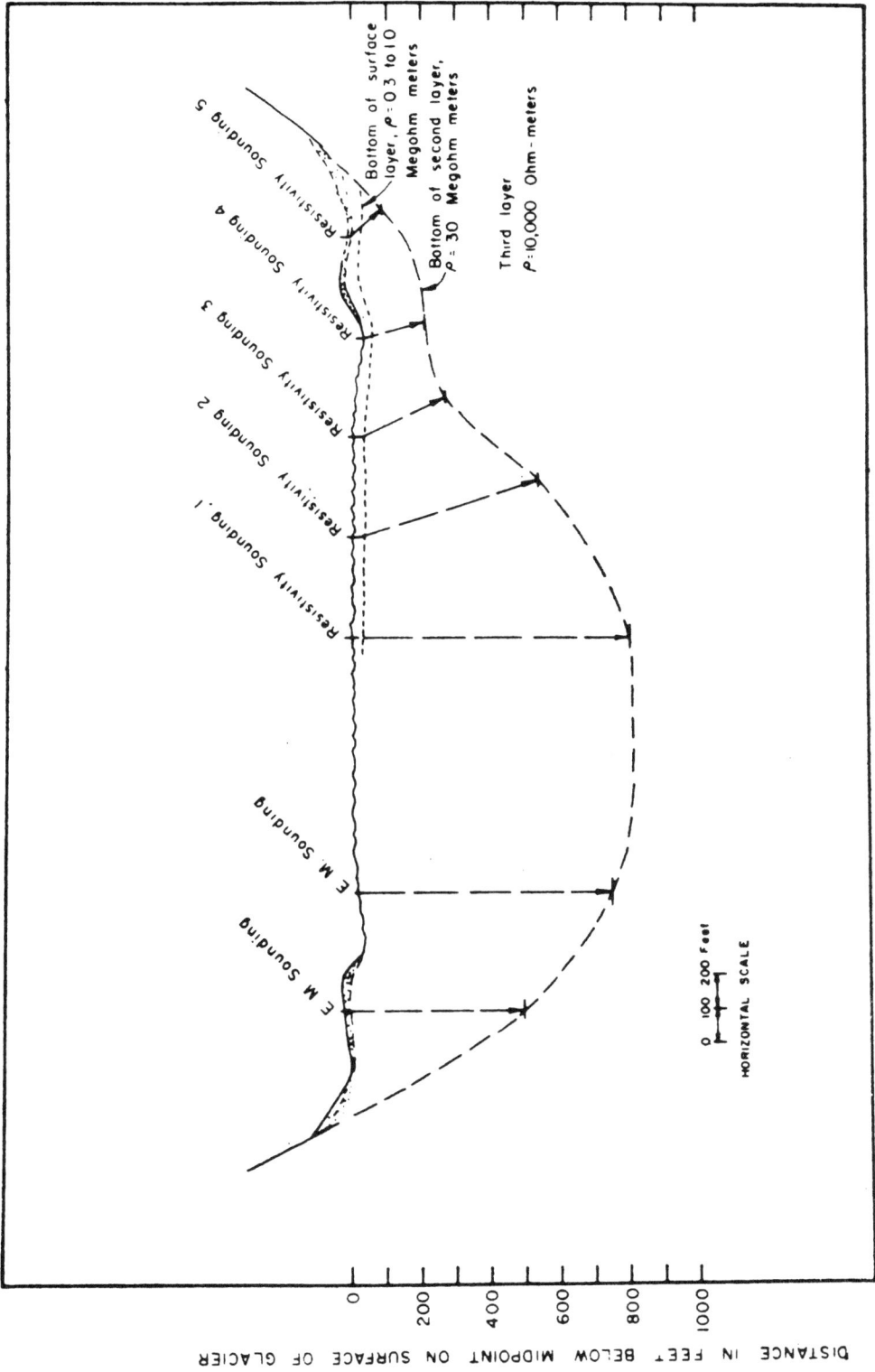

Bottom of surface layer, $\rho = 0.3$ to 10 Megohm meters

Bottom of second layer, $\rho = 30$ Megohm meters

Third layer $\rho = 10,000$ Ohm-meters

Resistivity Sounding 5

Resistivity Sounding 4

Resistivity Sounding 3

Resistivity Sounding 2

Resistivity Sounding 1

E M Sounding

E M Sounding

0   100   200 Feet

HORIZONTAL SCALE

DISTANCE IN FEET BELOW MIDPOINT ON SURFACE OF GLACIER

0
200
400
600
800
1000

FIGURE 7.   Postulated cross-section of Athabasca glacier, lower part of the lower step.

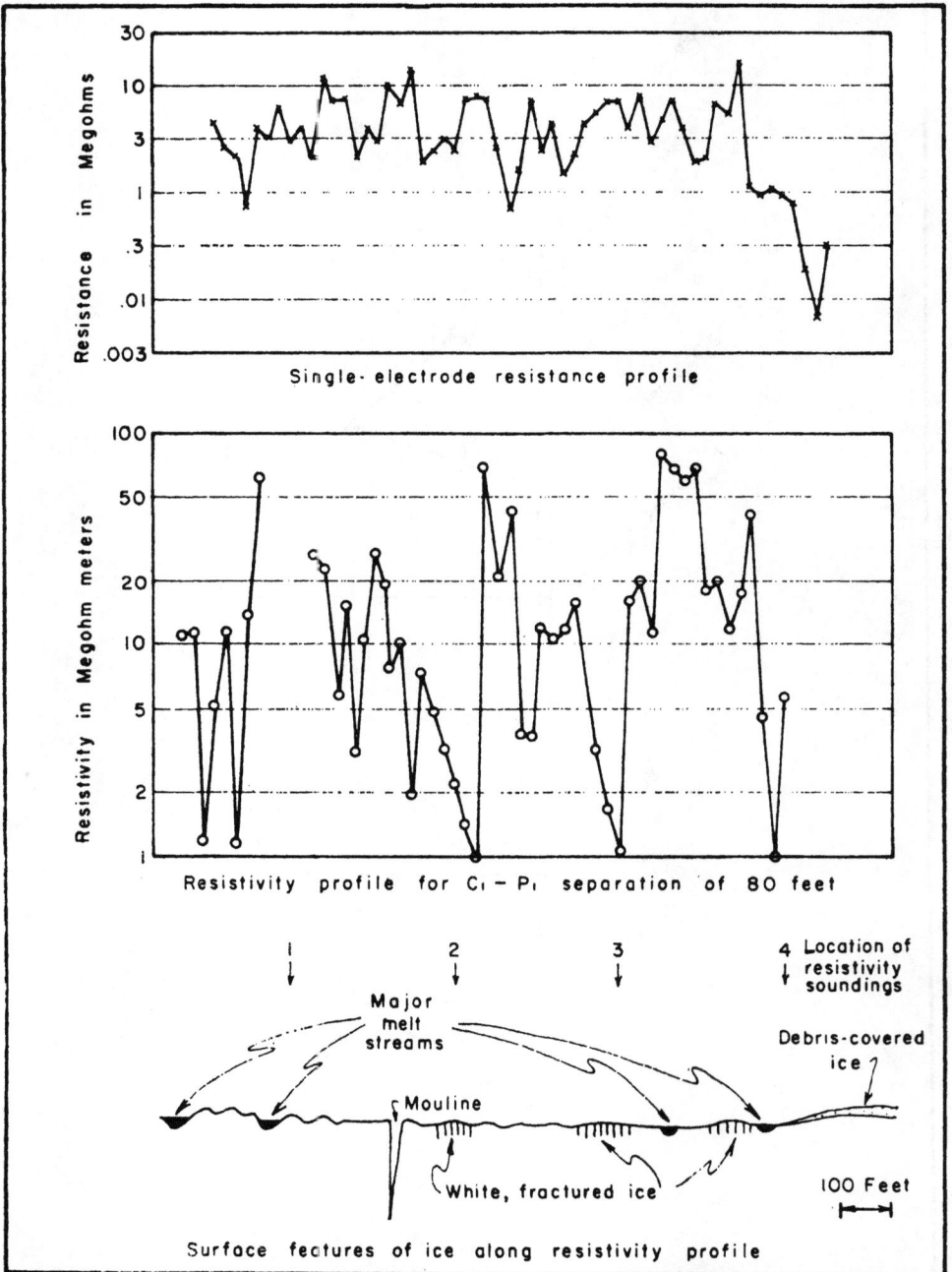

FIGURE 8.   Resistivity profile from the midpoint of Athabasca glacier to the northwest edge.

suitable for the interpretation of these data is shown on Figure 6. Since the points on the field curves are scattered, there is some uncertainty how best to match the data with theoretical curves. A range of possible solutions is indicated for each sounding.

A cross-section of the glacier as indicated by resistivity data is shown in Figure 7. The glacier appears to consist of a surface layer 50 to 100 feet thick, with a resistivity in the neighbourhood of 0.3 megohm metres. Beneath this surface layer, the ice has a resistivity of 30 megohm metres or greater. The lowermost layer is believed to be bedrock, and has an indicated resistivity of 10,000 ohm-metres or less. The exact value for the resistivity of the underlying rock cannot be determined because of the very high contrast with the resistivity of the overlying ice.

A resistivity profile was made across the glacier from its mid-point to the north-east edge to study the near-surface resistivity changes indicated by the depth soundings. Three moving electrodes were used: one current electrode and two pickup electrodes placed at distances of 80 and 100 feet from the current electrode. All three electrodes were oriented in a line across the glacier, normal to the drainage pattern. All three were moved in twenty-foot increments in measuring resistivity along the profile. The resistivity profile and the single-electrode resistance of the current electrode are plotted on the graph shown in Figure 8.

The apparent dielectric constants calculated for the depth sounding made at the centre of the glacier are shown on Figure 9. Unfortunately, no theoretical curves are available for the interpretation of such curves for a capacitive layer overlying a conductor. However, the general shape of such curves might be inferred from data presented by Zablocki (1957). He presents curves for a conductive surface layer overlying a capacitive medium. It is reasonable to assume that our case would resemble these curves inverted. If so, the dielectric constant should behave as indicated by the dotted lines in Part B of Figure 9. For small resistivity contrasts between the surface layer and the underlying medium, the apparent value of the dielectric constant will decrease monotonously from the true value for the surface layer to the true value for the lower medium. However, if the surface layer is much more resistant than the lower medium, the apparent dielectric constant will first increase over its true value for the surface layer and then decrease to the true value for the lower medium.

The field data indicate that the glacier has a dielectric constant of approximately 140. The higher values indicated for $C_1 - P_1$ separations between 100 and 1000 feet are caused by current flowing into the underlying conductive rocks, and so, probably do not represent true values of dielectric constant.

Electromagnetic Methods

The electromagnetic work consisted of five measurements of mutual coupling between loops spaced at fixed separations of from 500 to 1820 feet (see Figure 10). Measurements were made over the frequency range from 100 to 10,000 cycles per second. The transmitting loop consisted of 1 to 3 turns of wire and was 100 to 300 feet on a side. The receiving coil consisted of a single turn of eight con-ductor shielded cable and was 60 feet on a side. Neither loop was tuned.

An oscillator and a 70-watt audio amplifier supplied several amperes of current to the transmitting loop. A reference voltage was induced in a small coil placed at one side of the transmitting loop and then carried to the measuring apparatus over a two conductor cable. The phase angle difference and the ratio of the amplitudes of the voltage induced in the receiving loop and the reference voltage were measured with a ratiometer and null detector. A variable frequency bandpass filter

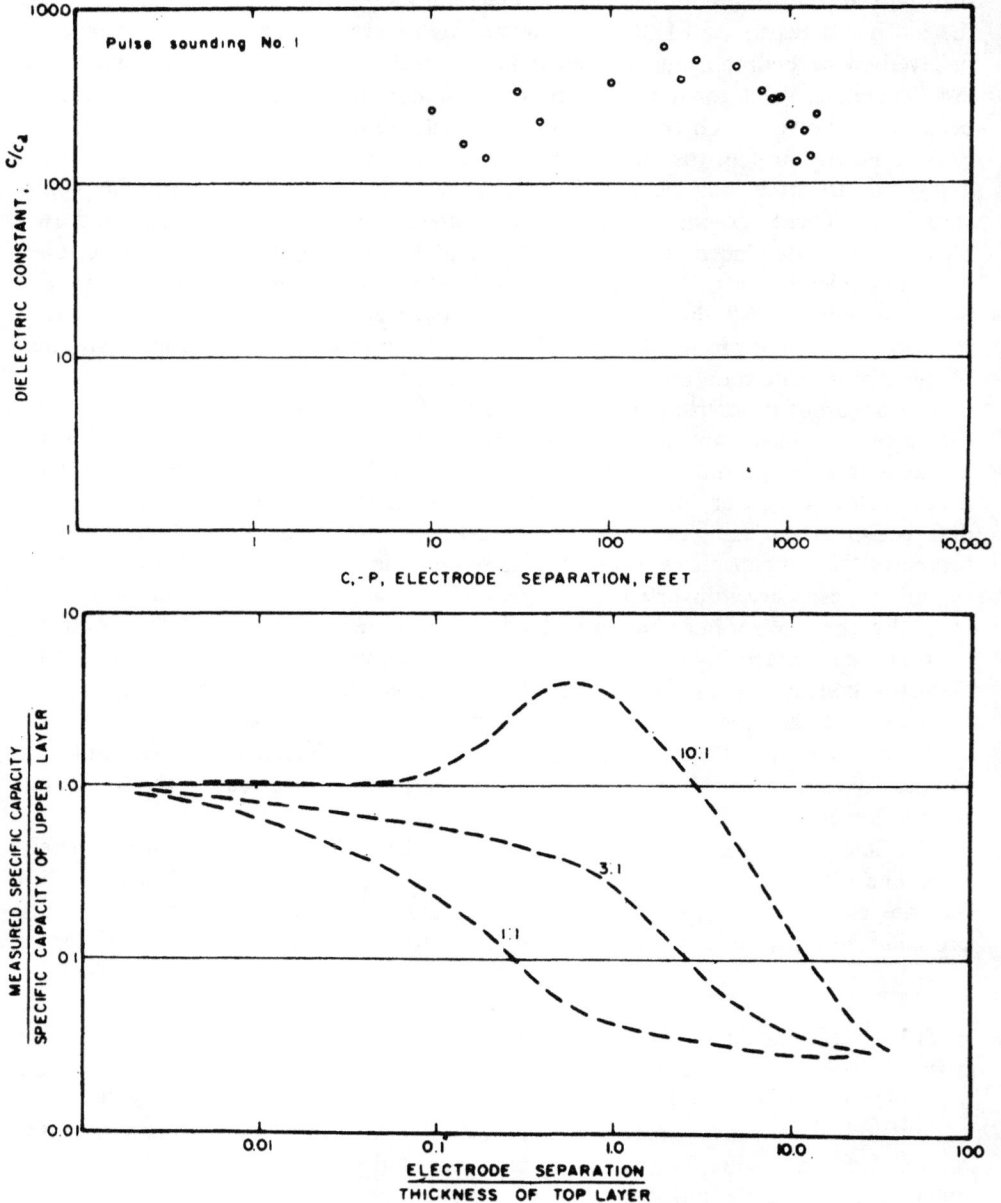

FIGURE 9.   Apparent dielectric constant as a function of electrode separation.

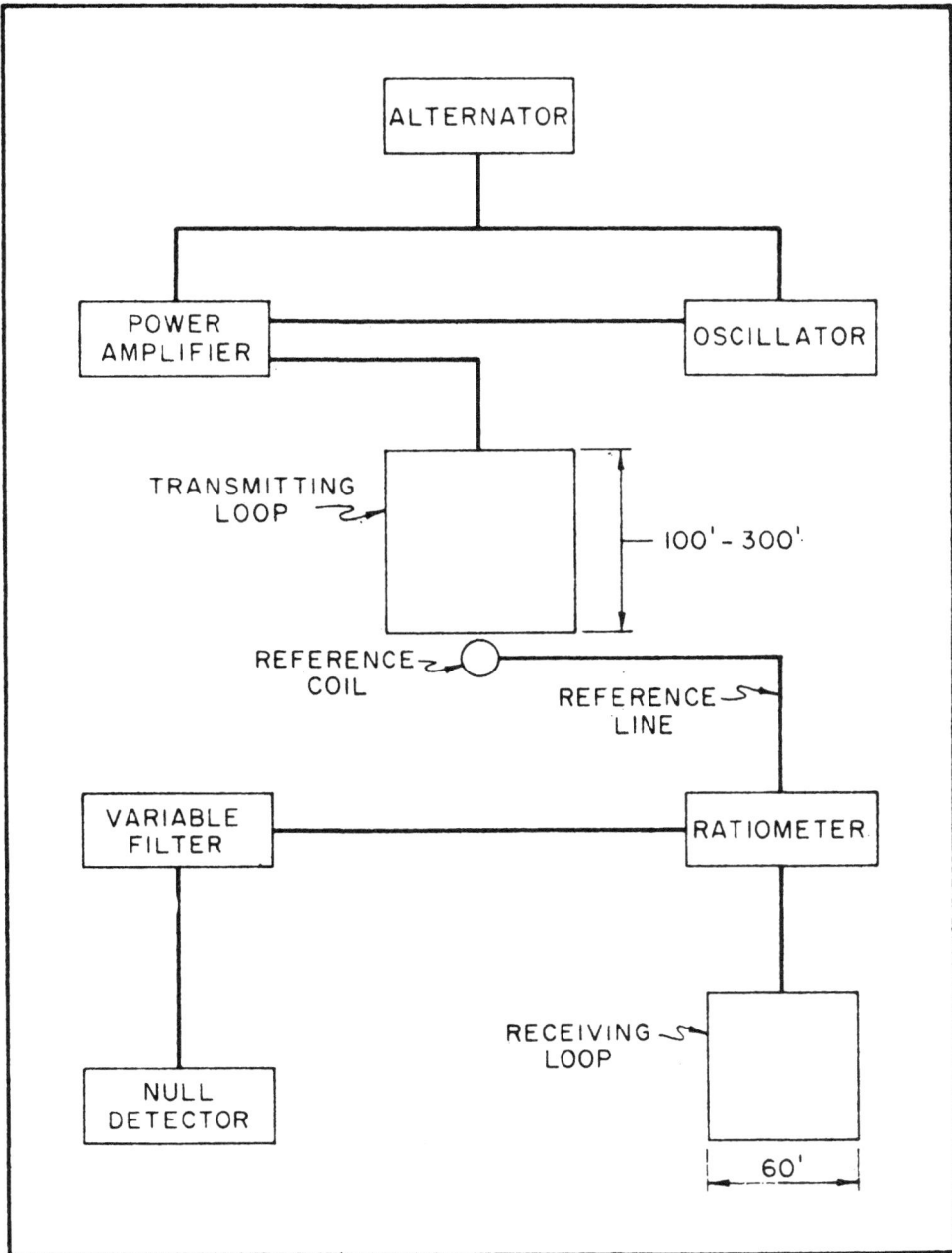

FIGURE 10.   Block diagram of variable frequency electromagnetic apparatus.

was placed between the ratiometer and the null detector to reduce interfering noise from sferics and signals from a powerful low-frequency radio station at Jim Creek, Washington.

The amplitude ratios and phase differences observed at different frequencies are a function of the impedances of the receiving and reference coils and the reference line. These effects were determined by measuring the frequency response of the system with the coils very close together and with the reference line extended to its full length. All subsequent measurements were normalized by the results of these first measurements.

The dial readings of the ratiometer are not expressed directly in terms of amplitude ratio and phase angle. After the field measurements were completed, the equipment was set up in the laboratory and the true ratio and phase angle were determined for each setting of the dials by measuring the unbalanced voltage to the null detector with no field through the receiving coil. Using this procedure, amplitude ratios were determined with an accuracy of about $\pm$ 2 per cent and phase angles with an accuracy of about $\pm$ 2 degrees at frequencies below 3000 cycles per second. Above 3000 cycles the data are of doubtful accuracy.

Instrumentally, the only problem in making the field measurements was in obtaining a sharp null at frequencies below about 300 cycles per second and above 3,000 cycles per second. The difficulty at the low frequencies was in the lack of signal strength, and the difficulty at the high frequencies was interference from sferics and signals from the Jim Creek station. The power amplifier and generator were difficult to move about over the glacier so they were placed in one location for all of the measurements. With the exception of transporting these heavy items of equipment, no significant field problems were imposed by conditions peculiar to the glacier.

Theoretical Curves

Equations for the mutual coupling between horizontal loops lying on the surface of a homogeneous flat earth are given by Wait (1954, p. 290-6) for the case where both the dielectric constant and the conductivity of the earth are important. Wait has calculated a family of curves of mutual coupling Z/Z_0 as a function of $B = ((\mu\omega/\rho)^{1/2}r$ for parametric values of the ratio $b = \omega\rho\epsilon$, where

$Z_0 =$ mutual coupling between loops in free space,

$Z = |Z|\ e^{i\theta} =$ complex mutual coupling between loops in presence of earth,

$|Z| =$ amplitude of mutual coupling in presence of earth,

$\theta =$ phase angle of mutual coupling in presence of earth,

$\rho =$ resistivity of the earth in ohm-metres,

$\mu =$ magnetic permeability of free space in henries/metre,

$\omega = 2\pi f =$ angular frequency,

$r =$ the spacing between coils in metres,

$\epsilon =$ dielectric constant of the earth in farads per metre.

These theoretical curves are useful in interpreting field curves of mutual coupling as a function of coil spacing for a constant frequency. As the measurements described here are of mutual coupling as a function of frequency for a fixed coil spacing, $\omega\rho\epsilon$ increases with frequency and it was necessary to calculate and plot curves of mutual coupling against B for constant values of $b = kB^2$. The families of curves obtained are shown in Figure 11. These curves are applicable to the present work only if the ratio of the thickness of the ice to the coil spacing is quite large, say $h/r > 1$ or 2, where h is the thickness of the ice or the height of the coils above bedrock.

Wait (1955, pp. 630-7) gives equations and some computations and curves for the response of loops raised above a conducting homogeneous earth where the dielectric constant may be neglected. Slichter and Knopoff (1959, pp. 77-88) present equations and computations for loops on the surface of two-layer conducting earth, which case degenerates to that of coils raised above a conducting homogeneous earth if the conductivity of the upper layer is zero. Again, these curves are plotted in such a way that they apply directly to measurements made by varying the coil separation at a constant frequency. The curves given by Wait and by Slichter and Knopoff may be replotted in terms of varying frequency, but they do not adequately cover the range of interest. Wait 1958, pp. 73-80) derives equations

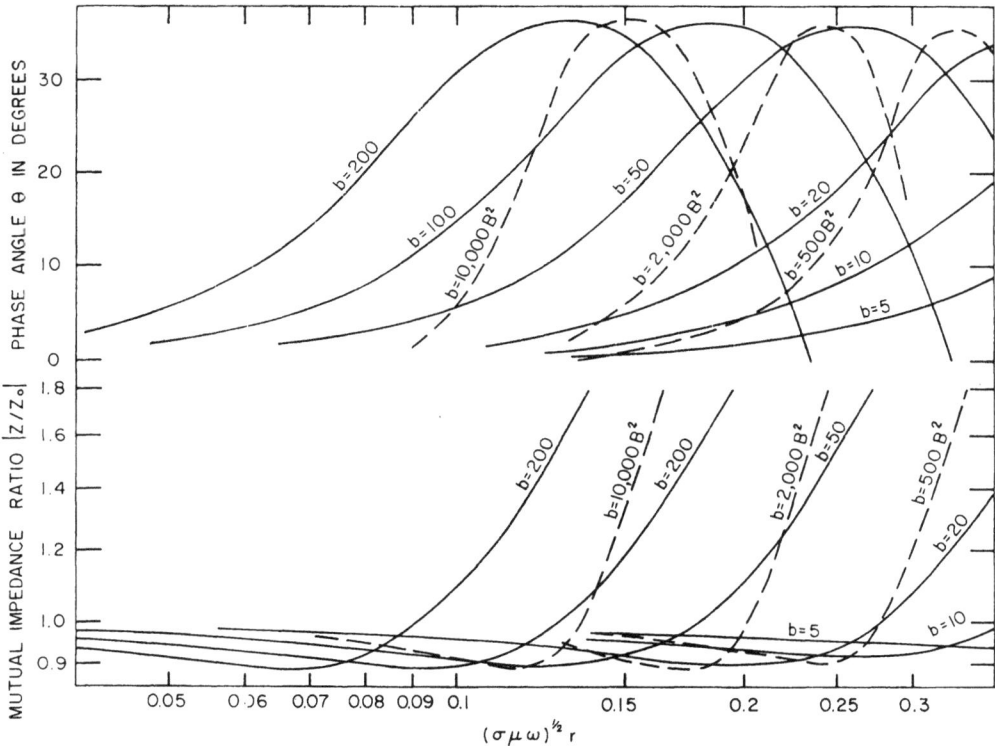

FIGURE 11.   Mutual impedance plotted as a function of the conductivity parameter for two horizontal loops on a lossy dielectric earth.

for the response of loops above a two-layer conducting earth. The United States Geological Survey has an extensive programme under way of evaluating these equations on a digital computer. The results are plotted as families of curves of mutual impedance versus B for various value of the ratios h/r and $K = ((\rho_1)/(\rho_2))$ where B, h, and r are as defined above and

$$d = \text{thickness of the upper layer},$$
$$\rho_1 = \text{resistivity of the upper layer},$$
$$\rho_2 = \text{resistivity of the lower semi-infinite layer}.$$

Families of curves calculated by the United States Geological Survey, as yet unpublished elsewhere, for horizontal loops raised above a homogeneous ($K = 1$) earth are shown in Figure 12. Also shown in Figure 12 (dashed curves) are portions of curves for $K = 0.3$ and $d/r = 0.25$. These curves are valid in measuring the thickness of ice only if the response from the ice is negligible. Neither set of theoretical curves presented here is valid for a lossy dielectric over a conducting earth, but each is a good approximation for the two limiting cases when the ice is either very thick or when its effect is negligible.

Both the abscissa and the ordinate of the field curve contain an undetermined constant multiplier. The process of normalizing the field curves by measurements made with the coils close together is equivalent to measuring the free space mutual coupling Z_0 for a very small value of r. Between dipoles, Z_0 varies as r^3 and it is not feasible to compute Z_0 at a large spacing from measurements made at short

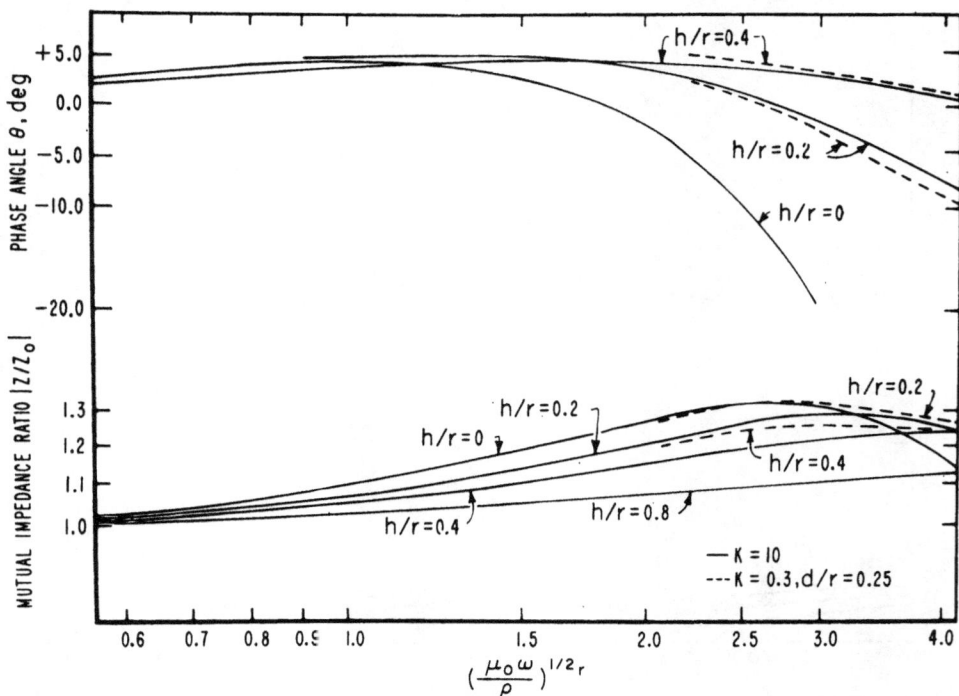

FIGURE 12. Mutual impedance plotted as a function of the conductivity parameter for two horizontal loops raised above a two-layer conducting earth.

spacings. Therefore, the ordinate of the field curves equals $\psi_1 \, Z/Z_0$ where ψ_1 is undetermined. Similarly, the abscissa of the field curves is $f^{1/2}$ where $\psi_2 \, f^{1/2} = B$ and ψ_2 is unknown. To facilitate interpretation, the field curves and the theoretical curves are plotted on log paper and interpretation is accomplished by the usual practice of shifting a field curve about over the family of theoretical curves until a match is found. If a valid fit between curves is found, ψ_1, ψ_2 and the other parameters are readily found. The position of the ordinate of the field curve relative to the field curve determines ψ_1; the position of the abscissa determines ψ_2 from which ρ may be calculated. The particular theoretical curve which is matched specifics $\omega\rho\epsilon$ or h/r depending on which type of curve is being used. If the field curve can be extrapolated to zero frequency, ψ_1 may be determined from the relation $\psi_1 \, Z_0/Z = 1$. This fixes the ordinate of the field curve which may be useful if the field data are rather inaccurate and scattered. If ρ is known by some independent mean ψ_2 and B can be determined and the abscissa is fixed.

Results of Electromagnetic Measurements

The first two measurements were made with the coils 500 feet apart in a locality where the average thickness of the ice is probably 800 feet or more. The results of one of these measurements, or depth soundings as they may be called, is shown as (a) in Figure 13. At most, (a) shows a response of about one-half of 1 per cent which is the order of magnitude of the scatter of the data. This small response may be attributed to the effect of bedrock at a depth of 800 feet. As the ice had

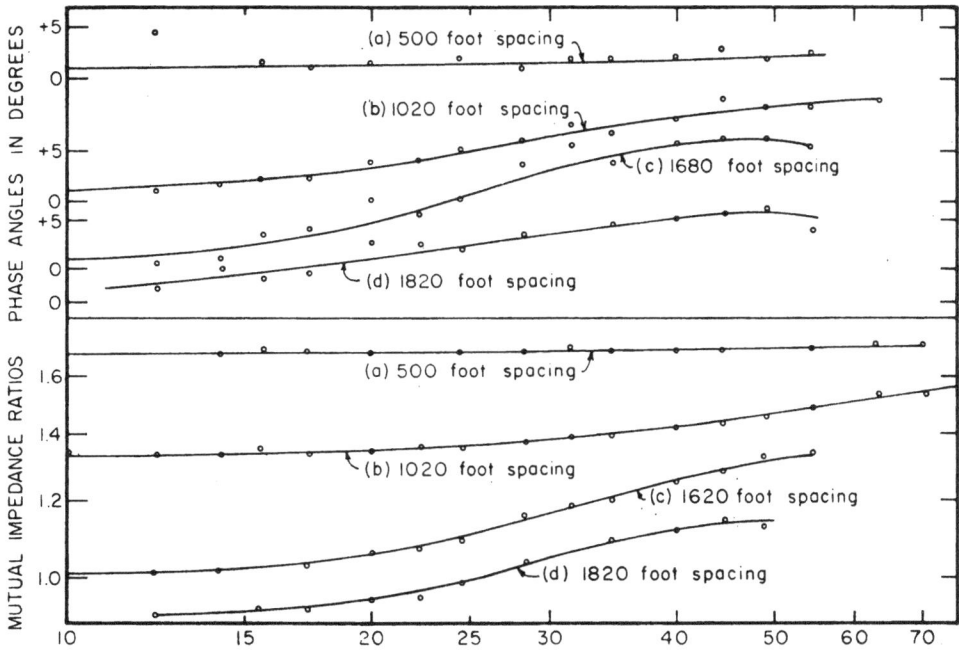

FIGURE 13. Changes in mutual impedance as a function of the square root of frequencies measured on the Athabasca Glacier.

an imperceptible effect on these first measurements, the remainder of the work was devoted to three depth soundings, using larger spacings with the line between the coils parallel to and near the edge of the glacier. The results of the measurements are shown as (b), (c), and (d) in Figure 13. Depth soundings (b) and (c) were made at the same distance from the edge of the glacier; the coil locations for (d) were about 200 feet farther from the edge.

In Figure 14 the results for depth sounding (b) are superimposed on theoretical curves. The amplitude ratio curve fits very well on the theoretical curve for $h/r = 0.4$, except for some uncertainty at the highest frequencies. The phase difference curve roughly fits theoretical curves at low frequencies but departs by a factor of two at the highest frequency.

The amplitude ratio curve for (c) (Figure 14) fits between curves for $h/r = 0.2$ and $h/r = 0.4$ at low frequencies but departs from all of the theoretical curves at high frequencies. The phase difference curve for (c) is even more extreme than the curve for (b). Depth sounding (d) (comparison not shown) is similar to (c) except that the phase differences are somewhat smaller.

Ignoring the rest of the data, a simple interpretation can be made from the amplitude ratio curve for depth sounding (b). For this curve $h/r = 0.4$, $r = 1,020$ feet, $h = 408$ feet. The abscissas are so matched that, for instance, the square root of 300 cycles per second on the field curve corresponds with $((\mu\omega)/\rho)^{\frac{1}{2}}r = 0.535$ on the theoretical curve. This gives the resistivity of the bedrock as 800 ohm-metres. A thickness of ice of 400 feet and a resistivity of 800 ohm-metres for the bedrock are reasonable values although there is no direct evidence to substantiate

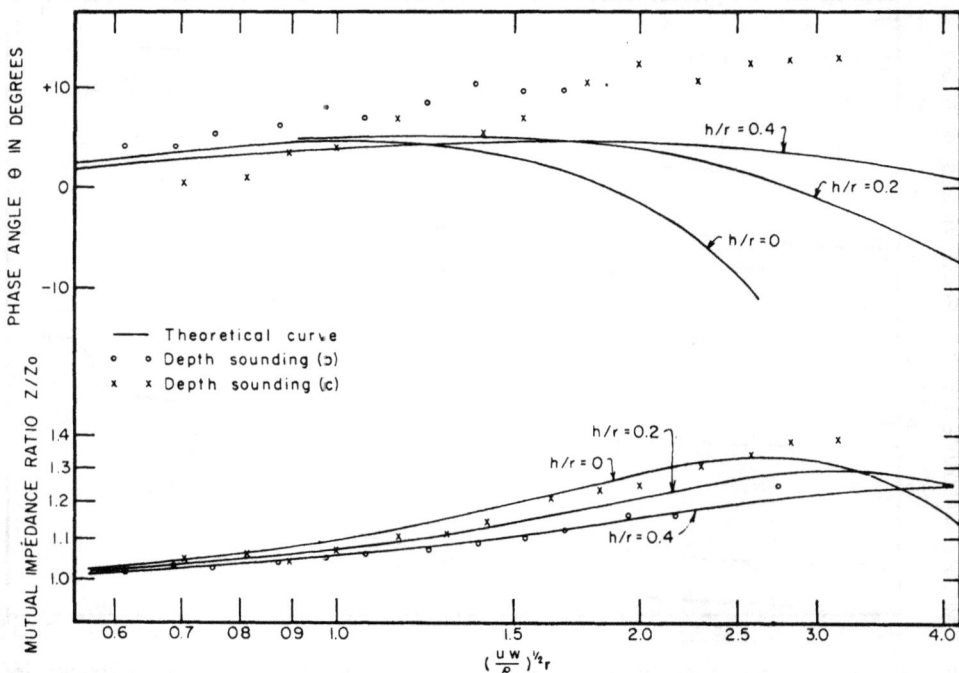

FIGURE 14.   Comparison of results from Athabasca Glacier with theoretical curves for horizontal loops raised above a two-layer conducting earth.

them. Actually, 400 feet and 800 ohm-metres should be considered minimum values, as the factor which causes the curves for (c) and (d) to be abnormally high probably raises (c) by a lesser amount.

Three possibilities have been considered for the departure of the phase curves and the amplitude curves for (c) and (d) from computed curves. They are layering within the bedrock, the steeply dipping surface of the bedrock, and a response from the ice. The curves for a layered earth for $K = 0.3$ and $d/r = 0.25$, shown dashed in Figure 12, come closer to fitting the field data than do the curves for a homogeneous earth. Possibly, some other combination of K and d/r, not yet computed, might give a better fit. However, it seems unlikely that the phase curve can be matched by considering a layered earth. Nearby exposures indicate that the rock beneath the Athabasca Glacier is a thick section of fairly homogeneous carbonate rock; consequently, there are probably no large conductivity changes with depth in the rocks beneath the glacier. Drainage beneath the edges of the glacier may cause a thin wet zone having low resistivity.

Beneath the coils the surface of the bedrock may be dipping as much as 30 or 40 degrees. If so, it is not a very close approximation to consider an ideal case of horizontal loops raised above a horizontal earth. A better approximation is actually to account for the angle between the plane of loops and the bedrock surface by replacing the loops with two sets of component loops, one set with its plane parallel, and the other set with its plane perpendicular to the bedrock surface. The response of vertical loops is at least of the correct sign to help account for behaviour of the field curves. Curves for loops having various inclinations are shown in Figure 15.

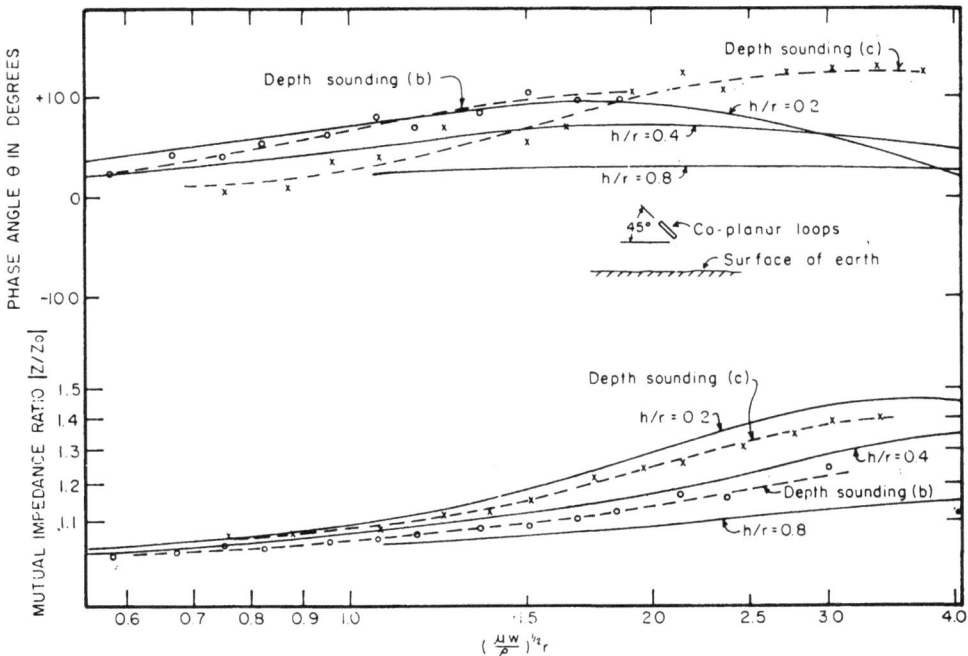

FIGURE 15.   Comparison of results from Athabasca Glacier with theoretical curves for co-planar inclined loops raised above a homogeneous conducting earth.

The portions of the field curves that cannot be accounted for by a conducting earth amount to only a few per cent in amplitude and about 10 degrees of phase angle. None of the curves for a lossy dielectric (Figure 11) can be used directly to account for such a small response. In the region where the response is small, the amplitudes are less than unity. The only way to simulate the observed effect is to assume that *b* decreases very rapidly with frequency. However, on the basis of the curves that have been computed, it is impossible to account for both the amplitude and the phase curve with the same values of *b*. As discussed in the next section, laboratory studies show that dispersion in ice near the freezing point takes place in the range from 5,000 to 50,000 cycles per second. If the electromagnetic data presented here are affected by the ice, dispersion is taking place at frequencies of a few hundred cycles. In view of the above discussion, it seems very unlikely that the ice had an appreciable effect on the electromagnetic measurements. The effect, if any, cannot readily be explained by the theoretical data presented in this report. Assuming that the theory here would be correct for the response from ice, upper limits can be set for the conductivity and dielectric constant of the glacier.

ELECTRICAL PROPERTIES OF ICE

Extensive laboratory studies of the electrical properties of ice have been reported in the literature (see, for example, Smythe and Hitchcock, 1932). The water, or ice, molecule is polar and exhibits dispersion in the values of dielectric constant at audio frequencies. A curve, relating dielectric constant and frequency for ice very slightly below its melting point, is shown in Figure 16. At very low frequencies, the dielectric constant is 73.7, while at high frequencies the dielectric constant is 4.0. The dispersion takes place in the frequency range 5,000 to 50,000

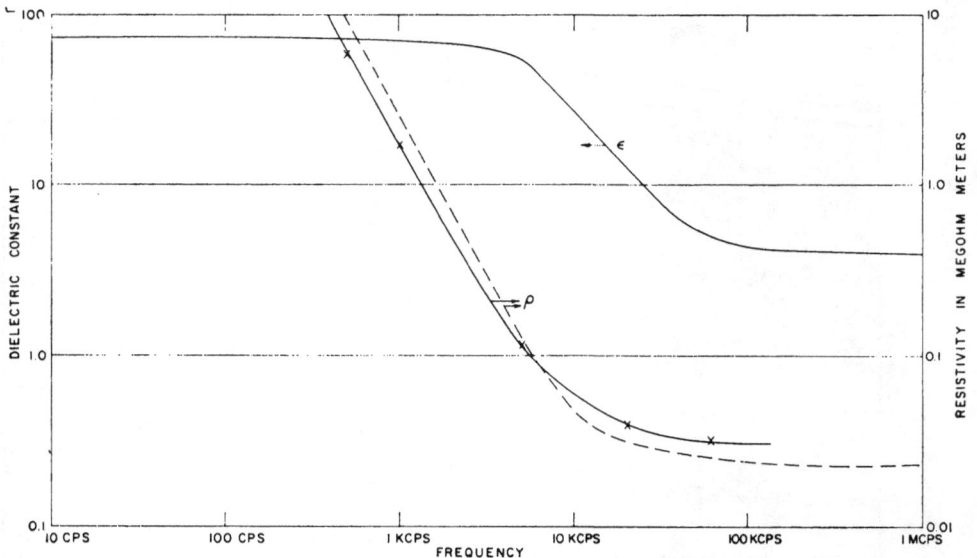

FIGURE 16.   Summary of the electrical properties of ice slightly below the freezing point (data from Smythe and Hitchcock 1932).

cycles per second. The relaxation frequency is 15.5 kilocycles per second. Smythe's and Hitchcock's values for resistivity of ice are shown on the same graph. The high-frequency resistivity approaches a low asymptotic value of 0.0335 megohm metres, but increases as the inverse square of frequency below the relaxation frequency. The dashed curve represents the values for resistivity calculated from the dielectric constant data assuming that the only conduction is Debye-type loss associated with molecular oscillations is shown as a dotted curve on Figure 16. The agreement is excellent, suggesting that for the ice studied by Smythe and Hitchcock, molecular resonance is the only significant conduction mechanism.

Smythe and Hitchcock also found that very small amounts of salt altered drastically the electrical properties of ice near the melting point. For ice containing 0.001 molar potassium chloride, a dielectric constant of 176 and a resistivity of 0.53 megohm-metres was measured at 300 cycles per second.

Glacial ice differs from laboratory ice in genesis. For the most part, glacial ice is compacted snow rather than frozen water. In this respect, ice resembles any other detrital rock. Crystal fabrics on the Saskatchewan Glacier have been described by Meier, Rigsby, and Sharp (1954), and their observations are probably applicable to the Athabasca Glacier as well.

According to Meier and his co-workers, much of the ice exposed on the surface of the Saskatchewan Glacier displays a well-developed flow foliation consisting of alternating layers a fraction of an inch thick of white bubbly ice and denser bluish ice. This foliation was noted over large areas of the Athabasca Glacier, also, but in addition, about 5 to 10 per cent of the surface is composed of white ice, well jointed with two sets of near-vertical intersecting joints. This white ice is not as hard as the rest of the surface, but it melts more slowly, forming prominent ridges. The lower melting rate is probably determined by the greater albedo, lower thermal conductivity, and better drainage. It was noted in the horizontal profile that these patches of white ice had a lower resistivity than the patches of dense blue ice. This lower resistivity may be a result of intergranular moisture films conducting electricity through the white ice.

In addition to this intergranular porosity, crevasses are probably important in determining over-all conductivity in the ice, even though the part of the glacier on which this work was done is the least crevassed of the whole glacier. The great transverse crevasses which develop at the ice falls apparently are healed within a few hundred feet from the foot of the ice fall. A poorly-developed crevasse system on the lower step consists of vertical fractures trending inward and upstream at a 45-degree angle towards mid-glacier. A few of these crevasses are open an inch or two, but most appear to be closed. In many places, melt streams fall into these crevasses, forming deep moulins typical of valley glaciers.

The fact that all of the melt streams drain into moulins, so that there is no drainage from the toe of the glacier except by marginal and subglacial streams indicates that the moulins must be connected laterally with the margins of the glacier. Very few of the moulins are observed to fill up, despite the very large run-off during the melt season.

An open crevasse might be expected to increase greatly the resistivity of the ice. Most of the fractures, however, are not open and probably are filled with a thin film of water, affording a path for conduction. If the Athabasca Glacier is at the melting point, there is no excess cold to freeze the water in the fractures. Even if the glacier is slightly below the freezing point, there may be water in these fractures. As the fractures take up the motion of the glacier, there must be pressure melting of the ice where irregularities on either side of a fracture bear the brunt of the downglacier pressure. The surface zone of relatively low resistivity found on Athabasca Glacier may represent the depth to which these fractures are water-bearing.

In many respects, the porosity of glacial ice appears to resemble the porosity in limestones, consisting of three types: inter-crystalline microporosity, vugs and solution cavities, and finally, joints. It seems reasonable that the equation relating water content and resistivity in limestones might be applied to ice. In limestones, resistivity and water content are related as follows:

(3) $$\rho = 1.4\rho_w s^{-1.8}$$

where

ρ_v is the resistivity of the water contained in the rock, and

s is the volume fraction of water.

Samples of water taken from run-off streams on the glacier have a resistivity of 650 ohm-metres at 0°C. Assuming equation 3 applies to glacial ice, one may calculate that the ice with a resistivity of 0.3 megohm-metres has a water content of 4.0 per cent by volume.

Fine-grained detrital rock material in ice can lower the resistivity considerably because such impurities can retain water in a liquid state even well below freezing. Parts of the Athabasca Glacier appear to include high concentrations of rock, notably along the lateral moraines where landslides have covered the margins of the glacier with rock. That this material is much more conductive than clean ice is seen from the electrode resistances recorded along the resistivity profile (Figure 8). The electrode resistance was only 80,000 ohms in the moraine, while it was about 8 megohms to the centre of the glacier.

In addition to the accumulation of debris along the margins, the entire surface of the glacier appears to be covered with rock dust in late summer. In many places this dust is concentrated in bands a fraction of an inch apart, apparently outlining sedimentary layers. However, this dust appears to be only a surface feature, caused by wind-borne particles being caught in riffles on the ice surface created by differential melting of the layers in the ice.

As the bulk of the ice in the glacier is dirt-free, it is felt that the moisture in crevasses and microporosity is far more important than rock dust in determining the electrical conductivity at low frequencies. At high frequencies, conduction through moisture is probably insignificant compared to volume conduction through the ice.

CONCLUSIONS

Resistivity studies on Athabasca Glacier have indicated that electrical methods may be useful in studying the thickness and texture of temperate glacial ice. Both galvanic and electromagnetic techniques for measuring resistivity appeared to be capable and effective in measuring ice thickness. Electromagnetic methods are probably preferable to galvanic methods if the primary interest is in the thickness of the ice and the nature of the material underlying it. The electromagnetic method has three advantages for such applications. The resistivity of the underlying rock can be determined without going to large coil separations, while if galvanic methods are used, impossibly large electrode separations are required to determine the resistivity of the rock, because it is so much lower than that of the ice. The electromagnetic method is also somewhat superior to the galvanic method if the surface beneath the ice is steeply dipping. The galvanic method is sensitive to boundaries parallel to the electrode spread whether they are horizontal or vertical. The electromagnetic method is sometimes less sensitive to boundaries perpendicular to the loops than to boundaries co-planar to the loops. Also, the interpretation for a dipping surface is, in this case, simpler for the electromagnetic method than for the resistivity method. The third advantage of electromagnetic methods is that they are insensitive to resistivity variations within the ice at the frequencies and coil separations used to detect sub-ice conditions.

On the other hand, if one is primarily interested in features in the glacial ice itself, galvanic resistivity methods are preferable. As is suggested by the present work, resistivity methods can be used to separate zones of massive ice from zones of compacted névé, and in so doing may be helpful in tracing structure in a glacier. Resistivity may be used to detect the depth to which the glacier contains liquid water. As a corollary, it should be possible to measure thermal layering in a glacier, if it contains some cold ice, by measuring the resistivity layering, because resistivity changes rapidly as ice temperature is lowered beneath the freezing point.

The most important question is if resistivity methods are worth-while when compared with the seismic method or other methods of studying ice thickness. Seismic methods generally are more precise in measuring ice depths, with accuracies of 1 or 2 per cent being within reason. On the other hand, the best accuracy that may be obtained with electrical methods is 5 to 10 per cent, principally because the results are affected by thickness over quite a large area.

However, seismic methods run into great difficulties if the ice is only a few hundred feet thick, or if it is highly crevassed, or if the bottom is irregular or steeply dipping, all circumstances which prevent the detection of a coherent reflection from the bottom. In rare circumstances, also, ice may overlay rock with a velocity such that there is no reflection coefficient to generate reflections. In all these cases electromagnetic methods will work.

The great advantage of electrical methods may be ultimately in the ability to identify, more or less, the subglacial rocks. Such identification is based either on a general resistivity-lithology correlation or on specific knowledge of the resistivity of rock units from drill holes or marginal outcrops. Seismic velocity might also be used for such rock identification (except that it is not so unique as resistivity identification), particularly because the high velocity of sound in ice makes it impossible to determine velocities less than ten to twelve thousand feet per second, and even with greater subglacial velocities tremendously long geophone spreads are required.

All in all, it appears that electrical methods may have an important role in glaciological studies and subglacial exploration geophysics.

REFERENCES

MEIER, M. F., RIGSBY, G. P., and SHARP, R. P.   1954.   Preliminary data from Saskatchewan Glacier, Alberta, Canada; Arctic, vol. 7, no. 1, pp. 3-26.

MOONEY, H. M., and WETZEL, W. W.   1956.   The potentials about a point electrode and apparent resistivity curves for a two-, three-, and four-layer earth; University of Minnesota Press.

SLICHTER, L. B., and KNOPOFF L.   1959.   Field of an alternating dipole on the surface of a layered earth; Geophysics, vol. 24, no. 1, pp. 77-88.

SMYTHE, C. P., and HITCHCOCK, C. S.   1932.   Dipole rotation in crystalline solids; J. Amer. Chem. Soc., vol. 54, pp. 4631-47.

WAIT, J. R.   1954.   Mutual coupling of loops lying on the ground; Geophysics, vol. 19, no. 2, pp. 290-6.

———   1955.   Mutual electromagnetic coupling of loops over a homogeneous ground; Geophysics, vol. 20, no. 3, pp. 630-7.

———   1958.   Induction by an oscillating magnetic dipole over a two-layered ground; Appl. Sci. Research, sec. B, vol. 7, pp. 73-80.

ZABLOCKI, CHARLES.   1957.   Analog studies of induced polarization over a layered earth (abstract); Geophysics, vol. 22, no. 2, p. 502.

A Distribution Study of

Abandoned Cirques in the

Alaska-Canada Boundary Range

MAYNARD M. MILLER

ABSTRACT

The distribution, orientation, and frequency patterns of 218 cirques in the northern Boundary Range, Alaska-Canada, are analysed. The mean elevation of the lowest ice-filled basins is 3500 feet. In the valleys not formerly glaciated by trunk glaciers from the Wisconsinan Cordilleran ice-sheet, abandoned cirques occur in a repeated five-fold sequence. The mean elevations of these cirques, or cirque-like basins, are: 350, 1100, 1750, 2450, and 3150 feet. The standard deviation in each system is small and the effect of orientation appears to be random. There is no apparent evidence of regional control by lithology, bedrock structure, or pre-Glacial topography. The frequency distribution, therefore, suggests that the cirque levels relate to former névé lines. An apparent correlation of cirque pattern with a sequence of other erosional forms in the main valleys of the range strengthens the inference of relation to Wisconsinan glacial stages. The study suggests that, during the late Pleistocene, glaciers in this maritime region were uniquely sensitive to climatic change.

WITH THE CERTAINTY of increased strategic and commercial activity in the far northern regions over the next several decades, it is natural that there is a corresponding interest developing in the various branches of arctic geology. One of the most important of these disciplines is that of polar and alpine geomorphology. As a knowledge of the nature and morphogenesis of arctic landforms and surface materials can yield information indispensable to operations in these regions, a full comprehension of arctic Pleistocene events and related processes in glaciers, sea ice, and permafrost is a pertinent and practical goal. In outlining the nature of former glacial ages, the classical approach has been to extrapolate from the evidence of relict depositional and erosional features, while regarding the régime and effects of modern ice masses as phenomena of post-Thermal Maximum time, essentially disconnected from the chronology of the Ice Age.

In this paper the problem is approached in the opposite way. Through selected locations of a specific landform, an attempt is made to work directly and sequentially into the past from the present position and activity of existing glaciers. The study bridges the last major glacial age, or Wisconsin Cordilleran glaciation, in the Alaska-Canada Coast Range. The region considered lies in the northern Boundary Range (Bostock, 1948, pp. 83-8), and particularly refers to that portion of the range inland from Lynn Canal in the vicinity of sixtieth parallel (Figure 1).

The field work for the present study was carried out in connection with a more comprehensive investigation of Alaskan and Canadian coastal glaciers in which the writer has been engaged over a number of years. The part of this programme discussed in the present paper was supported by a grant from the Resa Fund of the Society of Sigma Xi and by the Expeditionary Research Fund of the Explorers' Club. The work is also related to the systematic regional studies of the Juneau Icefield Research Program, under the aegis of the foundation for Glacier Research, Inc. For assistance in the field, the writer is indebted to Dr. T. R. Haley, Mr. W. W. Miller, and Mr. Barry Prather. Appreciation is also extended to Mr. Hunt Gruening of Alaska Coastal Airlines and Mr. Kenneth Loken, Juneau Air Taxi Service, for competent and cheerful co-operation during the extensive flights made to observe many of the geomorphological features here discussed.

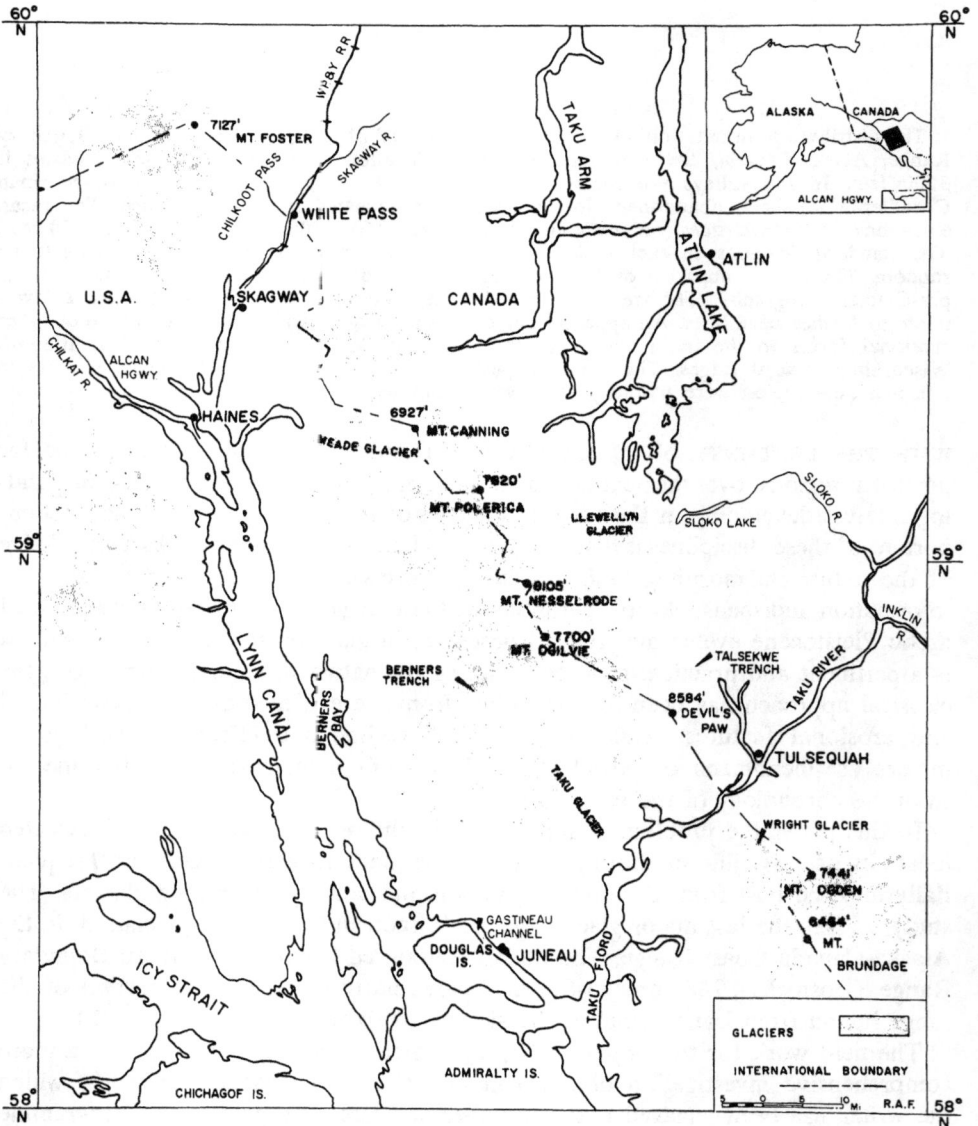

FIGURE 1.   General map of the northern Boundary Range, Alaska-Canada.

LOWER GLACIAL LIMITS FROM THE POSITION OF ABANDONED CIRQUES

The general levels of forest-covered cirques in present non-glaciated valleys of the maritime flanks of the northern Boundary Range provide some useful information on the depression of the Cordilleran ice cover during the Pleistocene. The position of cirques has been used by Ljungner (1948, p. 36) to show the lower limit of ancient glaciation in Scandinavia. Flint (1947, pp. 95 and 230) has also considered the elevation of old cirques as representing excavation during the maximum of glacial ages in the western United States. He suggests that, as the same cirques could be occupied by ice from successive glaciations, it may be difficult to assign a set of cirques to a particular age, but that "future refinements of study may show that some correlation is possible."

In this paper, such a refinement is attempted in an effort to understand a regional pattern of cirque development and to aid in the interpretation of related geomorphic features. From this, a working hypothesis is evolved which is in accord with available evidence and which can be tested by further, more detailed, research. The fundamental assumption is that the elevation of the cirque floors gives a rough estimate of the former regional snowline. As will be discussed, they more likely approximate to the level of former permanent névé lines, a proposition originally suggested by Hann (1903, p. 310).

Also to be considered is the possibility that the distribution and character of these cirques may have economic significance. The main area of the present study lies in the Taku district, within a radius of sixty miles of Juneau, Alaska's capital city. About 60 per cent of the cirques investigated are carved into bedrock areas on the maritime margin of the Juneau ice-field. The other 40 per cent are on the slopes of ridges above the main westerly-trending river valleys and on island massifs near Taku and Gastineau fjords (Figure 1).

A FIVEFOLD TANDEM SEQUENCE

In the northern Boundary Range, 218 distinct cirques were studied. In these, there are five noteworthy cirque systems, one of which is distinctly old. These cirques are numerous and well displayed everywhere around the edge of the present ice-fields. To illustrate their relation, a listing of cirques found on the flanks of the main valleys and slopes on the southern and western sides of the Juneau ice-field is given in Table I. Although many are composite and multiple, their forms are usually well defined and, as will be shown, the elevations of the different sets are unexpectedly accordant.

To simplify the discussion, reference is made to five groups of moderate-sized cirques in small, separated valleys leading from ridges not directly connected with the main ice-field area. As indicated in Table I, their elevations appear to be representative. By selecting these smaller valleys, the erosional effects of large through glaciers which have modified many other cirque systems in the region are of less concern. The valleys of Nugget Creek, Salmon Creek, and Gold Creek drain into Gastineau Channel from the area north and east of Juneau; they are noted within the rectangles labelled A, B, and C on the Juneau quadrangle map of Figure 2. The other two valleys are tributary to the inner end of Taku Fjord; the cirques in them are marked by smaller rectangles in the map of Figure 3.

TABLE I

ELEVATION OF CIRQUE FLOORS FROM THE TAKU VALLEY TO BERNER'S BAY

Location and map quadrangle	Elevation in feet and orientation of cirque, as noted					
Taku River, B6						
Mouth of Taku R.:						
E. Side	550NW	950NW	1550W	2300N	3100W*	—
	—	—	1700NW	2700NW	—	—
Johnson Cr.	—	—	1700NW	2150NE	3100N*	3600NW*
Davidson Cr.	—	1250N	1600NW	2400SW	—	—
	—	1200NW	—	2200NW*	—	—
	—	1200N	—	2300N	—	—
	—	1200NE	—	2450N	—	—
	—	—	—	2300N	—	—
	—	—	—	2250NW	—	—
	—	—	—	2350NW	—	—
Turner Lake Area	350N	1150NW	1550N	2300N*	3300NW*	3600W*
	—	900W	1600N	2200NW	3100SW	3600NW
	—	—	1450W	2200W	—	—
	—	—	1800NW	2550W	—	—
	—	—	—	2200W*	—	—
	—	—	—	2200NW	—	—
	—	—	—	2300SW	—	—
	—	—	—	2200E	—	—
Juneau, B1						
Norris Peak:						
N. Side	—	—	—	2600W	3200N	—
S. Side	—	1100S	—	—	—	—
S.W. Br. Norris	—	—	1600SE	2700S	3000E*	—
Taku Inlet:						
Scow Cove	—	—	1500E	—	—	—
W. of Flat Point	—	1150S	1500SE	—	—	—
Annex Cr. valley	—	900SE	1600SE	2250W	—	—
Carlson Creek:						
Sunny Cove	300SE	1050E	—	—	—	—
Salmon and Gold forks	—	1200NE	1600NE	2800SE	3200N	3500E*
Sheep Fork	450NE	900NE	1800NE	2300NE	3300E*	3600NE*
	—	—	1800NE	—	—	—
Hawthorne Peak	300NE	900NE	1800E	2300SE	3100NW*	3600N*
N.E. valley	—	1150SE	1700E	2500SE*	—	—
Rhine Cr.	—	—	1800SE	2500SE	—	—
	—	—	—	2400E	—	—
Thane Area	—	1000SW	1500W	—	—	—
Sheep Cr.	550SW	900W	1500NW	—	3100NW*	—

*These containing ice at present.

TABLE I—*continued*

Location and map quadrangle	Elevation in feet and orientation of cirque, as noted					
Juneau, B2						
Lemon Cr. valley	300W	800W	1900N	2350W	—	—
	—	—	—	2450W	—	—
	—	—	—	2600SW	—	—
	—	—	—	2600N	3000N	3400N*
Mendenhall valley:						
Steep Cr. Sector	—	1300N	2000NW	2650N	—	—
McGinnis Cr.	350SW	1150W	1850S	2700S	3200SW*	—
Douglas Island						
N. Side	—	1200NE	1550NE	—	—	—
Fish Cr.	450N	1000N	1650N	2300NE	—	—
W. Side	—	1150W	1750W	2550NS	—	—
Juneau, B3						
Auke Bay area:						
Peterson Cr.	350N	900N	—	—	—	—
Montana Cr.	—	—	1850W	—	—	—
Juneau, C3						
Herbert R. valley	—	—	1650N	2350N	—	—
Husky Cr.	—	—	—	2600NW	—	—
Eagle R. valley	300SW	1200W	1850NW	2350NE	—	—
	—	—	—	2300SE	3100E*	3500NE
Cowee Cr. area:						
Yankee Cove	450W	1200W	1850NW	2300NW	—	—
Canyon Cr.	—	1300N	1850W	2350W	—	—
Cowee Cr. (main)	300SW	1000SW	1600SW	2400SW*	—	—
Davies Cr.	350S	1200W*	1600S	2200S	—	—
Berner's Bay:						
Sawmill Cr.	—	—	1600N	2500N	—	—
	350W	1150SW	1700SW	2700W	—	—
Juneau, D3						
N.E. Side of Bay	—	1250W	1800W	2200W	3200W	—
S. Br. Antler R.:						
S. Side	300NW	1250N	—	2350S	3000N*	3400N*
	—	1000N	1900N	2500N	—	—
N. Side	—	—	1750S	2200N	—	—
	—	—	1900S	2600S	—	—
Antler R. valley	—	—	—	2200N	3100NW	—
S. Side	—	1400S	1850NW	2750N	—	3500N*
N. Side	250SE	—	1750SW	2800W*	3200S	—
	—	—	1800S	2850SE	—	—
Lace R. valley	—	1250W	—	2200N	3000W	3500N*
	—	1200W	—	2300W	2900W	3450N
	—	1100N	—	2650N	3100W	—
	—	1000S	1800SE	2750S	3200W*	—
	—	1200SE	—	2550W	—	—
	—	—	—	2600W	—	—
	—	—	—	2350SW	—	—
Mean elevation of the major cirque systems (to nearest 50 feet):	350	1100	1750	2450	3150	3500

*These containing ice at present.

Some of the cirques with the lowest elevation have been so altered by headward erosion and valley fill that they are more in the form of elongated basins and, in some instances, where mining operations have taken place, have been formally designated on maps as basins. For convenience, the different sets of cirques are

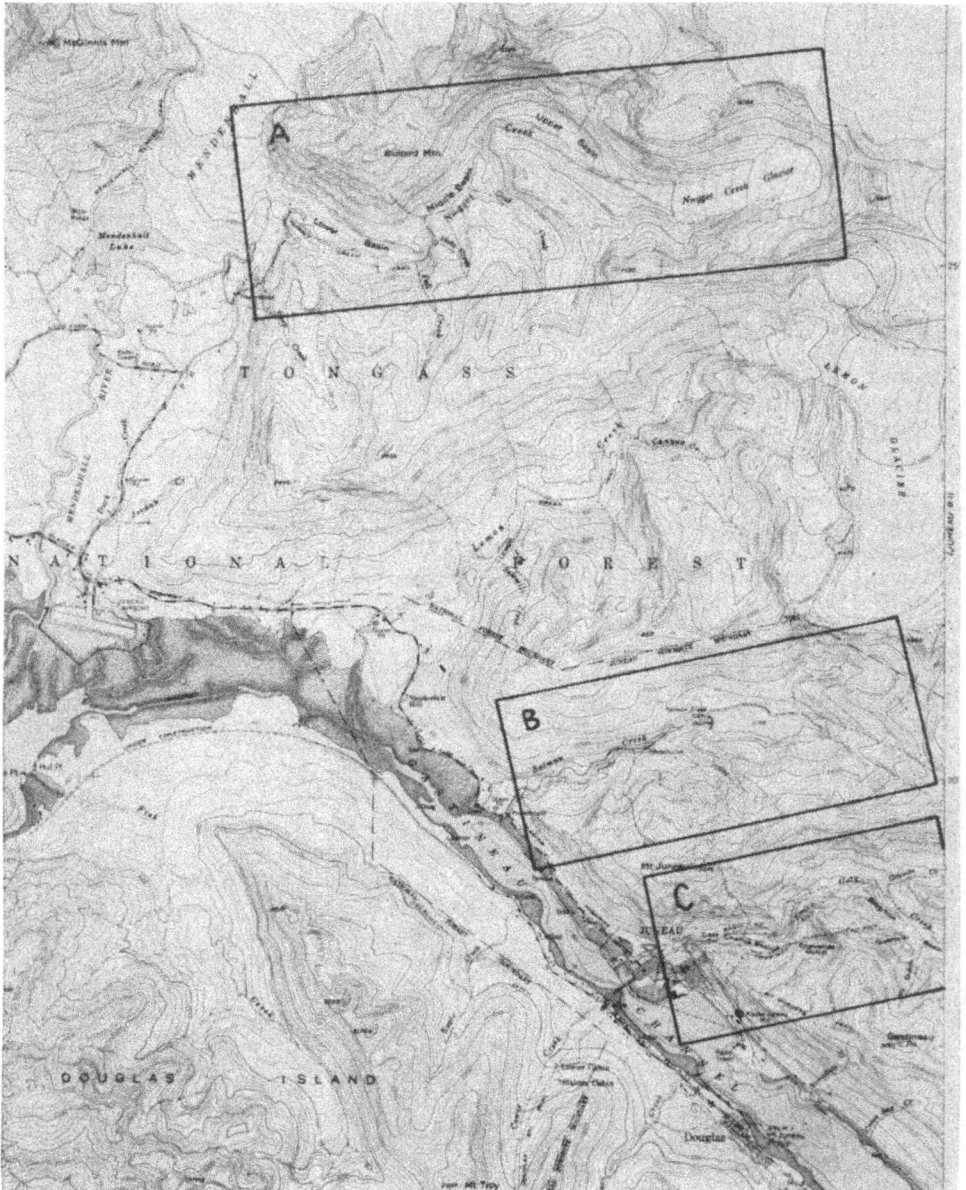

FIGURE 2.   Portion of Juneau B-2 quadrangle showing topographic configuration of tandem cirques in selected valleys of the Juneau area. Scale: length of Juneau Airport runway, 1 mile.

referred to in Table II as C1 to C5, beginning with the lowest and oldest and grading upward to the highest cirque at present containing remnant bodies of ice.

In view of the great diversity of rock types in which they occur, grading from low-grade metamorphics to granodiorites in different parts of the area, the general conformity of level indicated by cirques in each system is particularly striking. To demonstrate this statistically, an elevation-frequency diagram has been prepared in

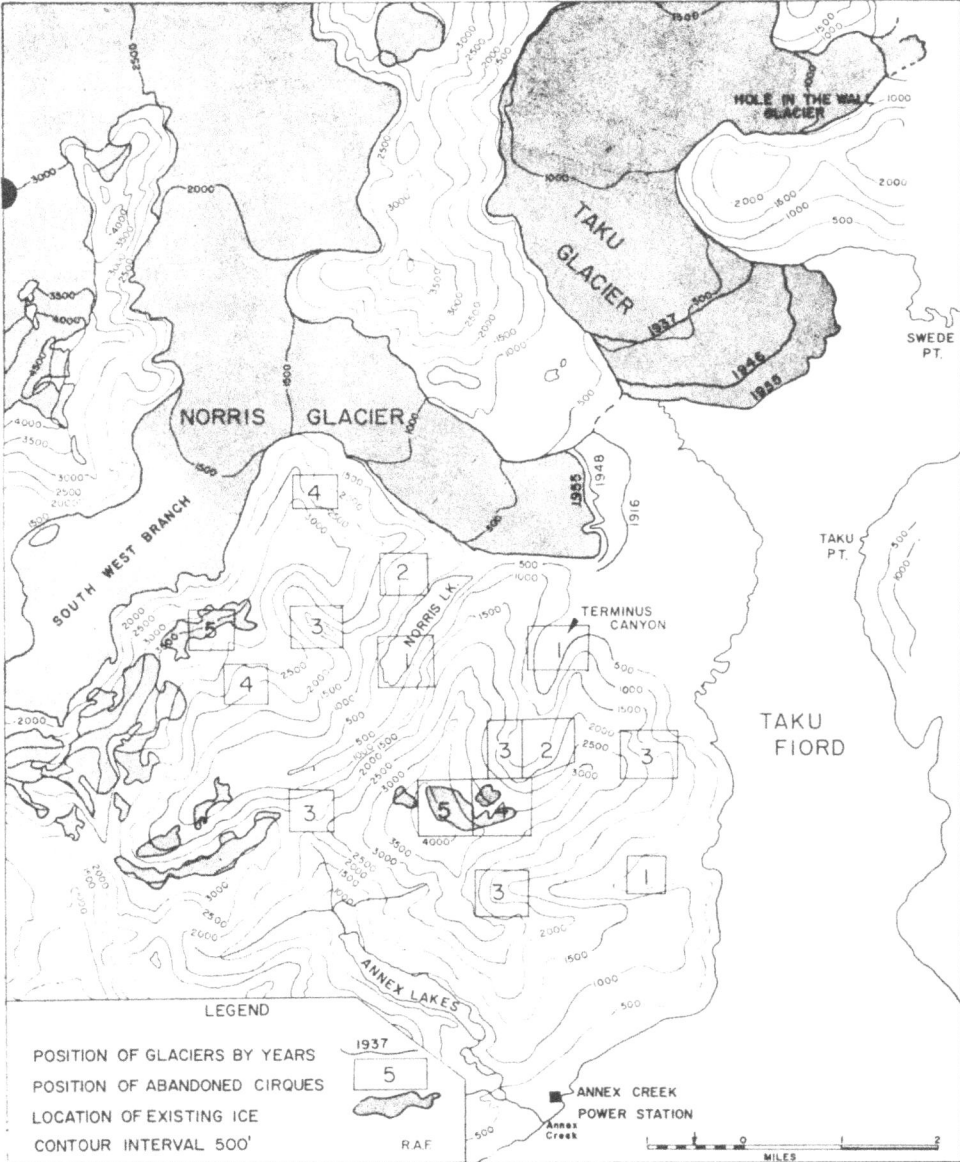

FIGURE 3. Section of inner Taku Fjord showing terminal positions of existing valley glaciers and fivefold pattern of abandoned cirques in the Norris Lake sector.

Figure 4. The diagram embraces the data in both of the table listings. The position of each of the 218 cirques is indicated.

These are not random samples, but are the result of systematic consideration of all cirques in every valley between Taku Fjord and Berners Bay (Figure 1). The elevations of the cirque floors are as inferred from the contours of the Juneau B1, B2, C3, and D3 sheets and of the Taku River B6 sheet of the 1:63,360 series of the topographical maps recently produced from aerial photographs by multiplex methods (United States Geological Survey, 1951-5). From field comparisons made on some of the cirques with aneroid elevations, these maps are believed to permit assessments to an accuracy of twice the contour interval, that is, ± 100 feet. The elevations of cirques in the Juneau area are the most certain, because they are also taken from available large-scale maps prepared for development purposes and mineral exploration.

In row A of Figure 4, the combined positions of all cirques in the district survey are noted, covering a zone some eighty miles in length around the southern and western edges of the ice-field. Although the fivefold distribution pattern may not be as clearly evident in the combined presentation, it is well shown in rows B, C, and D of the figure. Row B reveals the more local distribution pattern in the Lynn Canal sector (Herbert River to Lace River); row C, the Juneau-Gastineau Channel sector; and row D, inner Taku Fjord. Where there is divergence of the pattern locally, it may be ascribed to orientation[1] and orographical differences, and on the regional basis it may be explained as due to positional influences.

The significant fact, therefore, is not that there is a mean level of cirques representing separate systems — the orographical differences and directions of exposure

[1]The orientation of each cirque is noted in the listing in Table I.

TABLE II

FLOOR ELEVATIONS OF ABANDONED AND PARTIALLY ABANDONED CIRQUES
(in feet above msl)

Reference designation	Nugget Glacier valley (Rect. A, Fig. 2)	Salmon Creek valley (Rect. B, Fig. 2)	Gold Creek Canyon, Juneau (Rect. C, Fig. 2)	Valley at delta, Norris Glacier (Fig. 3)	Norris Lake Canyon (Fig. 3)
C1	500 (Lower Basin)	250 (Flume Pond)	300 (Last Chance Basin)	300	300 (Norris Lake)
C2	1300 (Middle Basin)	1000 (Salmon Creek Reservoir)	1050 (Silverbow Basin)	1100*	900
C3	1800 (Upper Basin)	1900	1800 (Granite Creek Basin)	1600*	1700 (Tarn Lake)
C4	2300† (Nugget Glacier tongue)	2600	2300	2600	2100
C5	3100 (Main Nugget Glacier)	3200 (Ptarmigan Glacier)	3100 (Mount Olds Glacier)	3300 (unnamed glacier)	3200 (unnamed glacier)

*Composite cirque with two levels as noted.
†Incised into headwall of C3 (Figure 2)

partly obscure this correlation — but rather that in local sectors and in almost any valley in particular there is a complete sequence of five abandoned cirques. This is also demonstrated in the breakdown of localities noted in Table I and is specifically illustrated by the type distribution pattern at Gold Creek near Juneau (row E of the frequency diagram). As in this typical case, the cirques in each sequence are usually tandem in arrangement. It is suggested that the pattern is developed in distinct régime stages, or sets of similar stages, for each level. Thus different névé lines are probably represented. As has been shown elsewhere (Miller, 1956, chapter VI) even over a period as short as ten years there is a great variation in névé-line positions — as much as 1200 feet vertically in such an interval. If the suggested relation does pertain, each level must concern the mean limiting position over very considerable lengths of time, that is, upwards of some thousands of years.

IMPROBABILITY OF RELATIONSHIP TO FORMER MARINE LEVELS

An alternative consideration might be that this conformity is a reflection of former marine surfaces related to pre-Pleistocene landscape with a eustatic relation considerably different than today. This appears unlikely, however, if we assume, as Johnson (1917), Antevs (1932), and others have, that cirque formation is the

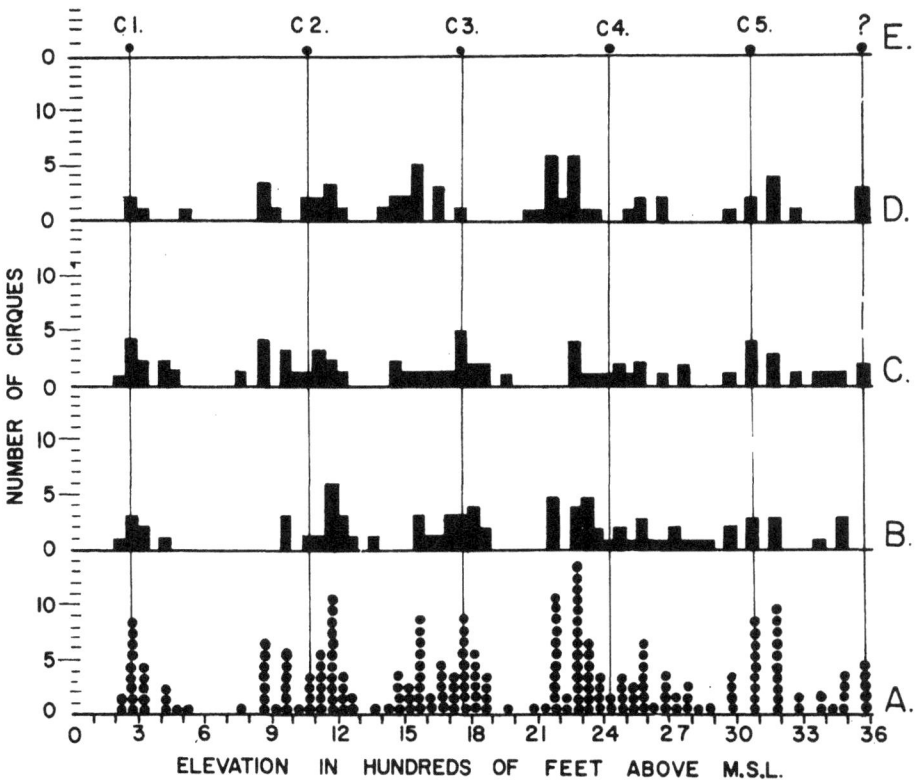

FIGURE 4.   Frequency distribution of abandoned cirques as a function of elevation in the maritime sectors of the northern Boundary Range.

last phase of glacial history in a mountain region. Beyond this theoretical assumption, there are the following evidences against the possibility: (1) that a patterned sequence of cirques, with definitely three levels and possibly up to five, is also found at somewhat higher positions on the flanks of the deep valleys on the interior side of the range where, as has been indicated by other studies (Miller, 1956, chapter III) no Tertiary marine deposits occur (also V. Kerr, 1948, p. 53); and (2) that similar cirque systems occur on the ice-carved slopes of large U-shaped valleys which trend towards the coast from positions well inside the Boundary Range. To all appearances, these valleys were incised by the maximum mountain ice-sheet of the Wisconsin age and, in some cases, by continental ice prior to the initiation of the cirques. Examples of these are Berner's Trench, the valley of Gilkey Glacier in Figure 1, and the Taku River valley sector typified by the view in Figure 5.

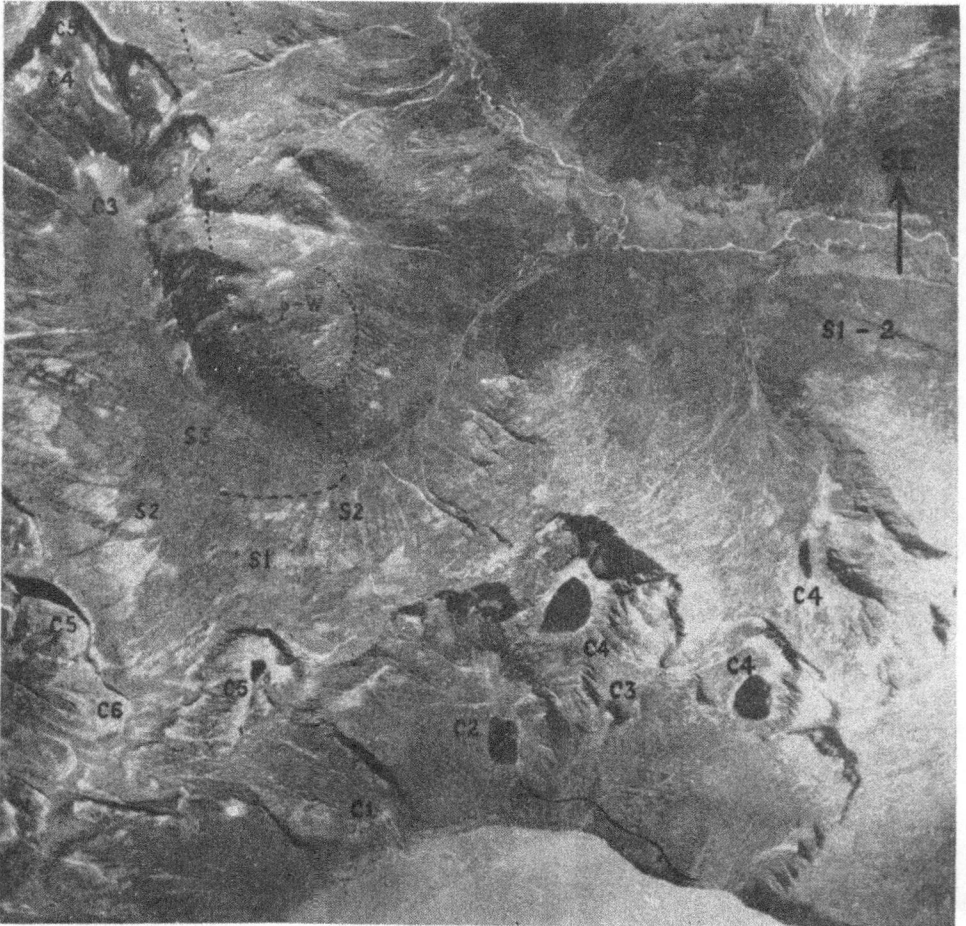

FIGURE 5.   Vertical aerial photograph showing development of tandem and composite cirques with associated tarn lakes and glacial valleys south of the Juneau ice-field. The bedrock is massive granodiorite. A portion of Taku River shows at bottom of the photograph.

Further to this, the constricted and protected nature of the narrow fjords and smaller valleys leading off the Juneau ice-field suggests that even if these had been inlets in pre-Pleistocene time they were so confined that wave sculpturing and other marine erosion processes would have been at a minimum. This suggestion is also borne out by the lack of strand flats, wave-cut benches, or sea cliffs, and the absence of marine deposits at elevations greater than 500 feet. Additionally, range upon range of lesser mountains up to 5000 feet high occur on the closely-spaced islands which extend for 80 to 100 miles west of the main Boundary Range, indicating that the open coast even in pre-Pleistocene time, was far removed from the Taku district.[2]

RELATIVE DIFFERENCES OF ALLUVIATION AND FORM

The lowest basins at 200 to 500 feet elevation are large, although not as well defined as the others. This is due not only to the longer period of erosion and alluviation to which they have been subjected, but to certain other modifying circumstances, noted below. They have also been made more indistinct by the relatively heavier afforestation near sea-level.

They are considered probably to be old cirques on the basis of the following criteria: (1) the semi-enclosed form and the arcuate nature of adjacent slopes; (2) the presence of a tarn lake, or of a downvalley projection of bedrock presumably representing a remnant threshold; and (3) the presence of cyclopean stairs above the basin area, and in most cases also a steep slope below it. Many have been subjected to late-Glacial marine sedimentation and others have received considerable deposits of outwash gravel so that the initial topography has been buried.

Most of the lower cirque-like depressions have also been modified at their outlets by ice passing over them laterally. This is illustrated in the case of both C1 and C2 cirques in the lower part of the photograph of Figure 5 (the dark line in the figure represents the trace of such an ice limit at the base of a truncated spur). And in some cases, glaciers have even pushed upslope into the cirques during stages when the trunk valleys to which they are tributary were channels for ice from other sources. This could only have occurred at times of raised névé limit when the lowest amphitheatres were no longer catchment areas. The faceting and rounding-off of the bases of their confining spurs supports this view and suggests that their present embayment form was developed during a separate later glacial stage, or in a retrogressive cycle of the same glaciation which produced the cirques. This would require the interesting circumstance of local glaciers in disconnected valleys being in a receded condition at the time of maximum advance of ice in the fjords. Such a situation has been described in Washington State (Mackin, 1941) where, during the Wisconsin maximum, the great lowland ice tongue of the Puget Glacier pushed up into the Cascade Mountains and filled some of the lower basins which had previously been occupied by local valley glaciers. The observation begs the fundamental question which has been brought to attention by the present vigorous

[2]On the western shore of the Alexander Archipelago and northward to Cape St. Elias, broad strand flats do occur at low and intermediate levels (D. J. Miller, 1953). These are quite in contrast to what is found in the Boundary Range, further suggesting that marine processes were not effective in this interior coastal area.

advance of the Taku Glacier in its sea-level channel at a time when local glaciers in the side valleys and at higher elevations are in a state of excessive retreat.

The cirques at the second level (C2) are the best defined and usually the deepest. They are also the most elongated, suggesting a considerable period of development with strong headward glacial erosion. They are heavily forested and would appear to represent a distinct stage in the valley development.

The cirques at the third and fourth levels (C3 and C4) are also well formed but usually not as long, suggesting a shorter period of development. They are mantled with successively less alluvial material and soil, and much less vegetation, a fact which is in part due to their elevation above the timber-line. The sharply delineated character of the fourth level of cirques indicates that they have largely resulted from the retrogressive phase(s) of the last major glaciation. An example of the C3 and C4 systems is seen in Figure 6, showing part of the sequence of tandem cirques in Salmon Creek valley. Cirques in the fifth system (C5), however, are usually shallow and are either barren or only partially filled with ice. From the assessment of the total number shown in the frequency diagram, the mean elevation of each key horizon on the west side of the range is taken as 350, 1100, 1750, 2450, and 3150 feet. The average position of another and higher system of broad basins filled with ice is approximately 3500 feet. The highest system (referred to below as C6) is morphologically not a true cirque system. Although its general elevation is close to the present mean névé line on the western side of the Juneau ice-field, the existence of these uppermost basins should not be considered in the same way — that is, as representative of higher snow lines — since they connect with the broad and presently-glaciated highland and hence have been more continuously affected by successive stages down through post-Wisconsin time.

IMPLICATIONS OF THE PATTERN

It is reiterated that the cirques represented in this study are those which were not occupied by glaciers in post-Glacial time, in spite of their positions relatively close to the present ice-field complex. The statistics in the tables reveal that the standard deviation of the floor elevations, especially of the higher cirques, is remarkably small with respect to the mean level for each system. Furthermore, the effect of geographical orientation on the elevation pattern appears to be random and so probably not significant. Added to this is the fact that on the maritime side of the northern Boundary Range a full fivefold sequence of cirques is found along the walls of each of the major deglaciated valleys. From these observations and the apparent lack of lithologic or structural control, the distribution pattern suggests regional significance. As no evidence has been found of control by pre-Glacial topography, there is the exciting implication that the levels represent former mean positions of the regional névé line which has raised and lowered cyclically during the glacial ages. On the basis of this interpretation, the frequency distribution would have major climatological significance.

A possible correlation may exist with other ice-carved forms on the valley walls and on flanks of nunataks in the interior of the range which is glaciated at present. A striking example is provided by a sequence of distinct rock shoulders on exposed bedrock separating surfaces of former erosion on the bordering ridges of the Taku Glacier. Such surfaces are also well displayed on the walls of bordering ice-free

FIGURE 6. Abandoned cirques at the 1900- and 2600-foot levels at the head of Salmon Creek valley (Rectangle B, Figure 2).

valleys, as illustrated by the designations S1 to S3 in Figure 5.[3] There is a continuous array of such berms, with intervening erosion surfaces, elsewhere in the range, adding weight to the interpretation suggested (Miller, 1956, pp. 116-22).

[3]The highest ridge crests in this sector may contain remnants of a pre-Wisconsin surface (p-W). The upper ice limit of major Wisconsin glaciation is interpreted from the highest berm. The trace of this berm is indicated in Figure 5 by dotted line.

The writer considers well that any attempt at broad correlation of ice-eroded features in a heavily-glaciated region must, by its very nature, be most cautious. This is particularly true in a region which may have been subjected to differential crustal warping during the Pleistocene. It is nevertheless of interest that four very prominent and one minor rock bench are found in sequence on rock exposures above the present glacier level in the maritime sector of this range. Geophysical measurements of depth transects in ice also reveal a lesser berm beneath the Taku Glacier in its lower-valley sector (Poulter, 1949, Figure 4). This lesser subglacial bench by position becomes sixth in the sequence. It may not correlate technically with the upper basin, or C6 "cirque" level at 3500 feet in the sequence of Table I, as the latter involves much more continuous glaciation. But at least it represents activity of a recent stage of very much retracted highland ice and hence provides one of the forms bridging the past to the present. As all the larger (older) berms appear to represent significant glacier-carved remnants of former valley floors, it is difficult to resist considering that they may have been sculptured in the same successive stages of glaciation which are suggested as responsible for the pattern of abandoned cirques in the peripheral zones. At least the striking similarity of sequence is tantalizing.

If the concordance of patterns in these two types of erosional features is verified elsewhere in the Boundary Range and if they can be shown to have a traceable relation to sequence of depositional forms in the peripheral lowlands, the inference of a causal relationship to Wisconsin glacial stages will be greatly strengthened. In a search for such corroboration, the study is currently being extended into the valleys and flanks of the adjacent middle and southern Boundary Range, particularly in the Stikine district. Although these other data are not yet available for processing, information to date suggests that the ice masses in the entire Boundary Range during the last major glaciation were unusually sensitive to climatic change. Possibly this seeming sensitivity is related to the unique position of the range, situated as it is in a climatologically-sensitive interaction zone between the cyclonic cellular circulation of air masses over the Gulf of Alaska and anti-cyclonic polar air masses of the continent. A discussion of this aspect is treated elsewhere, particularly with respect to the glacial fluctuations of the "Little Ice Age" (Miller, 1958, 1961).

ABANDONED CIRQUES AS LOCI OF PLACER GRAVELS

As some of the lowest cirques have received considerable deposits of marine and terrestrial sediments, part or all of their initial floor topography has been buried. This, of course, must be considered in attaching any significance to the apparent conformity of level in the lowest system. By assuming the degree of marine sedimentation as comparable in each of the key lower cirques described, and by considering the spur remnants as bedrock thresholds, one at least has a clue to the bedrock positions. Behind the outlet thresholds of the C1 and C2 depressions, glacio-fluvatile materials have been trapped. This is particularly true in those valleys which serve as drainage from existing glaciers. Since the valleys have been cut into the metamorphic zone along the edge of the axial crystalline core of the range, the associated basins have been carved mainly out of lode-injected schists.

These basins are, therefore, loci of gold-bearing gravel. Clays deposited by marine transgressions form the base upon which most of these auriferous gravels have been laid down (Knopf, 1912, p. 33). This probably explains the slightly higher than average elevation of the floor in the Lower basin of the Nugget valley (Table II), and of the outlet basin at Sheep Creek in the Thane area (Table I).

It is of interest that the first discovery of "colours" in the Juneau gold belt was in the outlet streams below the thresholds of several of these abandoned cirque valleys. These findings led to the first rush of gold seekers to Alaska in the 1870's, a quarter of a century before the Klondike strike. This resulted also in the founding of the city of Juneau. On the United States Geological Survey map (Figure 2), it may be seen that in the canyons east of Juneau two of the lowest amphitheatres have been given the descriptive names Last Chance basin and Silverbow basin. Other such terms used locally are Perserverance basin, Flume basin, Lower basin, Middle basin, Upper basin, and Nugget basin. There is, of course, the possibility that a detailed knowledge of the sequence, distribution, and morphology of abandoned cirques, and a recognition of their regional pattern, could serve as criteria in the search for new ore veins. It is possible that such knowledge could also prove to be useful in evaluating the potential of placer operations in zones of auriferous gravels impounded behind the lips of some of the lower cirques.

REFERENCES

ANTEVS, E. 1932. The alpine zone of Mt. Washington; Auburn, Maine.
BOSTOCK, H. S. 1948. Physiography of the Canadian Cordillera, with special reference to the area north of the fifty-fifth parallel; Geol. Surv., Mem. 247, Canada Department of Mines and Resources.
FLINT, R. F. 1947. Glacial geology and the Pleistocene epoch; John Wiley and Sons, Inc.
HANN, J. 1903. *In* R. and DEC. WARD, eds., MacMillan Co.; Handbook of climatology.
JOHNSON, D. 1917. Date of local glaciation in the White, Adirondack, and Catskill Mountains; Bul. Geol. Soc. Amer., vol. 28, pp. 543-52.
KERR, F. A. 1948. Taku River map area, British Columbia, Canada; Geol. Surv., Mem. 248, Department of Mines and Resources, pp. 1-84.
KNOPF, A. 1912. The Eagle River region, southeastern Alaska; Bull. 502, U.S. Geol. Surv.
LJUNGNER, E. 1948. East-west balance of the Quaternary ice caps. in Patagonia and Scandinavia; Bull. Geol. Inst. Upsala, vol. XXXIII, pp. 12-96.
MACKIN, J. HOOVER. 1941. Glacial geology of the Snoqualmie-Cedar area, Washington; J. Geol., vol. XLIX, no. 5, July-August, pp. 449-81.
MILLER, D. J. 1953. Late Cenozoic marine glacial sediment and marine terraces of Middleton Island, Alaska; J. Geol., vol. 61, no. 1, pp. 17-40.
MILLER, M. M. 1956. The glaciology of the Juneau icefield, S. E. Alaska; Office of Naval Research, Final Report, Task Order 83001, 2 vols., 800 pp.
——— 1958. Glaciers on the rampage; Science World, vol. 3, no. 8, pp. 4-7.
——— 1961. Casual factors in the fluctuation of Arctic glaciers and sea ice; Trans. New York Acad. Sci. (in press, report of paper delivered before the Society, May, 1958).
POULTER, R. C., ALLEN, C. F., and MILLER, S. W. 1949. Seismic measurements on the Taku Glacier; Stanford Research Institute, Stanford, California.
UNITED STATES GEOLOGICAL SURVEY (1951-5). Alaska Topographic Map Series, at scale of 1:63,360.

Glacial Geology of the Mount Chamberlin Area, Brooks Range, Alaska[1]

G. WILLIAM HOLMES AND
CHARLES R. LEWIS

ABSTRACT

Glaciers originating in the Franklin Mountains, northeastern Brooks Range, advanced in four major Pleistocene glaciations and in two minor Recent fluctuations. The oldest advance, the Weller glaciation, moved across the foothills to the south flank of the Sadlerochit Mountains; it is recorded by scattered quartzite and granitic erratics on bedrock ridges, and smooth sheets of till on the middle and lower slopes that have been profoundly modified by frost action and gravity and mixed with silty colluvium. Next, the Chamberlin glaciation deposited till sheets and outwash aprons with well-defined topographic expression. Although the end moraine of this glaciation is mappable and the moraine-outwash relation is generally preserved, minor topographic features have disappeared. The less extensive Schrader glaciation is represented by ridged, lobate end moraines, some of which enclose Lake Schrader. These moraines are marked by a few shallow ponds and low hillocks, and by a widely-distributed, well-developed assortment of frost features. The last major advance, the Peters glaciation, did not extend beyond the mountain front and formed fresh, bouldery, steep-sided lateral and end moraines. Frost features are rare or absent on Peters moraines. Modern glaciers have formed two groups of moraines: a slightly weathered single or double moraine several hundred yards from the ice front, and a very fresh moraine in contact with the ice.

THE MOUNT CHAMBERLIN AREA, which includes Lakes Peters and Schrader, is located in the eastern Brooks Range, northeastern Alaska. The area mapped includes the Mount Michelson B-2 quadrangle and the southern part of the C-2 quadrangle (Figures 9* and 10*). Mount Chamberlin (Figure 1) is about 60 miles south of the Arctic Ocean, 100 miles west of the Canadian border, and 30 miles north of the crest of the range. The area is drained by the Sadlerochit River and its tributaries. The large Hulahula River valley lies immediately to the east, and the Canning River drains the region to the west.

Investigations were conducted by Holmes from June 21 to August 29, 1958, and in April and May 1959, and by Lewis in June 1959. Traverses were made throughout the northern half of the area, down the Sadlerochit River, and up the major valleys and tributaries in the mountains. The section immediately east of Mount Chamberlin was mapped by photogeologic methods, aided by light aeroplane reconnaissance. The project was conducted by the United States Geological Survey under contract to the Terrestrial Sciences Laboratory, Geophysical Research Directorate, Air Force Cambridge Research Center. Fernand de Percin, meteorologist, Quartermaster Research and Engineering Center, and John E. Hobbie, limnologist, University of California, taking time from their own studies, assisted Holmes on

[1]Publication authorized by the Director, United States Geological Survey.

*See separate container of figures.

several traverses. Livingston Chase, geologist, and Lloyd Spetzman, botanist, both of the Survey, accompanied Lewis on the Sadlerochit River traverses. Their assistance is gratefully acknowledged. The base camp on Lake Peters and other logistic functions were under the expert management of Major Frank Riddell, Royal Canadian Army (retired).

Compared to other parts of the eastern Brooks Range, the Mount Chamberlin area has received considerable scientific attention. Leffingwell (1919) spent three seasons in the region, including two months' study of the area around lakes Peters and Schrader. He described, mapped, and named many of the major bedrock units, and devoted some attention to the glacial deposits and the frost features. He noted especially the glacial deposits around Lake Schrader (1919, pp. 136-8). Whittington and Sable (1948), and Brosgé, Dutro, Mangus, and Reiser (1952) re-examined the bedrock in the central and northern part of the area. Their interest was in extending the stratigraphic and structural investigations of Naval Petroleum Reserve No. 4, which lies along the north slope of the range to the west. These investigations also resulted in two recent papers on glaciation of the north-central Brooks Range (Detterman, 1953; and Detterman, Bowsher, and Dutro, 1958) which are especially germane to the present study. During the International Geo-

FIGURE 1.   Mount Chamberlin (left) with Lake Peters and part of Lake Schrader, looking south, showing large alluvial fans in Lake Peters and delta between the lakes. Kettle pond in foreground, right, is on a Schrader moraine, and kettle at the northwest end of Peters Lake is on Peters moraine.

physical Year studies were conducted in the area east of the Hulahula River, centring on McCall Glacier in the headwaters of the Okpilak drainage (Mason, 1959; Sater, 1959; Keeler, 1959).

TOPOGRAPHY AND DRAINAGE

The region surrounding the area, specifically the Franklin and the Romanzoff mountains, is as rugged as any portion of the Brooks Range, and includes the highest peaks in the North American Arctic. Mount Chamberlin (Figure 1), and to the east, Mount Michelson and unnamed peaks in the drainage of the Okpilak and Jago rivers are slightly less or slightly more than 9,000 feet above sea-level. The general level of the peaks is 6000 feet, and local relief is typically 3000 to 4000 feet. The mountains are sharp crested and steep, and are deeply dissected by closely-spaced streams.

North of the Franklin Mountains is a double pair of ridges and valleys, beyond which are the outlying Shublik[2] and Sadlerochit mountains. The Arctic foothills province borders the Sadlerochit Mountains on the north and extends to the basin of Lake Schrader. North of the foothills is the Arctic Coastal Plain, which is narrow in this region. Descriptions of the physiographic subdivisions of Arctic Alaska appear in Payne and others (1951), Black (1955, p. 118), and Reed (1958, pp. 5-10).

The Sadlerochit River drains most of the area by means of its main branch on the west and the lake branch (Leffingwell, 1919, p. 56) or "Neruokpukkoonga Creek"[3] (Whittington and Sable, 1948). "Katuk Creek" drains the southeastern part of the quadrangle and empties into the Hulahula River.

Streams in the area rise from small glaciers or flow from melting snow and seasonally frozen ground. Most of the streams from the Brooks Range have flood-plains of boulders and gravel, whereas the streams originating on the rolling tundra of the Arctic foothills flow over silt- and peat-mantled valley floors. Although total rainfall is light, the streams respond quickly and strongly to rainstorms.

SUMMARY OF BEDROCK GEOLOGY

Lithology

The following summary is based on the work of Leffingwell (1919), Whittington and Sable (1948), and Brosgé, Dutro, Mangus, and Reiser (1952).

The rugged mountains of the southern part of the area are composed of the Neruokpuk formation (late Devonian or older), a sequence of quartzites, phyllites, schists, and argillites. Younger rocks lie to the north in east-west trending belts, appearing as hogbacks or anticlinal hills and low mountains. In stratigraphic succession, these are: unnamed conglomerate (Mississippian?); Kayak shale (Mississippian); Lisburne group (Mississippian), limestone and dolomite; Sadlerochit formation (Permian and Lower Triassic), sandstone, siltstone, and shale; Shublik

---

[2]Spelled Shubelik on U.S.G.S. Alaska Reconnaissance Topographic Series, Mount Michelson sheet, scale 1:250,000, but spelled by Leffingwell as written here in reference to the mountain range and the stratigraphic unit.

[3]Place names not known to be accepted by the Board of Geographic Names are enclosed in quotation marks.

formation (Middle and Upper Triassic), dark shale, siltstone, sandstone, and limestone; Kingak shale (Jurassic), black shale and dark siltstone; and the Ignek formation (Cretaceous), gray shale, siltstone and sandstone.

Structural Geology Related to Geomorphology

The Franklin Mountains are part of the Romanzoff uplift, a major structural element of the Brooks Range (Payne, 1955). The last major tectonic phase apparently was during the Pliocene epoch, and continued into the Pleistocene. The accompanying erosional episode probably produced the major valley systems in this area.

The present elevation of the Franklin Mountains is attributable in part to resistance of the Neruokpuk meta-sediments. This relation may have been accentuated by movement on major thrust faults along the north flank of the range.

The anticlines and synclines north of the Franklin Mountains, which include simple and overturned folds, produce an outcrop pattern of alternating ridges of resistant limestone of the Lisburne group and sandstone of the Sadlerochit formation, and lowlands underlain by shales of the Sadlerochit, Shublik, and Kingak formations (Whittington and Sable, 1948). The valleys and ridges influenced paths of north-flowing glaciers and allowed them to coalesce as piedmont sheets through the east-west troughs.

QUATERNARY GEOLOGY

Weller Glaciation

The oldest glaciation recognized in the region is here named for Mount Weller, a prominent peak in the Sadlerochit Mountains, which was named by Leffingwell (1919, p. 100) for Professor Stuart Weller.[4] The existence of this glaciation was suggested by Leffingwell (1919, p. 137) who speculated that glacial ice reached the lower, southern slopes of the Sadlerochit Mountains; he described erratics of possible glacial origin along "Camp 263 Creek."[5]

Deposits of the Weller glaciation include scattered erratics, a few gravel benches, and several short, low, till ridges at elevations of 1750 to 2150 feet along the southern slopes of the Sadlerochit Mountains. The subdued, rounded ridges slope downvalley and resemble remnants of recessional lateral moraines. Above these ridges are several small gravel benches that may be kame terrace remnants of the Weller glaciation. These benches are about 300 feet above the present river bed. Weller drift also appears as scattered erratics and till patches on bare ridges of Kingak shale north of "Neruokpukkoonga Creek," and as strongly reworked or colluviated till with scattered boulders south of the Sadlerochit River. Along the south side of the Sadlerochit valley, in the northeast part of the area, east-striking ridges of soft shale show evidence for rounding by ice of the Weller glaciation. These ridges are 300-400 feet above the river. Deposits of the Weller glaciation are not recognized south of the till border of the next younger glaciation.

[4]Leffingwell also immortalized other professors of geology at the University of Chicago, which he attended, by naming peaks for them, e.g. Professors A. A. Michelson, R. D. Salisbury, and T. C. Chamberlin.

[5]The stream called "Camp 263 Creek" by Leffingwell (1919, p. 136) is *west* not east of Sunset Pass.

Sections of Weller deposits are rarely exposed. Scattered erratics are two to six feet in diameter; most are quartzite, sandstone, or limestone, but a few are schist or gneiss. The recessional ridges are well drained and have a firm surface of gravel and silt with little vegetation and rare surface boulders. The till in these ridges is best exposed where one is breached by "Camp 263 Creek"; even here, however, the till is badly slumped and mixed with alluvium of the former stream. In this stream cut, the material consists of poorly-sorted, angular gravel and small boulders in a light brown sandy matrix. Although Leffingwell reported granitic erratics along this creek, only sandstone and limestone with minor fragments of schist and gneiss were noted by Lewis. Small frost features occur on these ridges and patterned ground is well developed on the tundra-covered, colluviated, till slopes.

In summary, the Weller glaciation is the oldest and most extensive glaciation recognized in the area. During this advance, a piedmont glacier fed by valley glaciers from the Hulahula and Sadlerochit valleys covered most of the foothills of the Franklin Mountains and reached northward to the lower slopes of the Sadlerochit Mountains. The extensive erosion shown by the discontinuity and fragmentary preservation of the Weller deposits indicate glaciation is pre-Wisconsin in age.

FIGURE 2.   West side of "Kikittut Mountain," from surface of Chamberlin ground moraine east of Lake Schrader outlet. Sloping channels on side of this mountain were probably formed during the Chamberlin glaciation.

Chamberlin Glaciation

The oldest glaciation represented by distinct end moraines in this area is here named the Chamberlin glaciation, from Mount Chamberlin, the dominating peak in the area. Leffingwell (1919, p. 137) described the terrain of the Chamberlin drift sheet, noting large, scattered boulders on the smoothly rolling tundra north of Lake Schrader. The maximum extent of the Chamberlin glaciation is indicated in places by a well-defined drift border, and elsewhere by lateral drainage channels and scattered erratics. Ice from the Hulahula valley moved laterally (westward) through the lowland south of "Kikittut Mountain" and joined the glacier that flowed from Lake Peters valley. The drift border on the south side of "Kikittut Mountain" is generally well defined, although in places colluvium and alluvium have modified the till. Lateral channels that cut across bedrock structure, and modern gullies on the west side of "Kikittut Mountain" (Figure 2), probably formed during the maximum of the Chamberlin glaciation. The ice surface apparently sloped steeply northward in this sector, towards the lobate terminus south of the main branch of the Sadlerochit River. North of "Kikittut Mountain" the former position of the ice border is indicated by scattered erratics on bedrock bluffs east of "Neruokpukkoonga Creek." Where "Neruokpukkoonga Creek" makes its sharp

FIGURE 3. Chamberlin end moraine with hills of Ignek shale (centre) and Kingak shale (right), at the big bend in "Neruokpukkoonga Creek." Smooth profile of Chamberlin end moraine in middle distance, left, merges with outwash apron that extends to the bank of the creek. Erratics of the Weller glaciation lie on the shale hills in the background.

swing to the west, the Chamberlin drift border is distinct, and a well-preserved outwash apron extends from the moraine front (Figure 3). The northernmost limit of Chamberlin glaciation is also marked by a subdued end moraine which merges with an outwash apron that extends nearly to the southeast bank of the Sadlerochit River.

Stream cuts and other natural exposures of undisturbed Chamberlin till are rare as a result of congeliturbation and mass movement. Turf-covered colluvium, derived from till and, in places, bedrock, normally extends to the banks of the streams. The few exposures, such as along the north-flowing creek immediately east of "Neruokpukkoonga Creek," may give some indication of the lithology of Chamberlin drift, but cannot be regarded as undisturbed till. Shallow permafrost limits the examination of the till, especially on the slopes and summits of the moraines.

Chamberlin morainal deposits rest on the rolling terrain of the piedmont apparently deriving their gross form from a pre-existing mature topography. Along the flat crests of the divides (Figure 4) are a few low mounds of till, but elsewhere the profiles are smooth and gently rounded. Streams from the Franklin Mountains and the foothills cross the Chamberlin moraine in valleys 600 to 800 feet below the ridge tops. Tributaries are widely-spaced, marshy rills, which flow through peat-

FIGURE 4.   Smooth surface of Chamberlin ground moraine on flat interfluve north of Lake Schrader. Schrader moraines occupy valleys on each side of this ridge. Sadlerochit Mountains are on the skyline.

and silt-covered swales. The smooth surfaces of the Chamberlin drift sheet are in part the result of long periods of frost action and mass movement. Surface boulders are rare, although a few boulder patches occur on the ridges, and frost action commonly brings isolated stones to the surface. Patterned ground is well developed on Chamberlin drift, as may be expected on terrain that has endured long periods of frost and glacial climates.

Chamberlin till consists of fragments of all bedrock types of the area notably quartzite, schist, sandstone, limestone, and conglomerate, and has a dark grey matrix derived in part from dark shales. Along its eastern border this drift also contains a small amount of granitic rocks (presumably from the Mount Michelson area) which were not observed in the younger tills. Few particle size analyses were made, and samples are not necessarily representative, yet the results seem to agree with field observations (Table I). Chamberlin till appears to be typical of alpine drifts from crystalline source rocks, having a relatively small fraction of silt and clay sizes, a moderately large sand fraction, and a large fraction of coarser particles. The till samples are very poorly sorted.

Weathering characteristics and soil profiles are weakly developed owing to churning and solifluction of the active zone (Tedrow and Cantlon, 1958, pp. 168-70) and to the slow rate of chemical alteration. The only obvious indications of weathering, other than the paucity of surface boulders, are that quartzite and schist fragments are commonly stained, and that a few schist and sandstone particles have partially disintegrated.

TABLE I

PARTICLE SIZE ANALYSES OF TILLS[a]

Glaciation	Sample number	Composition[b]			Median size mm	Sorting coefficient[c]	Remarks
		Gravel per cent	Sand per cent	Silt and clay per cent			
Peters	1	60	31	9	8.0	4.7	Lateral moraine; derived from Neruokpuk formation only
Peters	2	33	40	27	0.9	8.2	End moraine from several bedrock sources and lake silt
Schrader	5	38	29	33	0.7	21.8	Probably enriched by lake silt
Schrader	6	36	44	20	1.2	9.5	
Chamberlin	3	61	28	11	12	7.2	Buried till
Chamberlin	7	44	38	18	3.2	9.0	

[a]Made by Quality of Water Branch, Water Resources Division, U.S.G.S., Washington, D.C. Samples are of particles less than 76.2 mm (3 inches) in diameter.

[b]Size limits are those used by United States Army Corps of Engineers; gravel, 76.2 mm (3 inches) to 5.76 mm (no. 4 United States Standard sieve); sand, 5.76 mm to 0.74 mm (no. 200 United States Standard sieve); and silt and "clay" size ("fines"), less than 0.74 mm.

[c]Sorting coefficient $\sqrt{(Q_1/Q_3)}$ where Q_1 is size (in mm) corresponding to first quartile (or 75 per cent finer than) and Q_3 is size corresponding to third quartile (or 25 per cent finer than).

Small outwash aprons of weathered, bouldery gravel flank the moraine on "Neruokpukkoonga Creek." Although these aprons have been modified by encroaching colluvium and by streams, the fact that they merge with the moraine front suggests that this glaciation is not extremely old. As stated above, a deeply trenched and much more extensive outwash apron extends northward from the moraine between "Neruokpukkoonga Creek" and the main branch of the Sadlerochit River. This outwash apron is about 150 feet above the stream, and consists of poorly stratified gravel and boulders in a matrix of angular sand and silt size shale particles. This apron may have been continuous with a valley train on the Sadlerochit River.

In review, the Chamberlin glaciation is the oldest and most extensive advance represented by recognizable moraines and outwash plains. During this advance large valley glaciers from the Hulahula and Sadlerochit drainages coalesced immediately north of the Franklin Mountains and flowed to within a short distance of the Sadlerochit Mountains. Smooth outlines, the absence of morainal hillocks and depressions, the high development of frost forms, and evidence for moderate chemical weathering show that a substantial interval has passed since this glaciation. On the other hand, the preservation of the moraine-outwash relation, the identification of a continuous moraine border, the close relation between the ground

FIGURE 5.   Lateral moraine of Schrader glaciation on Lake Schrader, composed of several steep-sided boulder-strewn ridges. Bedrock knobs rise above this moraine.

moraine and the present topography, and the fact that the till is in continuous or nearly continuous sheets indicates that it is considerably younger than the Weller glaciation.

Schrader Glaciation

The Schrader advance, younger and less extensive than the Chamberlin glaciation, is here named for the large lake in the centre of the area. The lake was named by Leffingwell for F. C. Schrader, an early explorer and investigator of the Brooks Range. This glaciation left well-defined lobate moraines, notably north of Lake Schrader. An end moraine on the main branch of the Sadlerochit River and one on "Katuk Creek" on the eastern edge of the area also are attributed to the Schrader glaciation. The largest ice mass of this glaciation flowed down the valley now occupied by Lake Peters, pushed through the basin of Lake Schrader and split into three lobes. The edges of this glacier were in places at least 1,000 feet above the present Lake Schrader or about 1,200 feet above its deepest part.

Unlike the older moraines, the Schrader drift is bordered in many places by well-defined ridges (Figure 5), which have relief of 50 to 100 feet and moderate to steep slopes. The lateral moraines generally are lower; in places the moraine border can be distinguished only by a sudden decrease in the frequency of surface boulders. Commonly small streams flow along the outer margins of the lateral

FIGURE 6. Earth flow on northwest shore of Lake Schrader in Schrader till, which occurred in August, 1958. Larger lateral moraine of Schrader glaciation is shown on opposite shore.

moraines. Schrader moraines are also distinctive with respect to their well-defined recessional ridges. Their relief is typically 30 to 50 feet, and they are separated by swales floored with silt and peat. Four to five sharp recessional ridges appear south of the front of the two major end moraines north and northeast of Lake Schrader. Other morainal features, such as kettles and knobs, are not common. Most of the ponds are drained by marshy water courses, and a widely spaced stream system drains the rolling ground moraine behind the terminal ridges.

Patterned ground and other frost features are well developed on Schrader moraines, and active earth flows (Figure 6) are common. Surface boulders are closely spaced on the flat ridge crests, but on many slopes or in swales stoney till is completely buried by colluvium, peat, or turf. A loess mantle was not observed on this or other drift in the area, probably because of remoteness from broad, bare, flood plains. However, Schrader till near the mouth of "Katuk Creek" has a more silty texture in the surface layer, which might be derived from loess blown from the Hulahula flood plain.

Schrader till is composed of particles of quartzite, schist, conglomerate, sandstone, limestone, and shale. It has a medium to dark gray, fine-grained matrix that reflects the dark shale source material. In sections where the permafrost table is low, as along "Katuk Creek," a faint weathering profile, approximately three feet thick, is exposed. In sections in the fresh earth flows along Lake Schrader, where the till has been churned, stained rock fragments occur in the upper two feet, but no distinct soil profiles were observed.

The few samples of Schrader till (Table I) and field observations indicate that this material contains considerably more fine material than is generally expected in tills derived from crystalline source rocks. This may be in part a result of local sources of loess, shale, or limestone, or of concentrations at the sample points of fine material by frost action, but could also reflect incorporation of proglacial or preglacial lake sediments in the drift. As the lake basins are to a great extent controlled by the configuration of the bedrock, it is possible that a lake or lakes filled the basins just prior to Schrader glaciation. Ice moving across lake sediments (or even silty deltas) would deposit till rich in fine-grained material. The two samples analysed are very poorly sorted, and one is an extreme example of poorly sorted material. This also suggests abnormal enrichment of the till by silt or clay size material.

Outwash aprons fronting Schrader moraines are narrow and low. The best developed apron, whose surface is about twenty feet above the stream, merges with the terminus of the large Schrader moraine due north of Lake Schrader. This outwash deposit is a recognizable terrace near the moraine; but a short distance downstream its outlines are obscured by colluvium from the valley sides. A smaller outwash plain extends downstream from the prominent moraine on "Neruokpuk-koonga Creek." Terraces on the Sadlerochit River below the Chamberlin outwash are probably valley train remnants of the Schrader glaciation.

Considering the distribution and scale of Schrader glaciation in relation to Chamberlin glaciation, it seems that the younger advance was in all respects a scaled down version of the older, and that the time interval between them was not extremely long. On the other hand, the two advances do not seem to be pulsations or phases of a single glaciation. The interval between the two, as indicated by degree of topographic modification and weathering, was apparently not as long as between the oldest and next oldest glaciations of the Alaska Range and the central Brooks Range (Holmes, 1959; Péwé, 1952; Detterman, Bowsher and Dutro, 1958).

Peters Glaciation

The youngest of the major glaciations is named for Peters Lake, which was named by Leffingwell for W. J. Peters, a United States Geological Survey geologist and early explorer of the Brooks Range. This glaciation contrasts sharply with the older advances in respect to the extent, size, lithology, and topographic expression of its deposits.

Moraines of the Peters glaciation are confined to the major valleys of the Franklin Mountains. The most distinct end moraine is on the creek one-and-a-half miles east of Lake Peters, which is regarded as the type moraine. The end moraine there consists of a group of very steep, boulder-strewn hillocks and ridges that cross the valley about two miles above its mouth. The outwash associated with this moraine apparently was swept down the bedrock gorge and was deposited as a large delta in Lake Schrader. A distinct lateral moraine (Figure 7) parallels the shore of Lake Peters approximately 200 feet above the water surface, and extends

FIGURE 7.   Southwest shore of Lake Peters, showing bouldery lateral moraine of Peters glaciation, alluvial fans, and fresh talus encroaching on fans.

up the valley south of the area boundary. This lateral moraine has steep slopes towards the lake and persistent reverse slopes towards the valley sides. The crests are boulder-strewn and somewhat hummocky, and the ridges are breached by tributaries or encroached on by alluvial fans. The lateral moraine on the west side of the lake merges with a remnant of the end moraine on the northern shore of Lake Peters, beyond which is a small outwash apron. The terminal moraine on the northeast side of the lake is less well defined, although remnants of the end moraine and a recessional moraine are preserved, and a small isolated outwash terrace lies a short distance north of the till border. At this point the Peters moraine lies against the valley wall which in turn is covered by Schrader till. Here the contrast in scale of the two glaciations is seen; the Schrader moraine, which is composed of several parallel ridges, is approximately 800 feet higher than the top of the single-crested Peters moraine.

The Peters moraine of "Katuk Creek" (in the Hulahula drainage) is a narrow elongated loop, indicating the ice tongue was thin and closely confined by the canyon. The terminal section is marked by small, unmodified knobs and closed depressions. The outwash of this glacier, as in the type valley, may have been flushed down the rock-walled gorge by melt-water.

The Peters moraines are strewn with closely-spaced, slightly weathered boulders, and, except for the encroachment by alluvial fans, are not mantled by surficial deposits. Frost features are absent or poorly developed. The till is composed largely of quartzite, schist, and phyllite of the Neruokpuk formation which give it a light grey-brown colour. Only the terminal moraines contain appreciable quantities of limestone or dolomite of the Lisburne group. Buried fragments in Peters till are not perceptibly weathered, although surface boulders are stained and lichen-encrusted.

Two samples of Peters till illustrate contrasting size characteristics of a single till deposit, and reflect the influence of local sources. A sample (no. 1) from the east lateral moraine on Lake Peters is a typical sandy, gravelly, alpine till, which may be expected from crystalline source materials such as the Neruokpuk formation. Its sorting index is low, suggesting some influence by melt-water, or a low percentage of fine particles in its source material. Till (sample no. 2) from the end moraine on Lake Peters contrasts sharply with the lateral moraine till, and resembles the poorly-sorted, silty, Schrader till samples. The fine texture of this sample of Peters till is attributed to incorporation of lake sediments by the advancing Peters glacier, and suggests that a lake existed in the valley prior to this advance. Other evidence for a lake in the Peters trough prior to this glaciation are: (1) the deep basin and shallow outlet threshold of this lake; (2) the bedrock control of the outlet of Lake Schrader which in turn would control the level of Lake Peters; and (3) delta deposits in Lake Schrader, if composed of Peters outwash, would indicate high lake levels just prior to this advance. If the Peters Glacier advanced into a pro-glacial lake, a distinct end moraine may not have formed; this may explain the poorly-developed moraine at the end of Peters valley.

Peters moraines contrast sharply with Schrader moraines in regard to thickness and volume of deposits. Moreover the older glaciation occurred as a series of pul-sations, and formed several morainal ridges, whereas the smaller Peters glaciers in this area built one or, rarely, two ridges. On the other hand, the contrast in degree of weathering and erosional modification, although conspicuous, is not extremely large, and is perhaps of the same order as the difference between Schrader and Chamberlin moraines. Outwash deposits are small, like those of the Schrader glaciation. However, much of the Peters fluvioglacial material may be in Schrader Lake or in the Hulahula valley.

Cirque Glaciations

Small fresh moraines lie in front of nearly all the glaciers in the mountains, especially those surrounding Mount Chamberlin (Figure 8). These moraines are exceptionally small in comparison with the youngest of the major moraines; they represent only minor Recent readvances of the alpine ice.

The small moraines of the glacier on the northwest flank of Mount Chamberlin ("Chamberlin Glacier") suggest that glaciation in Recent time was compound. In contact with the ice is a very fresh moraine covered by large angular boulders ("Cirque Moraine II"). Downstream from this ridge is a flat, boulder-strewn outwash apron, crossed by a braided stream. Beyond the outwash is a pair of steep-sided moraines ("Cirque Moraine I"), which are covered by very slightly stained boulders. A few alpine plants have become rooted in the sandy matrix, and the boulders are somewhat more firmly placed than in the moraine in contact

FIGURE 8.   Moraines of the Cirque glaciations on the west slope of Mt. Chamberlin. A fresh moraine lies against the glacier, and two slightly weathered moraines rest a few hundred yards downstream.

with the ice. A narrow valley train leads from the outer moraine into the rocky, V-shaped canyon below. On the valley side wall beyond the outer loop is a lateral moraine remnant which may represent an earlier readvance or possibly a recessional moraine of the Peters glaciation.

These cirque moraines are composed entirely of Neruokpuk meta-sediments, and have a light grey, sandy matrix. Except for the stained boulders on the outer ridges, there is no evidence of weathering.

GLACIAL HISTORY

Major uplift of this part of the range in late Pliocene or early Pleistocene (Payne, 1955) apparently set the stage for a series of glacial advances. This contrasts with the Rocky Mountains in the continental United States, the equivalent of the Brooks Range, where fewer glacial advances are clearly recorded. The distribution of the oldest glacial deposits in the Brooks Range suggests that the general topographic relations have remained essentially the same since the first recorded glaciation. This also contrasts with the situation in the Rocky Mountains of the United States, and with the Alaska Range (Wahrhaftig, 1958; Péwé, 1952) where the earliest glaciers moved over a landscape different from the present one, after which the area was uplifted.

During the Weller glaciation, the oldest and most extensive glacial advance in the area, ice from valleys in the Franklin Mountains coalesced into extensive piedmont glaciers that covered much if not all of the foothills south of the Sadlerochit Mountains. The piedmont ice was deflected eastward by the Sadlerochit Mountains and terminated in the lower course of the river. Lateral drainage during this glaciation may have escaped through Sunset Pass to the north as suggested by Leffingwell (1919, p. 137) although no erratics have yet been found in this broad valley. A considerable interval elapsed between the Weller and Chamberlin glaciations, as indicated by the contrast between the severely eroded oldest drift, and the modified, but recognizable, moraines of the Chamberlin glaciation.

The Chamberlin glaciation, like the oldest, was a major event. Ice from the major valleys coalesced and formed very broad, thick piedmont glaciers. These glaciers moved northward over the piedmont and overrode low foothills on a terrain that was not radically different from the present one. Drainage from these glaciers deposited well-preserved, extensive outwash plains, indicating a long period of ice-front stability. On unglaciated piedmonts, broad sheets of alluvium were deposited. It is probable that lakes formed in the Peters and Schrader basins after the Chamberlin advance.

The Schrader glaciation was also a major advance, but was on a smaller scale than earlier glaciation. Schrader moraines still enclose large lakes and are marked by a few kettle ponds, suggesting that their age is Wisconsin. Evidence is strong that the present Lake Schrader was formed after this glaciation, and that a separate or contiguous body of water occupied the Lake Peters depression after the glacier retreated.

It is possible that the Peters glacier advanced into a lake to its terminal position. Although Schrader moraines are considerably larger than Peters moraines, by a factor of more than 10:1, the younger advance is thought to have been a Pleistocene event, probably a late phase of the Wisconsin. It has been suggested (Henry W. Coulter, personal communication) that the reason for the great disparity in mass between the Peters (presumably latest Wisconsin) and older moraines is that the Peters advance flowed only from cirques on the north side of the Brooks Range, whereas earlier glaciations were nourished by large ice-caps or highland ice that centred south of the crest of the range. At present the south slope receives more precipitation than the arctic side, and this condition may have obtained during the Pleistocene. The weakness of the Peters glaciation may also be an indication of less precipitation than during the preceding glaciations. Lower precipitation on the arctic slope may be related to the formation of a continuous ice cover on the Arctic Ocean.

The cirque moraines are at least one order of magnitude smaller than the Peters moraines, suggesting that they represent Recent advances, for which there is abundant evidence elsewhere in Alaska and in many alpine regions.

Although correlations of the stages described here with others in the Brooks Range, the Alaska Range, or elsewhere are little better than guesses, it is believed useful to state the writers' opinion of their chronology. These tentative correlations are made with full consideration of the problems of evaluating the age of the deposits. No radiocarbon data are available, and weathering characteristics and soil profiles are of limited value. Moreover, the reduction of Pleistocene landforms in the arctic may be less a function of absolute age than in the subarctic temperate regions, and more closely related to the number, length, and intensity of periods of warm climate that they have survived. During warm periods the permafrost table would be lower, and mass movement, congeliturbation, and stream action would be more effective than at present or during glacial advances. Conversely, during cooler phases of the Pleistocene, frost-churning and mass movement would be minimal.

Tentative correlation with glaciations reported in the north-central Brooks Range is possible through the descriptions by Detterman, Bowsher, and Dutro (1958) and personal conferences with Dutro (Table II). Correlation with the youngest glaciations is fairly certain; Fan Mountain and cirque moraines occupy analogous positions in high cirques, and Alapah Mountain and Peters moraines are comparable in respect to size, distance from cirques, and degree of modification.

TABLE II

QUATERNARY CHRONOLOGIES IN THE BROOKS AND ALASKA RANGES

	Alaska Range		Brooks Range	
	Péwé, 1952 Holmes, 1959		Holmes and Lewis	Bowsher, Detterman and Dutro, 1958
Recent	Black Rapids II Black Rapids I		Cirque Moraine II Cirque Moraine I	} Fan Mountain
Pleistocene	(Post-Donnelly) Donnelly Delta	? ? ?	Peters Schrader Chamberlin	Alapah Mountain Echooka Itkillik
	Darling Creek	?	Weller	{ Sagavanirktok Anaktuvuk

Echooka and Schrader moraines are similar, each having modified knob and kettle topography and each damming large lakes at the present. Itkillik and Chamberlin moraines are apparently equivalent, although in the Sadlerochit area the Chamberlin moraine appears to be more modified than Itkillik drift to the west. Less certain is the correlation of the Weller glaciation with older advances. Detterman, Bowsher, and Dutro name two early glaciations, but state (1958, p. 51) that Anaktuvuk and Sagavanirktok moraines have not been recognized together in the same area, and may represent a single advance. Uncertainties also arise from discrepancies in the topographical form of the Weller and Anaktuvuk-Sagavanirktok moraines, but this may be the result of local differences in mass-wasting, in the development of thermokarst features, or in lithology.

A more tenuous correlation is attempted between the Brooks Range and the Alaska Range, with which the senior author is familiar. This correlation is based primarily on similar sequences of deposits. Correlation based on degree of erosional modification is probably more certain in respect to the youngest glaciations than for the older advances.

REFERENCES

BLACK, ROBERT F., in HOPKINS, DAVID M., KARLSTROM, THOR N. V. et al. 1955. Permafrost and ground water in Alaska; U.S. Geol. Surv., Prof. Paper 264-F, pp. 113-46.

BROSGÉ, WILLIAM P., DUTRO, J. THOMAS, JR., MANGUS, MARVIN D., and REISER, HILLARD N. 1952. Stratigraphy and structure of some selected localities in the eastern Brooks Range, Alaska; U.S. Geol. Surv., Geological Investigations Naval Petroleum Reserve No. 4, Alaska, Preliminary Report 42, 28 p.

DETTERMAN, R. L., in PÉWÉ, TROY L., et al. 1953. Multiple glaciation in Alaska; U.S. Geol. Surv., Circ. 289, pp. 11-12.

DETTERMAN, R. L., BOWSHER, A. L., and DUTRO, J. THOMAS, JR. 1958. Glaciation on the Arctic slope of the Brooks Range, northern Alaska; Arctic, vol. 11, pp. 43-61.

HOLMES, G. WILLIAM. 1959. Glaciation in the Johnson River-Tok area, Alaska Range; Bull. Geol. Soc. Amer., vol. 70, p. 1620.

KEELER, CHARLES M. 1959. Notes on the geology of the McCall Valley area; Arctic, vol. 12, pp. 187-97.

LEFFINGWELL, ERNEST DE K. 1919. The Canning River region, northern Alaska; U.S. Geol. Surv., Prof. Paper 109, 251 p.

MASON, ROBERT W. 1959. The McCall Glacier project and its logistics; Arctic, vol. 12, pp. 77-81.

PAYNE, THOMAS G. 1955. Mesozoic and Cenozoic tectonic elements of Alaska; U.S. Geol. Surv., Misc. Inv., map I-84.

PAYNE, THOMAS G., et al. 1951. Geology of the Arctic slope of Alaska; U.S. Geol. Surv., Oil and Gas Inv., Map OM-126.

PÉWÉ, TROY L. 1952. Preliminary report of multiple glaciation in the Big Delta area, Alaska; Bull. Geol. Soc. Amer., vol. 63, p. 1289.

REED, JOHN C. 1958. Exploration of naval petroleum reserve no. 4 and adjacent areas of northern Alaska, 1944-53, Part 1, History of the exploration; U.S. Geol. Surv., Prof. Paper 301, 192 p.

SATER, JOHN E. 1959. Glacier studies of the McCall Glacier, Alaska; Arctic, vol. 12, pp. 82-6.

TEDROW, J. C. F., and CANTLON, J. E. 1958. Concepts of soil formation and classification in Arctic regions; Arctic, vol. 11, pp. 166-79.

WAHRHAFTIG, CLYDE. 1958. Quaternary geology of the Nenana River valley and adjacent parts of the Alaska Range; U.S. Geol. Surv., Prof. Paper 293, pp. 1-70.

WHITTINGTON, CHARLES L., and SABLE, EDWARD G. 1948. Preliminary report on the geology of the Sadlerochit River area, Alaska; U.S. Geol. Surv., Geological Investigations Naval Petroleum Reserve No. 4, Alaska, Preliminary Report 20, 18 p.

Glacial Marine Sedimentation

S. WARREN CAREY AND

NASEERUDDIN AHMAD

ABSTRACT

Processes and terminology of ancient glacial sedimentation have been based on Quaternary terrestrial glacial phenomena, although such terrestrial deposits are likely to be geologically ephemeral. The contemporary marine glacial sediments on the shelves that may be buried in due course under other sediments are more likely to survive through geological periods, but those incorporated in currently active geosynclines (such as those in Alaskan waters) have the best chance of survival through geological eras. Hence study of marine glacial sedimentation, rather than current terrestrial forms, is essential for the recognition and analysis of ancient glaciations.

Profoundly different types of behaviour and marine sedimentation are predicted according as the base of the sea-going glacier is colder than freezing point, or, is permeated by melt-water in the region where it becomes buoyant. Wet-base glaciers have no marked change of surface altitude where they become buoyant; they produce great thicknesses of unfossiliferous tills that may have occasional sand or mud interbeds, and these give place seawards to marine sands and silts with few erratics, interbedded with frequent turbidity current deposits, and occasional submarine flows of till-like material containing abundant glacially-striated erratics. Dry-base glaciers have a marked change of surface altitude, forming ice mountains rising from the line of buoyancy; they produce little in the way of tills, but yield marine pebbly mudstones and siltstones, and the associated sediments have a high lime content grading into limestones that may contain dropped erratics; they may be associated with dark pyritic mudstones with calcareous concretions and large glendonites (calcite pseudomorphs after glauberite). Glaciers of these two types are identified in Antarctica, and sediments of the two types are found in the Tasmanian Permian system, and also in the marine Pleistocene sediments of Middleton Island in the Gulf of Alaska.

INTRODUCTION

Selectivity of the Geological Record

The study of glacial sediments has grown out of the study of modern glaciers and of the terrestrial deposits left by the Pleistocene glaciation. Until recently comparatively little attention had been paid to glacial sedimentation at sea. From the point of view of the palaeogeographer, however, marine glacial sediments may be more important in that they are much more likely to survive in the geological record. Today the most obvious and widespread glacial sediments are the terrestrial deposits–moraines, tills, and outwash. However, on the scale of geological time all these deposits are likely to be completely destroyed. But in areas where the glacial sediments are included in a continuing cycle of subsidence and sedimentation, the glacial record will be preserved, perhaps for several geological periods. Such areas occur in the Arctic, Alaska, and off the Ross Shelf. The most permanent records will be those where the glacial sediments have been enclosed in a currently active geosyncline, where strong folds will, in due course, take the glacial formations deep into the crust, and where several cycles of peneplanation will be necessary to erase the last vestiges of the record.

The remains of the "Eocambrian" glaciation are largely of this type—in folded geosynclinal piles usually resting on other marine strata, with rarely any sign of a glaciated bedrock pavement. There were of course contemporaneous moraines, tills, and outwash, but they were the first to be wiped out. No doubt there were shelf deposits, but they, too, have largely disappeared.

The record of the late Palaeozoic glaciation is partially geosynclinal but there are still many shelf deposits, often unfolded, often associated with glaciated bedrock pavements, but usually leading up into a following succession of marine or lacustrine sediments that served as a protecting blanket during the erosion cycles of the Mesozoic and Tertiary. Terrestrial moraines and tills are relatively rare in the Palaeozoic, depending for their preservation, as they would, on penecontemporaneous burial through tectonic subsidence of the shelf on which they lay.

In spite of this difference in the state of the record in Pleistocene, Palaeozoic, and Precambrian glaciations, there is still a strong tendency to look to terrestrial glaciation for analogies for ancient glacial sediments, rather than to marine environments. Some geologists have gone so far as to discredit ancient glaciations unless they find preserved a striated pavement or other terrestrial attributes of the ice-sheet. This limitation is clearly observed in the nomenclature, where rock names are carried over from terrestrial environments. There is a dearth of names to designate marine glacial sediments in spite of the frequency of their occurrence, and cumbersome circumlocutions are often used (e.g. Miller, 1953, p. 26).

Scope of Present Contribution

This investigation was inspired by the presence of marine glacial sediment in the Tasmanian Permian system whose characteristics did not fit in clearly with commonly understood processes. We therefore set out to examine theoretically the physical conditions which should be expected under glaciers that extend below sea-level and become floating ice-barriers, and to forecast the sedimentation process and kinds of sediment that should result. This study led to the conclusion that two radically different types of floating glacier and kinds of glacial marine sedimentation should result, according to whether melt-water is present or absent at the base of the glacier. We are able to identify the two predicted types of floating glacier in Antarctica, and to identify the two predicted types of sedimentation in the Permian systems of Tasmania. Both types seem to have been present in the Arctic during the Pleistocene, but only one type (the higher temperature one) is found in the Arctic region at present.

Conditions for a Floating Ice-Sheet

Any glacier may protrude to sea and become buoyant if its rate of alimentation is sufficient to maintain it against the processes of melting and spreading. Because the Ross Barrier is in a bight some 400 miles wide and of similar length, it has been suggested that a somewhat restricted re-entrant of this type is a necessary condition for an ice-barrier. The Gulf of Alaska is much more open yet a floating ice-sheet developed there during the Pleistocene and is demonstrated by the glacial marine sediments folded up and exposed in Middleton Island (Miller, 1953) at present. The wider the angle of the coast, the more rapidly the floating ice will spread, but we do not agree that a restricted bight is in any way an essential condition. Around the Antarctic coast a number of glaciers protrude to sea and become buoyant. The Dronning Maud Land coast is broadly convex and has a fjord-like ice-sheet flowing out to sea, forming a floating ice-shelf for many hundreds of miles. Entrenched within this general ice-sheet, there is a fjord-filling glacier that drains part of the interior ice plateau of the continent (Figures 1A, 2, and 3). There are many others.

The most important condition would seem to be an abundant flow of terrestrial ice. Without this contribution a substantial thickness could never be built up because the low conductivity of snow insulates sea ice against atmospheric cold, and bottom melting balances surface accretion after only a small thickness has been attained. The thickness of the floating ice-sheet depends on a number of variables: the rate of outflow of ice from the feeding terrestrial glaciers, the mean temperature of this terrestrial ice (which determines the volume of sea water which will eventually be frozen by it), the rate of snow accretion on the surface, the rate of basal melting, and the rate of spreading which is strongly influenced by any lateral restrictions of the coast. Robin (1953) has suggested a stable equilibrium thickness for any given conditions.

Environmental Zones

We may recognize the following environments of sedimentation in relation to a glacier that extends to sea (Figure 1).

A. Terrestrial — where the base of the glacier is above sea-level.

B. Grounded shelf — where the base of the glacier is below sea-level, but not floating.

C. Floating shelf — where the glacier is floating.

D. Inner iceberg zone — from the ice-barrier to the limit of winter pack ice.

E. Outer iceberg zone — beyond limit of pack ice but within limit of icebergs. We propose to seek criteria for recognizing the sediments of each of these zones. Before attempting this, however, it is necessary to make another distinction which is of profound importance in every aspect of glacial sedimentation — namely wet-base and dry-base glaciers.

Wet-Base and Dry-Base Glaciers

Glaciologists have long discriminated between "arctic" glaciers and "temperate" glaciers. The distinction which we find to be critical in respect to sedimentation is similar to, but not identical with, Ahlmann's division. A wet-base glacier is defined as one whose base is at melting temperature, and, as a corollary, has basal melt-waters. A dry-base glacier is defined as one whose base is below melting temperature. A wet-base glacier may of course be a dry-base glacier further inland. We have selected the Penchsokkia Glacier of Dronning Maud Land and its floating ice-shelf as an example of a dry-base glacier, and the Ross Barrier ice-sheet with its feeding glaciers as an example of a wet-base glacier, at least in its seaward end (Figures 1-4; these figures are based on seismic data from Robin (1954) and Poulter (1947) respectively). Superficially a wet-base glacier differs from a dry-base glacier in that there is little or no surface indication of where the floating

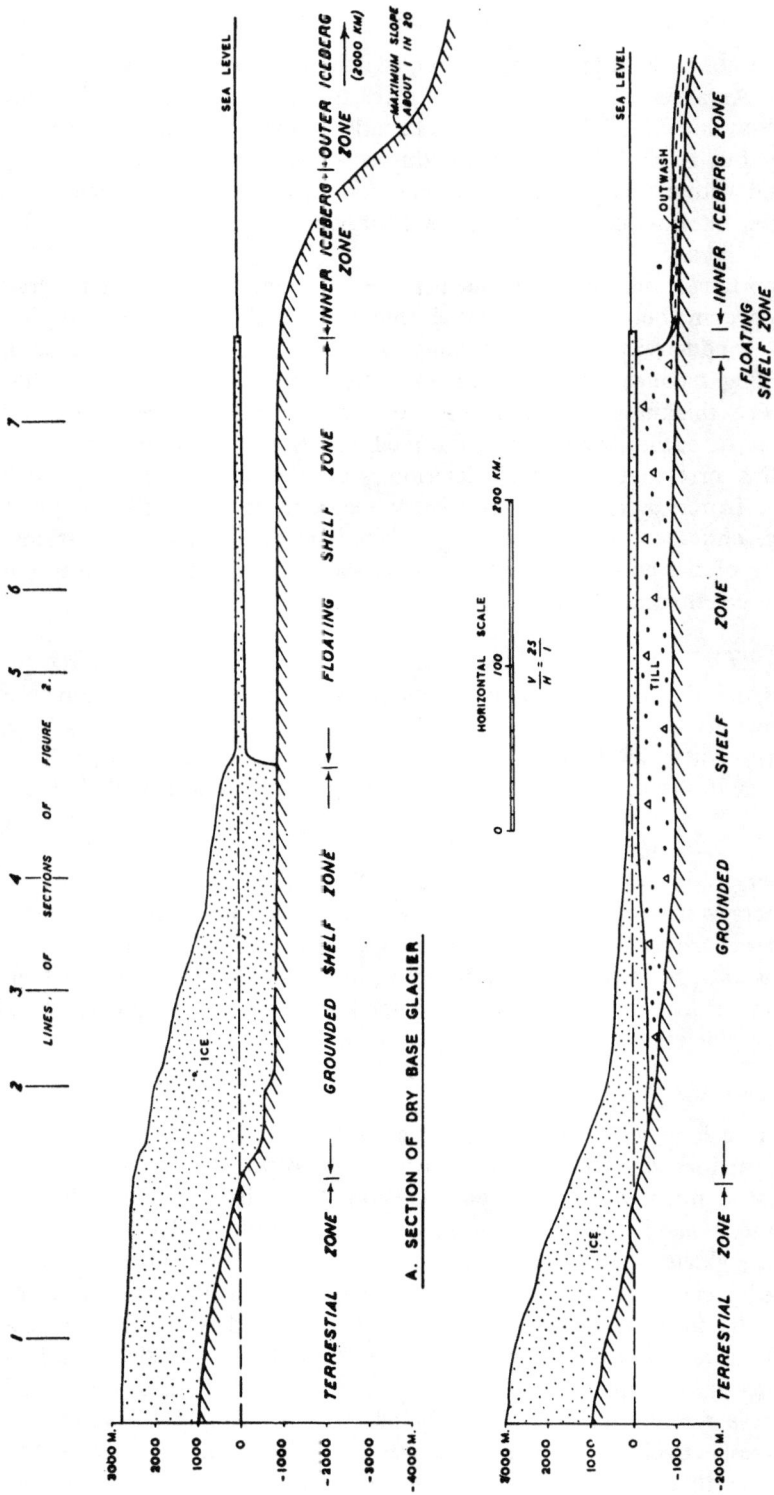

FIGURE 1. Profile sections through dry-base and wet-base glaciers with 25 times exaggeration of vertical scale. Profile A (dry-base glacier) is the profile of the Pencksøkka glacier of Dronning Maud Land as determined seismicly by Robin (for map see Figure 3). Transverse sections of this glacier are given in Figure 2. Profile B (wet-base glacier) is the profile of the Ross Barrier as measured seismicly by Poulter.

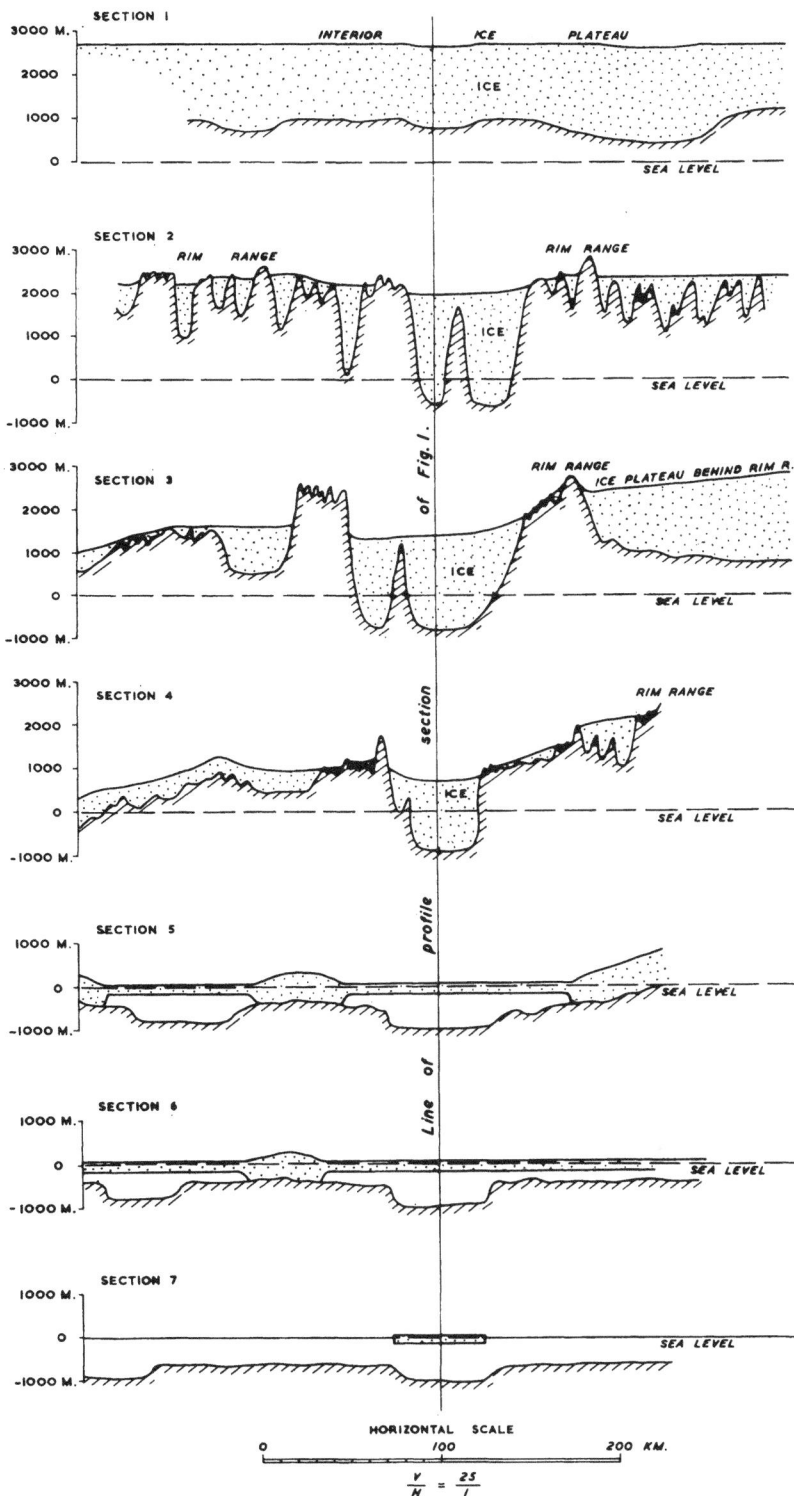

FIGURE 2. Transverse sections of a dry-base glacier (based on seismic measurements by Robin). Compare Figures 1 and 3.

shelf touches down and becomes grounded. A dry glacier on the other hand has a sharp break of surface slope at this line, and is found to increase in thickness by a factor of four or five. Its floating shelf may be very much wider. In place of the sudden increase in ice thickness, the wet-base glacier has a sudden increase of subjacent till thickness at the point of grounding. However, these contrasts are merely symptoms of very different behaviours.

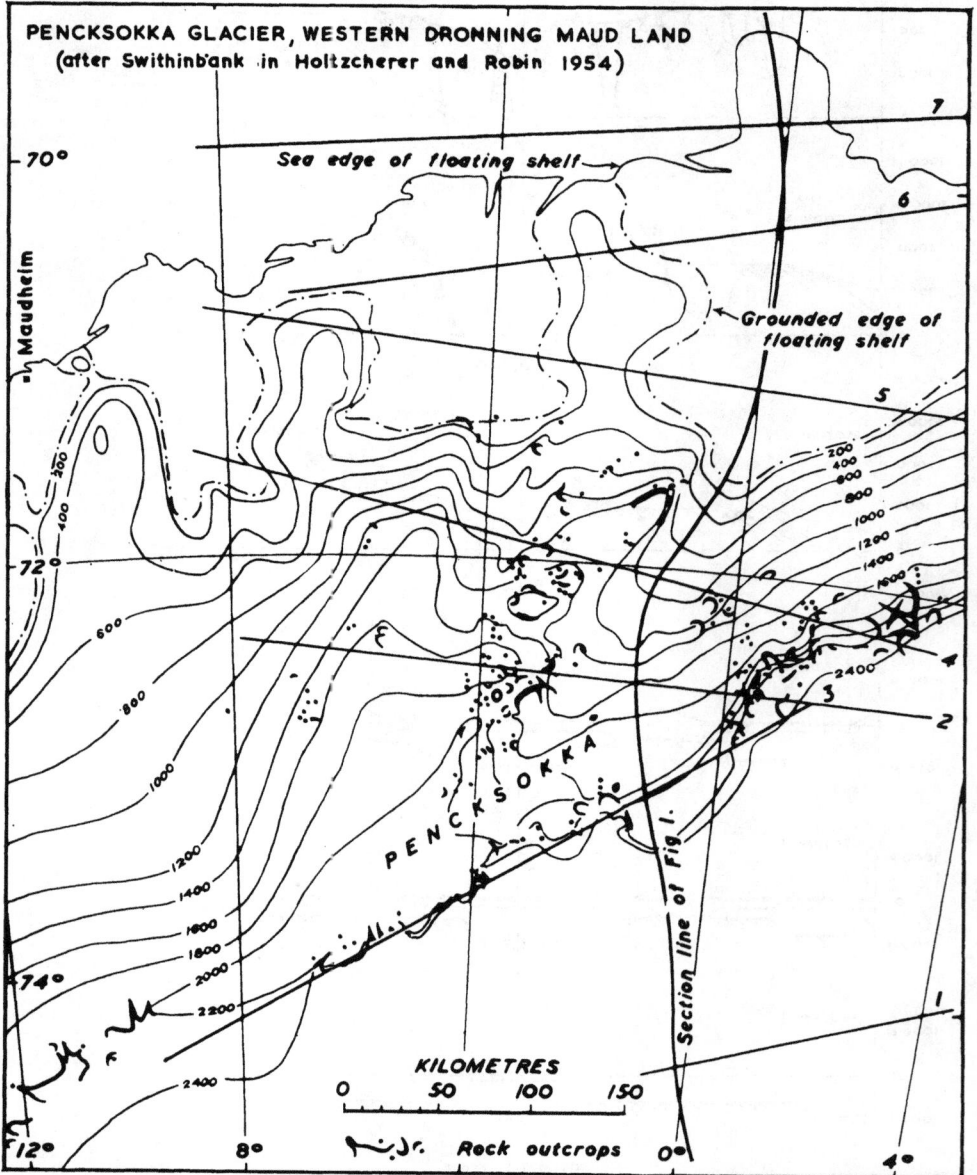

PENCKSOKKA GLACIER, WESTERN DRONNING MAUD LAND
(after Swithinbank in Holtzcherer and Robin 1954)

FIGURE 3. Pencksokka glacier (western Dronning Maud Land). Contour intervals 200 m on ice surface.

In the ensuing discussion we will attempt to show that dry-base glaciers should produce marine sediments with many dropped erratics, whereas wet-base glaciers should produce well-bedded marine sediments with much rarer erratics, but often containing interbedded layers from a few inches to several feet thick of till-like material, without disturbance of the underlying sediments; that the dry-base glaciers are associated with abundant thick-shelled fossils and limestone containing many erratics, whereas the wet-base glaciers produce more silty, non-calcareous sediments with thin-shelled fossils; that the dry glaciers may produce glendonites and large calcareous concretions and saline specialized faunas; that the wet glaciers below sea-level produce thick non-fossiliferous tills which pass rather abruptly laterally into very different sediments; that in a particular facies the wet-base glaciers result in strongly dragged and rolled structures; and that dry-base glaciers are more likely than the wet to produce favourable environments for petroleum accumulation. Moreover we have seen phenomena matching all these characteristics in the marine glacial sedimentation of the Permian system of Tasmania. Let us then examine systematically the physical behaviour of these two glacier types and their geological consequences.

Heat Flow through Glaciers

The variables in the thermal régime in a dry glacier are shown qualitatively in Figure 5. The surface temperature might fluctuate between the extremes AA', but melting prevents the temperature of the ice itself ever rising above 0°C. The temperature fluctuation at the top of the ice is therefore between A' and 0°C (J). These temperatures are ephemeral and no equilibrium gradient can be established

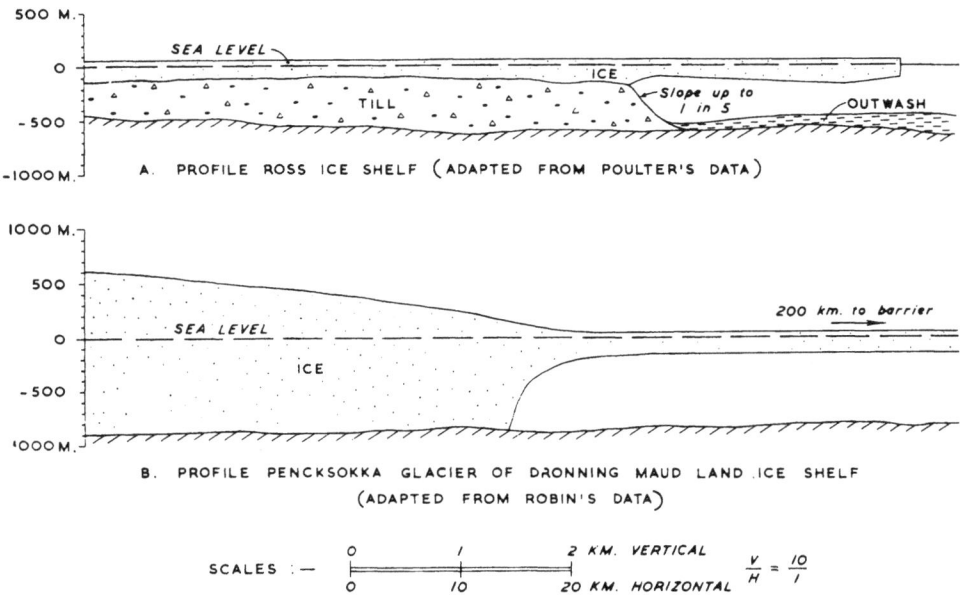

FIGURE 4.   Contrast of wet-base and dry-base glaciers at buoyancy line (ten times vertical exaggeration).

with them. The ranges of seasonal temperature variations must therefore become rapidly less with depth owing to the very low conductivity of loose snow, and the annual fluctuation vanishes at a shallow depth (B). This has been confirmed by field measurements; Wade, for example (1945), found constant temperature below forty-five feet. The firn is a very poor conductor indeed, and a good deal of the heat flow in this zone is convective through air circulation. Thus Poulter (1947, p. 372) reports feeling an air draught two feet above the end of a bore ending sixteen feet below the snow surface. The gradient at any instant may be anywhere in the figure JBA'. Below B the temperature gradient shows no seasonal variation.

A static ice-sheet in thermal equilibrium would have to conduct the normal terrestrial heat flux to the surface, and at the same time the surface would be subjected to the annual range of seasonal temperature fluctuation. The slope ED is determined by the conductivity of the rocks and the regional heat flux. The conductivity of ice is of the same order as that of rocks, but the conductivity of compact snow is an order of magnitude less. Hence, DC (the lower part of the gradient in the ice) has much the same slope as in the underlying bedrock (a degree or so every 100 feet) but changes rapidly to about a degree every ten feet as the firn becomes less compact.

There are, however, further variables. The flow of the glacier has a convective effect equivalent to bringing the very cold surface temperatures nearer to the basement, thus depressing the basal temperature below D, and so on for all points between D and B. Hence the temperature curve would be displaced to the position FB. The shape of this curve is a function of the rate of glacier flow, of the rate of change of this rate with depth, and of the rate of change of conductivity of the ice with depth. Also, the bedrock segment of the gradient EF is not in equilibrium. It tends towards a uniform gradient (for uniform rocks) but the longer the glaciation continues the further F moves to the right towards lower temperatures (but always,

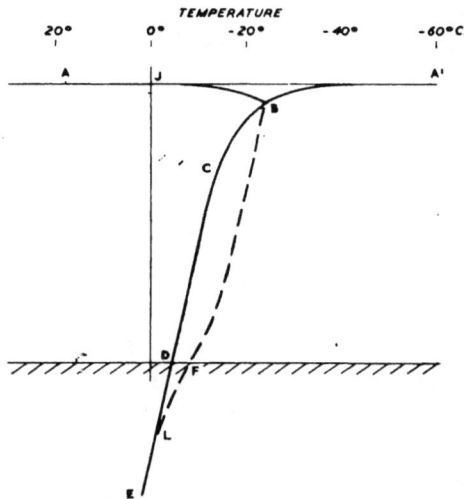

FIGURE 5.   Thermal gradient through a
dry-base glacier.

of course, above the temperature of B). Heat is flowing upward through the rocks near F at a faster rate than near E, and this leads to a progressive downward migration of the point L as the chilling effect of the glacier extends more deeply into the bedrock. Many thousands of years are required to approach an equilibrium gradient.

A complicating feedback relation exists between the rate of flow of the glacier and the thermal gradient. The flow-rate of ice is sharply affected by temperature. For example, the temperature dropping from 0°C to −5°C halves the rate of flow, other conditions being equal. Hence a very cold, slow glacier receives more terrestrial heat during its passage than a warmer, faster one. A portion of the gradient below B may become very steep at nearly constant temperature, receiving virtually no summer heat by conduction from the surface, and virtually no heat from the earth's heat flux owing to the convective removal of this heat by ice flow to further down within the glacier. The ice also receives heat from conversion of mechanical heat by viscous flow and bottom friction slippage. The whole of the potential energy lost by the ice commencing to flow at say 3000 m altitude and ending at sea-level is converted to heat. This is enough to raise the ice temperature by about 12°C, but by itself could achieve little melting.

The three contributors of heat — friction, terrestrial heat flux, and heat exchange from basement to ice with depression of temperature of the former — all lead to a progressive rise in temperature towards the base. Further, the same three contributors cause the temperature at the base to rise progressively downflow. Hence the thicker the glacier and the greater the distance downflow, the more the probability of a melting zone appearing at the base, even though the upper zones of the glacier may be typically arctic. This introduces a further complicating factor of melt-water.

In the zone of surface seasonal fluctuation (ABA′, Figure 5), melt-water will

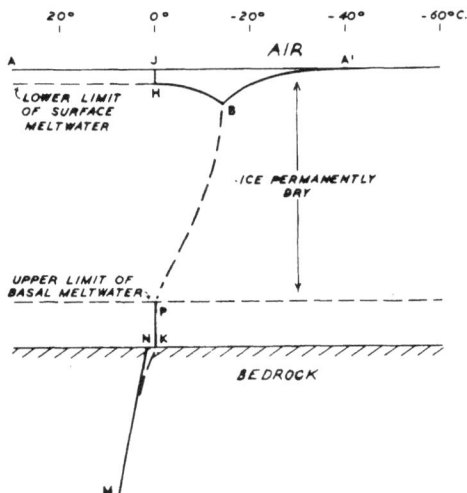

FIGURE 6.   Thermal gradient through a wet-base glacier.

always be present where the surface air temperature is above 0°C. Owing to the time required for heat conduction through the boundary layer of water adsorbed to the surface of the ice crystals, surface melt-water can be heated above 0°C, and although it is in the process of losing this excess heat by melting adjacent ice, it is at the same time percolating downwards, so that the zone of melting is not confined to a surface skin. Hence the ice temperature is 0°C down to the limit of melt-water (JH, Figure 6). Melt-water percolating more deeply is frozen by the cold ice while its surrendered latent heat raises the temperature of the ice towards zero. Likewise the temperature at the base cannot rise above 0°C, and if the initial rock temperature (N) before the onset of glaciation was above 0°, the gradient MN would be rapidly offset to K at 0°C and a layer KP in the ice would be at 0°C. Assuming equal conductivity of rock and ice, the position of P would be below the prolongation of MN because of the removal of heat by glacier flow. The constant temperature segments KP and JH do not mean the absence of heat flow but that heat flow is absorbed in melting ice or freezing water until the ice becomes dry at H and P. Interstitial melt-water is therefore confined to the zones of the glacier above H and below P respectively. The increasing compactness of ice downwards means that the permeability of the basal ice must be very low. Hence the PK may be very close to K. This means that the upper limit of melt-water may not be far above the limit of complete melting. This question will be examined further when we consider deposition of till below glaciers.

All glaciers have a surface melt-water zone at some time of the year. The bases of many Arctic and Antarctic glaciers are, however, well below freezing point so that there is no basal melt-water zone. In some temperate glaciers B and P meet and J, P, B, K, are all at 0°C so that there is interstitial melt-water throughout. Such glaciers do not differ hydraulically from sedimentary rocks saturated with ground water, and interstitial flow and the seepage pressures behave as in ground-water flow.

VERTICAL SUPPORT OF THE GLACIER

While the sole of the glacier is above sea-level, the whole weight of the glacier is borne by the underlying rocks and sediments. This is also true for dry glaciers when the sole of the glacier is below sea-level, but not yet buoyant. In a wet glacier, on the other hand, the subglacial sediments are saturated with ground water that may exert a powerful uplift pressure on the base of the glacier, and partly support its weight, just as do uplift pressures on the base of a concrete gravity dam. A longitudinal section along the glacier changes little in respect to load distribution from year to year, hence an equilibrium pressure gradient is attained in the interstitial melt-water and in the ground water below the glacier. The equilibrium gradient relates the load of the ice, the head of water on the submarine seepage outlet, the permeability of the saturated beds, and the rate of flow. Because the landward-saturated sediments are confined under the load of the ice, a strong hydraulic gradient exists in a seaward direction. Hence the uplift pressure in the grounded shelf zone is greater than the hydraulic head at an equivalent depth below sea-level, and the proportion of the glacier's weight supported by the ground water is greater than would be the case with an equivalent column of ice freely submerged to the same depth below sea-level. The subglacial sediments therefore do not bear the full load of the ice, and the ice load on the sediments diminishes progressively to zero at the point of flotation.

Thus a wet-base glacier has a gradual load transition that is distributed all the way from where the sole is a little above sea-level to where there is sufficient depth of water to float the glacier. A dry-base glacier, on the other hand, suffers an abrupt change in the nature of its support when it becomes buoyant. Inside this transition its full weight is borne by stresses in the frozen sediments. There is no hydraulic contribution. Across the line of transition its load is borne entirely by flotation. The water below the floating section is subject to tidal rise and fall. The abrupt transition means that there must be a hinge zone at the transition to buoyancy, even though this be hundreds of miles from the ice-barrier. Any lifting of the margin of the grounded zone along the main shear plane during an exceptionally high tide would result immediately in a wedge of sea ice frozen in the gap, so that the falling tide would leave the whole of the buoyant glacier cantilevered from the edge of the wedge. As the semi-diurnal tidal cycle is too rapid for the ice to flow to equilibrium, folds and fractures appear at the surface along the hinge zone. This phenomenon is well known in Antarctica (Robin, 1954, p. 199). The inner edge of the water under a dry glacier should normally terminate abruptly at an ice wedge connecting the sole of the glacier to the sea floor. The slope of this abrupt contact is greatly increased by the rapid spreading of the floating segment owing to the absence of basal friction.

As soon as a glacier becomes buoyant there is virtually no frictional resistance to its horizontal movement. The weight of the upper portions bear downwards opposed by the buoyancy of the displaced water, so that the glacier is squeezed laterally to cover a larger area with a lower centre of gravity for the whole system. Both the originally dry and wet types spread similarly in the buoyant section. Both receive flow contribution from the land, and receive surface precipitation. Only the dry may receive addition to the base from freezing of sea water. Both may produce floating shelves of comparable thickness. But the spreading of the wet glacier begins long distances back from the actual line of buoyancy. There it is already almost supported by the ground water seepage pressure, and the load on subjacent sediments is very low. The frictional resistance to spreading (the product of the glacier's effective weight on the bottom and the coefficient of friction of wet ice on water-lubricated clay) is a negligible one in contrast to the dry glacier. Here the *full* load bears on the bottom until the actual line of buoyancy, and the coefficient of friction is that between dry ice welded to dry cold rock. The weight times coefficient product is very high. Hence the cross-section of ice flowing over the basement may be ten times that necessary to discharge the same volume through the corresponding buoyant zone.

This means that the thickness of a dry-base glacier must change abruptly at the buoyancy line, by a fraction of ten or so, whereas the change is quite gradual in a wet-base glacier and may be distributed over a hundred miles or more. Hence there is no notable change of surface elevation at the line of buoyancy of a wet-base glacier whereas in a dry-base glacier the surface rises abruptly along the buoyancy line from the flat floating shelf to ice mountains.

These expectations are completely vindicated in the field. Robin (1954) found the Dronning Maud Shelf to be remarkably level for hundreds of kilometres and then to rise abruptly into ice mountains (Figures 1-3). His seismic reflection measurements revealed that the flat portions were floating over deep water, and that the ice mountains were grounded ice, also in relatively deep water. At the transition zone, ice about 1000 m thick becomes buoyant in water 800 m deep and thins rapidly by spreading to less than 200 m thickness, leaving 760 m of open water beneath it, while the 1000 m thick grounded ice is still resting firmly on an only slightly shallower bottom. The top surface of the grounded ice is at more than 800 m above sea-level, whereas the top surface of the floating ice is at 30 m. An equilibrium gradient between these two levels is determined by the glacial flow-rate (e.g. Nye, 1951, 1952; Robin, 1953). This may result in even more of the sheet becoming buoyant until a gradient of equilibrium is reached. The ice first becomes buoyant where the ice-sheet has over-deepened valleys under the more active flow channels. The buoyant zone rapidly spreads and the consequent fall of its surface level to that of the ice shelf leaves the ice on the interfluves standing as ice ranges well above the re-entrant embayments of the shelf ice, for the interfluve ice almost maintains the original level even though the rock interfluves are far below sea-level (see Figures 2 and 3). This has been demonstrated by seismic work (Robin, 1953) and is beautifully revealed by the glaciological map of western Dronning Maud Land prepared by Swithinbank (Figure 8 of Robin, 1954).

By contrast (see Figures 1 and 4) Joulter (1947) found the surface of the Ross Shelf to be remarkably uniform for hundreds of kilometres even though seismic reflections proved that the shelf is grounded to within 13 km of the outer edge. Also his seismic measurements show that the thickness changes little. The changes that occur are due to two other causes: (1) there is a steady increase in thickness from 500 feet twenty-three miles from the edge, to 735 feet three miles from the edge, owing to the increased precipitation gradient towards the edge; (2) progressive thinning from 755 feet to 525 feet in the outermost three miles owing to rapidly increasing bottom melting by inflowing sea water, and a similar thinning by bottom melting thirteen miles in where a three-knot "warm" current under the shelf follows the outer edge of the subglacial sediments.

The only surface indication of the buoyancy line in a wet glacier is the tidal cracking. The semi-diurnal tidal movements are too rapid a cycle for the sub-glacial seepage pressures to remain in equilibrium with them. Hence long straight fractures develop along the buoyancy line, and may eventually lead to the calving of tabular bergs many tens of miles in length (Poulter, 1947).

Even far out on this self-spreading floating shelf, however, the flow pattern may be dominated by the active trunks of the feeding glacier. For example, Robin (1954, p. 198 and Figure 8) described a tongue projecting from the free margin of the Dronning Maud Land Shelf, the tongue apparently being the terminal outflow of a main trunk glacier that had maintained its identity right across the shelf. For several hundred kilometres in both directions the coast is a fjord coast with a mountain ice-sheet extending to below sea-level and continuing out over the continental shelf as a floating ice-shelf from 50 to 100 more km wide. However, at the Greenwich meridian there is the outlet of the Pencksokka Glacier flowing from the high continental ice-sheet 1000 km inland (Figure 3). The floating ice-shelf at the mouth of this fjord is about 200 km wide and of about the same length, but in the line of flow of the trunk-fjord glacier the floating ice-shelf protrudes a further 50 km over a width of 50 km (Robin, Figure 8). Such phenomena may substantially affect the pattern of contemporaneous sedimentation. In constructing Figure 1 of this paper the dimensions of this Pencksokka Glacier were used as a basis.

SEDIMENTATION BELOW GLACIERS

Terrestrial Zone

The factors that cause deposition of sediment by a glacier are necessarily very different from those causing deposition by a stream.

The principal cause of deposition in streams, reduction of velocity, has no effect in a glacier, which can continue to transport its full load no matter how slowly it moves. The second critical threshold to water transport, size of body, is also irrelevant in a glacier, which can carry huge blocks as readily as clay— hence the characteristic lack of size-sorting in till. Conversely, bottom melting, the most significant factor in glacial deposition, has no counterpart in streams. Because the principles of fluvial and glacial deposition are dramatically different, geologists need to make a conscious effort to free their minds of fluvial models when contemplating glacial deposition.

There are two critical surfaces in a terrestrial glacier, the upper limit of basal melt-water and the lower limit of interstitial ice, which may be taken as the boundary between glacier and deposited sediment (Figure 7). Traced downflow, both surfaces should rise with respect to the basement because the extent of offset of K with respect to N (Figure 6) should diminish in that direction. Traced upflow, first one and then the other surface (if the glacier is cold enough) should reach the basement. These surfaces are almost parallel to the flow lines but cross them at very gentle angles. (The angle depends on the balance of heat convection and heat conduction in the ice, hence the faster the glacier the flatter the angle.)

Because the till is deposited under the full load of the glacier it is born as a consolidated sediment, which is another point of sharp contrast with water-laid sediments which are "normally loaded" (in the soil mechanics sense) as they are deposited. The englacial flow fabric is preserved for the till develops *in situ* by the replacement of interstitial ice by melt-water, through a transitional zone of considerable width (compare Harrison, 1957, p. 296).

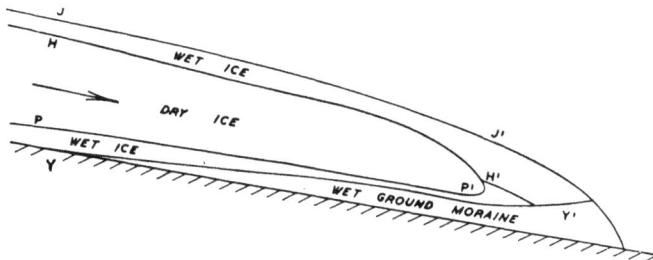

FIGURE 7.   Melting zones of a glacier.

The upstream limit of melting (locus of Y, Figure 7) would not be regular. Glacier ice is far from isotropic. It consists of many ice-streams convergent from different névé fields at different elevations, conveyed thence by feeder routes of different thickness and different rates of flow. Different streams may converge with quite different basal temperatures. These differences gradually vanish, but only by conduction, which is slow. Again the detritus carried by the different streams differs widely in amount, composition, and grain size. This may greatly affect the rate of melting, because in the zone of melting the ascending terrestrial heat is absorbed only as latent heat (gradient KP, Figure 6), hence the rate of upward migration of the surface of complete melting is directly proportional to the ice-sediment ratio. The sediment variation of the different streams also affects the glacial flow parameters both in the subzero and in the water-saturated states. Again the stream with the higher sediment content usually (but not always) has the slower velocity and hence receives more terrestrial heat per mile of flow and melts first. In addition there are irregularities in the basement. Quartzites are better heat conductors than basalts. Basement projections cause upward and lateral deflection of the flow lines with reduced rate of flow in their ice, and increased rate of flow elsewhere.

All these variations cause the upper limit of complete melting to be a wavy surface very dentate in transverse section although smooth in longitudinal section in the direction of flow, and the line in which this surface meets the bedrock basement is extremely dentate with very long upstream projections and outliers.

Long before permanent sedimentation becomes general it might well occur in the lee of basement projections and form lee drumlins (rock drumlins or "crag and tail"). In the threshold region where sedimentation is about to become general, melting occurs on the most favourable lines. Deposition here would inevitably cause deflection of the flow lines upwards and laterally, so that a drumlin is born. The favourable ice-stream would continue to arrive at this point, the drumlin would grow in size, with successive additions continuing in the streamline shape. As the melting becomes more general, continuous ground moraine develops, but its upper surface would still be drumlinoid, as is beautifully illustrated in the United States Army Air Force air photograph of the Carp Lake area of British Columbia (published as Figure 12-12, Longwell and Flint, 1955, p. 196). The deflection of the flow lines by the drumlin deposit increases as the drumlin grows. This deflection causes crowding of the flow lines at the front, which implies greater shear, and widening of them behind the drumlin, which implies less shear. Material is therefore dragged from the upstream end and deposited at the rear slope. Probable bottom slippage, between glacier and basement in this zone of deposition, may accentuate the process.

One of us (Carey, 1953, pp. 76-8) has drawn dimensional analogies between dunes and drumlins, and between lee shadow sand drifts and lee drumlins. In both cases we are dealing with a fluid (wind and ice respectively) flowing over an inert basement on which there is loose material being shaped and transported by the moving fluid. The viscosity of ice is some 10^{18} times that of air, but wind velocities are 10^9 times as great as glacier velocities. A more viscous "wind" would need to move more slowly to produce a similar drag, and if the viscosity of the "wind" were progressively increased until it reached the viscosity of a glacier, the velocity required to mould the "dune" into a comparable shape might be of the order of glacier flow.

Although the most important cause of deposition is basal melting, the question arises as to whether deposition can occur upstream from where wet ice is in contact with the rock floor (Y, Figure 7), and also whether deposition can occur above where PP' rests on the basement, that is, when dry, sub-zero ice rests on the basement. Let us first clarify what we mean by "deposition" under these circumstances. Can subglacial sediment be said to be deposited when its voids are occupied by ice continuous with the ice of the glacier above? The whole of the shear involved in the flow of the ice above must be transmitted through this material, and it must surely flow, even if comparatively slowly. Deposition then begins at the line of complete basal melting (Y, Figure 7). The most that could occur elsewhere would be stagnation of rock-loaded ice to await the ultimate melting of that part of the glacier.

Let us now examine the question of bottom slippage of the glacier in the different cases of a wet glacier on its already deposited till, a wet glacier on bedrock, and a dry glacier on bedrock. In all three cases the total shear involved in the flow of any column cannot exceed the cohesion between the glacier and its bedrock, for if it reaches that level slippage occurs. In a dry glacier the cohesion between dry ice and subzero rock is high. However, the shear needed to cause flow in a subzero glacier is also very high. Hence the thickness of ice must increase until the total flow through the cross-section balances the total precipitation. The thicker the ice, the greater the shear at the base, but as this cannot exceed the cohesion, the latter determines the maximum permissible thickness. If the precipitation rate is such that a greater thickness is required to remove it by flow alone, then slippage must occur at the base.

An equivalent wet glacier will flow at a very much faster rate under the same load, therefore a very much thinner section will transmit the same precipitation with a correspondingly smaller shear load on the base. However, the shear cohesion of wet ice on wet rock is also very much smaller, so again slippage could occur under appropriate conditions. Similarly in the case of the wet glacier on its own till, the flow-rate in the ice will be large and the corresponding cross-section necessary to transmit a given flow will be small. However, we have to consider the relative rates of shear under given loads in the wet ice, in the wet till, and the limiting friction between the wet ice and the top of the wet till. It seems that in all cases the deposited till will be much more resistant to shear flow than the wet ice. The lower part of the ice contains the same constituents as the till but they are separated by ice wet with its own liquid phase. The ice has mechanisms of flow not possessed by the till, for example crystal glide plane slippage, two-phase deformation, energy transfer from crystal to crystal. On the other hand, the till has a wide range of grain size coupled with load precompression so that it will contain little interstitial water and should have relatively high shear strength. Although it contains clay size-grade material it contains little clay in the sense of phyllosilicate minerals. Little if any deformation of the till should take place as the ice flows over it. The most rapid rate of shear might be in the lower part of the ice, and there might also be slippage between the ice and deposited till perhaps resulting in rolling of some of the surface layer of the till. The till should preserve the flow fabric of the basal portion of the glacier, that is with long axes of pebbles in the transverse direction or parallel direction or both (see Glen, Donner, and West, 1957).

The empirical geological evidence seems to indicate that slippage at the base of the glacier is the general rule. Wherever glaciated surfaces are revealed, whether from the Pleistocene or Palaeozoic glaciations, striations, grooving, roches moutonnées, and plucking of the basement seem to be universal. From such evidence we would judge that both wet and dry glaciers normally slide along their floors in addition to flowing within their mass. The empirical evidence is less conclusive for wet glaciers sliding on their own tills, than for dry glaciers moving over bedrock.

Transition to Buoyant State

A glacier that contains interstitial melt-water at its base, and rests on water-saturated sediment, suffers no abrupt changes during the transition to buoyancy. There is no change of basal temperature, there is a gradual change of vertical support, and there is a transition from distributed flow throughout the ice to nil drag on the base of the glacier.

In contrast a dry-base glacier suffers abrupt changes in the heat flow, in the nature of support (from solid to hydraulic), and in the conditions of horizontal shear. Each of these changes has marked effects on both the behaviour of the glacier and the nature of the sediments deposited.

If a dry-base glacier has a temperature below the freezing temperature of sea water (about $-1.9°C$), and flows out to sea and becomes buoyant, sea water rapidly freezes and welds onto the bottom of the glacier, forming an insulating layer. The temperature gradient now has the form shown in Figure 8. The temperature at the base of the new ice is at freezing point. There is a steep gradient to S on the original gradient inherited from the non-buoyant stages. Heat flows rapidly across the steep gradient RS so that S moves up towards B and eventually equilibrium is reached in a form like RTB, whose curvature is due to the decreasing conductivity of the upper levels of ice. Such warming of the glacier as is implied

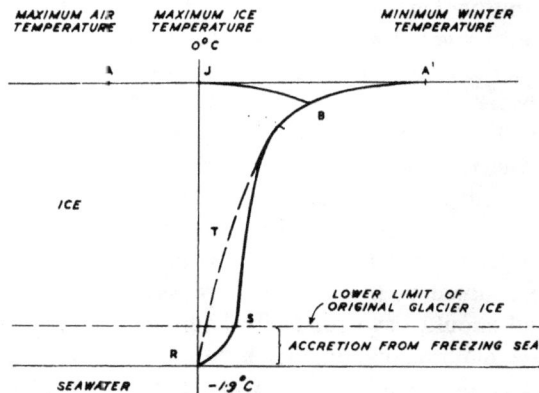

FIGURE 8.   Thermal gradient through a floating ice-sheet derived from a wet-base glacier.

by the change from gradient RSB to RTB is achieved by heat derived from the underlying sea water and results in progressive freezing of more sea water and thus the addition of more ice to the bottom of the glacier. Before this equilibrium curve is reached, melting will have commenced at the base (for sea water is normally above melting temperature). Even a subzero glacier that adds sea ice to its base at its landward end will begin to melt from the base at a point well inside the terminal barrier. In such a subzero glacier, the underlying rock floor is also in a frozen permafrost state (Figure 5) and will freeze any sea water that may be brought in contact with it.

Sea water may be brought in contact with new surfaces of subzero ice by one of three processes — the outward movement of the glacier, tidal movements, and (on a much larger scale) by landward extensions of the buoyant area. Such landward extensions must occur during a time of climatic amelioration due to reduced alimentation, thinning through surface melting, and eustatic rise of sea level.

During the retreat stages of a large marine ice-sheet, many hundreds of square miles of ice previously grounded become buoyant and considerable volumes of dense brines must be produced. Not only are new, subzero surfaces of glacier bottom and bedrock bottom brought into contact with sea water, but the buoyant ice-sheet spreads very rapidly to a fraction of its thickness, bringing very large areas of subzero ice against sea water. Such ice is much colder than that at the original base. Much sea ice must be frozen onto the bottom, forming an insulating blanket or filling joint cracks. Resultant cold brines (being the partly-dehydrated sea water) would sink to the bottom and form outward gravity currents moving to the deep ocean floor beyond the shelf. If barred basins exist, either under the floating ice-barrier or beyond the ice-barrier, the cold brines flow into such depressions and lie there perhaps to crystallize slowly with rising temperature. This will be further discussed in later pages.

SEDIMENTS PRODUCED BY SUBGLACIAL MELT-WATER

Subglacial melt-water sediments are conspicuous features of terrestrial glaciation and as such have been studied and described at length, but they are not directly party of this study. However, subglacial melt-water drainage and sedimentation may produce sediments in association with submarine products. Such submarine deposits if not properly understood and interpreted, could lead to a quite erroneous concept of the climatic sequence recorded by the sediments. It is therefore necessary to examine critically terrestrial subglacial drainage processes and sedimentation, particularly the long sinuous ridges called eskers (in the sense used by Flint, 1947) or osar (in the sense used by Charlesworth, 1957, not eskers of Charlesworth). In this paper we propose to follow Flint's usage.

The origin of eskers has produced a long controversy which is summarized comprehensively by Charlesworth (1957) with extensive bibliography, and thus needs no repetition. Sufficient to say that for the following reasons we are satisfied that eskers (or perhaps only some of them, these being the ones that will concern the rest of the discussion) are formed in subglacial streams flowing under the ice with or without connection with the surface melt-water of the glacier.

No adequate explanation seems to have been published concerning the mechanism of development of open subglacial channels. The mere opening of a crevasse right through to the sole of the glacier cannot produce such a channel. If part of the glacier is subzero, the crevasse will freeze; if the whole of the crevasse is water-saturated or at least not subzero, the water head at the bottom of the crevasse, even when the crevasse is filled to overflowing, is little more than the hydraulic head previously existing in the melt-water below the glacier, and may be even less because artesian pressures may exist in subglacial melt-water. Subglacial channels can only develop from the seepage outlet of the melt-water at the terminus of the glacier. The subglacial till and the melt-water–saturated part of the glacier are simply water-saturated rocks and behave so hydraulically. Seepage can produce open channels through water-saturated sediment well below the surface. This is a process well known in soil mechanics under the name of "piping" or "underground erosion" (see, for example, Terzaghi and Peck, 1948, p. 230) and has led to the failure of many earth dams. We suggest that esker channels are merely examples of this process. Such channels cannot develop in a glacier which is subzero throughout, nor in permafrost sediments. They commence at the seepage outlet of the melt-water, which is either under the terminus of the glacier on land, where the melt-water seepage enters the sea below the end of a grounded glacier, or where a floating glacier becomes buoyant. The seepage pressure at the outlet is sufficient to wash out the finest particles, thus enlarging the water-bearing spaces and hence increasing the rate of flow. Sand particles can next be washed out, and as the process develops, sand and gravel, so that an open channel appears. This channel takes the steep hydraulic gradient farther back into the water-bearing material, so that the process of underground erosion works continuously back from the terminus of the glacier. The ground water bears the full static load of the overlying ice, and thus an extremely steep hydraulic gradient is always present at the head of the piping. In fact the thicker the glacier above the channels, the more rapidly they may work back. Underground channels may therefore develop back from the seepage outlet as interstitial water is present in the subglacial till, irrespective of any contribution of melt-water from the surface. Eskers as long as 150 miles are known. The development of channels will stop where the freezing isotherm passes below the rock floor, or, alternatively, at the plucked crag of a large *roche moutonnée*, or, where sediments are met that are of such a size-sorting that they form a natural filter (see Terzaghi and Peck, 1948, p. 50), or that are too impermeable to allow seepage at a sufficient rate. Herein perhaps lies the reason why eskers are not universally developed (see Charlesworth, 1957, pp. 421-3). They require a suitable grain size distribution in the till. It would be useful to compare the grain size histograms of tills associated with eskers with those of tills that are not. The grain size of the esker sediment is not relevant because all the fines are washed out and the coarse material resorted.

Eskers develop on the top of till already deposited. The till below is already pre-compressed and though water-saturated has low permeability. The basal part of the saturated glacier is most vulnerable, for a considerable proportion of its ice is already melted. The interstitial melt-water is milky with fine sediment, and if located near the head of an esker, will seep rapidly towards the esker. The weight of the glacier is bridged across the developing channel and at times the seepage pressure may remove more clay than sand as fast as the material is freed by melting, and so the esker extends. Water pressure within the tunnel is of course enough to keep the tunnel full, and may even contribute substantially to the support of its back. The pressure is sufficient to drive the water up grades, "over hill and dale." Erosion into the underlying till occurs to some extent. The channels widen downstream as melting progresses and rise upward with the melting. Tributaries may develop. During the engineering investigation for fifty-foot deep foundations of a heavy structure, one of us (S.W.C.) found an unsuspected complex of dendritic drainage pattern of underground channels developed in silty sand below a cemented conglomerate. The open channels had worked back hundreds of yards from an original ground water seepage outlet below low water level of a nearby river.

It is abundantly clear that terrestrial eskers, as we know them, have suffered little deformation by glacier flow. Any channels oblique to glacier flow are rapidly closed by such flow. Moreover in their lower reaches eskers are often associated with undisturbed crevasse fillings. Hence the eskers finally deposited must be the product mainly of the moribund melting stage of the glaciation. But esker-like water channels are not incompatible with glacier movements, which are necessarily slow, too slow to forbid piping processes. The glacier may flow and alter the form of the channels to streamline shapes. But all voids of both till and glacier are saturated, and water compresses little, even under the total weight of the ice above. Hence the water-filled openings remain, though deformed, and a hydraulic gradient must build up quickly across any constricted links, sufficient to widen them rapidly or to erode the underlying till enough to pass the water flow. The ispatinows of Saskatchewan (Tyrrel, 1895; and Figure A, plate XIII of Charlesworth, 1957) probably belong here. These are ridges of drumlin form but of esker material, and were regarded by Charlesworth (p. 423) as related to eskers.

The conclusion important to the present study is that there seems no reason why eskers should not form in the transition zone below sea-level, extending right out to where the glacier becomes buoyant, which might be at a depth of 1000 feet or more. Eskers so formed could be preserved during ice retreat. Any warming should lead both to thinning of the glacier and to rise of sea-level, each of which tends to lift the glacier off the bottom, leaving the esker to be covered by marine glacial sediments.

Such deposits should be looked for in ancient glacio-marine sediments. The ice-sediment contact zone is the locus of unusually strong shear and the esker would probably be remoulded into the ispatinow form. Eskers are unmistakeable on young glaciated surfaces because of their form in plan. But, in a Permian cliff section, they might easily pass unrecognized or be misinterpreted.

SEDIMENTS OF THE GROUNDED SHELF ZONE

The grounded shelf zone of a dry glacier does not differ in any significant way from the terrestrial zone. The full load of dry ice loaded with debris grinds the pavement. There can be little sedimentation, if any.

A wet glacier is profoundly different. We have already shown that the weight of the glacier borne by the subjacent rock or sediment diminishes progressively to zero, and that this process is accompanied by progressive thinning of the glacier as the frictional inhibition to spreading declines. Further, the till is deposited at the base of the glacier as the level of complete melting rises. The conditions in such till are those of low vertical load but very great horizontal shear. Hence a great deal of shear will occur at the top of the till. The lighter the load, the shallower is the effective depth of penetration of this shear into the till and the greater its horizontal component. The till still bears part of the weight of the glacier and will be dragged or rolled forward to be left where the water depth is just sufficient to give complete buoyancy. Here a submarine slope will develop at the angle of rest for the material. Under steady conditions material will be constantly dragged or rolled forward to the edge of this slope and pushed over the edge as a foreset deposit. This slope will move steadily forward. A warming climate will cause thinning of the ice and a rise of sea-level so that the buoyancy line will retreat landwards. A new foreset slope will develop and advance over till previously deposited. Such conditions may produce melt-water sediments interbedded with till. Melt-water outwash silts deposited ahead of this foreset slope may develop strong drag, slump, or roll structures. One of us (Ahmad) has studied balled and rolled structures at Wynyard at the top of the till just at the transition into submarine drift sediments. These might have developed as an ice-sheet that was previously partially resting on the till, gradually lifted as the buoyancy line moved back.

Climatic cooling would produce thicker ice, and thus more weight on the base of the glacier, and great dozing of the upper sediments, which would be pushed forward to a new foreset slope front. Unlike truly terrestrial tills, the tills and melt-water sediments deposited below sea-level would suffer only partial load compression, or even none. Hence, during increasing glaciation, a thickening ice-sheet emerges to a fallen sea-level and the extra ice load would consolidate the sediments to lower void levels.

Whether the ice is stationary, advancing, or retreating, these processes have the effect of determining a base-level for till accumulation analogous to wave-base for marine settling. Till builds rapidly up to the base of the ice or is cut down to that level, then stays at that level while its foreset slope is advanced. The top of the till and the thickness of the ice-sheet may remain uniform for hundreds of miles. Such conditions were found by Poulter in the Ross sheet. In this example the gradient of the foreset slope was surprisingly steep for marine conditions. Seismic reflections indicate that the foreset slope opposite Lindbergh Inlet drops from 800 feet to 1700 feet depth in about two miles, with a maximum slope approaching one in five.

Submarine slopes of such steepness must be very unstable, and as fresh loads of till are dozed over, must from time to time inevitably produce mud slides down the slope. Studies of submarine slump phenomena have now established that such mud flows may roll out for many miles from their starting point. Because the slumping material is till the resulting deposit would be difficult to distinguish from till. It might occur as a thin bed or as a thick bed within stratified marine glacial drift. It might be completely unstratified with abundant erratics in an unsorted matrix. The erratics would show no signs of dropping through water. The under-lying sediments could be undisturbed. Such flows might well pick up marine fossils while eroding channels in the sediments over which they move. They might over-run and incorporate living shell beds, crinoid beds, or the like. They would, however, lack the characteristic fabric of true till, and with increasing distance of transport would resemble tills less and less as sorting and stratification become more effective, but pebbles would still show glacial striae. A very important conclu-sion is that such submarine flow "tills" could easily be mistaken for true tills, leading to a very different interpretation of the palaeoclimatic sequence. Several "tillite" bands interbedded with well-bedded siltstones have been described from the marine Permian of Tasmania. Thus from Woody Island, Banks, Hale, and Yaxley (1955, p. 224) have described the D'Entrecasteaux tillite, six inches thick, interbedded between stratified siltstones that contain occasional erratics. The tillite is rich in erratics and rarely contains *Spirifer* shells, but the beds above and below contain marine fossils. Interpreted as a tillite, we can offer no feasible explanation for its occurrence as such a thin bed in undisturbed marine strata, nor for its contained fossils. Interpreted as a submarine flow from a till foreset deposit, its occurrence and characteristics are completely satisfactory and logical.

Quite as important as the till flows is the role of melt-water that appears on the foreset slope at the embouchures of esker streams at the zone where the ice-sheet becomes buoyant. The total flow of subglacial water laden with clay and silt must be very great, and turbidity currents must surely be a dominating phenomenon, flowing down the foreset slope and thence far out to sea. Hence the sediments below the floating ice-shelf and the iceberg zones should be dominantly of this type. Such sediments should consist originally of unweathered, finely-ground rocks and minerals, of sand to clay grade. There are many processes whereby these sediments should develop laminar rhythmic, or thick bedding. Annual changes in current circulation could easily produce bedding, as could annual or longer range patterns of glacier thickening or thinning altering the bottom load in the sediments.

We have seen that, under a wet-base glacier, eskers may develop at the glacier-till contact far below sea-level, and if there is a rise of sea-level owing to partial amelioration of climate, these river-washed channel sediments might be buried under firstly submarine outwash sands and silts with interbedded till-flows; then under foreset "tills"; then covered by a substantial thickness of normal till; all with a continuously warming climate and representing a single retreat phase. Yet such a section would probably be misinterpreted as a record of complex, oscillating glacial and interglacial conditions.

None of these till-flow and turbidity current phenomena should be expected with dry-base glaciers. Nor does the contrast end there. The pattern of iceberg transport is quite different. The wet-base glacier suffers basal melting for possibly hundreds of miles back from the buoyancy line. The lower, sediment-rich, layers of the glacier drop their sediment by melting miles inland and if such sediment reaches the open sea it does so by dozing action of the glacier to the foreset slope, by melt-water transport of fines, or by turbidity currents. By the time the ice reaches the ice-barrier where icebergs are calved, little sediment may be left in the ice. Many observers have reported lack of sediment in the Ross Barrier icebergs, even those tipped by differential melting to expose a complete section. This has been a puzzling feature but might have been expected. As a result in front of wet glaciers, the iceberg zone sediments are siltstones, sandstones, and claystones, with only a few dropped erratics. In contrast the dry-base glacier carries all its sediment right to the buoyancy line. There it lifts and spreads rapidly, adding frozen sea water to its base because of its subzero temperature. Thus erratics and clay alike are trapped in the floating ice-shelf, which, for comparable ice flow, is many times wider than the buoyant shelf zone of wet glaciers (Figure 1). The dry glacier, lacking melt-water turbidity currents and foreset till flows, produces instead mudstones with abundant erratics, all of which have dropped through water. These two contrasting types of sediment association are both characteristic of parts of the Tasmanian Permian succession.

Still another sharp contrast exists with respect to the salinity of the waters. The wet-base glacier discharges large volumes of melt-water into the sea at its buoyancy line, quite comparable to the discharge of a major river. No sea water is frozen onto the glacier base where it becomes buoyant. The net effect is to *reduce* the salinity of the sea water in the vicinity.

When a dry-base glacier becomes buoyant both ice and exposed bedrock are below freezing temperature, and sea water is frozen onto them, forming an insulating layer at freezing temperature. However, as lower temperature exists within, heat continues to flow from the sea water producing slow addition to the frozen layer. The newly-buoyant glacier rapidly spreads to a fraction of its original thickness (as has been previously explained) and thus the cold heart of the glacier is continually brought very close to the sea water so that a large surface area is exposed to a steep freezing gradient. These processes result in the production of cold brines in the sea, that sink because of their density to flow along the bottom as cold salinity currents, and might flow out and over the continental shelf to be lost, or, on the other hand, might be trapped in any barred basins that exist. The cold brines would be especially characteristic of times of general glacial retreat, when dry glaciers previously grounded have their thickness reduced and their depth of water increased with the result that many square miles of subzero ice and bedrock are exposed to the sea within a short time. During advancing glaciation, the phenomenon would be much reduced but still present, because wherever the dry glacier eventually becomes buoyant it quickly spreads, exposing its cold heart to the sea.

There are at least two important consequences to this process, the precipitation of calcite, and the precipitation of sulphates. Most sea water is nearly saturated in $CaCO_3$, but the solubility of $CaCO_3$ increases as the temperature of the water drops towards zero. The effect of the freezing of a portion of the sea water is to increase the titre of $CaCO_3$ but precipitation may not occur immediately from the very cold water. As the brine flows outward into a warmer environment $CaCO_3$ is inevitably precipitated, either directly or *via* organisms building shells. Hence the first consequence is that dry-base glaciers have either an off-shore calcareous facies, or a calcareous facies in their glacial retreat sequence, whereas wet-base glaciers produce environments unsaturated with lime. The limestones associated with dry-base glaciers will contain abundant erratics, for, as we have shown, icebergs from this kind of glacier still carry their erratics and sediment when they become buoyant whereas ones from wet-base glaciers do not. The Darlington limestone and the Berriedale limestone closely fit these environments in the Tasmanian Permian, and we shall show that fossil ecology concurs.

Precipitation of sulphates is the next possible consequence of the salinity currents associated with dry-base glaciers. We offer the suggestion that the large glendonites characteristic of certain horizons of the Permian sediments of New South Wales and Tasmania were probably formed in this way.

Glendonites, calcite pseudomorphs after glauberite (Na_2SO_4. $CaSO_4$), were first described by David, Taylor, Woolnough, and Foxall (1905) from New South Wales, and have subsequently been reported from many localities in the two states. Raggatt (1937) has summarized the New South Wales occurrences, and Banks and Hale (ms) have recently collected the information on Tasmanian occurrences and discovered several new localities. Characteristic of these occurrences are: (1) host rock commonly dark siltstone or sandstone; (2) glacial erratics present; (3) large calcareous concretions commonly present and in some examples the concretions grew around the glendonites; (4) pyrite nodules commonly present; (5) marine fossils commonly present, mostly sparse but rarely abundant, and in some specimens the glauberite seems to have grown on fossils; and (6) characteristic of particular horizons.

These conditions all fit the environment that we now suggest. The presence of dropped erratics implies floating icebergs or a floating ice-shelf. The fossils imply marine conditions. The dark colour, and the pyrite, suggest the enclosed basin environment necessary to trap the salinity currents. The calcareous concretions suggest water of a high salinity and rising temperature because the solubility of $CaCO_3$ decreases with rising temperature. The crystallization of sodium sulphate suggests abnormally high salinity.

However, there has been much discussion about the enigmatic presence of glauberite instead of mirabilite ($Na_2SO_4.10H_2O$). The evaporation of sea water produces mirabilite, and no glauberite has been crystalized experimentally below 25°C. Freezing of sea water likewise forms mirabilite, as has been shown experimentally by Nelson and Thompson (1954) and observed in nature by Debenham when pools of sea water were frozen on the surface of the Antarctic ice. However the experimental evidence so far as it goes is in agreement with the process herein suggested by us, although further work is necessary for its unquestioned acceptance.

Thompson and Nelson (1956) have investigated the partial freezing of sea water to various salinities with withdrawal of the partially concentrated brine. They report that at moderate subzero temperatures the first precipitate is a very small quantity of calcium carbonate. This is followed on further freezing by a crop of mirabilite. The precipitate of the $CaCO_3$ occurs with rising temperature after the brine has been separated from the ice. The less the degree of refrigeration and the lower the salinity the longer the time delay. "Thus for brines with temperatures somewhat below the freezing point of the original sea water, a period of several days at room temperature was required before precipitation became noticeable" (Thompson and Nelson, 1956, p. 232). This time interval was decreased to several hours by refrigeration to $-22.9°$ C and to less than an hour by refrigeration to $-36°$ C.

Under conditions of ice formation at the base of a subzero glacier where sea water, having yielded some ice, and actually crystallizing or about to crystallize mirabilite, would sink to the bottom and flow down the ocean floor with a temperature rising very slowly towards a maximum well below "room temperature," many days or weeks would be required for the $CaCO_3$ precipitation. Conditions then existing would be appropriate for the further reaction described by Thompson and Nelson (1956, p. 237) in which saturated Na_2SO_4 solutions react very slowly with calcium carbonate to form sodium carbonate with anhydrite, gypsum, or glauberite, according to the conditions. The slowness of this reaction is indicated by the observed rate of pH increase (owing to sodium carbonate replacing sodium sulphate in solution) — 7.51 initially, 8.99 after several days, 9.31 after three months (1956, p. 237, n.).

Thus, where mirabilite is formed, glauberite is a most unlikely product from a surface-closed system that freezes sea water. However, glauberite is a likely product of slow crystallization in sea-floor basins receiving brines derived from partial freezing at the base of a subzero glacier. These conditions are likely to be fulfilled only during fairly rapid retreat of a sea-going subzero ice-sheet. We suggest that glendonite horizons in dark marine sediments containing erratics might be so interpreted.

Finally we ask whether the occurrence of pyritic nodules characteristic of these glendonite horizons is not itself a consequence of the high sulphate content in a reducing environment.

Contrast between wet- and dry-base glacial sedimentation is also predicted from palaeotemperature measurements derived from O^{16}–O^{18} ratios. Near wet-base glaciers sea water, in contact with benthonic organisms, is strongly diluted by melt-water that increases the O^{16} ratio and produces an anomalously high apparent temperature compared with the arbitrary "standard." Adjacent to dry-base glaciers, benthonic water is brine concentrated by partial freezing, with no addition of melt-water.

FOSSILS ASSOCIATED WITH MARINE GLACIAL SEDIMENTS

No life of geological consequence can exist below either the dry- or the wet-base glaciers in their non-buoyant states. Even with a buoyant ice-sheet there are severe limiting factors such as light, salinity, and oxygen.

Under an extensive ice-sheet, such as that of Dronning Maud Land, complete and utter darkness would prevail at relatively short distances in from the ice-barrier. An ice-barrier only seventy feet high implies that its base is some 500 feet below sea-level. Light is severely filtered at this depth, and under the ice transmission is only by means of light scattering. Even were this not so and full daylight was projected downwards from the base of glacier, the absorption of the water itself would bring on total darkness before a mile had been penetrated. There can be no question that Stygian darkness prevails in the water beneath buoyant glaciers of any areal extent. All life depends directly or indirectly on photosynthesis for food, and thus absence of light is an important restriction.

It has already been established that an animal fauna exists on deep ocean floors well below the limit of penetration of daylight. Overhead there are lighted seas teeming with life of various kinds and an ultimate source of food is therefore available. Moreover turbidity currents bring sediments to the ocean depths, and no doubt supply some organic matter that can start another food chain for the bottom dwellers. Beneath a glacier the only source of food would seem to be currents moving under the ice. Such currents have been shown to exist under the Ross Shelf as far in as the buoyancy line (Poulter, 1947), and favourable bottom configuration may possibly result in tidal flow taking the form of a wide circuit under the ice. There might also be convective circulation — cold briny waters sinking and flowing outwards along the bottom, with warmer water flowing in just under the ice. Therefore a sparse food supply is possible for organisms able to function without light and remaining in the path of current circulation.

Other restrictions exist. Any barred basins associated with dry-base glaciers would be highly saline and also would lack a food supply. In addition an oxygen deficiency is likely, partially because there would be no plants for its generation. Most of the minerals of the sediments would be fresh, unoxidized, and hence reducing, and any animal life would reduce oxygen concentration and check its own multiplication. However, currents could bring in some oxygen.

To sum up, whereas we may not be able to maintain that life is impossible far in beneath a floating ice-sheet, there are stringent limitations. We would expect sediments of this zone, although marine, to be largely if not entirely unfossiliferous. Abundance of life appears in the iceberg drift zones.

THICK-SHELLED FAUNAS AND THIN-SHELLED FAUNAS

Many visitors to the Tasmanian Permian sections have been puzzled by the abundance of glacial erratics in limestones that are full of very large, thick-shelled fossils. The common view has been that thick-shelled fossils suggest the tropics. Thus Murphy (1928) and Kirk (1928) state that organisms secreting large amounts of lime are most abundant in the tropical seas and at a minimum in cold waters.

What then is the explanation of the Tasmanian Permian where glacial erratics are associated with fossils that have calcite shells, quite exceptionally thick by present standards or in comparison with fossil faunas throughout time? We suggest that this may be another consequence of the subzero, dry-base glacier.

A wet-base glacier causes dilution of the sea by large volumes of fresh water. A dry-base glacier, by contrast, causes saturation or actual precipitation of lime along bottom zones in the path of outward salinity currents. This may affect shell thickness in three ways: (a) the physiology of some organisms may be such that they precipitate more $CaCO_3$ into their shells in waters in which these ions are most concentrated; (b) any organisms selectively adapted to high $CaCo_3$ may multiply at the expense of others which can dominate them in less saline waters; and (c) the lime saturation of the bottom water inhibits resolution of shells and produces a limestone.

We should not therefore be surprised to find erratic-bearing coquina limestones with abundant thick shells sited off the embouchure of a dry-base glacier during a time of glacial retreat. The Darlington and Berriedale limestones of the Tasmanian Permian are examples. Here we find riotous multiplication of large *Eurydesma*, with shells approaching an inch in thickness, large pectens, gastropods, thick spiri-fers, and *Martiniopsis*. The bryozoans (such as *Stenopora* and the fenestellidae) show prodigious development with very calcareous colonies. It is perhaps signifi-cant that all these are benthonic forms, for the high salinity should be essentially confined to salinity currents flowing outwards along the sea floor. This being so we should not be surprised if the calcareous facies changed rapidly into a silty facies along the boundary of the salinity current. Brill's mapping (1954) of the isopachs of the Berriedale limestone suggests the embouchure of a dry glacier to the south-west of Hobart and a salinity flow northeastward. There is a sharp facies change from limestone to siltstone, and the rocks show a conspicuous limestone to cal-careous siltstone rhythm (Brill, 1954). This is not unexpected, if our interpretation is correct, because the amount of brine produced depends on the area of the ice-sheet that becomes buoyant, with the result that the bottom salinity current would necessarily be very sensitive to any climatic pulse superimposed on a general retreat.

Thick shells have also commonly been interpreted as indicative of shallow water and strong waves. But we question the validity of this assumption. The characteris-tic environment off a subzero glacier has water several hundred feet deep if not deeper. The rather rare association of high $CaCO_3$ content with fairly deep water may be the reason for the special developments of the large calcareous bryozoans characteristic of these beds.

The dry-base glacier also produces an environment that might well be favourable for petroleum source rocks — abundant life, high salinity, and a likelihood of barred basins. Perhaps by no mere coincidence the Berriedale limestone has a marked foetidity and a bituminous odour.

Should barred basins occur in this zone, they are likely to become filled with highly concentrated brines, especially during a time of retreating glaciation. These brines being derived from the freezing of large volumes of sea water as an ice-sheet previously grounded becomes buoyant, owing to reduction in thickness and rise of sea-level, and exposes large areas of subzero glacier and of subzero sea bottom (see previous discussion). Apart from producing saline precipitates, any such brine pools lying on the bottom beyond the ice barrier might produce freak faunas if any organism succeeded in adapting itself to the high salinity. The tasmanite beds of the Permian of northern Tasmania could perhaps be an example of this kind.

In northern Tasmania basal tillites are followed by a short interval of non-fossiliferous beds with a size distribution close to tillite but containing erratics showing evidence of having been dropped through water, next by a considerable thickness of dark, poorly bedded and poorly fossiliferous mudstones with dropped erratics. Within this group occur a number of lenses of tasmanite, a variety of oil shale made up almost entirely of small yellow bodies, which have been called *Tasmanites punctatus*. The systematic position of these organisms is not known, although they have been examined in detail by several specialists. Schopf (1957, pp. 712-13), in an annotated review of the literature, states that "one may agree that these spore, cyst or egglike forms could be placed with almost equal justification in either the animal or plant kingdoms. The truth is that they are still extremely problematical. . . ." He recommends that if they be regarded as plants they be classified as *Tasmanites* plantae *incertae sedis* (Norton, 1875) but if they be regarded as animals they should be referred to as *Leiosphaera* following Eiseneck (1938) who first proposed a name within the animal kingdom. Schopf's review makes it clear that all authors who have studied these organisms agree on the generic identity of the North American Devonian forms and the Tasmanian forms, and this identity probably extends to Brazilian Devonian forms (Sommer, 1951, 1953). Schopf points out that Ransch (1884) and Singh (1932) both report the presence of some trilete (land plant) spores interspersed with the *Tasmanites*. The following facts concerning tasmanites are established.

(1) Banks (1956) has cleared up the stratigraphic anomalies concerning the occurrence and has established that tasmanite occurs within the Quamby group, a marine sequence of poorly fossiliferous dark siltstones containing sporadic glacial erratics. Earlier suggestions that the tasmanite might be a marine variant of the Mersey coal measures have been shown by Banks to be based on erroneous correlation.

(2) The environment of deposition of the tasmanite itself was marine. There are marine fossils enclosed in the tasmanite (Twelvetrees, 1911; Voisey *et al.,* 1938).

(3) These beds occurred during a time of general amelioration of climate but the sea in which the tasmanite formed contained numerous floating icebergs. The tasmanite itself contains occasional erratics.

(4) The glaciers came from lands to the west and south, but the sea floor also rose to the north and east of the tasmanite area as proved by overlap of progressively younger Permian sediments on to the basement.

(5) The depth of deposition was below wave-base, and deep enough to float drifting icebergs without grounding. If the hypothesis herein advanced is to stand, the depth of water must have been more than sufficient to float the waning ice-sheet at its lifting inner margin and allow a down gradient for the brines, that is, a depth much less that 1000 feet would be improbable. The depth is not likely to have been much greater than this because of the overlap relations of the rising basement mentioned above.

(6) The sediments associated with the tasmanite contain glendonites (pseudomorphs after glauberite), calcareous concretions, and pyrite.

(7) The *Tasmanites* organisms multiplied to prodigious numbers so that their bodies formed many million tons of rock. (The possible reserves still remaining have been estimated at over thirty million tons.)

A possible hypothesis is that heavy brines formed by freezing of the glacier flowed outwards along the bottom until they met the rising basement floor. Here the brines formed pools, far too saline to allow any normal life. However, the *Tasmanites* adapted itself to these conditions and multiplied prodigiously, with few predators, and its remains accumulated as tasmanite.

Contrary perhaps to this suggestion is the occurrence of organisms identified with *Tasmanites* in the Silurian and Devonian of North America and in the Devonian of Brazil. These may not in fact be the same organism, but specialists who have examined them have found no significant difference. Accepting their identity, the Brazilian Devonian could perhaps have glacial affinities, but such phenomena can scarcely be suggested for North America. A fact common to all the occurrences is the barred basin pyrite "black shale" environment. There is some suggestion of coeval salinity of a tropical type, so perhaps saline bottom flow into barred basins might produce a similar environment. However, we realize that we have stretched our speculation far beyond our facts, but have done so in the hope that a common factor in the occurrence of *Tasmanites*, if it exists, might be brought to light.

STRATIFICATION AND FABRIC IN THE TERRESTRIAL ZONE
OF A SEA-GOING WET GLACIER

We are concerned only with subglacial till and possibly some contained melt-water channels carrying water-worn and water-sorted melt-water sediments. The till retains the characteristic flow fabric of the glacier (Holmes, 1941; Glen, Donner, and West, 1957; Seifert, 1954).

In the grounded shelf zone of the wet glacier we again meet unstratified till. However, intermittent lifting of the shelf bottom during a retreat sequence is very likely to cause interstratification of beds deposited under the floating shelf zone. We may find evidence of dragging, rolling, and slumping both of till and of these interbedded sediments, and foreset beds of cross-bedded "till" may occur.

In the floating shelf zone of a wet glacier, stratification should normally be well developed. Stratification is caused by four independent processes taking part in the transport and distribution of sediment in this zone.

(1) Subglacial melt-water carries clay and silt-grade material and forms turbidity currents down the foreset slope, whence more sediment is winnowed and carried far afield under the floating shelf and beyond as bedded sands, silts, and muds.

(2) Till mud flows from the unstable foreset slopes would commonly erode broad channels in the finer sediments. Such flows also would be capable of spreading far afield, although their character would change progressively with distance. Near their source they could easily be mistaken for till although their interbedding with siltstones in quite thin beds, as is common in the Tasmanian Permian, is scarcely compatible with a true till. Such flows may possess true flow fabric with the long axes of the pebbles oriented in the direction of flow. However, the transverse maximum characteristically developed in a long glacier (see Glen, Donner, and West, 1957) should not appear in such a mud flow. With increasing distance from the source the mechanical analysis becomes progressively narrower in size-grade spread, and less till-like.

(3) Erratics are dropped from the still melting ice-shelf above. In a wet glacier, as we have seen, this is not the major source of sediment in the floating shelf zone. The resulting product is bedded siltstones and sandstones with occasional erratics.

(4) Stratification is also caused by ocean bottom currents other than turbidity currents. These are likely to be quite strong and to be responsible for a good deal of redistribution and winnowing of sediments. Poulter (1947) reported a three-knot current under the floating shelf on the Ross Barrier.

The second source of stratification in the floating shelf zone is climatic change. Even the annual rhythm might be well developed. The pattern and strength of bottom currents could be very different between summer and winter, for the annual freezing of pack ice (perhaps for hundreds of miles from the edge of the floating barrier) not only alters the thermal distribution, but also protects large areas from wind drive. The melt-water contribution might also be substantially affected, for, although the rate of melting at the base would show no seasonal variation, nevertheless the surface precipitation of snow on the grounded ice-sheet would vary the pressure of the sheet on the sediments below and cause variation both in the rate of spreading of the sheet and in the hydraulic gradient in the sediments, the latter governing the seepage pressure of the melt-water. Under favourable circumstances therefore, well-developed, varved silts might appear in the floating ice-shelf zone. These might be difficult to distinguish from subaerial terrestrial varves. However, the presence of products from the foreset slope, such as thin beds of mud flow "tillite" with signs of channel erosion at their base, turbidity flow horizons, and even rolled or balled concretionary masses, should help to identify marine varves below the floating shelf. Longer climatic cycles should be clearly recorded as stratification in this environment.

The sediments below the floating shelf of a dry glacier should be rather different. There are no tillitic mud flows, and no melt-water turbidity currents. The main source of sediment is the melting of the shelf above, which in this case is rich in erratics as well as in rock flour. The principal product should be pebbly mudstones. Stratification should be much poorer than in the case of the wet glacier. There should be substantial changes in the strength of bottom currents, and hence in the degree of winnowing and redistribution. The longer climatic cycles would change the thickness of the glacier and its bottom temperature. The latter would modify the amount of sea ice frozen to the base of the glacier and vary the area that would receive the melt products from the most heavily-loaded part of the glacier. It would vary the strength of the bottom salinity currents. However, the long-range rhythms, although present, should not be revealed as prominently as by the sediments in the floating shelf zone of the wet-base glacier. The fabric of these pebbly mudstones should not be expected to have any relation to the flow of the glacier, unless statistical orientation of pebbles in the floating shelf could result in such orientation after having dropped through the water, an occurrence which we do not consider likely.

REFERENCES

AHMAD, N., BARTLETT, H. A., and GREEN, D. H.   1959.   The glaciation of the King Valley, western Tasmania; Roy. Soc. Tasm., vol. 93, pp. 11-16.
BAKER, W. E., and AHMAD, N.   1959.   Re-examination of the fjord theory of Port Davey, Tasmania; Roy. Soc. Tasm., vol. 93, pp. 113-15.
BANKS, M. R., HALE, G. E., and YAXLEY, M. L.   1955.   The Permian rocks of Woody Island, Tasmania; Roy. Soc. Tasm., vol. 89, pp. 219-29.
BRILL, K. G.   1956.   Cyclic sedimentation of the Permian system of Tasmania; Roy. Soc. Tasm., vol. 90, pp. 131-40.
CAREY, S. W.   1938.   The Carboniferous sequence in the Werrie Basin; Proc. Linn. Soc. N.S.W., LXII, parts 5-6.
——   1954.   The Rheid concept in geotectonics; J. Geol. Soc. Aust., vol. 1, no. 1, pp. 67-117.
CHARLESWORTH, J. K.   1957.   The Quaternary era, with special reference to its glaciation; Arnold.
DAVID, T. W. E., TAYLOR, T. G., WOOLNOUGH, W. G., and FOXALL, N. G.   1905.   Occurrence of the pseudomorph glendonite in New South Wales, with notes on the microscopic and crystallographic characters; Rec. Geol. Surv. N.S.W., vol. VIII, pp. 161-79.

FLINT, R. F.   1947.   Glacial geology and the Pleistocene epoch; Wiley.
GLEN, J. W.   1952.   Experiments on the deformation of ice; J. Glaciology, vol. 2, no. 12, pp. 111-14.
————   1954.   The creep of polycrystalline ice; Proc. Roy. Soc., Ser. A., vol. 228, no. 1175, pp. 519-38.
GLEN, J. W., DENNER, J. J., and WEST, R. G.   1957.   On the mechanism by which stones in till become orientated; Amer. J. Sci., vol. 255, no. 3, pp. 194-205.
HARRISON, P. W.   1957.   A clay-till fabric: Its character and origin; J. Geol., vol. 65, no. 3, pp. 275-308.
HOLTZSCHERER, J. J.   1934.   In HOLTZSCHERER, J. J., and ROBIN, G. DE Q., Depth of polar ice caps; Geog. J., vol. CXX, pp. 193-202.
JENNINGS, J. N., and AHMAD, N.   1957.   The legacy of an ice cap: The lakes of the western part of the central plateau cf Tasmania; Aust. Geographer, vol. VII, no. 2.
KIRK, E.   1928.   Fossil marine faunas as indicators of climatic conditions; Ann. Rep. Smithsonian Inst., 1927, pp. 299-307.
LONGWELL, C., and FLINT, R. F.   1955.   Introduction to physical geology; Wiley.
MENARD, H. W.   1933.   Pleistocene and Recent sediment from the floor of the north-eastern Pacific Ocean; Bull. Geol. Soc. Amer., vol. 64, no. 11, pp. 1279-94.
MILLER, D. J.   1953.   Late Cenozoic marine glacial sediments and marine terraces of Middleton Island, Alaska; J. Geol., vol. 61, no. 1, pp. 17-40.
MURPHY, R. C.   1928.   Antarctic zoogeography and some of its problems; Amer. Geog. Soc., Spec. Pub., no. 7, pp. 355-79.
NELSON, K. H., and THOMPSON, T. G.   1954.   Deposition of salts from sea water by frigid concentration; J. Marine Research, vol. 13, pp. 166-82.
NYE, J. F.   1951.   The flow cf glaciers and ice-sheets as a problem in plasticity; Proc. Roy. Soc., Ser. A., no. 207, pp. 554-72.
————   1952.   A method of calculating the thickness of ice-sheets; Nature, vol. 169, p. 529.
POULTER, T. C.   1947.   Seismic measurements on the Ross Shelf ice; Trans. Amer. Geophys. Un., vol. 28, pt. I, pp. 162-70; pt. II, pp. 367-84.
RAGGAT, H. G.   1937.   On the occurrence of glendonites in N.S.W., with notes on their mode of origin; Proc. Roy. Soc. N.S.W., vol. 71, pp. 336-49.
ROBIN, G. DE Q.   1953.   Measurements of ice thickness in Dronning Maud Land, Antartica; Nature, vol. 171, no. 4341, pp. 55-8.
————   1954.   in HOLTZSCHERER, J. J., and ROBIN, G. DE Q., Depth of Polar ice caps; Geog. J., vol. CXX, pp. 193-202.
SWITHINBANK, C.   1955.   Ice shelves; G. J., vol. CXXI, pt. 1, pp. 64-76.
TERZAGHI, K., and PECK, R. B.   1948.   Soil mechanics in engineering practice; Wiley.
THOMPSON, T. G., and NELSON, K. H.   1956.   Concentration of brines and deposition of salts from sea water under frigid conditions; Amer. J. Sci., vol. 254, no. 4, pp. 227-38.
WADE, F. A.   1943.   The physical aspects of the Ross Shelf ice; Proc. Amer. Phil. Soc., vol. 89, pp. 160-73.

Pleistocene Physical and Biologic Environments of Pacific Southcentral and Southwestern Alaska[1]

THOR N. V. KARLSTROM

GLACIAL DEPOSITS of Pacific coastal Alaska record five major Pleistocene glaciations. Each largely covered the mountainous coast and followed an interglacial in which alpine ice was probably at least as contracted as today. Morainal distribution indicates that the Pleistocene glaciers repeatedly (1) fed from the same higher alpine areas which today contain the most glaciers, and (2) were systematically larger near the coast and on Pacific sides than on interior sides of the coastal ranges. These relations conform with present climatic zonation orographically produced by predominant precipitation supplies from the Pacific, and reveal no significant differential uplifts between alpine areas, nor profound regional atmospheric circulation changes since early Pleistocene.

During each glaciation, the extensive cover largely destroyed existing vegetation and created severe barriers to biota migration into or across the region. Repopulation by displaced or new forms necessarily occurred during interglacials. Two main ice-surrounded, but unglaciated, mountain areas bordering proglacial lakes in upper Cook inlet and southwest Kodiak Island persisted as significant biota refuges during the last two glaciations. Such coastal refuges have been previously inferred, but imprecisely located, from distribution studies of regional biota. Because of insular location between the mainland and the Aleutian Islands, detailed ecological studies of the Kodiak refuge should prove particularly worthwhile.

[1]Abstract only.

Sequential Development of Surface Morphology on Fletcher's Ice Island, T-3[1]

DAVID D. SMITH

ABSTRACT

Fletcher's Ice Island is composed primarily of iced firn containing numerous elongate, lensoid bodies of old lake ice formed during the build-up of the parent Ellesmere ice-shelf. The parallelism of these bodies imparts a pronounced structural grain to the island. Overlying the two bedrock ice types are modern lake ice pads formed at the close of an unusually short, cool summer within the last five years.

Ablation causes a progressive, though non-cyclic, sequential development of surface forms characterized by repeated inversion of relief. Ice units having higher albedo values and more favourable crystal structure are etched into prominence by differential solar weathering.

The surface of the island has a gently undulating topography made up of numerous broad, parallel ridges and intervening narrow but shallow valleys; together, these forms comprise the first-order relief. Second-order relief features are present on the first-order ridges. These consist of (1) long narrow ridges underlain by bodies of resistant old lake ice, (2) inter-ridge valleys and lowlands underlain by non-resistant iced firn, and (3) mesa-like forms which develop on the modern lake ice pads. A variety of third-order or microrelief forms is also present.

First-order ridges develop in areas where resistant second-order ridges are concentrated; first-order valleys form in intervening areas where second-order ridges are more widely spaced. Second- and third-order relief is the direct topographic expression of differential ice resistance; first-order relief is the indirect topographic expression of this differential resistance. Geomorphically, all three orders of relief are a product of structural control.

FLETCHER'S ICE ISLAND, T-3, is an 18-km long floating mass of freshwater ice which is currently adrift in the Arctic Ocean. Its drift course between the North Pole and the Beaufort Sea has described a crudely elliptical orbit, the long axis of which is oriented about N 50°-60° E. The island's movement along this course is in a clockwise direction. Since discovery in 1947 near Ellef Ringnes Island at about 80° N, 104° W, T-3 has made one complete transit of the orbit and is currently underway on a second.

First occupied by a United States Air Force party in March 1952, the island has been studied by scientific parties of various types intermittently ever since (Bushnell, 1959, pp. 1-2). In the early spring of 1957, International Geophysical Year Drifting Station Bravo was established on T-3 and this camp has been occupied continuously to date. As part of the IGY Arctic Basin programme, the writer spent three-and-one-half months on the island during the summer season of 1958 conducting a surface morphology investigation as part of a research contract between Dartmouth College and the Terrestrial Sciences Laboratory, Geophysics Research Directorate, Air Force Cambridge Research Center.

[1]Based on research sponsored by the Geophysics Research Directorate of the Air Force Cambridge Research Center, Air Research and Development Command, under Contract AF19(604)-2159 with the Department of Geology, Dartmouth College.

After the discovery of Ice Island T-1 by the United States Air Force in 1946, an extensive search revealed the existence of several other large ice islands (T-3 among these) and many smaller ones. About sixty ice islands are now known (Koenig and others, 1952; and Montgomery, 1952), most of which occur in or near the Canadian Arctic archipelago. The origin of these islands was the subject of rather widespread speculation until a comprehensive paper by Koenig and others (1952) indicated that their most probable source was along the northwest coast of Ellesmere Island. In that paper, Hattersley-Smith proposed (pp. 95-100) that the islands result from fragmentation of the Ellesmere ice-shelf.

Field work by Crary (1952, 1958) and Marshall on Ice Island T-3, and by Hattersley-Smith (1955, 1957a, 1957b), Crary (1958), and Marshall (1955) on the Ellesmere ice-shelf has confirmed the shelf as the source of T-3 and the other ice islands on the basis of similarity in ice types, ice structures, and surface characteristics.

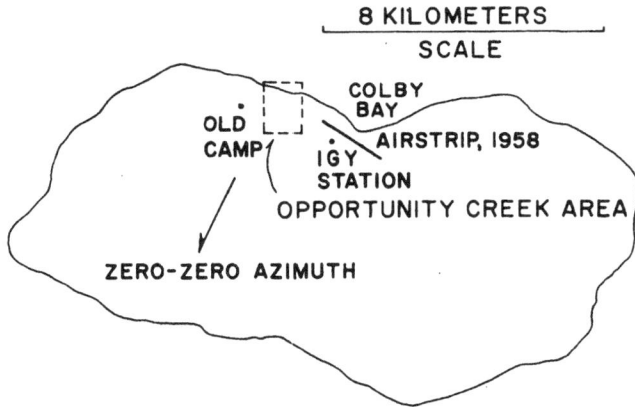

FIGURE 1.   Outline map of Fletcher's Ice Island, T-3.

Although more than twenty-five papers have been published as the result of scientific work on Ice Island T-3 by various parties, only four of these (Crary, 1958; Crary et al., 1952 and 1955; and Stoiber et al., 1956) deal primarily with geologic problems of the ice island proper. The surface of the island is described briefly in Crary (1958, pp. 7-8) and Crary et al. (1952, p. 212) but the problem of the origin of surface features is not considered. The origin of similar features on the parent Ellesmere ice-shelf is discussed by Hattersley-Smith (1957a, and in Koenig et al., 1952, pp. 100-1) and by Zubov (1955, pp. 5-6). In addition Marshall's description (1955) of the structure and stratigraphy of the shelf is fundamental to the T-3 problem.

The results described in this paper are based on detailed mapping in the Opportunity Creek area (Figure 1) and a number of reconnaissance traverses elsewhere on the island. Other aspects of the writer's work on T-3 are presented in Smith (1960a and 1960b).

METEOROLOGIC SUMMARY

In comparison with records of preceding summers, the 1958 season on T-3, and apparently throughout much of the Polar Basin, was unusually warm and long. The eighty-seven-day melt season began June 8 and ended September 2; fifteen days during this period, however, had an average daily temperature below freezing. The effective ablation season was thus roughly seventy-two days.

As expected, the pattern of daily high and low temperatures was closely related to elevation of the sun; warmest temperatures generally occurred in the six- to eight-hour period bracketing solar noon. Superposed on this cycle was the modifying effect of cloud cover, ground fog, and wind. A brief summary of meteorological data for the 1958 melt season is presented in Table I.

TABLE I

SUMMARY OF METEOROLOGIC DATA FOR 1958 MELT SEASON*

	June 8 through 30	July	Aug. 1 through Sept. 2	Season
Air temperature: (in degrees Farenheit)				
High	43	40	40	43
Low	22	25	24	22
Average	33.2	32.4	32.1	32.5
Cloud cover:				
Clear†	4 days, 17.4 per cent	8 days, 25.8 per cent	3 days, 9.1 per cent	15 days, 17.4 per cent
Partly cloudy†	5 days, 21.8 per cent	5 days, 16.1 per cent	8 days, 24.5 per cent	18 days, 20.6 per cent
Cloudy†	14 days, 60.8 per cent	18 days, 58.1 per cent	22 days, 66.6 per cent	54 days, 62 per cent
Ground fog:	8 days, 35 per cent	7 days, 22.6 per cent	25 days, 76 per cent	40 days, 46 per cent
Percentage of total possible sunshine	43 per cent	55 per cent	27 per cent	41.6 per cent
Average wind velocity	12.5 kph	15.5 kph	12.9 kph	13.5 kph

*Based on data from United States Weather Bureau station on T-3.

†Clear: 0–3/10's cloud cover; partly cloudy: 4–7/10's cloud cover; cloudy: 8–10/10's cloud cover.

DESCRIPTION OF THE ISLAND

Ice Island T-3 (Figure 2) is roughly oval to kidney-shaped, has a length of 18 km, a width varying from 6½ to 9½ km, and according to Crary (1958, p. 12) a thickness of about 60-65 m. It is composed of freshwater ice of several types; each has a characteristic lithology, structure, and surface expression. In air views, the parallel pattern of melt-water lakes and low, intervening ice ridges is the most striking feature of the gently undulating surface. Over much of its area the island rises 4.5-6 m above the surface of the adjacent pack ice. Along part of its perimeter a cliff-like scarp is present; elsewhere the island surface slopes very gradually to the level of the pack.

Because a detailed description of the 1958 snow cover, snow melt, run-off, and ice ablation is presented in Smith (1960a), only a brief summary is given here. The initial snow cover, which ranged from 15 cm to 1.7 m in depth and which virtually masked the undulating topography of the ice surface, began thawing on June 8 and was completely melted or converted to slush by July 2 when run-off began along the edges of the island as a sluggish, over-bank stage slush flow. Normal stream run-off developed five days later and was characterized by an initial peak discharge which diminished rapidly to a relatively steady summer low-water stage that prevailed for the rest of the melt season except for scattered periods of higher discharge correlated with high temperatures and minimum cloud cover. Although the amount of ice lost during the melt season varies with the ice type (as presented later), an average value for total ice loss for the island as a whole during the 1958 melt season is roughly .75 to 1 m.

Records of the amount of ice lost in previous melt seasons indicate that the surface of T-3 generally has a net annual loss except for occasional short, cool summers when little or no ice is ablated. Even if an unusually cool summer results in accretion at the surface, the long term net surface budget seems clearly negative.

Inasmuch as the island rotates slowly as it drifts, a previously-established zero-zero reference azimuth (Figure 1) was used as north rather than the true geographic direction. Accordingly, all compass points and directional terms used in description of the island in this paper refer to this azimuth.

ICE LITHOLOGY AND STRUCTURE

The ice types exposed at the surface of Ice Island T-3 (Smith, 1960a) consist primarily of two bedrock units, iced firn and old lake ice, overlying which is a modern lake ice unit formed within the last five years. The iced firn, interstratified with linear bodies of the old lake ice, makes up at least the upper half of the

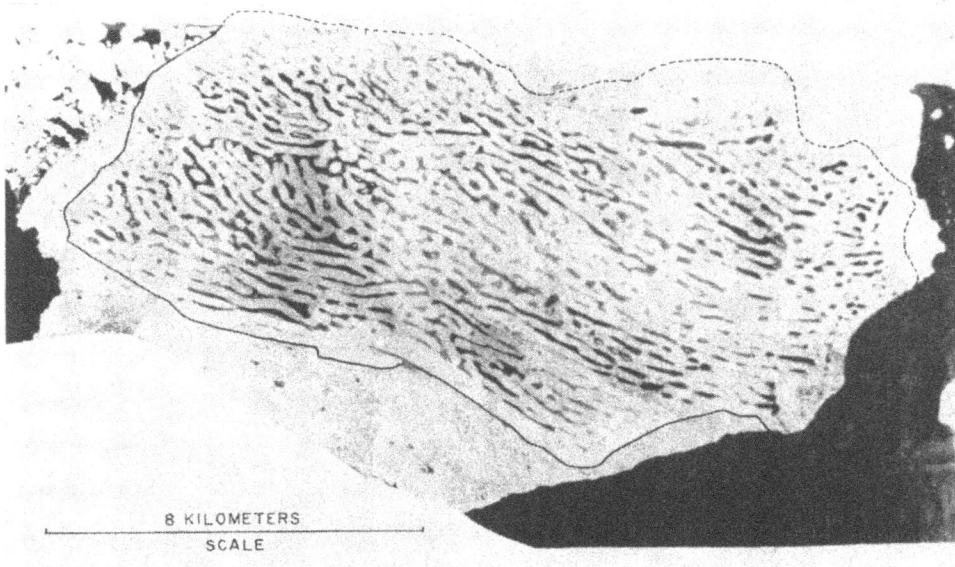

FIGURE 2.   Photo mosaic of Ice Island T-3 prepared by United States Air Force from photography flown in late July 1952 at 5500 metres. The linear black bands on the island are meltwater lakes which occupy first-order valleys; the intervening white areas are the first-order ridges. The parallelism of the latter is the expression of the parallel structural grain of the island's bedrock ice types.

island (Weeks *in* Crary, 1958, p. 35). The modern lake ice is analogous to surficial deposits overlying the bedrock. A few other ice types of limited areal extent, such as the banded ice shown in Figure 6, are present but have no apparent bearing on the problem at hand. A brief summary of the three units follows.

The iced firn consists of a crudely stratified mass of subangular to subspherical ice grains ranging from 0.5 to 1.5 cm in diameter (Figure 3). It contains thin dirt layers 15 to 20 cm apart which Marshall (1955, p. 110) interprets as annual increments of windblown sediment swept onto the shelf in the course of shelf build-up during summers when the adjacent land mass was snow free. This dirt is now being concentrated at the surface as a lag deposit which accelerates ablation of the iced firn.

The old lake ice units interbedded with the iced firn are narrow, elongate, parallel to subparallel masses 1.5-5 m thick, up to tens of metres wide, and hundreds of metres long, and typically lensoid in cross-section (Figure 4). They are composed of ice crystals 15-30 cm in diameter and 30-45 cm long. No dirt bands are present with the result that the surface of the mass is dirt free.

The modern lake ice occurs as thin but extensive pads overlying the bedrock ice types. The pads are 1-2 m thick, generally elongate in plan, and have a long dimension of as much as several hundred metres. They are composed of two or more layers of closely packed, vertically oriented, candle-like crystals 2.5-5 cm in diameter and 10-40 cm long (Figure 5), and are generally dirt free. The pads apparently formed at the end of the 1953 or 1955 melt season when the melt-

FIGURE 3. Weathered surface exposure and core of iced firn. The granular texture of the firn is clearly shown in the core which has been exposed to the sun. Note the "crystal gravel" and concentration of dirt which are typical of iced firn surfaces. The rule is 15 cm long.

water lakes, which developed on the island as snow melt progressed, did not drain but rather froze in place because of the short, cool character of the summer.

The parallelism of the elongate bodies of old lake ice (Figure 6; and Smith, 1960a, Figure 6) over large parts of the island results in a pronounced east-west structural grain in the bedrock ice. The parallelism of the melt-water lakes (and the bodies of modern lake ice) is a response to this structural control (Smith, 1960b).

SURFACE MORPHOLOGY

The surface of T-3 and of the parent Ellesmere ice-shelf is characterized by a very gently undulating topography made up of numerous broad, parallel to sub-parallel ridges alternating with shallow valleys. During the melt season, the valleys are occupied by elongate melt-water lakes and drainage channels (Figure 2). The parallel pattern of these lakes and ridges is a striking feature in any air view of the island. On the ground, however, the ridges are scarcely detectable.

The margins of most of the island have gradual slopes; in the northeast quadrant, however, the edge of the island is a sharply defined scarp 3-7.5 m high (Figure 7). Pack ice has been driven aboard the island along much of its perimeter, locally forming hummocks of considerable size (Figure 8). The northeastern edge of the island, in particular, is lined by an almost continuous ridge of hummocked pack ice 4.5-14 m high. Along the southern edge of the island where extensive areas of old sea ice appear to be firmly attached to the island mass, almost no hummocks are present.

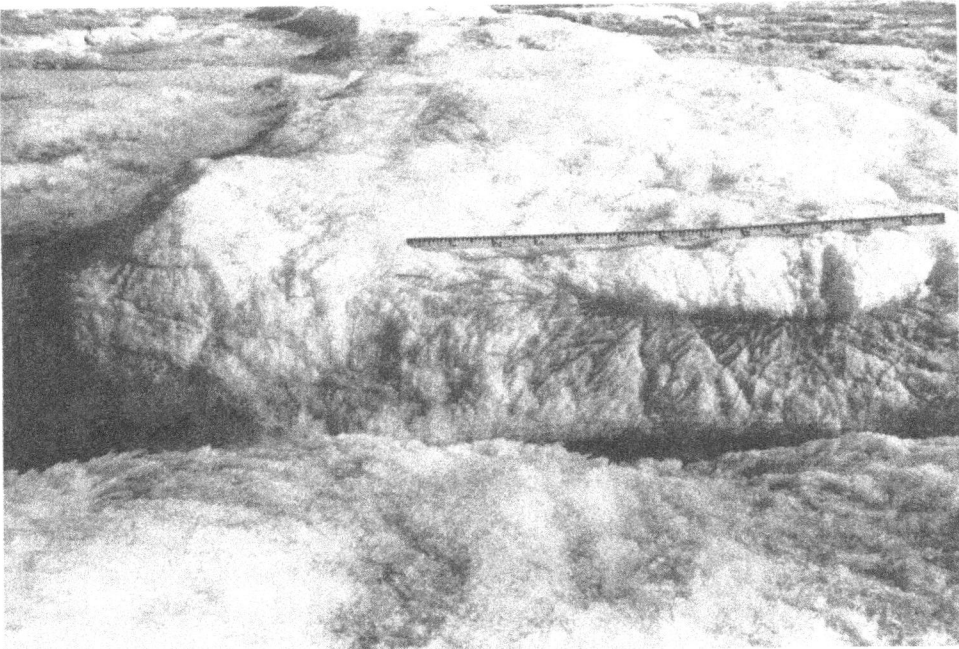

FIGURE 4. Cross-section of narrow, lensoid body of old lake ice exposed in stream channel. Note interfingering of left edge of lense and enclosing iced firn. Pseudo-synclinal jointing is well developed. Stadia rod is 4 m long.

Three Orders of Relief

Relief features on T-3 are readily divisible into three orders of magnitude. First-order relief forms consist of the broad, parallel ridges and gentle but narrow, melt-water-filled valleys so prominent in air views of the island; as will be developed shortly, these forms are the indirect topographic expression of different ice lithologies. Second-order relief forms are the direct topographic expression of individual ice types and consist of narrow ridges underlain by bodies of old lake ice, relatively broad inter-ridge valleys or lowlands underlain by iced firn, and the mesa-like features underlain by pads of modern lake ice. Third-order or microrelief features are widely varied in character and generally are the expression of small ice bodies formed in ponds and channels at the close of the preceding melt season. In a geomorphic sense, all three orders of relief are the result of structural control.

First-order relief. The gently undulating topography of the surface of the island and of the Ellesmere ice-shelf (described as "rolls" by Hattersley-Smith, 1957a, pp. 32-3; and as "corrugations" by Zubov, 1955, p. 5) is made up of an alternating series of broad, low ice ridges and relatively narrow, shallow valleys (the "troughs" of Hattersley-Smith, 1957a, p. 33). These features together comprise the first-order relief forms.

Although locally near the east and west edges of the island, the ridge and valley topography is prominent (Figure 9), it should be noted that for most of the island the difference in elevation between adjacent ridges and troughs is so slight, and as a result the undulating topography so subdued, that in the writer's opinion the terms "ridge" and "trough" are rather misleading and might better be replaced.

FIGURE 5.   Cross-section of modern lake ice pad in channel wall exposing two layers of candles.

On the other hand, because their usage in this sense is well established, no other terminology is proposed; however, "valley" is used interchangeably with "trough" in this paper.

Except around the periphery of the island, almost all first-order valleys are occupied by melt-water lakes during the ablation season. As apparent in Figure 2 the ridges separating the valleys are generally parallel but locally merge, thus either pinching out the intervening low and the lake occupying it or so reducing the width that only a narrow drainage channel remains.

First-order ridge systems on T-3 (Figure 2) in general trend parallel or sub-parallel to the long axis of the island. Distance between ridge crests averages about 300 m. Difference in elevation between ridge crests and the floors of adjacent valleys ranges from about 1 to 3 m and averages about 1.5 m. Crary states (*in* Hattersley-Smith *et al.*, 1955, p. 27) that ridges on the parent Ellesmere ice-shelf have "an average wave length of approximately 760 feet [230 metres] . . . and . . . the difference in height between ridge and trough averaged about 7 feet [2.1 metres]." The ridges and troughs of the shelf trend parallel to the coastline and the shelf edge.

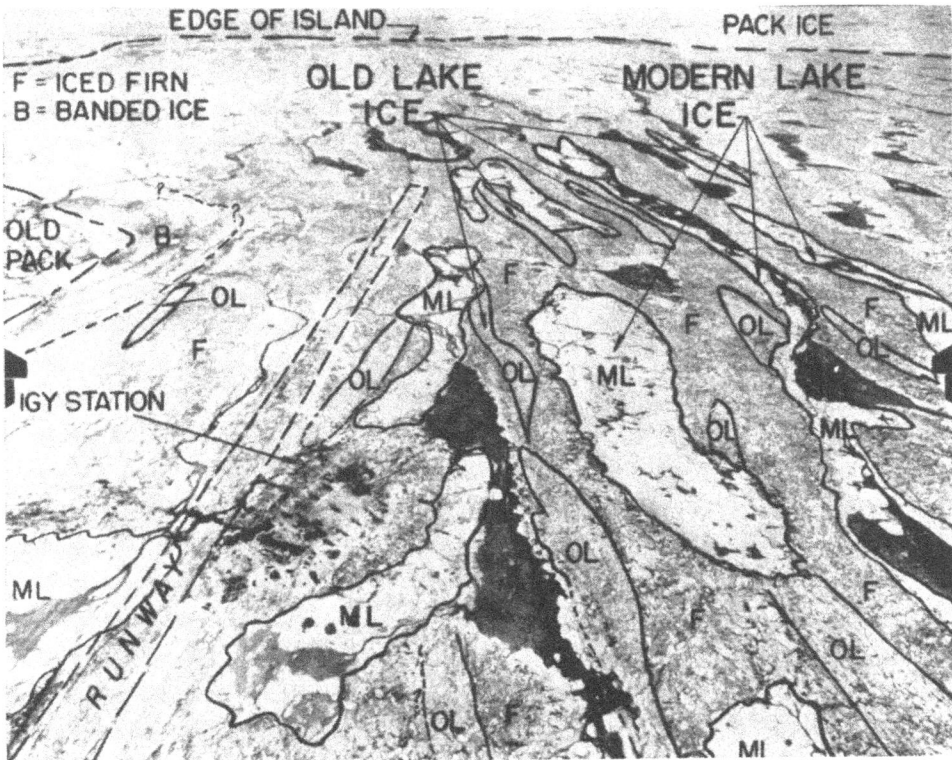

FIGURE 6.   Air view west from IGY station area. The parallelism of the old lake ice bodies (OL) is the fundamental control of the island's structural grain and thus of the parallel first-order ridges and valleys. Only the more prominent OL bodies are outlined in the area west of the camp. First-order ridges shown are atypically narrow. The runway is about 60 m wide. R.C.A.F. photo, July 30, 1958.

FIGURE 7.   The cliff along the island's edge is typical of its northeast quadrant but uncommon elsewhere. Open water adjacent to the edge was rare.

FIGURE 8.   Hummock of pack ice along the northeastern edge of the island. This mass of ice, more than 15 m high, was the largest of the many hummocks on T-3.

Detailed examination of these first-order relief forms on the ground and in large-scale air photos reveals that the ridges are actually made up of a series of relatively closely spaced, resistant, second-order ridges. First-order valleys, on the other hand, generally occur in areas where the resistant second-order ridges are more widely spaced. This relationship between first- and second-order ridges is shown in Figure 10 which is a diagrammatic transverse profile and cross-section showing two typical first-order ridges and the intervening valley. It is based on mapping and a series of profiles levelled in the Opportunity Creek area.

Second-order relief. When examined in detail, most first-order ridges are seen to have undulating to irregular surface topography composed of a series of second-order relief features (Figure 10): narrow linear ridges, relatively broad valleys or lowlands, and extensive mesa-like features. These second-order relief forms are the direct topographic expression of the major ice types which make up T-3. The second-order ridges are underlain by moderately resistant bodies of old lake ice. The adjacent inter-ridge valleys and lowlands are developed on areas underlain by more easily ablated iced firn. The mesa-like features are the expression of pads of modern lake ice which is highly resistant to ablation.

The old lake ice ridges (Figure 11) are relatively narrow, elongate forms which range from a few metres to tens of metres in width and from 275 to 1825 metres in length. They rise .6-1.5 metres above the adjacent second-order valleys and lowlands. These ridges commonly have steep sides and their upper surface is irregular to jagged. Distance between adjacent second-order ridges varies greatly. In and near the Opportunity Creek area they are typically 90-165 m apart. Where second-order old lake ice ridges are closely spaced, broad first-order ridges develop.

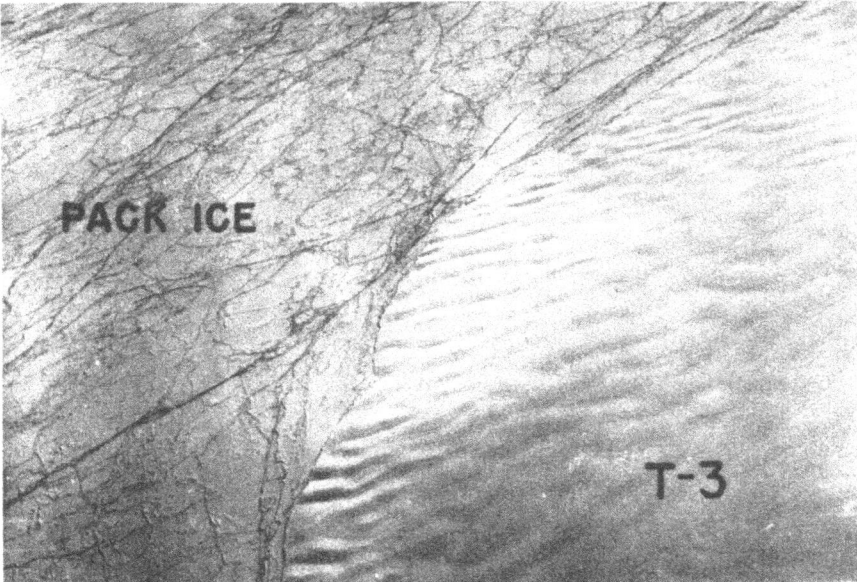

FIGURE 9.   Air view of prominent ridge and valley topography along the eastern edge of the island. The undulations were much more subdued over most of the island's surface. U.S.A.F. photograph.

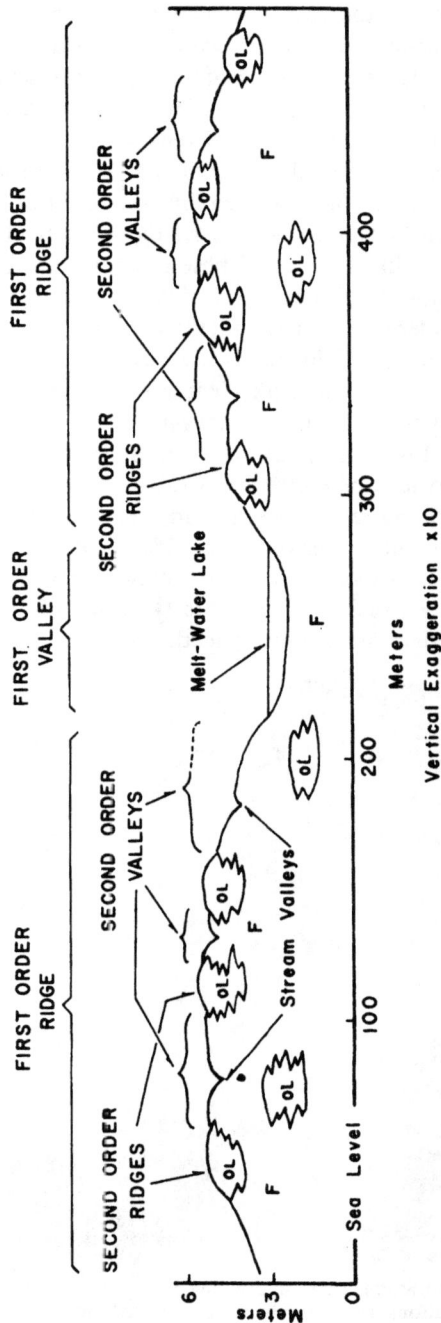

FIGURE 10. Diagrammatic transverse profile and cross-section showing typical relations of first- and second-order ridges and valleys, and underlying ice types. Note the relative concentration of resistant old lake ice bodies (expressed topographically as second-order ridges) in the first-order ridge areas and their sparseness in the first-order valleys.

Where iced firn lowlands are dominant and old lake ice ridges are more widely scattered, first-order valleys tend to develop (Figure 10).

The second-order valleys (Figure 11) develop between the old lake ice ridges on areas underlain by iced firn. They are generally elongate but at many localities their width is such that they are more aptly termed lowlands. The slopes of these valleys or lowlands are extremely gentle except near the axes of some valleys where they steepen abruptly along the edges of an incised channel in which the local drainage flows (Figure 10). Because these valleys are underlain by iced firn which disaggregates into individual pebble-like crystals as melting occurs, their surface is covered with a thin grus-like layer of "gravel" (Figure 3).

The mesa-like features (Figure 12) are underlain by pads of modern lake ice. They are elongate to irregular in plan, have a long dimension of as much as several hundred metres and have strikingly flat surfaces broken locally by slight valley-like depressions which mark the position of former drainage lines. These mesas stand 1-1.5 m above the adjacent areas of iced firn and, at the end of the 1958 ablation season, generally about .5 m above the adjacent old lake ice ridges.

The relative elevation of second-order forms with respect to each other is a function of their differential resistance to ablation, which in turn is dependent upon a number of factors such as albedo characteristics, crystal structures, and contained dirt. The relative importance of these factors will be considered shortly. Of the three second-order forms just described, the mesa-like features are underlain by the most resistant ice type and thus have the highest relative elevations.

FIGURE 11.   Narrow, elongate second-order ridge underlain by a body of old lake ice. The second-order valleys adjacent to it are developed on iced firn. This ridge is considerably narrower than similar features shown in Figure 6. The stadia rod is 4 m long.

It should be emphasized that the second-order and not the first-order forms are the striking relief features of the island as viewed by the ground observer.

Two interesting contrasts exist between first- and second-order forms. (1) First-order ridges are broad and gentle; the valleys are relatively narrow and have poorly defined margins; they are generally little more than the width of the lakes or channels which occupy them. Second-order ridges on the other hand have abrupt sides and are quite narrow with respect to the intervening valleys which are broad and gentle (Figure 10). (2) In first-order relief, valleys are virtually one and the same with the drainage channels and lake basins that occupy them, that is, except for the lake basin or the drainage channel, almost no valley floor exists (Figure 10). In second-order relief, on the other hand, the stream occupies a narrow channel which more or less follows the axis of a broad topographic low; in this case, the channel is thus a relatively small part of the valley.

Third-order relief. Third-order or microrelief forms of a variety of types are present on the island surface. Most types develop on the second-order iced firn valley or lowland areas. These features are the topographic expression of small bodies of more resistant ice which generally formed in small ponds, dirt wells, and channels at the close of the preceding melt season. They include (1) small mesas which appear to be almost scale-model replicas of the second-order mesas described above; (2) small subconical features about the size and shape of a beehive (Figure 13); and (3) narrow, elongate to sinuous flat-topped ridges held up by a protective ice-cap which formed as a stream channel-fill. These and other micro-

FIGURE 12.   Second-order mesa-like feature underlain by pad of modern lake ice. Note .75 m relief along the left edge at contact with adjacent second-order valley developed on iced firn. Ice axe is .9 m long.

relief features are the subject of a paper currently in preparation. Not only are the microrelief forms interesting in themselves, but they also serve as valuable aids to developing an understanding of the occurrence and sequence of development of the full-scale first- and second-order features.

Mode of Development

Differential weathering caused by solar radiation is the fundamental process of relief development on T-3. As ablation progresses during the summer season the response of the individual ice types varies. Lithologies which have higher albedo values, more favourable crystal structures, and little or no contained dirt are more resistant to ablation and thus weather into topographic prominence as compared with their less resistant neighbours. Thus second- and third-order relief features are the direct topographic expression of individual bodies of ice of different lithologies and dimensions. First-order relief features, on the other hand, are the indirect topographic expression of resistant ice bodies in that first-order ridges develop in areas where second-order ridges are more closely spaced. In the geomorphic sense, all three orders of relief are the product of structural control, that is, resistant ice masses.

From a consideration of the slope of the curves and the total seasonal ablation for each of the three ice types as shown in Figure 14, it is apparent that modern lake ice is the most resistant to ablation and that iced firn is the most susceptible. Although the curve for old lake ice indicates that it is roughly intermediate in

FIGURE 13. Small, subconical third-order (microrelief) feature probably formed by the inversion of a dirt well. The flat upper surface apparently marks the water level in the dirt well at the time of the 1957 freeze-up. Rule is 15 cm long.

resistance to the other two, widespread topographic evidence suggests that the curve should lie somewhat closer to that of modern lake ice.

Analysis of typical sequence. Ablation during successive seasons causes a progressive sequential development of surface forms. The sequence is characterized by a repeated inversion of relief. An analysis of this sequence follows.

For clarity of description assume a more or less level initial surface underlain by iced firn that contains lensoid bodies of old lake ice a short distance below the surface. As the initial surface melts downward, the upper parts of the old lake ice masses are exposed. Because of greater resistance, they ablate more slowly than the surrounding iced firn and thus develop into topographic highs (second-order ice-ridges). The surface now consists of a series of interspersed second-order highs and lows. As ablation progresses, areas with closely spaced, old lake ice ridges lag behind areas in which these ridges are more widely scattered. The former areas become topographic highs, the latter topographic lows. More and more melt water from the highs concentrates in the lows tending to perpetuate them. Thus first-order ridges and valleys develop.

With further melting some second-order ridges disappear as the lensoid ice masses melt away; others appear, however, as bodies deeper in the stratigraphic section are exposed. Thus a second-order ridge lasts for a given number of seasons and then is replaced by part of a second-order valley. This constitutes one form of inversion of relief.

As long as the relative areal concentration of old lake ice bodies remains the same through successively lower stratigraphic horizons, the relative positions of

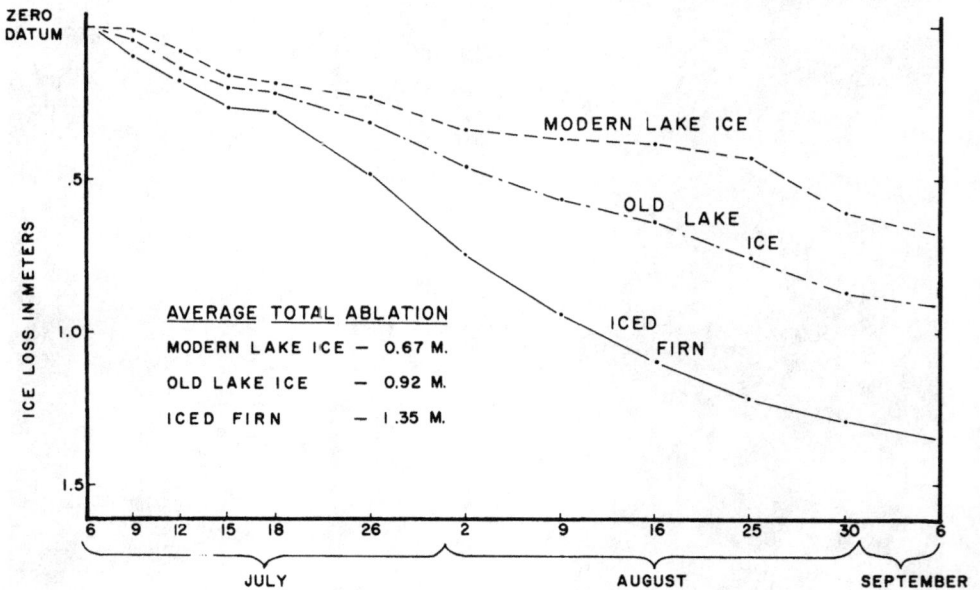

FIGURE 14.   A comparison of average ablation rates of iced firn, old lake ice, and modern lake ice in the Opportunity Creek area for the summer of 1958. Each curve is based on average readings for a series of ablation poles considered representative of the ice type indicated.

first-order forms will remain unchanged. If the areas of closely spaced, old lake ice bodies shift laterally in successively deeper horizons then the location of the first-order features will also shift. This constitutes a second-form of inversion of relief.

Superposed on the fundamental sequence is the complicating factor of wide variation in character of successive ablation seasons. Not only does this factor have the key role in determining the rate of development of the sequence, but it modifies development in the following important way. At least two summers in the last eight have been so short and cool that the melt-water lakes developed on T-3 as a result of snow melt did not drain but rather froze in place. It was during one or both of these summers that the present modern lake ice pads formed. By the end of the 1958 ablation season, they were generally the most prominent second-order topographic highs on the marginal slopes of the island. Yet these features record the former position of lake basins. Obviously, then, since their development they have reversed their topographic expression. Apparently because of their protective effect the second-order mesas tend to block or destroy the valley character of the first-order troughs by developing gradually a topographic high which in turn tends to divert drainage away from the area of the former low. If their presence is of sufficient duration these mesas may initiate a shift in position of first-order features. This constitutes the third type of inversion of relief. In general the modern lake ice mesas occur within or immediately adjacent to first-order valleys showing that there has been little shift in position of first-order features since this ice formed some three to five years earlier. At a few localities, generally near the edge of the island, however, the presence of these mesas on first-order ridges suggests that a shift in position has occurred. Thus a short cool summer can disturb the fundamental sequence by producing extensive pads of moderately long-lasting modern lake ice in first-order valleys with the resulting inversion of at least some of the lows.

Aside from the possible cyclic character of the probable recurrence of short, cool summers that produce pads of resistant modern lake ice, the outlined sequence of development is non-cyclic. The writer does not consider the repeated sequential development of old ice ridges by destruction of exposed lenses and uncovering of deeper ones as the surface progresses down the stratigraphic section to be cyclic in the geomorphic sense of the word.

Finally, it is interesting to note that streams on T-3 play a surprisingly small part in the development of the surface morphology. They seem to serve primarily as sluiceways for the removal of melt-water rather than as major erosional agents. Although ice-temperature data indicate that they serve as heat sources for the surrounding ice (Smith, 1960a) and thus accelerate melting in their immediate vicinity, the total amount of ice so removed is virtually insignificant in comparison with the over-all lowering of the island surface by solar weathering.

Life Span

The life span of a given first-, second-, or third-order relief feature is dependent upon the duration of existence of the causal ice mass which in turn is a function of a number of factors. These include (1) the colour, structure, albedo characteristics, and thickness of the given ice type as well as (2) the amount and arrangement of contained particulate matter, (3) the variation in microclimate (principally incoming radiation) related to the topography of the ice surface, and (4) rate of rotation of the island. Thorough quantitative evaluation and assignment of rank as to relative importance of these factors is not possible at this time, although work on some of these problems is now being done by Larsson (1959, pp. 26-31) and Nakaya (1959, pp. 43-4). A general qualitative consideration of these factors, however, suggests that the initial thickness, crystal structure, albedo characteristics, and lack of contained dirt are probably the most important.

In general the following life expectancies seem to be approximately the right order of magnitude. Most third-order or microrelief features apparently develop within one season although some types of abandoned channel-fills may be three or more seasons old. Second-order ridges and mesas probably require about five seasons to develop their maximum relief and to begin to decay. It seems probable that although mesas probably attain the greatest elevation, the ridges, because of greater initial thickness of the old lake ice mass, last longer. First-order ridges and valleys probably require even longer to develop maximum relief and eventually to decay. Comparison of air photos of the island taken at an interval of six years reveals little shift in position of first-order ridges in the inner three-quarters of the island. Mapping indicates that such shifts in position apparently are more common along the margins of the island, perhaps because the slope is steeper, drainage is better, and ablation seems to proceed somewhat more rapidly.

Comparison With Other Areas

If the preceding generalizations concerning the mode of development of surface morphology on Ice Island T-3 are correct, they should also obtain for the development of surface forms on the parent Ellesmere ice-shelf and on the other ice islands, providing that the net surface budget of these other areas is also negative, and that a generally similar stratigraphic horizon is currently exposed at their surface.

A personal communication from Dr. G. Hattersley-Smith indicates that the mode of development described above applies in general to the Ellesmere ice-shelf. The results of current work on the Ellesmere ice-shelf by an Air Force Cambridge Research Center party (Anderson, 1959, pp. 78-9) should provide additional data for comparison between the two areas as well as an evaluation of the applicability of the T-3 work to the shelf. As to the other ice islands, to the writer's knowledge, only one (T-1) has definitely been examined on the surface (Crary et al., 1952, p. 222), and the period of time spent there was so brief that little is known about it. It is uncertain whether the Russian drifting station NP-6 is actually an ice island (Bushnell, 1959, p. 1) as previously reported (Burkhanov, 1957, p. 9).

Summary

Relief features on Ice Island T-3 are readily divisible into three orders of magnitude. In the geomorphic sense all three orders are the result of structural control, that is, resistant ice masses. Second- and third-order forms are the direct topographic expression of ablation-resistant ice masses etched into prominence by differential solar weathering. First-order forms, on the other hand, are the indirect topographic expression of these resistant ice masses in that first-order ridges develop where second-order positive forms are relatively closely spaced.

Ablation during successive seasons causes a progressive, though non-cyclic, sequential development of surface forms. The sequence is characterized by repeated inversion of relief of at least three different types. The life span of a given feature depends upon a large number of factors; some of the more important are crystal structure, albedo characteristics, and thickness of the ice mass underlying the feature in question. In general, most third-order features develop within one melt season, second-order positive forms probably require about five seasons to attain maximum relief, and first-order forms apparently last even longer.

ACKNOWLEDGMENTS

The writer gratefully acknowledges the support and encouragement of numerous individuals at Dartmouth College, at Air Force Cambridge Research Center in Boston, and on Ice Island T-3. In particular he wishes to thank David F. Barnes of the Research Center's Geophysics Research Directorate, who suggested the problem and arranged the logistic support; and Dr. Arthur E. Collin, Fisheries Research Board of Canada, whose discussions and suggestions in the field are the basis for many of the ideas in this paper.

In addition, Dr. Geoffrey Hattersley-Smith, Defence Research Board of Canada, and Dr. Albert P. Crary, United States Antarctic Research Board, Professor John B. Lyons and Professor A. Lincoln Washburn, Dartmouth College, and Professor Wilford F. Weeks, St. Louis University, Mr. John Murray, United States Weather Bureau, and the writer's wife, Myrtle N. Smith, all offered helpful comments and suggestions. Special thanks are due Lt. Col. James J. Giles, U.S.A.F., former Commander of T-3, for his friendly co-operation.

REFERENCES

ANDERSON, D. G. 1959. Ellesmere ice shelf investigations in BUSHNELL, V., ed., Proceedings of the second annual Arctic Planning Session, October, 1959; United States Air Force, Air Research and Development Command, Air Force Cambridge Research Center, Geophysics Research Directorate, GRD Research Notes no. 29, pp. 78-86.

BURKHANOV, V. F. 1957. Achievements of Soviet geographic exploration and research in the Arctic; Defence Research Board of Canada, Trans. 253R by E. R. Hope.

BUSHNELL, V., ed. 1959. Scientific studies at Fletcher's Ice Island, T-3, 1952-1955; United States Air Force, Air Research and Development Command, Air Force Cambridge Research Center, Geophysics Research Directorate, Geophysical Research Papers no. 63, AFCRC-TR-59-232(1), Bedford, Mass.

CRARY, A. P. 1958. Arctic ice island and ice shelf studies, part I; Arctic, vol. 11, pp. 2-42.

CRARY, A. P., COTELL, R. D., and SEXTON, T. F. 1952. Preliminary report on scientific work on "Fletcher's Ice Island," T-3; Arctic, vol. 5, no. 4, pp. 211-23.

CRARY, A. P., KULP, J. L., and MARSHALL, E. W. 1955. Evidences of climatic change from ice island studies; Science, vol. 122, no. 3181, pp. 1171-3.

HATTERSLEY-SMITH, G. 1957a. The rolls on the Ellesmere ice shelf; Arctic, vol. 10, no. 1, pp. 32-44.

——— 1957b. The Ellesmere ice shelf and the ice islands; Canadian Geographer, no. 9, pp. 65-70.

HATTERSLEY-SMITH, G., et al. 1955. Northern Ellesmere Island, 1953 and 1954; Arctic, vol. 8, no. 1, pp. 2-36.

KOENIG, L. S., GREENAWAY, K. R., DUNBAR, M., and HATTERSLEY-SMITH, G. 1952. Arctic ice islands; Arctic, vol. 5, no. 2, pp. 67-103.

LARSSON, P.   1959.   Studies of micrometeorology at Fletcher's Ice Island, T-3, *in* BUSHNELL V., ed., Proceedings of the second annual Arctic Planning Session, October, 1959; United States Air Force, Air Research and Development Command, Air Force Cambridge Research Center, Geophysics Research Directorate, GRD Research Notes no. 29, pp. 26-31.

MARSHALL, E. W.   1955.   Structural and stratigraphic studies of the northern Ellesmere ice shelf; Arctic , vol. 8, no. 2, pp. 109-14.

MONTGOMERY, M. R.   1952.   Further notes on ice islands in the Canadian Arctic; Arctic, vol. 5, no. 3, p. 183-7.

NAKAYA, U.   1959.   Future plans for oceanographic and glaciological studies at T-3, *in* BUSHNELL, V., ed., Proceedings of the second annual Arctic Planning Session, October, 1959; United States Air Force, Air Research and Development Command, Air Force Cambridge Research Center, Geophysics Research Directorate, GRD Research Notes no. 29, pp. 43-4.

SMITH, D. D.   1960a.   Development of surface morphology on Fletcher's Ice Island, T-3; Scientific Report no. 4, Contract AF19(604)-2159, Dartmouth College, Hanover, N.H., and United States Air Force, Air Research and Development Command, Air Force Cambridge Research Center, Geophysics Research Directorate.

——   1960b.   Origin of parallel pattern of meltwater lakes on Fletcher's Ice Island, T-3; Proceedings, XXI International Geological Congress, Copenhagen, Denmark, Pt. XXI, pp. 51-9.

STOIBER, R. E., LYONS, J. B., ELBERTY, W. T., and McCREHAN, R. H.   1956.   The source area and age of Ice Island T-3; Dartmouth College, Hanover, N.H., and United States Air Force, Air Research and Development Command, Air Force Cambridge Research Center, Geophysics Research Directorate, AFCRC-TR-57-251.

ZUBOV, N. N.   1955.   Arctic ice islands and how they drift; Defence Research Board of Canada, Trans. 176R by E. R. Hope, pp. 1-10, Priroda, no. 2, pp. 37-45.

The Major Factors of Arctic Climate

H. P. WILSON

ABSTRACT

The climate of the Arctic Basin is described by explanation of the more significant physical, geographical, and meteorological factors involved including radiation, properties of water substance, distribution of land, sea, and ice, mountain ranges, prevailing winds, and storm tracks.

ANYONE WHO IS ASSIGNED to planning operations in the Arctic must take into account the environmental factors of weather and climate. Although there is a great deal of information available, the problem of assimilating the data is not an easy one unless the complex physical processes involved are clearly understood. The objective of this paper is to explain as simply as possible how these processes operate and how they affect the climate of the Arctic.

THE ENVIRONMENTAL FIELD

The surface of the Arctic and sub-Arctic is composed of approximately equal proportions of land, water, and ice. Because of the annual variation of the area covered by the Arctic ice pack, these proportions vary significantly with the seasons. In winter, because both land and ice are covered with snow, and also because land and ice have similar thermal properties, the land masses of Siberia, Alaska, Canada, and Greenland are effectively joined together into a vast continental area. In summer, after the covering of snow has disappeared, and after the Arctic pack has broken up somewhat, the Arctic Ocean has much the same effect on weather as any area of very cold water. Therefore, the ratio of marine area to continental area in the Arctic varies widely over the year, and this fact is of major importance with respect to the seasonal variations of weather in the Arctic.

SOLAR RADIATION

The sun transmits vast amounts of energy in the form of radiation with wavelengths between 0.15 and 4 μ[1]. The intensity of radiation at the outer limit of the atmosphere measured in heat units is roughly 2 cal cm^{-2} min^{-1}. About 10 per cent is absorbed by the atmosphere and 45 per cent is reflected back to space by clouds; the remaining 45 per cent or 0.8 cal cm^{-2} min^{-1} reaches the surface of the earth. At noon on a clear day the solar radiation reaches the surface at a rate of 1.8 cal cm^{-2} min^{-1} in the tropics and about 0.7 cal cm^{-2} min^{-1} at midsummer at the north pole. Actually at that season the energy received per day is slightly greater at the pole than at the equator because the sun is continuously

[1]Note: 1 $\mu = 10^{-6}$ m.

above the horizon. On a yearly basis the Arctic receives 0.14 cal cm^{-2} min^{-1} while the tropics receive 0.35 or two-and-a-half times as much.

ABSORPTION

As most substances are completely opaque to solar radiation, the absorption and conversion to thermal energy takes place at the immediate surface. Water is significantly different in this respect as the absorption is spread through the top three or four feet. Snow and ice also have this property but to a lesser degree. Sea ice absorbs thermal energy faster than freshwater ice, because of impurities, and faster than snow, because of multiple reflection between the snow crystals.

REFLECTION

In general, any substance which appears white to the eye in sunlight is a poor absorber whereas dark objects are likely to be good absorbers. Table I gives the reflectivity of the substances with which we are most concerned.

TABLE I
ALBEDO OR REFLECTIVITY OF VARIOUS SURFACES

Surface	Albedo (per cent)
New snow	85
Old snow	70
Ice	70
Clouds	45–80
Water	5
Water, solar elevation 5°	45
Green forest	3–10
Grass	15–35
Bare ground	10–20
Desert sand	25
Average for whole earth	40

MECHANISMS FOR THE VERTICAL EXCHANGE OF HEAT

In general, the transfer of heat is governed by the equation $dQ/dt = K\ dT/dx$ which means that the rate of transfer is proportional to the gradient of temperature and that the heat flux is from hot to cold, or in the direction which will tend to equalize temperatures. The constant K is assigned a value appropriate to the process.

The major mechanisms for the vertical exchange of heat are: (1) long-wave radiation, (2) eddy diffusion or turbulent exchange, (3) convection, and (4) conduction. These are discussed in the following sections.

Terrestrial or Long-wave Radiation

Although solar radiation is familiar because we can both see its effects and feel it, it is not easy to realize that we are constantly being bombarded by radiant energy from terrestrial sources. Because of its importance to the heat budget of the Arctic a few statements are in order.

(1) The range of wavelength is from 4 to 80μ and the peak of energy for the meteorological range of temperatures is near 10μ.

(2) All liquid or solid substances with which we are concerned, including clouds, are nearly perfect absorbers of long-wave radiation, and also efficient radiators. This is not true of many other substances. For example, brass has a high albedo with respect to long-wave radiation and is a poor radiator.

(3) Oxygen and nitrogen are practically transparent. Carbon dioxide absorbs and radiates at wavelengths for $12\text{-}16\mu$ but is otherwise transparent. Water vapour is transparent from $8\text{-}16\mu$ but for most of the remainder it is an effective absorber and radiator. The net result is that, except for clouds, the atmosphere is transparent to long-wave radiation only between 8 and 12μ.

(4) The maximum rate of emission of energy over the long-wave spectrum is given by $E = CT^4$, where C is a constant and T the temperature expressed in degrees absolute.[2] For the range of temperatures with which we are concerned the emission rate varies from 0.2 to 0.8 cal cm^{-2} min^{-1} and is therefore comparable to the solar radiation rate. Water, snow, ice, soil, and vegetation radiate at rates that are very close to the maximum possible. Water vapour emits at a rate which is about 85 per cent of maximum. The rate for carbon dioxide is 18 per cent.

(5) The flux of heat in the atmosphere by long-wave radiation is a complex process. To gain an understanding of how it operates we can consider the atmosphere to be divided into a number of layers, each of which contains enough water vapour to absorb completely the radiation from the layers immediately above and below. The number of layers will be proportional to the water vapour content of the atmosphere. The flux will, of course, be from warm to cold. As the exchange between adjacent layers is time-consuming, the rate of flux through the atmosphere will be inversely proportional to the number of layers, or in other words the total vapour content.

(6) As temperatures and vapour pressures normally decrease with height the flux of heat by long-wave radiation is usually directed upward, and the rate of flux increases upward. This would result in a cooling of the atmosphere of roughly 1.5° C per day if the air were not gaining heat by other processes.

(7) The exchange between the lowest layer and the surface is different only with respect to the 15 per cent of the emission from the surface which is not absorbed by the atmosphere, and which escapes to outer space unless intercepted by clouds.

Turbulent Mixing or Eddy Diffusion

First it is necessary to define dry adiabatic and wet adiabatic lapse rates.[3] When a parcel of air is lifted, it expands and cools because of the work done in expanding. The rate of cooling for dry air is 5½°F/1000 feet or 1°C/100 m. If the air is saturated, the cooling is retarded because of the liberation of the latent heat of vaporization. For this reason the wet rate is less than the dry. As the rate of

---

[2]T(absolute) $= T$(centigrade) 273. The constant C has a value of 8.2 × 10^{-11} cal cm^{-2} min^{-1}.

[3]The lapse rate is the rate of decrease of temperature with height, and is usually expressed in terms of °F/1000 ft. or °C/100 m.

condensation decreases sharply with decreasing temperatures, the wet adiabatic lapse rate increases from 2˚F/1000 feet for very warm air to practically the dry rate at very cold temperatures. Because water is an incompressible fluid, there is no similar parameter for that fluid.

The horizontal motion of air over surface irregularities causes eddy motions which lead to mixing. Eddies may also develop because of horizontal or vertical variations of wind speed in the atmosphere well above the surface. As a parcel of air with a motion different from the average in its general vicinity must cause other parcels in its immediate vicinity to deviate also, the turbulence tends to spread out from its point of origin and persist until destroyed by friction. As turbulence is being created frequently, the atmosphere is generally turbulent to some degree.

The amount of turbulence varies with the lapse rate. If the lapse rate of the environment is negative, a parcel of air displaced up or down is immediately subject to a force which acts to restore it to its original level. As the lapse rate increases, this restoring force decreases in strength, and when the lapse rate is adiabatic, the restoring force is zero. Under adiabatic conditions only inertia and friction must be overcome to lift an air parcel. From this it will be obvious that turbulence is likely to increase sharply as the lapse rate approaches the adiabatic value. When the lapse rate is less than adiabatic, a parcel of air displaced downward is warmer than its environment. Because some mixing must occur, the net result is that some heat is transferred downward. Similarly when the air parcel is lifted it will be colder than its environment and some cooling will occur because of mixing. This means that turbulence causes a downward flux of heat as long as the lapse rate is less than the adiabatic value, and can cause an upward flux of heat only when the lapse rate is greater than the adiabatic. As vertical turbulent motions are increasingly inhibited as the lapse rate decreases, turbulent mixing is most effective in transferring heat when the lapse rate is close to but not equal to the adiabatic. Under favourable conditions this mechanism is 100 times as effective as the radiative transfer process.

In water, the turbulence is provided by wave action and horizontal flow in currents. Because of the far greater inertia of parcels of water, turbulent mixing is much slower in water than in air. Vigorous turbulent mixing of air results in an adiabatic lapse rate if continued long enough, but in water an isothermal lapse rate is produced.

Convection

Convection can occur when the lapse rate is greater than the adiabatic and when the air has full three-dimensional freedom of motion.

When the lapse rate is unstable, that is, greater than adiabatic, an air parcel displaced upwards is warmer and therefore lighter than its environment, and consequently is subject to a buoyant force directed upward. Obviously under these conditions heat can be transferred upward with great rapidity. However, close to the surface, convection is somewhat inhibited because the air required to replace the ascending air parcel cannot move with complete freedom because of friction with the surface.

In the Arctic, convection is not of prime importance because lapse rates do not often exceed the adiabatic value, except over larger land masses in summer. Convective clouds are often observed in July and August over the northern islands, but thunderstorms are very rare north of latitude 70. In winter when very cold air flows over open water, violent convection results but only through a layer about 200 feet thick.

Convection is capable of transferring heat about 100 times as effectively as turbulent mixing. It is a faster process, but its greater effectiveness is related to distance rather than to speed. In water, convection can operate whenever the temperature decreases upwards provided that the impurities such as sea salts in solution are uniformly distributed. In either air or water, convection may be initiated by warming at the bottom of the layer or cooling at the top.

Conduction

Conduction in water or air is so slow that it may be neglected. However, it is the only process available for heat transfer in solids, and is therefore of considerable importance in weather and climate. Actually, there is not much variation in conductivity among the various soil materials in solid form. The conductivity of a soil depends largely on how the intergranular spaces are filled. As the conductivity of ice is three times that of water and 100 times that of air, a frozen soil is a better conductor of heat than a wet soil which in turn is a much better conductor than a dry soil. Conduction is always from hot to cold and if continued long enough produces an isothermal temperature distribution. Because it is more than 1,000 times slower than radiative transfer which is the slowest of the other three processes, only a shallow layer of soil or ice is involved in the distribution and redistribution of the annual supply of heat from the sun.

PRECIPITATION

As great quantities of heat are required to evaporate water at the surface, and as this heat is released to the atmosphere where condensation occurs, the hydrological cycle is of major importance with respect to heat transfer processes. The water vapour evaporated at the surface may be transported upward by either convection or turbulence. Both processes tend to produce a vertical moisture distribution that is uniform with respect to ratio of water vapour to air. On the other hand, as vigorous mixing results in an adiabatic decrease of temperature with height, the relative humidity must increase with height. If a deep enough layer of the air is involved in the mixing, the air must become saturated at some level above the surface, and condensation and possibly precipitation will occur at higher levels.

When convection is the vertical transport vehicle, the upward motions of air parcels must be balanced by downward motions of neighbouring air parcels. For this reason convection is cellular in nature. Because eddy turbulence is usually limited to much shallower layers than convection it is not as effective as a precipitation mechanism. In the Arctic where convection is less frequent, turbulence assumes greater importance with respect to condensation and precipitation.

SUMMARY OF PHYSICAL PROCESSES AND PROPERTIES

TABLE II
PHYSICAL PROPERTIES

	Air	Water	Ice	Soil
Heat capacity $= \rho c$	0.0002	1	0.45	0.6
Conductivity $(\times 10^{-3}) = k$	0.05	1.5	5	2.5
Molecular transfer efficiency $= k/\rho c \, (\times 10^{-3})$	250	1.5	11	4
Radiative trans. eff.	10^3	—	—	—
Turbulent trans. eff. $= K/\rho c$	10^3 to 10^5	10	—	—
Convective trans. eff.	10^7	300	—	—
Storage eff. $= \rho c \sqrt{K \text{ or } k}$	0.1	7	0.05	0.04

$\rho =$ density, $c =$ heat capacity per gram, k or $K =$ coefficient of heat transfer.

TABLE III
DIMENSIONS OF FIELD OF CLIMATE

	Air	Water	Soils
Height or depth in feet[1]	35,000	300	20
Relative storage capacity	1	50	1

[1]The figures for height or depth indicate the penetration limits of the annual heat cycle.

The storage capacity of the air-water combination in marine areas is about twenty-five times greater than that of the air-soil combination in continental areas. The amplitude of the diurnal or annual temperature cycles would be expected to vary according to the following expression:

$$\text{Temperature amplitude} = \frac{\text{amplitude of variation of heat supply}}{\text{storage capacity}}$$

Table IV gives average values of the amplitudes for marine and continental areas.

TABLE IV
AMPLITUDES OF TEMPERATURE VARIATIONS

	Marine	Continental
Annual cycle	20°F	90°F
Diurnal cycle	0.5°F	20°F
Air mass changes with weather systems	5°F	25°F

HORIZONTAL AIR MOTIONS

So far we have treated the air and water as if they were stationary while actually, of course, they are free to move horizontally. Although ocean currents are of considerable significance, we can neglect them in this discussion because on the average they are in harmony with the mean motions of the atmosphere.

The local temperature change that can be brought about by horizontal advection in the atmosphere is greater than that which can be produced by vertical exchange processes. Mathematically, the rate of change is equal to the product of the wind speed and the component of temperature gradient that is parallel to the wind direction. For short periods this product can amount to 25°F per hour.

From a wider point of view the pattern of wind circulation over the Arctic may be considered to be composed of several parts. First, we have the westerlies of mid-latitudes which usually extend northward to about latitude 60. North of the Arctic circle the pattern is commonly cellular, consisting of several nearly station-ary lows and highs. Superimposed on this pattern there are likely to be from five to fifteen migrating lows, most of them in the westerly belt but probably one or two in the Arctic. Both the slowly changing and the more rapidly changing systems act to equalize the temperature distribution by advecting heat in both sensible and latent forms towards the Arctic. This condition raises a point which may be of interest. We are familiar with the fact that cyclones, tornadoes, and hurricanes can develop in air but not in water. This means that the atmosphere, but not the ocean, can act as an engine to convert heat to mechanical energy. The oceans contribute to the process only by acting as a heat reservoir. Although advection can readily transport great quantities of heat into the Arctic, it will not be realized as a benefit unless the imported heat is transferred to the surface from the higher layers where advection is most active. As heat can be transferred more rapidly upwards than downwards, the Arctic is less affected by warm invasions from the south than more southerly regions are affected by cold invasions from the Arctic.

LAG OF TEMPERATURE CYCLES

Annual — Mid-Latitudes

The variation of heat supply from the sun in the mid-latitude belt may be represented with sufficient accuracy by a sine curve of the form $R = K \sin t + H$ where R is the rate at which heat is being received at a given latitude and for a given day, K is the amplitude, t is the time, and H is the average received per day for the year. The maximum, $H + K$, is received on June 21, the minimum, $H - K$, on December 21, and the mean, H, on March 21 and September 21.

If the heat stored is very small, the temperature curve would have much the same shape as for the incoming heat, and the lag would be slight. On the other hand, if the storage is very large so that the surface temperature amplitude is small, the temperature can continue to rise until nearly September 21. Actually it is not difficult to show mathematically that the lag is proportional to the heat storage. Therefore, the lag is considerably greater for marine than for continental areas.

Diurnal — Mid-Latitudes

For the diurnal cycle the heat supply cannot be represented by a sine curve because no direct radiation is received during half of each cycle. Therefore, although the maximum occurs about three hours after the time of maximum insulation, the minimum is delayed until the end of cooling period or roughly until seven hours after midnight.

Arctic Annual

The annual heat supply cycle is similar to the diurnal cycle of mid-latitudes because no direct heating is received from the sun during the winter months. However, the delay of the seasonal minimum at high latitudes is not as great as would be expected by comparison with the diurnal cycle, mainly because of advectional exchanges with neighbouring regions with small lag characteristics, and also because the seasonal dawn and sunset are not as abrupt.

There is another factor which must be mentioned because of its importance to water transport, namely the delay caused by time-consuming processes of freezing and melting. If you begin with a gram of ice at a low temperature, and heat it at the rate of 1 cal per second, the temperature will rise 2° per second until it begins to melt. The temperature for the next 80 seconds will remain constant at 0°C while the ice is being converted to water, then rise at the rate of 1°C per second.

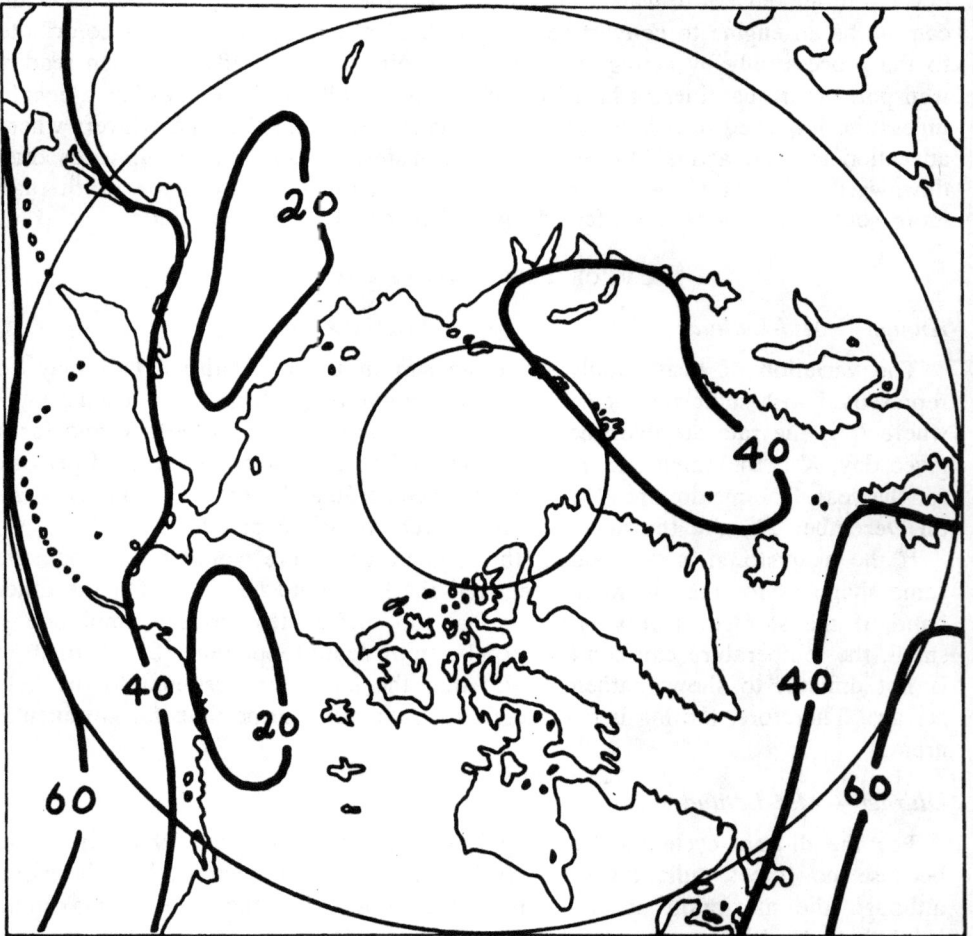

FIGURE 1.   Summer lag in days.

The minimum of Arctic ice should occur at the end of the melting period or more accurately when the seasonal decline of temperatures reaches 29°F which is the freezing point of Arctic sea water. Because of the time required for freezing, the navigation season is extended a week or more. This melting and freezing lag is more pronounced along the edge of the Arctic than farther north. For example, Great Bear and Great Slave lakes and Hudson Bay break up in June and do not freeze over completely until mid-December. This indicates a lag of three months. Figures 1 and 2 which are based on limited data, show the variation of lag of seasonal extremes of temperature over the Arctic.

EVAPORATION, CONDENSATION, AND PRECIPITATION

As nearly 600 calories of heat are required to evaporate 1 gm of water, great quantities of heat must be readily available for the evaporation process. For this reason evaporation is likely to be slow on land except when the sun is shining.

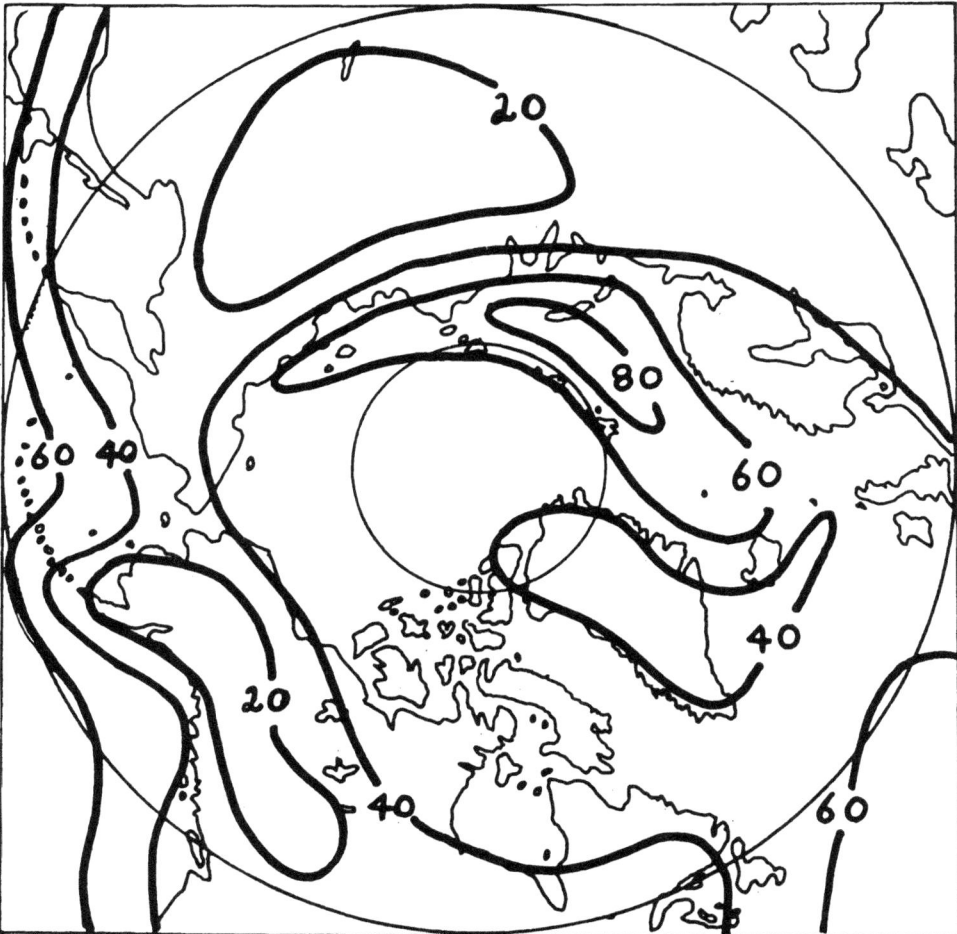

FIGURE 2.   Winter lag in days.

Because of the heat storage capacity of water, evaporation can continue day or night over the oceans, lakes, and rivers.

The evaporation process is quite different from the melting process in that it can operate over a wide temperature range whereas melting occurs only above 0°C. Evaporation depends essentially on the fact that water or ice evaporate as rapidly as the required heat can be supplied provided that the vapour is removed as fast as it is formed. If the vapour is allowed to accumulate, an equilibrium is soon established between the escape pressure at the water surface and the pressure of the water vapour. This pressure of equilibrium between a water surface and the water vapour in contact with it is called the saturation vapour pressure. It increases

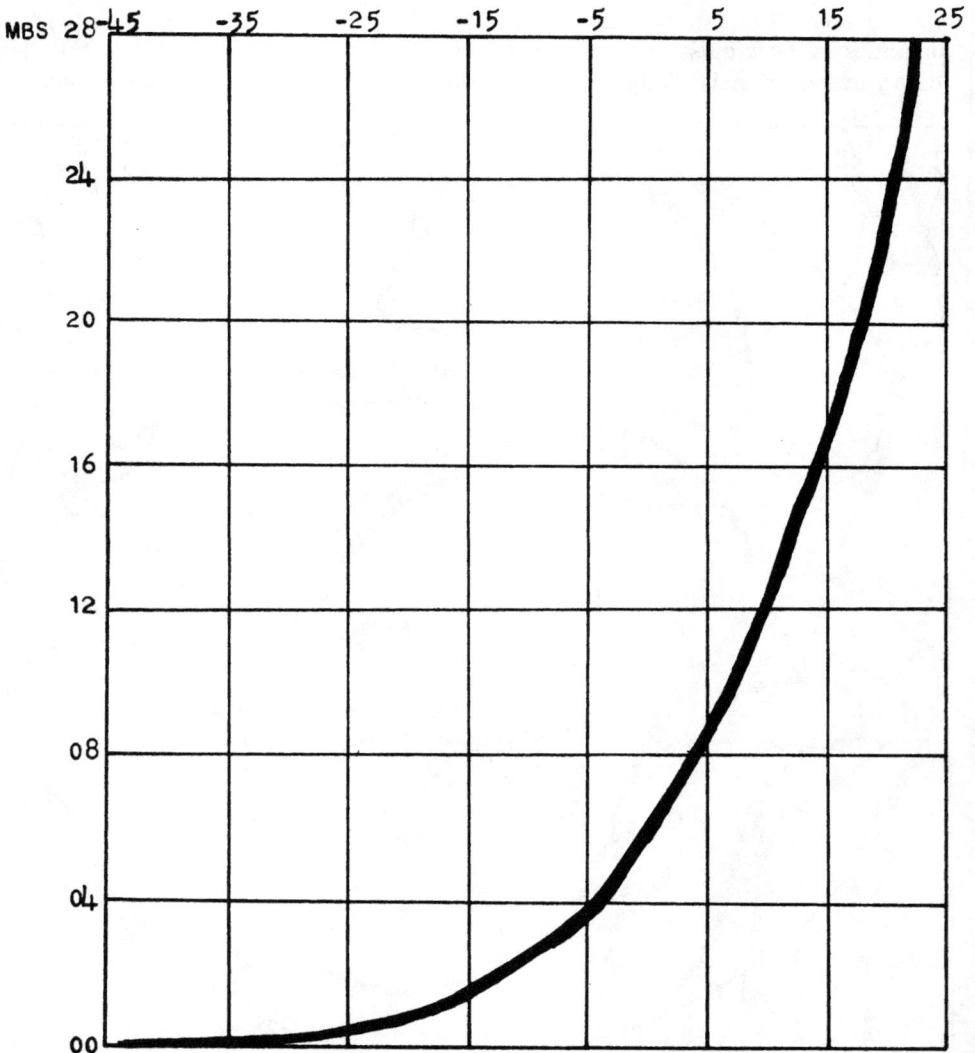

FIGURE 3.   Saturation vapour pressure curve °C.

with increasing temperature. Figure 3 shows the variation of saturation vapour pressure with temperature.

Ordinarily the vapour pressure in air is less than the pressure of saturation. In meteorology and climatology, two measures of moisture content are used. The relative humidity is the percentage ratio of the actual vapour pressure to the saturation pressure corresponding to the temperature of the air. The dew-point is the temperature that would be reached if the air were cooled until saturated.

The rate of evaporation is proportional to the difference of vapour pressure between the water surface and the air in contact with it. As the common measure of the humidity of air is its dew-point temperature, we must be able to obtain the evaporation rate from the difference between the temperature of the water and the dew-point of the air. From Figure 3 it can be seen that the vapour pressure differential corresponding to a $10°C$ temperature differential increases sharply with increasing temperature. Mathematically, the evaporation rate varies with the slope of the curve, or de/dT, where e is the vapour pressure and T is the mean of the water and dewpoint temperatures. Because the de/dT vs T curve has very nearly the same shape as the e vs T curve of Figure 3, we can with little error use Figure 3 to obtain directly a measure of the variation of evaporation rate with temperature for a water-dew-point differential of $1°C$. The unit rate would then be multiplied by the actual differential to obtain the corresponding rate.

On the other hand, Figure 3 can also be used to help explain the processes of condensation and precipitation. First of all, if the figures for vapour pressure in millibars[4] are reduced by one-third, the curve gives us the variation of saturation vapour density with temperature in terms of grams per kilogram, or parts per thousand. With a little juggling of dimensions, we can also estimate probable precipitation amounts. If a layer of air 3000 feet thick and saturated at a given temperature were lifted 3000 feet, the figures on the left would be very close to the precipitation that would result in hundredths of an inch.

From these facts we can understand why precipitation is so low in the Arctic, and also why the landscape is wet in spite of the lack of precipitation. Actually the Arctic ranks high with regard to the number of days with precipitation, but on the majority of these days the amount is one-hundredth of an inch or less. If the air is cooled below its saturation temperature, condensation occurs. The required cooling is usually provided by lift which may result from flow over hills or mountains, from convergent flow in cyclones, or from convective activity. These may act singly or in concert. Radiational cooling is always a factor at the tops of a cloud or fog or near the surface of the earth. Precipitation does not necessarily result if the air is cooled far below its saturation temperatures, as a cloud or fog can become quite dense without precipitation occurring. The reason for this is that initially only very small ice crystals or water droplets are formed, and they fall so slowly that they can be considered to be suspended in space. Therefore, they must be able to grow in order to develop an appreciable rate of fall. The following factors are involved in the growth of ice crystals and water droplets.

(1) The saturation pressure on the surface of a water droplet is lower than indicated by Figure 3 which applies to a plane water surface. The equilibrium

[4]A millibar (mb) is the meteorological unit of pressure .1013 mb = 760 mm of mercury.

vapour pressure increases with the radius of curvature of the droplet surface. Droplet formation begins when the relative humidity is about 75 per cent, and growth continues until the equilibrium size is reached.

(2) The temperature of freezing of a droplet also depends on its size. In still air a very small droplet can resist freezing down to about —40°C. The freezing point increases with increasing radius of curvature. Turbulence raises the freezing point.

(3) The saturation vapour pressure for a plane ice surface is lower than for a plane water surface. The maximum differential of 0.25 mb occurs at temperatures near —15°C. Because of this factor and also because of its irregular shape an ice crystal can grow faster than a neighbouring water droplet. Further, if the vapour pressure is intermediate between saturation for the droplet and saturation for the crystal, the crystal can grow at the expense of the droplet.

Because of these physical factors, the conditions favourable to precipitation are

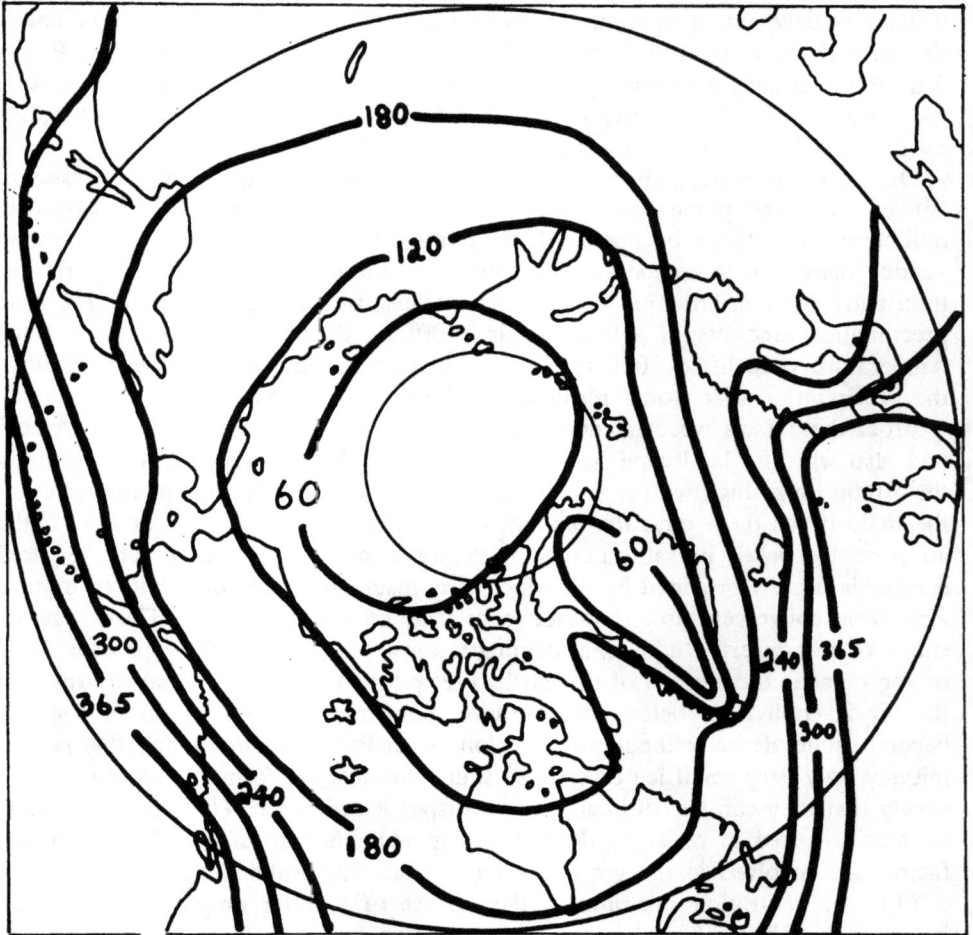

FIGURE 4.   Number of days temperature above freezing

(i) active lift, (ii) cloud must extend above freezing level, (iii) depth of cloud sufficient to reach —15° C, and (iv) turbulence within cloud. Although precipitation in the Arctic is usually small, it is noteworthy that heavy falls can occur. In 1937, when the Papanin party was at the North Pole, a twenty-inch fall of wet snow was observed. A few years ago, Eureka received 1.7 inches of rain over two days. This is nearly as great as the normal yearly total for that station. Also about six years ago, Norman Wells measured a total of four inches over a period of three days.

It is also noteworthy that along the edge of the Arctic there are some of the wettest places in the world. The southeast coast of Alaska, the southern tip of Greenland, Iceland, Norway, and southern Kamchatka all receive very heavy

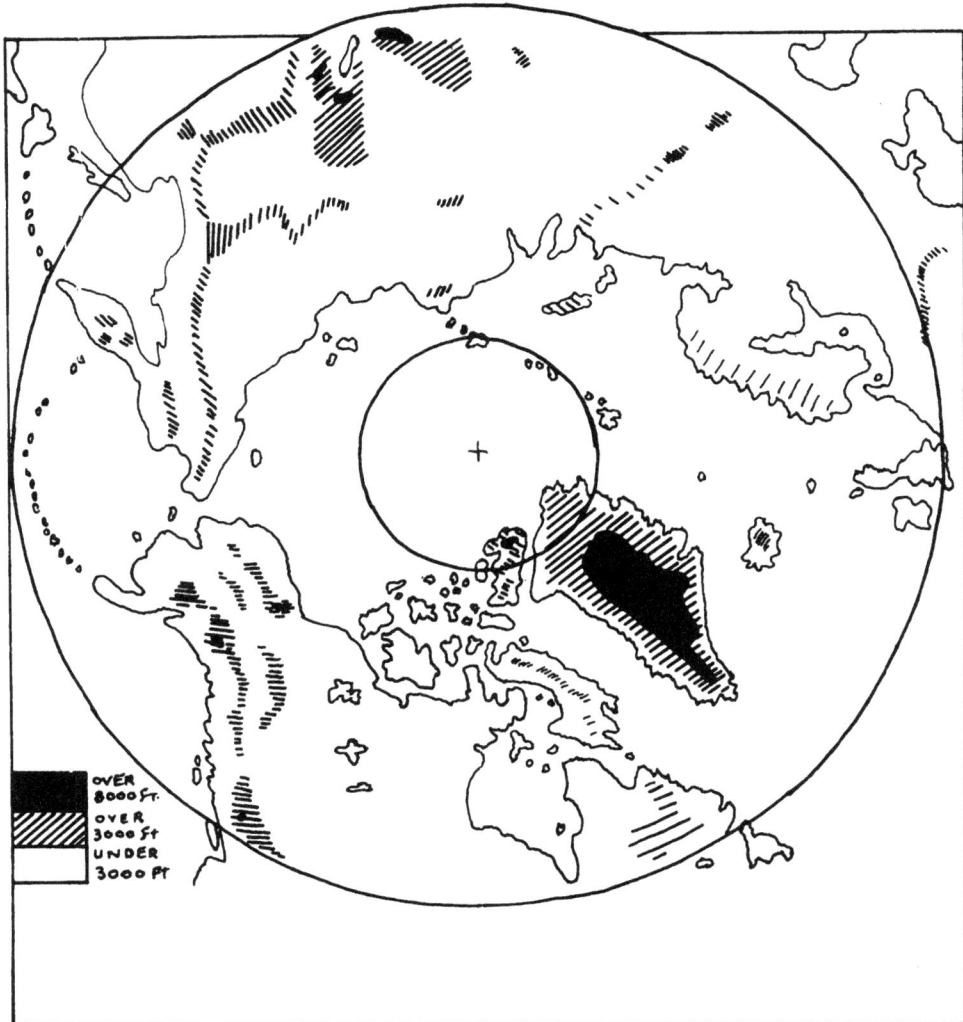

FIGURE 5.   Major mountain barriers.

annual precipitation. In each case a mountain barrier is located on or close to a major storm track.

STORM TRACKS

The development or increase of intensity of a cyclone of the type that is most frequent at mid-latitudes depends on the existence of a strong horizontal gradient of temperature for a source of energy. In winter, the greatest contrasts are found along the east coasts of Asia and North America. In summer, the strongest gradients are found along the north continental coasts, but the summer contrasts

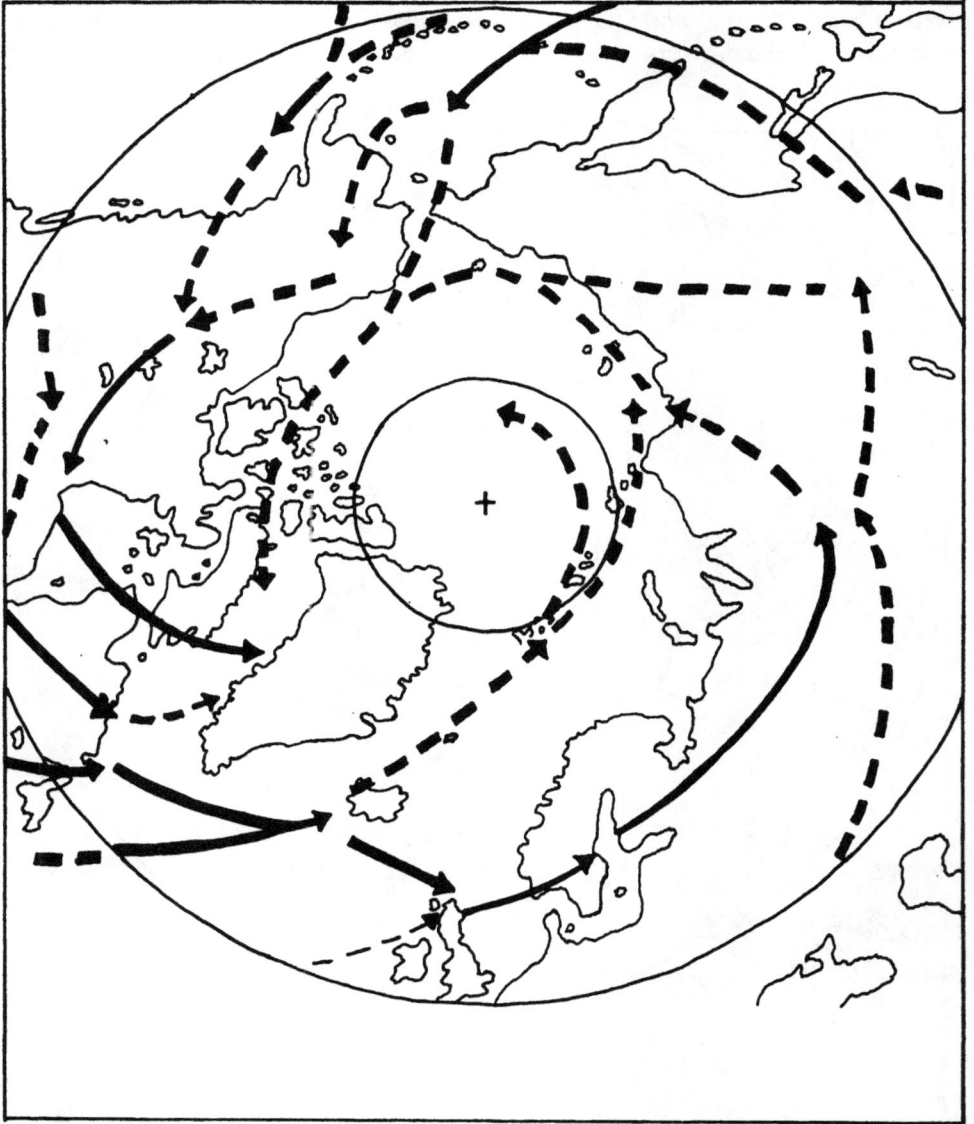

FIGURE 6.   Principal tracks of lows in July

are not as strong as those of winter. Figure 4 shows the variation of the number of days with mean temperature above freezing over the Arctic and sub-Arctic. This chart is roughly equivalent to a chart of mean annual temperature. The movement and growth of cyclones are greatly influenced by mountain barriers. Cyclones do not readily cross mountains, and if forced over mountains they usually weaken rapidly. Figure 5 shows the areas where elevations are above 2500 feet.

Figures 6 and 7 show the major tracks for the northern hemisphere. It will be noted that the patterns for the two seasons agree reasonably well with what would

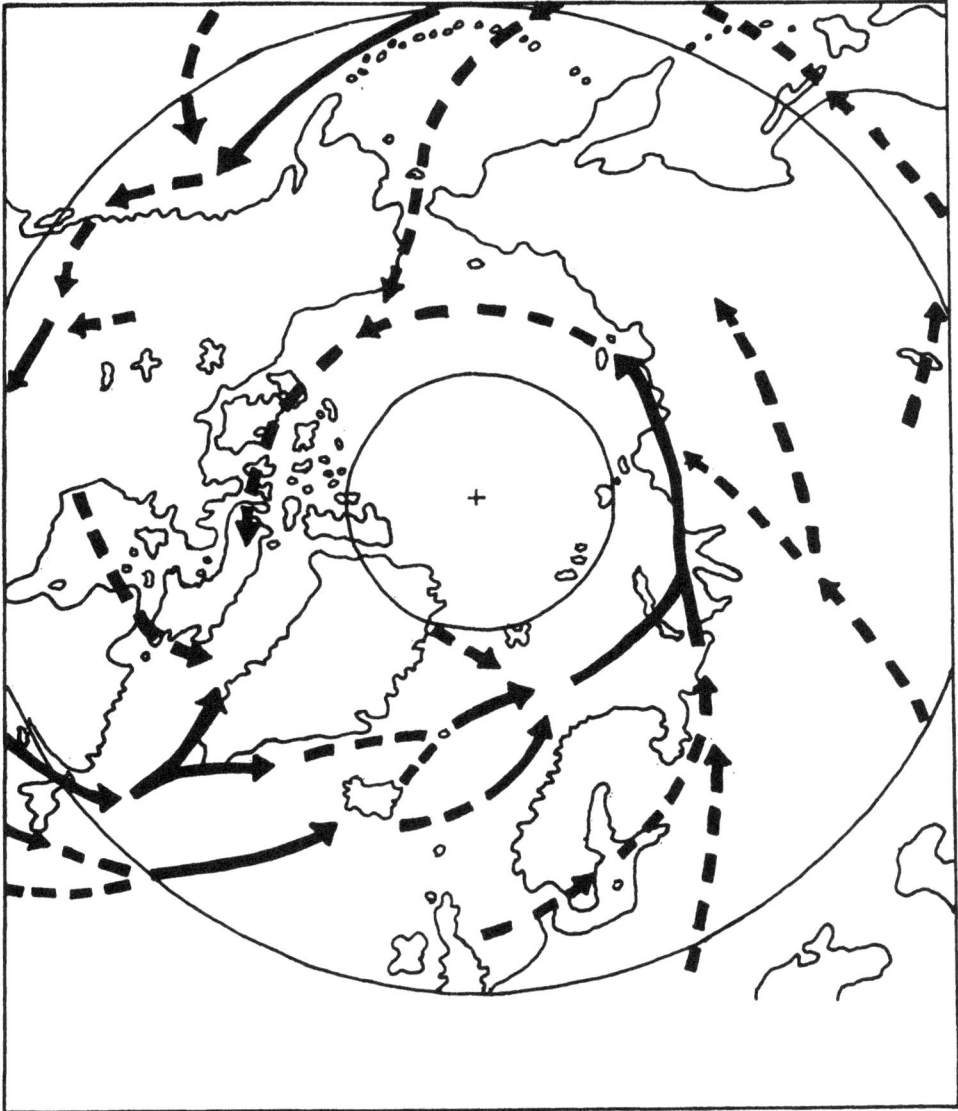

FIGURE 7.   Principal tracks of lows in January

be expected from the charts of thermal contrast and mountain barriers. Not shown on these charts are the major stall areas, where the storms become nearly stationary when they reach their maximum intensities and then decay. The stall areas correspond closely with areas of minimum pressure on mean pressure charts. The two major stall areas are located along the Aleutians and just west of Iceland.

The main tracks combined form a giant spiral, beginning in Japan and extending across the Pacific to the Gulf of Alaska, over Canada *via* Alberta, the Great Lakes, and Newfoundland to join the procession of lows from the east coast of the United States towards Iceland, then onwards north of Norway to the final graveyard north of Cape Chelyuskin. This main track accounts for a majority of the lows that migrate into the Arctic Ocean area. Another heavily travelled track lies northward along the west coast of Greenland, but most of the lows that follow this track die in Baffin Bay. The few that persist beyond Thule usually move around the north coast of Greenland then eastward to Spitsbergen. The continuation of the main track beyond Cape Chelyuskin is of considerable importance with respect to northern Alaska and the Arctic islands of Canada. The straggling procession of offshoots from Cape Chelyuskin is augmented in winter by irregulars from the Bering Sea. In summer, there are additions from the heated interior valleys of eastern Siberia and Alaska. In addition to the migrating lows, there are the deep cold-core lows that often dominate the circulation of the Arctic. These may be found anywhere in the Arctic but are most frequently located in the vicinity of northern Baffin Island or over the mountains of eastern Siberia. They are not always evident on surface charts and are slow-moving but they persist for long periods of time, sometimes for as long as six weeks.

REFERENCES

B. Haurwitz and James M. Austin.   1944.   Climatology; McGraw-Hill.
Clarence Eugene Koeppe and George C. DeLong.   1958.   Weather and climate; McGraw-Hill.

Pleistocene Climate Changes[1]

MAURICE EWING AND

WILLIAM L. DONN

ABSTRACT

It is postulated that the ice-free condition of the Arctic Ocean is the direct cause of Pleistocene ice ages in the northern hemisphere. Evidence indicating an open Arctic ocean is found from a study of thermal gradients, postglacial uplifts, Arctic seismicity, and the heat budget of the Arctic area. An attempt is made to explain the Pleistocene climate changes in terms of the alternations of ice-free with ice-covered states of the Arctic Ocean.

IT IS FIRMLY ESTABLISHED that the temperatures of the surface water of a large part of the Atlantic Ocean increased abruptly by 6° C to 10° C about 11,000 years ago (Emiliani, 1955). This change is clearly recorded in the deep-sea sediments of the north and equatorial Atlantic and of the connecting seas—the Mediterranean, the Caribbean, and the Gulf of Mexico (Schott, 1935; Bramlette and Bradley, 1942; Phleger, Parker, and Pierson, 1947-8; Ovey, 1950; Schott, 1952; Ericson and Wollin, 1956; Parker, 1958). The implication of a temperature increase shown by the change in planktonic fauna recorded in the sediments, has been confirmed by oxygen isotope ratios (Ericson, Broecker, Kulp, and Wollin, 1956; Ericson and Wollin, 1956; Ericson and Wollin, 1956a; Emiliani, 1955, 1957). Based on radiocarbon measurements from many different sources and investigators, Broecker, Ewing, and Heezen (1960) have established the date of this temperature change as 11,000 years before the present, confirming the earlier estimate of Ewing and Donn (1956). The principal sources used are Rubin and Suess (1955, 1956), Broecker, Kulp, and Tucek (1956), and Broecker and Kulp (1957).

Various other changes occurred in the same part of the ocean at the same time. For example, there was a decrease in the rate of clay and carbonate deposition in the equatorial Atlantic (Broecker, Kulp, and Tucek, 1956; Broecker and Kulp, 1957; Broecker, Turekian, and Heezen, 1958), a change from well-ventilated to stagnant conditions in the Cariaco trench just north of Venezuela (Heezen, Menzies, Broecker, and Ewing, 1959), a sharp drop in the rate of delivery of silt to the deep part of the Gulf of Mexico (Ewing, Ericson, and Heezen, 1958), and other changes in deep sea sedimentation which have been interpreted to indicate an abrupt general rise in sea-level.

The changes which occurred 11,000 years ago, and the abruptness with which they occurred have been taken by Ewing and Donn (1956) to mark the end of the Wisconsin glacial stage and have been discussed more fully by Broecker, Ewing, and Heezen (1960). The change was probably complete within a few centuries because all effects which disturb the sedimentary record would make the change appear more gradual.

In view of the great heat capacity of the ocean and the abruptness of the Atlantic temperature change, we have attempted to explain this change on the basis of a

[1]Lamont Geological Observatory (Columbia University) Contribution no. 413.

difference in oceanic circulation. This would alter the distribution of available heat, rather than change the total available heat. We have suggested that the ice-sheet which now covers the Arctic Ocean was formed about 11,000 years ago and that its presence caused the observed warming of the Atlantic, and a corresponding cooling of the Arctic Ocean, by reducing the interchange of water between the two oceans (Ewing and Donn, 1956). Many difficult questions are raised by this suggestion. Would the presence of the ice-sheet actually reduce the Arctic-Atlantic interchange? What caused it to form? What other changes would it have induced? Was it a result of the disappearance of the Wisconsin ice-sheets, or was it the cause of the change from a glacial to an interglacial stage? Unfortunately the atmosphere-ocean system is so complicated that we cannot give positive answers to any of these questions at present; we can only give opinions about them.

There is no doubt that the Arctic ice-sheet is a severe barrier to the transfer of moisture, heat, and momentum between ocean and atmosphere. The ice sheet is responsible for the continentality of the Arctic climate and prevents the wind from cooling, mixing, and stirring the water. It is our opinion that removal of the ice-sheet would cause four direct changes in the ocean:

(1) a marked increase in transfer of heat from the ocean to the atmosphere; this increase would be most pronounced in winter in view of the present very low Arctic air temperature in winter;

(2) a mixing of the water to the extent that the warm, saline Atlantic water would not plunge several hundred metres beneath the colder fresher Arctic, as at present (Worthington, 1953), but would tend to mix with it;

(3) a replacement of the present feeble, density-driven Arctic circulation by a vigorous, wind-driven circulation, probably in such a direction that the Atlantic-Arctic interchange of water would be greatly increased. This might well enable the Arctic Ocean to remain ice-free even while supplying great quantities of heat and moisture to the polar atmosphere. For example, Wexler (1958) has calculated that the average radiation loss from the present ice surface is 5.5×10^{18} cal/day during the four-month winter dark interval. Our estimate of 10^{19} of cal/day for the heat transport by ocean currents into the Arctic Basin is supported by the results of Jung (1955) giving 10^{18} cal/day and Timofeev (1957) giving 6×10^{18} cal/day for the heat transport into the Arctic basin between Spitsbergen and Greenland;

(4) a cooling of the Atlantic and a warming of the Arctic from the increased interchange of water.

According to this reasoning we might well explain the 11,000-year-old warming of the Atlantic by the assumption that the Arctic Ocean was formerly ice free. The abrupt warming of the Atlantic at this time was then the result of the cutting off of Arctic-Atlantic interchange. What conditions would have prevailed on land? According to our estimate they would have closely resembled those usually attributed to the Wisconsin glacial stage.

CONSEQUENCES OF AN ICE-FREE ARCTIC OCEAN

Some of the significant and possible consequences of an ice-free Arctic Ocean are:

(1) an increase in precipitation, particularly in winter, in a broad circumpolar zone, with the probable accumulation of ice sheets in Eurasia and North America.

(2) a warming effect on the lands immediately adjacent to the Arctic from the influence of the relatively warm water. We can thus imagine an ice-free border zone similar to that along much of the coast of Greenland today. (Field study is probably the only means of estimating the position of former ice limits.)

(3) a major change in the pattern of Arctic atmospheric circulation because the warming by the open ocean would probably change the present polar high to a polar low causing the present clockwise circulation to reverse to counterclockwise; this would be in the proper direction to increase, rather than oppose, the present circulation of Atlantic water in the Arctic Basin. Figure 1, showing an example of

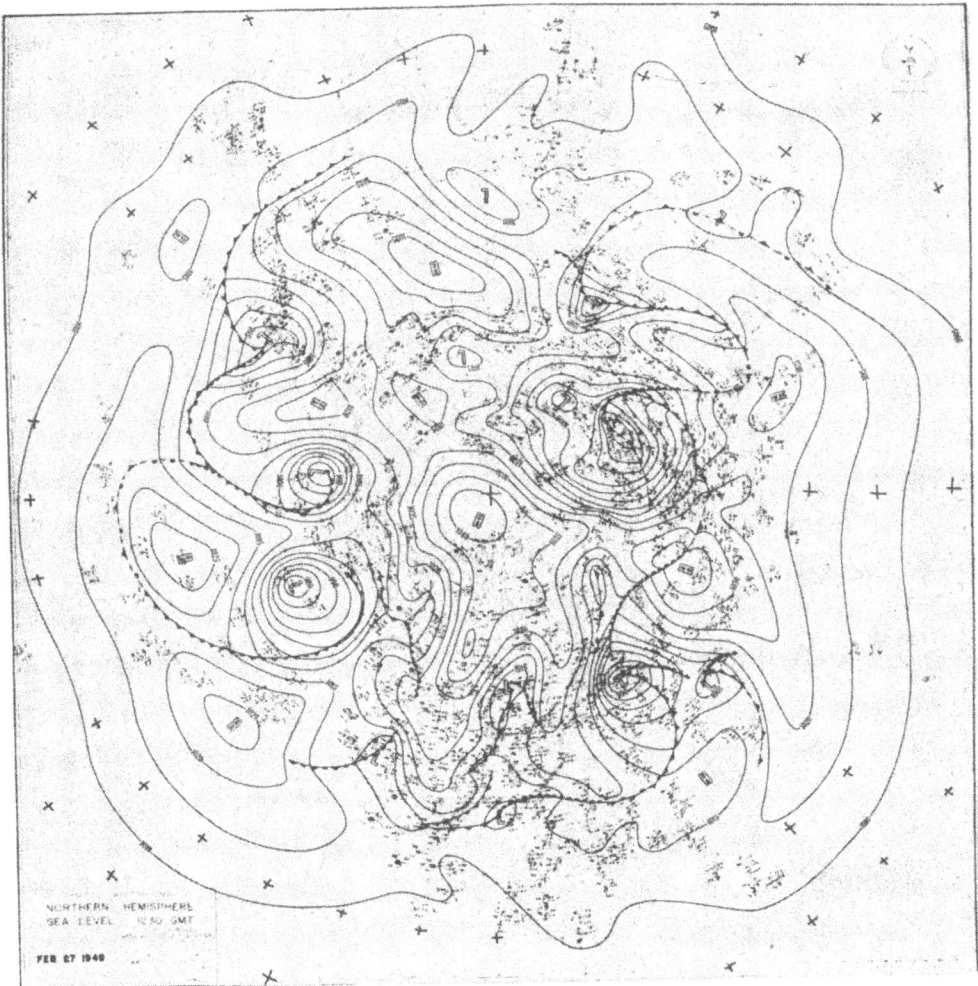

FIGURE 1.   Present-day pressure and wind pattern for the northern hemisphere showing the north polar high pressure centre (from United States Weather Bureau Historical Map Series— February 27, 1949).

present-day clockwise circulation in the polar atmosphere on a typical winter day, may be compared with Figure 2, showing the assumed counterclockwise circulation about a low-pressure area over the warm Arctic Ocean. Certainly the circulation over the open Arctic would reinforce the entrance of warm North Atlantic waters into the Arctic Basin (see Figure 3) and reverse the present-day surface circulation within the Arctic (Figure 4). But even a less profound circulation change would produce greatly increased precipitation simply from the presence of the open and relatively warm Arctic.

(4) It has been pointed out by Brooks (1949) that the cooling effect of an ice-cap once started, would increase geometrically with growth until some equilibrium

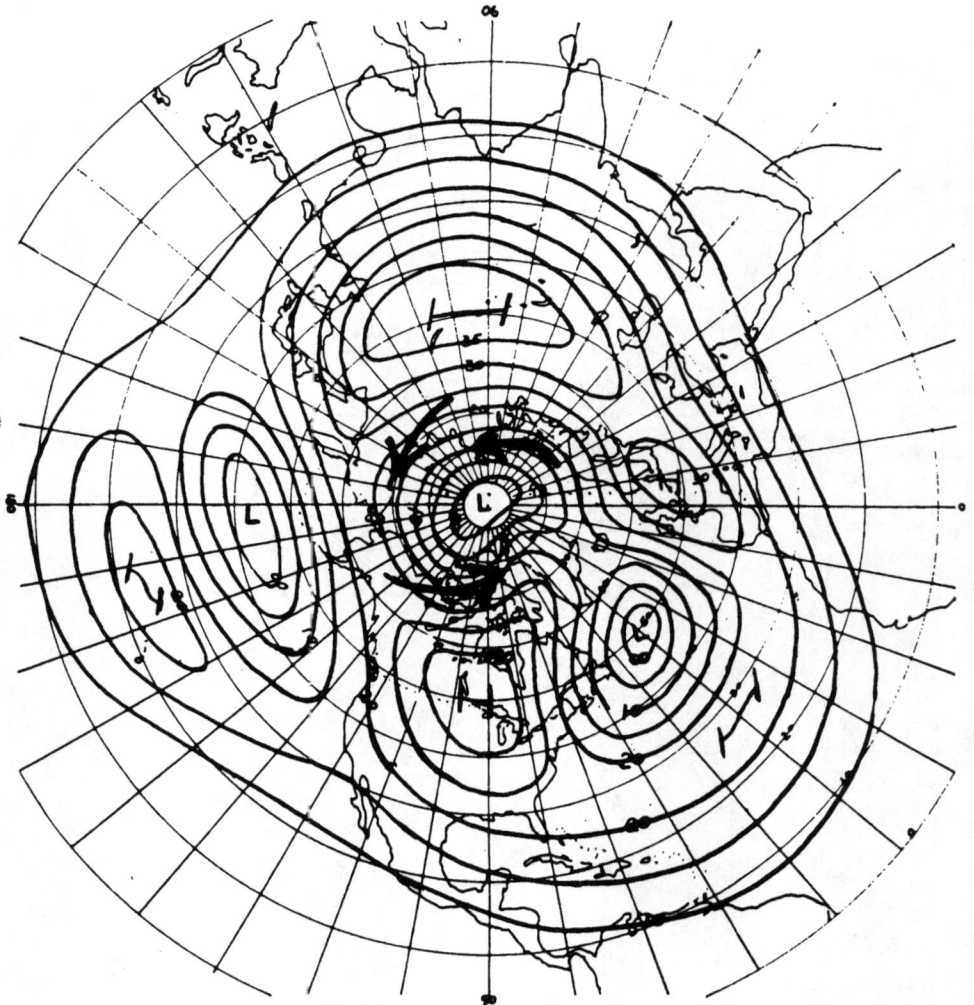

FIGURE 2. Assumed pressure pattern for the northern hemisphere with a relatively warm, ice-free Arctic Ocean. The heavy arrows show the counterclockwise air circulation about the polar low pressure centre.

latitude was reached. We propose that in the case of North America, the cooling effects of an ice-cap accumulated initially from an Arctic Ocean source would, as it grew southward, soon obtain greatly increased snow precipitation from the larger moisture supply available. The fact that the southward extent of the glaciated areas in Siberia is far less than that in North America may be explained by the mountain barriers which block the moisture-laden winds from the west and south. This suggests that the extent and thickness of the Siberian ice-cap is about the maximum to be expected from an Arctic Ocean source alone.

(5) The mid-latitude cooling resulting from the growth of the ice-caps would narrow the present climatic zones, thus producing greater temperature and moisture contrasts and would shift the present mid-latitude storm belt into the subtropic zone. The greatly increased precipitation in this zone could account for the very extensive pluviation of the large areas which are at present arid deserts.

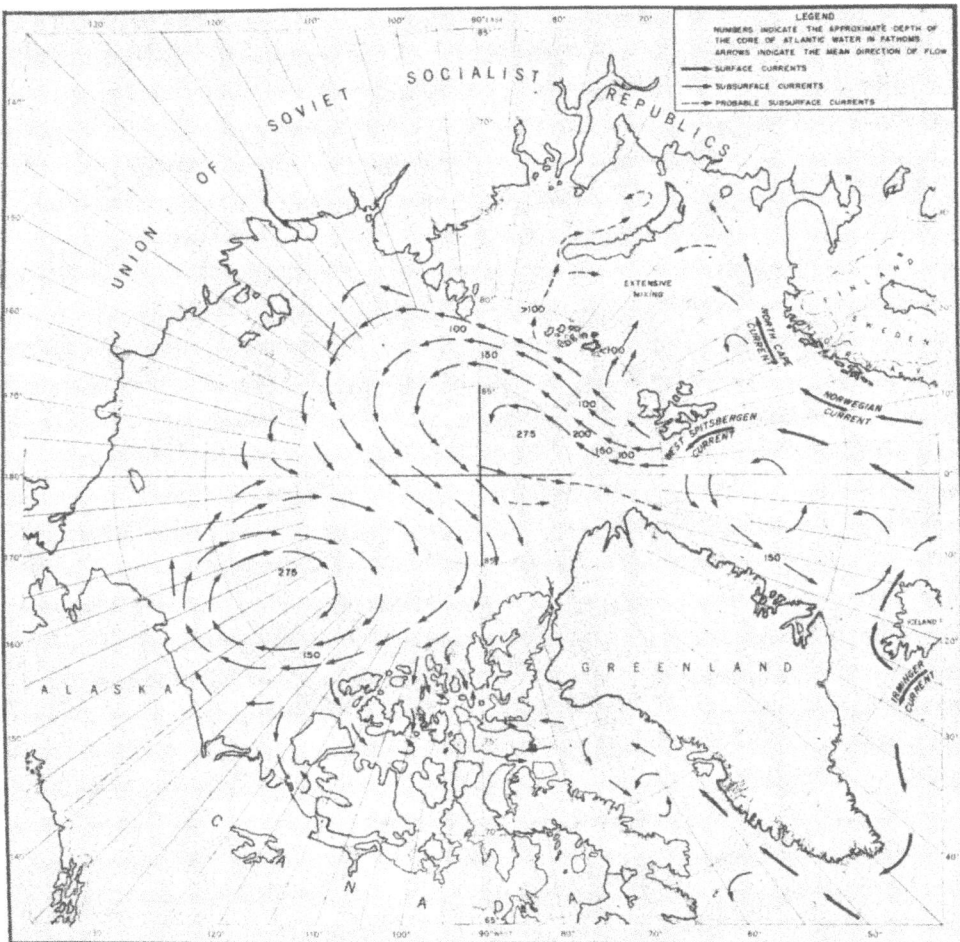

FIGURE 3. Circulation of Atlantic water in the Arctic Ocean and adjacent seas (from Figure 13—Oceanographic Atlas of the Polar Seas, United States Hydrographic Office, 1958).

(6) As the primary control of glaciation would be the Arctic-Atlantic inter-change, the cooling effect would decrease southward in the Atlantic Ocean and would be very slight in the other oceans as has been reported by Emiliani (1955). The general global refrigeration resulting from the increased albedo in the ice-covered glaciated areas and the cloud-covered pluviated areas would produce a slight decrease in temperature with a consequent increase in Alpine glaciation in areas remote from the Arctic region, particularly in the southern hemisphere;

(7) A gradual lowering of sea-level would occur as the ice-sheets built up. This would decrease the volume of interchange over the submarine sill which crosses the North Atlantic between Scandinavia and Greenland. This decreased inter-change plus the cooling produced by the ice-caps would gradually lower the tem-perature of the Arctic surface water until freezing could occur. At this point the

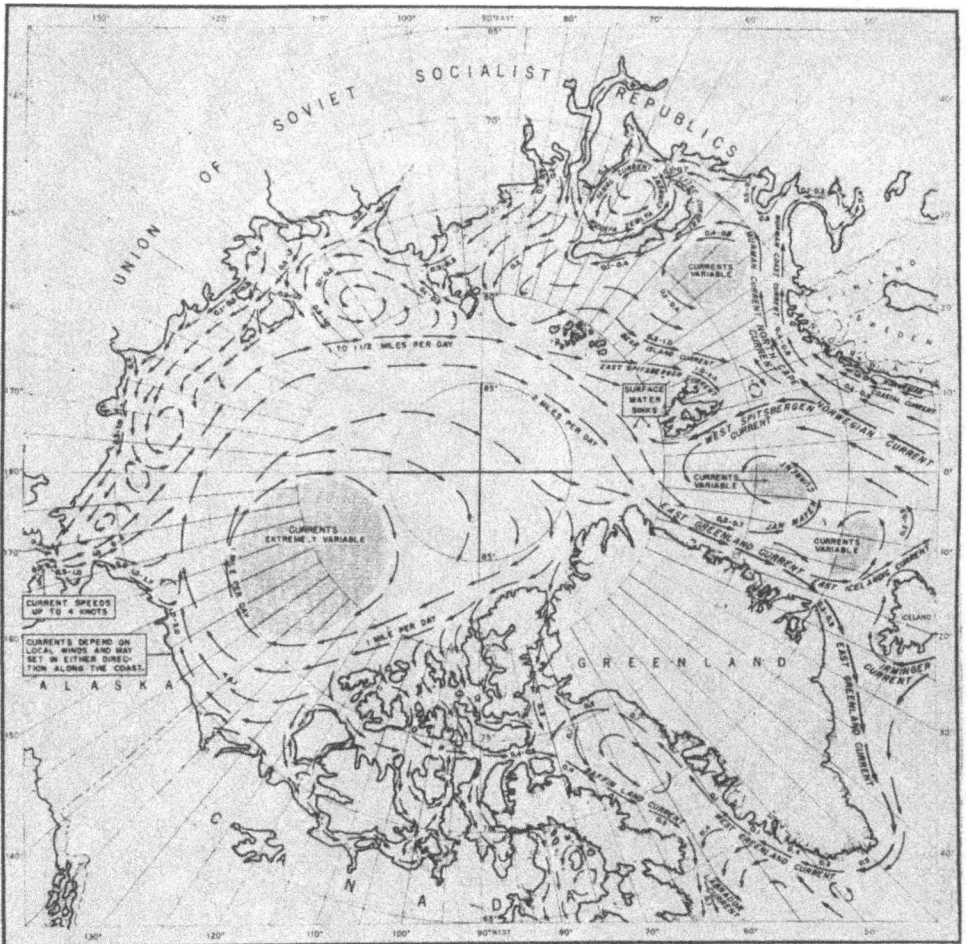

FIGURE 4.   General surface circulation of the Arctic area (from Figure 4—Oceanographic Atlas of the Polar Seas, United States Hydrographic Office, 1958).

interchange would abruptly diminish with a consequent rapid warming of the Atlantic water and cooling of the Arctic water. The meteorological model would revert to that of the present causing an immediate cut-off of glacial nourishment and a consequent wastage of the ice-caps.

(8) The proposed model can probably oscillate, explaining the glacial-interglacial stages of the Pleistocene while allowing external conditions to remain constant. On this model, the continental ice-sheets grow because there is heavy precipitation in circumpolar regions when the Arctic Ocean is ice free. The necessary heat to keep the ocean ice free is supplied by an enhanced Arctic-Atlantic interchange. But the growth of the ice-sheets would lower sea-level and thereby constrict the channel through which interchange occurs. Finally, when the ocean freezes over, the glacial stage would be ended; as the ice-sheets waste away through lack of nourishment sea-level would gradually rise, permitting Atlantic-Arctic interchange to increase until the Arctic ice would again be removed. The cycle may repeat until external conditions change.

The proposed model differs significantly from previous ones in that it provides an easy method for large accumulation of ice in high latitudes. It seems quite probable that observations north of 60° will be quite critical to its final evaluation. It would be a serious consequence, for this model, if there are large areas north of 60° which offer evidence against an augmented precipitation in Wisconsin time. It would be highly favourable if evidence can be found that heavy glaciation extended well up towards the Arctic Coast. The raised beaches and marine shells found at high elevations may ultimately show a consistent picture of isostatic adjustment from heavy glacial loading. Postglacial marine and terrestrial sediments cover some of the evidence but increasingly more of the land north of 60° is recognized as having been glaciated. Ultimately we may hope that the various stages may be identified and delineated.

GLACIATION AROUND THE ARCTIC OCEAN

The proposed model differs from previous ones in that it provides a direct source for a large accumulation of snow in the regions peripheral to the Arctic Ocean. In fact the model seems to require a substantial ice accumulation in these regions, many of which have been considered to be relatively unglaciated.

Northern Canada

Recent observations in central and northern Canada indicate strongly that the principal ice-divide lay well to the north of earlier estimates. Several Canadian publications (e.g., Wright 1956; Fyles, 1955; and others) show the major Keewatin ice-divide trending southwest from Wager Bay (northwestern Hudson Bay) to the Kazan River near Ennadai. Areas of Pleistocene marine and lacustrine submergences are reported in Lee (1959) at elevations as great as 612 and 1319 feet, respectively. Bird (1953) believes that, although the initial divide was in the northeast Keewatin province, the divide at maximum lay in the western part of Hudson Bay. He also minimizes the Labrador ice-sheet as a major center of glaciation which argues against primary nourishment from Atlantic moisture.

According to Downie, Evans, and Wilson (1953) the western shore of Hudson Bay is emphasized as a most important centre of glaciation. Although the indications on which the Keewatin divide is based have been quoted as referable only to the region of end-stage glaciation, we believe the end stage would be situated in the region where the ice was thickest in maximum stage. This is supported by independent geodetic studies of Fischer (1959) which show an area of maximum geoidal depression covering Hudson Bay and extending northwards over an approximately equal area of land. Fischer correlates this region with the region of maximum depression under the loading of glacial ice.

Recently increased exploration of the Canadian Arctic islands has resulted in observations which imply stronger glaciation in this area than has been recognized. Elevated beaches and strand lines, often with marine shells, have been found reaching to the northernmost of the Arctic islands as shown on the 1958 glacial map of Canada and in later reports. Shells on these raised strands or beaches are from hundreds to thousands of feet above present sea-level; the highest, with a Lamont age of about 22,000 years B.P. having been found by V. Sim at Eureka on Ellesmere Island at an elevation of 2,100 feet. Further, a recent report (1959) gives "the first clear evidence that glaciers crossed the Viscount Melville Sound and extended into the Queen Elizabeth Islands."

Admittedly, the evidence is not yet complete and much more is to be collected and interpreted but the observations made so far strongly indicate uplift of many hundreds to as much as 2000 feet in the northern islands. The absence of seismicity in this region leads to the conclusion that such uplift arises from glacial rebound, as accepted for the Hudson Bay, New England, and the Great Lakes areas, rather than from tectonic activity.

The present precipitation over northern Canada decreases northwards from less than fifteen inches at 60° N to fewer than four inches at the coast. An ice-free Arctic seems to be the most reasonable source of moisture supply for the thick northern Keewatin ice-sheet which probably extended over many if not all of the Canadian Arctic islands.

Northern Siberia

It has been generally believed that Siberia has suffered very little glaciation; this belief has been rationalized as the necessary consequence of the topographic barrier to storms and moisture from the south and west. However, recent Soviet reports and maps (Markov and Popov, 1959; Markov and Nalivkin, 1957; Geology of U.S.S.R., 1958) have extended the region of the known glaciation of Siberia to include a belt about 600 miles in width peripheral to the Arctic Ocean (Figure 5). In part of this region the glacial drift was obscured by a cover of 100 to 200 m of recent sediment. This buried drift was outlined after a thorough programme of exploration by means of bore holes about 1947-9. With a small exception, all of the area north of 60° N, including the Arctic islands, is now recognized as glaciated. The exception is a pocket of low-lying watery terrain extending from 30° to 160° E and from the coast to 66° N. Although this region is covered with recent lacustrine and alluvial sediment it also contains thin deposits of Wisconsin outwash (Resolution of Inter-agency Conference on Stratigraphy, 1959).

In view of the admitted lack of a moisture supply from the south or west, the ice-free Arctic postulated in our model seems to be the most reasonable source. We further suggest that the 600-mile extent of the Siberian ice-sheet is probably a good criterion of the width of a glacial belt nourished by an Arctic source alone and the width to which the North American ice-sheet would have been restricted had it not been augmented by precipitation from the moisture-laden southerly winds.

The restricted Arctic coastal plain of Alaska remains as the only Arctic region, which, although small compared to Eurasia and northern Canada, appears to be unglaciated. A possible explanation has been offered earlier by Ewing and Donn (1959).

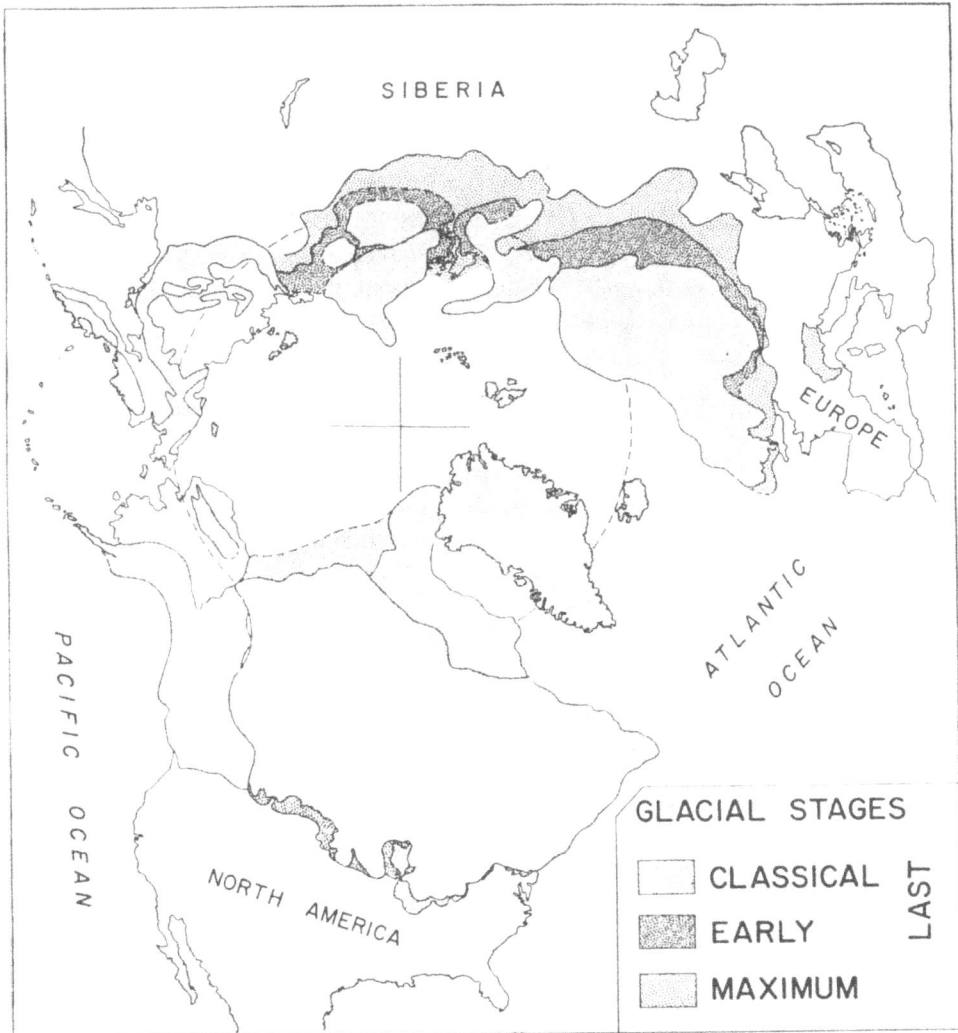

FIGURE 5. Distribution of areas of Pleistocene glaciation in the northern hemisphere for the maximum stage (third or Illinoian) and the last stage (Wisconsin) including the early and classical subdivisions (from Donn, Farrand, and Ewing, in press).

GLACIATION OF THE SOUTHERN HEMISPHERE

With the exception of the vast area of thick glaciation on Antarctica which has continued until the present, the Pleistocene glaciation of the southern hemisphere has been relatively minor. It is apparent from the summaries given by Charlesworth (1957) and Flint (1957) that although ice-sheets were present on Tasmania and South Island, New Zealand, south of 40° S where none exist at present, the other glaciers of the Pleistocene were essentially extensions of either high altitude or high latitude glaciers that are still present.

According to our model, the glaciation in the southern hemisphere, other than Antarctica, is explained as a simple consequence of the global refrigeration produced by the extensive glaciation of the northern hemisphere. This effect would be added to the cooling which still exists from the Antarctic ice-cap. Further cooling would be a consequence of the decreased radiation budget resulting from the increased albedo in the glaciated and cloudy pluviated areas of the northern hemisphere. We have estimated a decrease of several per cent (Ewing and Donn, 1958).

THE INITIAL STAGE

It has been postulated by us that the initial glacial stage was brought on by the strong polar cooling associated with the migration of the North Pole from the open Pacific Ocean system to the thermally isolated Arctic Ocean. This might have been accomplished by the movement of some outer shell of the earth over an interior zone.

The South Pole would have reached Antarctica about the same time that the North Pole entered the Arctic Ocean. The presence of the South Pole on land would result in very extreme thermal isolation and polar cooling with the consequent growth of an Antarctic ice-cap. The surrounding relatively warm ocean bodies would provide a very adequate moisture source for the nourishment and maintenance of this ice-cap. It is also a consequence of this model that the Antarctic ice-cap, having a continuous source of nourishment, like that on Greenland, would persist throughout interglacial as well as glacial stages.

Recent studies by Zeller (1959) of the thermoluminescence of limestones from Antarctica appear to give a minimum age since the last major temperature change in that region. This age is determined to be 170,000 years ago and certainly supports the consequence of our model that Antarctica would be under the same climatic régime since the initiation of the Pleistocene.

REFERENCES

BRAMLETTE, N., and BRADLEY, W.   1942.   Geology and biology of Atlantic deep sea cores; U.S. Geol. Surv., Prof. Paper 196, pp. 1-32.

BROECKER, W., and KULP, J.   1957.   Lamont natural radiocarbon measurements IV; Science, vol. 126, pp. 1324-34.

BROECKER, W., EWING, M., and HEEZEN, B.   1960.   Evidence for an abrupt change in climate close to 11,000 years ago; Amer. J. Sci. (in press) and 1959, International Oceanographic Congress, Preprints Amer. Assoc. Adv. Sci., 87.

BROECKER, W., KULP, J., and TUCEK, C.   1956.   Lamont natural radiocarbon measurements III; Science, vol. 124, pp. 154-65.

BROOKS, C. E. P.   1949.   Climate through the ages; McGraw-Hill Book Co.

CHARLESWORTH, J. K.   1957.   The Quaternary Era; vol. 1, p. 44 and vol. 2, p. 1322, Arnold.

DOWNIE, M., EVANS, A., and WILSON, J. T. 1953. Glacial features between the Mackenzie River and Hudson Bay plotted from air photographs (abstract); Bull. Geol. Soc. Amer., vol. 64, pp. 1413-14.

EMILIANI, C. 1955. Pleistocene temperatures; J. Geol., vol. 63, pp. 538-78.

——— 1957. Temperature and age analysis of deep sea cores; Science, vol. 125, pp. 383-7.

ERICSON, D., BROECKER, W., KULP, J., and WOLLIN, G. 1956. Late Pleistocene climates and deep sea sediments; Science, vol. 124, pp. 385-9.

ERICSON, D., and WOLLIN, G. 1956. Correlation of six cores from the equatorial Atlantic and the Caribbean; Deep Sea Research, vol. 3, pp. 104-25.

——— 1956a. Micropaleontologic and isotopic determinations of Pleistocene climate; Micropaleontology, vol. 2, pp. 257-70.

EWING, M., ERICSON, D., and HEEZEN, B. 1958. Sediments and topography of Gulf of Mexico; Habitat of Oil Symposium, Amer. Assoc. Petrol. Geol., pp. 995-1053.

EWING, M., and DONN, W. 1959. Theory of ice ages; Science, vol. 129, p. 464.

FISCHER, I. 1959. A tentative world datum from geoidal heights based on the Hough ellipsoid and the Columbus geoid; J. Geophys. Res., vol. 64, pp. 73-84.

——— 1959. The impact of the ice age on the present form of the geoid; J. Geophys. Research, vol. 64, pp. 85-8.

FLINT, R. F. 1957. Glacial and pleistocene geology; Wiley.

FYLES, J. G. 1955. Pleistocene features; in G. M. WRIGHT, 1955; Geol. Surv., Canada, Paper 55, 17.

GEOLOGY OF THE SOVIET ARCTIC. 1957. Ed. P. MARKOV and D. NALIVKIN, Trans. Arctic Geology Research Inst., Ministry of Geology and Mineral Conservation, no. 81, Moscow.

GEOLOGY OF U.S.S.R. 1958. All Union Geol. Research Inst. of Ministry of Geol. and Mineral Conservation (see Geomorphologic map), Moscow.

HEEZEN, B., MENZIES, R., BROECKER, W., and EWING, M. 1959. Stagnation of the Cariaco Trench, southeast Caribbean; International Oceanographic Congress Preprints, Amer. Assoc. Adv. Sci., pp. 99-102.

ICE AGE IN THE EUROPEAN SECTION OF THE U.S.S.R. AND SIBERIA. 1959. Ed. K. MARKOV and A. POPOV, Stato Lomonosov University of Moscow, Moscow.

JUNG, G. 1955. Heat transfer in the north Atlantic Ocean; Agric. and Mech. Coll. of Texas, Tech. Report 55-34T.

LEE, H. A. 1959. Surficial geology of the southern district of Keewatin and the Keewatin ice divide, N.W.T.; Geol. Surv., Canada, Bull. 51.

OVEY, C. 1950. On the interpretation of climatic variations as revealed by a study of samples from an Atlantic deep sea core; Meteorol. Soc. Centennial Proc.

PARKER, F. 1958. Eastern Mediterranean foraminifera; Rep. Swedish Deep Sea Exped. VIII, vol. 4.

PHLEGER, F., PARKER, F., and PIERSON, J. 1947-8. North Atlantic foraminifera; Rep. Swedish Deep Sea Expedition.

RESOLUTION OF INTER-AGENCY CONFERENCE ON UNIFIED STRATIGRAPHIC FRAMEWORK FOR NORTHEASTERN U.S.S.R. 1959. Ministry of Geol. and Mineral Conserv. and Soviet Acad. Sci., Moscow.

RUBIN, M., and SUESS, H. 1955. U.S. Geological Survey radiocarbon dates, II; Science, vol. 121, pp. 481-8.

——— 1956. U.S. Geological Survey radiocarbon dates III; Science, vol. 123, pp. 442-8.

SCHOTT, W. 1935. Die foraminiferen in den aequatorialen teil des Atlantischen Ozeans, Wiss. Ergeb. deutsche atlantischen expedition; Meteor 3, 3, pp. 43-134.

——— 1952. Ueber stratigraphische untersuchungsmethoden in rezenten tief see sedimenten, Göteborgs Kgl; Vetenkaps-Viterhets-Samhäll Handl. Sjätte Földgen Ser. B.6.

THORSTEINSSON, R., and TOZER, E. 1959. Western Queen Elizabeth Islands, Dist. of Franklin, N.W.T.; Geol. Surv., Canada, Paper 59-1.

TIMOFEEV, V. 1957. Atlantic water in the Arctic basin; Problemy Arktiki, no. 2, pp. 41-51.

WEXLER, H. 1958. Modifying weather on a large scale; Science, vol. 128, pp. 1059-63.

WORTHINGTON, L. 1953. Oceanographic results of project skijump I and skijump II in the Polar Sea, 1951-52; Trans. Amer. Geophys. Un., vol. 34, pp. 543-51.

WRIGHT, G. M. 1956. Map showing Pleistocene features indicating direction of ice movement; Geol. Surv., Canada, Paper 56-10.

ZELLER, E., and PEARN, W. 1960. Determination of past Antarctic climate by thermoluminescence of rocks; IGY Bulletin, Natl. Acad. Sci. no. 33, pp. 12-16.

A Survey of Temperatures in the Canadian Arctic[1]

M. K. THOMAS

ABSTRACT

The temperature régime of the Canadian Arctic is discussed and compared to conditions in the other sectors of the Arctic. Data are given outlining normal annual and diurnal variations in temperature, while temperature extremes are noted. Possible temperature trends in the Arctic are examined. Wind chill and heating degree days are examined in relation to temperature.

FOR THE PURPOSES OF THIS PAPER, the Arctic is defined as the barren grounds, tundras, and ice-caps north of the tree-line. The sub-Arctic consists of those areas to the south in the taiga or sparsely treed, boreal forest areas along the northern edge of settlement. The Arctic region is roughly delineated by the 50° isotherm of the warmest month of the year, which is usually July. In addition to outlining the Arctic region, Figure 1 has also marked on it the locations of meteorological observing stations which are referred to in this paper.

Before temperatures are discussed, the physical processes involved in producing and maintaining this northern hemisphere "icebox" are considered briefly. Although terrestrial radiation loss is, in general, fairly uniform at all latitudes, the tropical latitudes receive an abundance of solar radiation in comparison to the Arctic. The general circulation of the atmosphere is the result of this temperature differential and the energy balance of the earth and its atmosphere is thus maintained.

Because of the peculiarities of geography, the Arctic region is not restricted to a fairly regular latitudinal belt around a heat sink at the pole. On the contrary, in summer the cold pole of the northern hemisphere is in central Greenland and in winter there are two cold centres, one in Greenland and one in northern central Siberia. Also, because of the influence of geography on the general circulation of the atmosphere, the true Arctic extends much farther south in eastern North America than it does in either western North America or in Eurasia where the Arctic boundary more nearly parallels latitude.

ANNUAL TEMPERATURE RÉGIME

Graphs showing the annual temperature régimes for a few selected stations are given in Figure 2. It will be noted at once that some stations exhibit a greater change from summer to winter than others. As a rule, the extreme cases represent data from more "continental" stations than the others where there is a maritime régime. With the exception of interior Greenland, most Arctic areas are located

[1]Published with the approval of the Director, Meteorological Branch, Department of Transport, Toronto, Ontario.

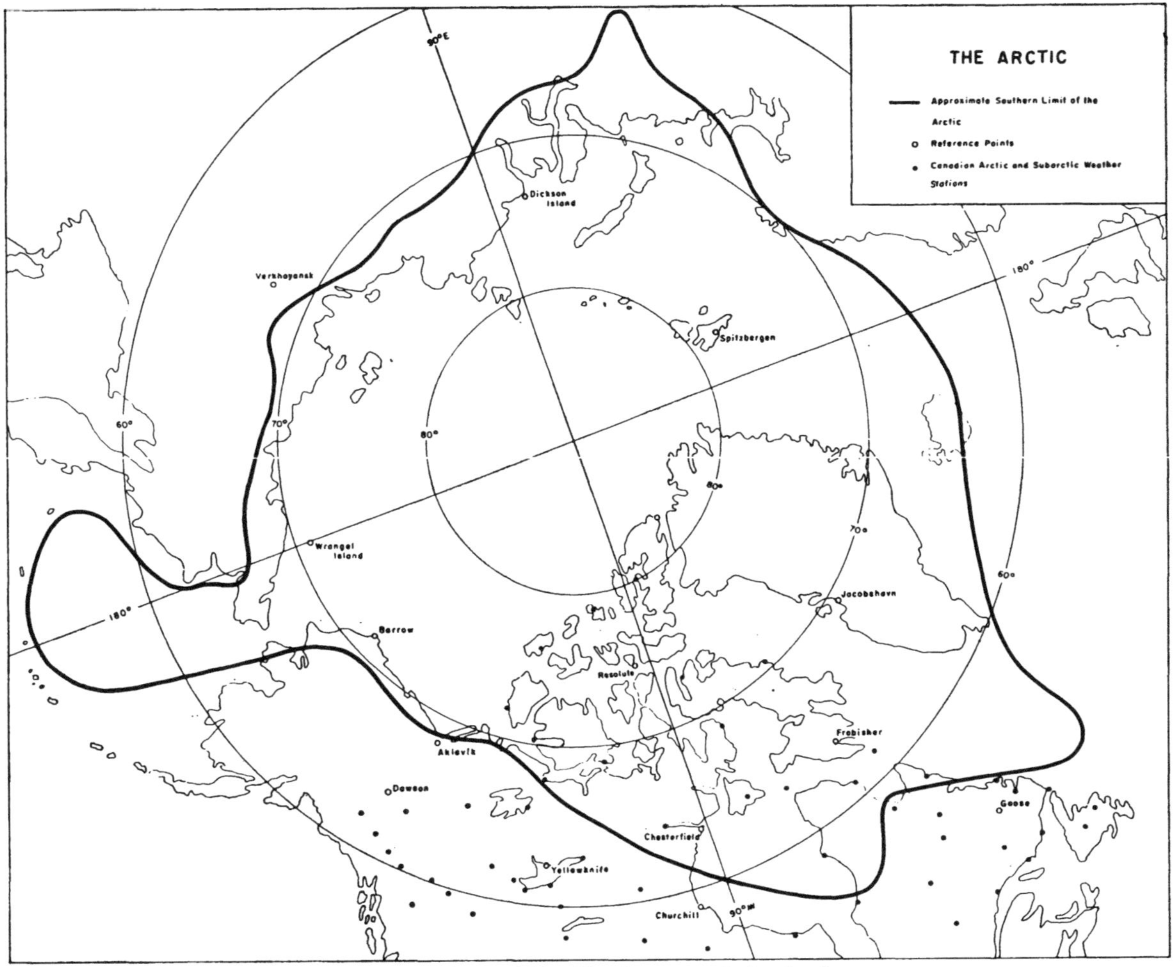

FIGURE 1. General identification map of the Arctic.

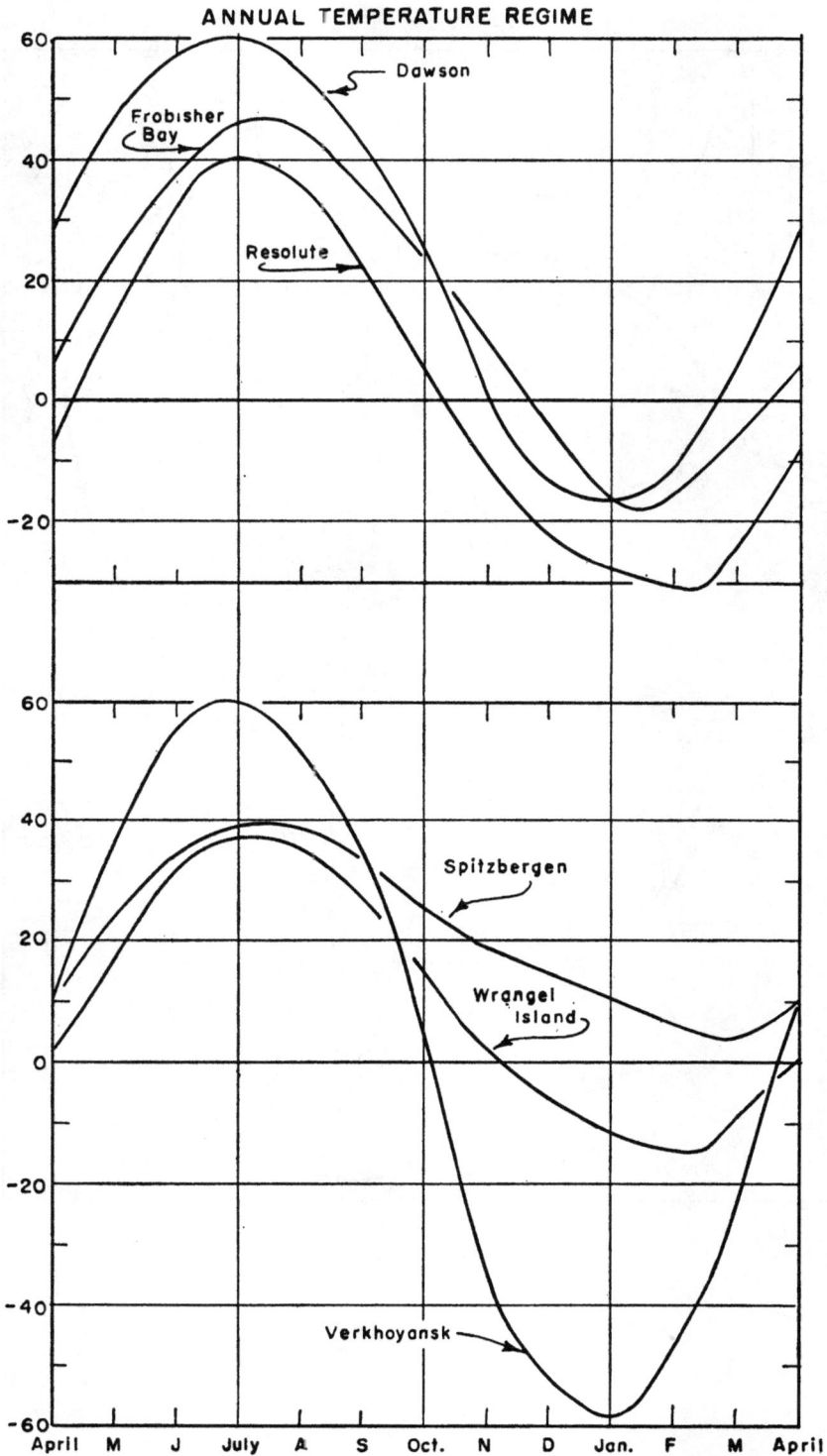

FIGURE 2. Annual temperature régimes showing mean monthly temperature (°F) for representative stations in the American and Eurasian Arctic and sub-Arctic.

fairly near the sea and so are more maritime than continental in nature. At Resolute there is a 70° difference in temperature from the warmest to the coldest month in a normal year, while at Dawson in a definite continental sub-Arctic type of climate, the range is 80°. In Eurasia the difference is even more pronounced. Along the Arctic coast at Wrangel Island, the range is about 50° in contrast to 120° range at Verkhoyansk in the sub-Arctic. Of all places on earth this area of the U.S.S.R. has the greatest variation in temperature from summer to winter.

Although the annual temperature cycle is a very marked and reliable phenomenon, there are sometimes wide fluctuations in both Arctic and mid-latitude regions from one winter to the next. Over a ten-year period at Resolute, for example, the coldest month on record averaged −36°, while in a relatively mild winter the coldest month averaged only −26°. It is interesting to note though that much wider variations do occur in more southerly latitudes. At Calgary, for example, one of the coldest winter months on record was January 1950 when the average temperature was −14°, but in the winter of 1930-1, March was the coldest month with the average temperature a remarkable 26° above, illustrating a spread of 40° compared to 10° at Resolute. Summer temperatures do not vary much from year to year either in the Arctic or in southern Canada.

Throughout all of Canada the coldest weather in any winter may occur from late November to early March. At Resolute, in January and February, the average temperature curve is relatively flat with slowly decreasing temperatures until a minimum is reached about March 1 just after the sun's heat begins to become effective but before it becomes appreciable. In summer, the time of maximum temperature is during the second week of July, a lag of only two weeks or so after the summer solar solstice.

WINTER TEMPERATURES

Mean daily temperatures in January are illustrated in Figure 3 which shows that the southern part of the Arctic islands is not as cold as the north-central mainland of Canada. The northwestern islands and northern Ellesmere Island are, however, the coldest areas in Canada in January. The southern Arctic islands are kept slightly warmer than the mainland by heat escaping through the ice from the water below. This effect is illustrated better by considering the lowest reported temperatures in Canada. The lowest temperature (−81° F) was observed at Snag in Yukon Territory; whereas −69° at Hazen Lake is the coldest temperature ever officially reported on the Arctic islands. In fact, temperatures as low as any reported from the Arctic have been observed in all the provinces except the Maritimes.

With the supply of heat largely cut off in midwinter, it is interesting to examine the diurnal variation of temperatures at Resolute as an example of conditions in that part of the Arctic. Here, it has been shown that any diurnal change in temperature from October 10 through to about March 10 may be attributed to random variations. As the atmosphere is usually completely stable, with a marked temperature inversion, an increase in wind speed from any direction or an increase in cloudiness will bring a temperature rise. Surprisingly, under these conditions in midwinter, the average daily range is as high as 10 to 15° F.

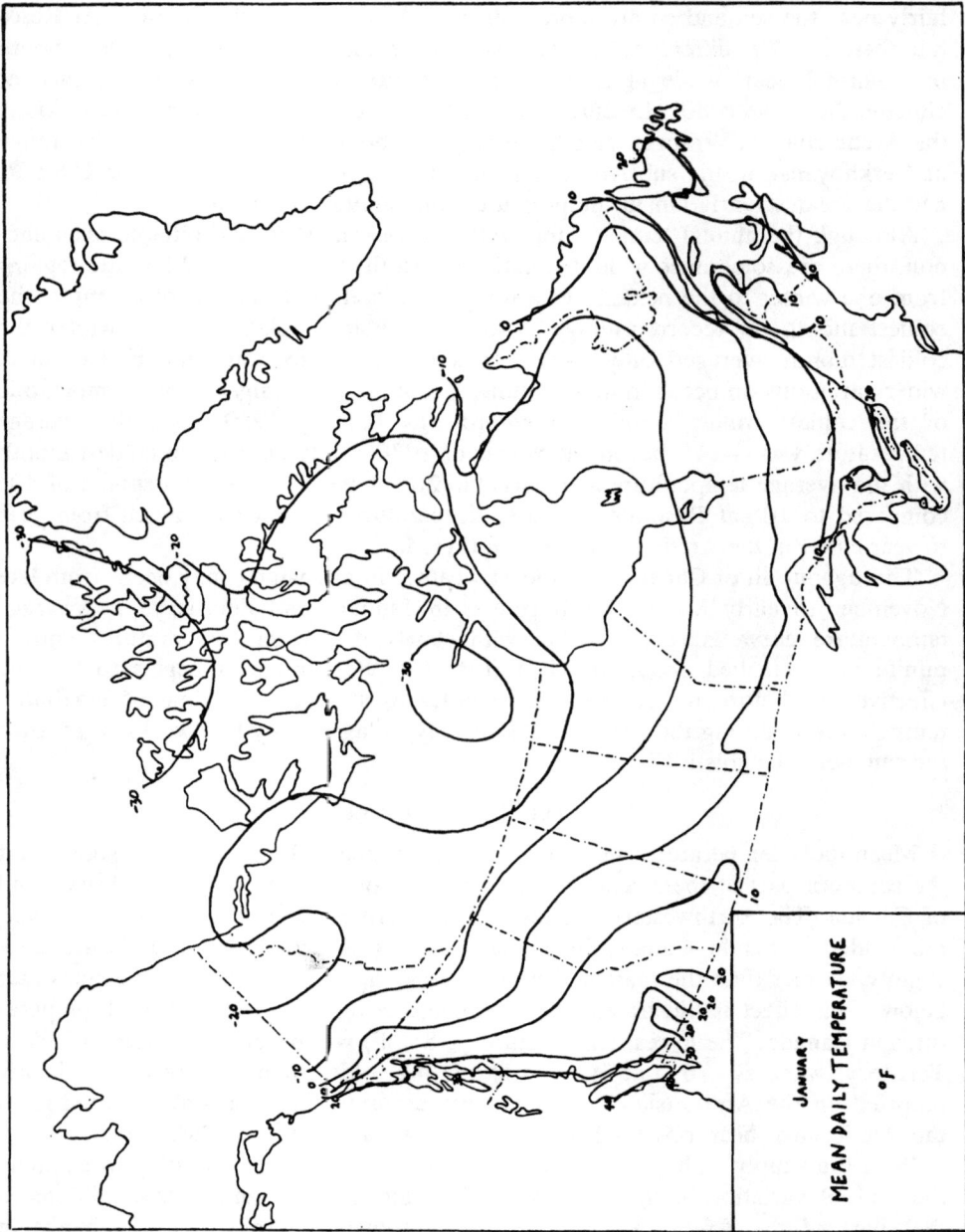

FIGURE 3. Mean daily temperature in January (°F).

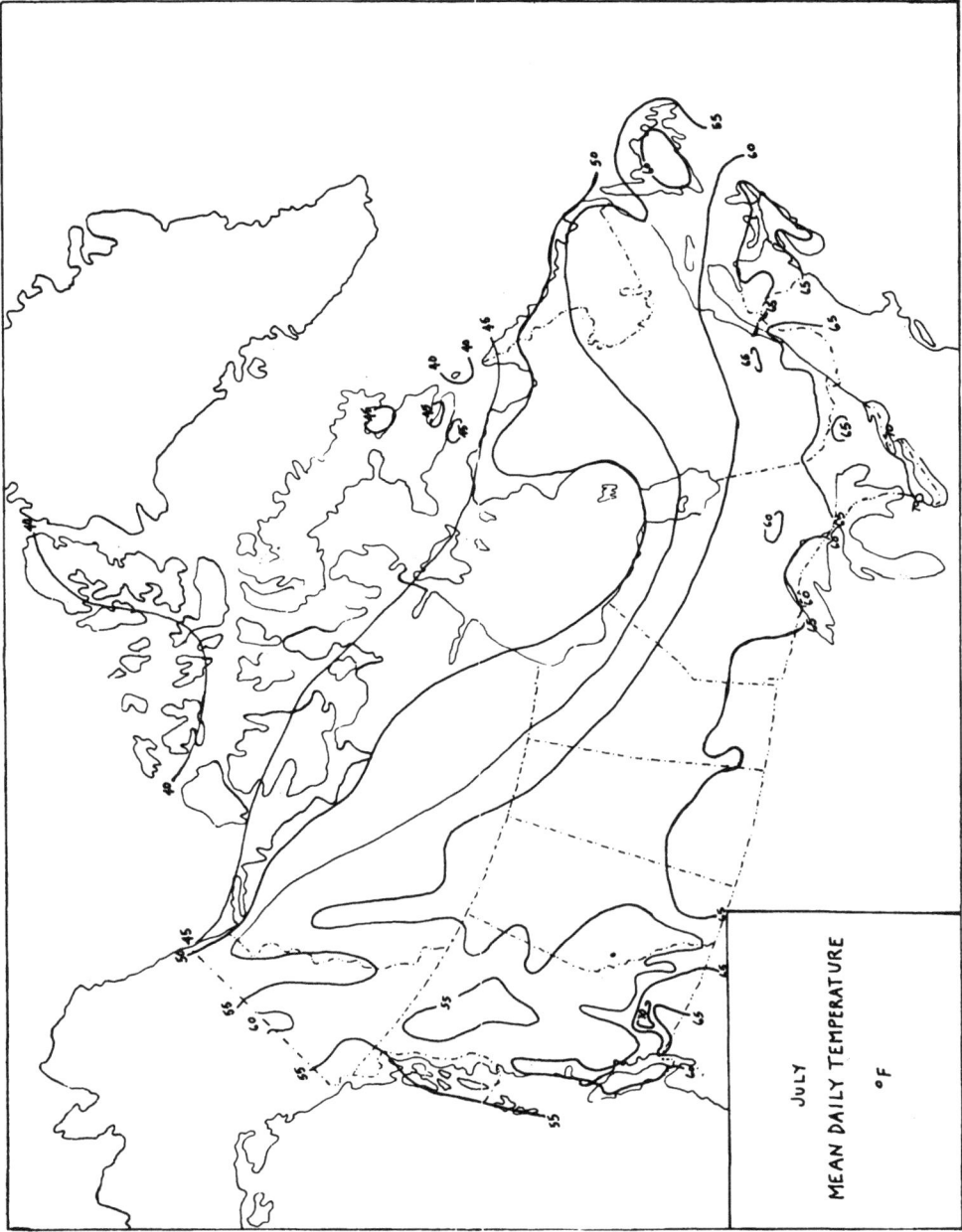

JULY

MEAN DAILY TEMPERATURE

°F

FIGURE 4. Mean daily temperature in July (°F).

SUMMER TEMPERATURES

Summer as we know it in the southern latitudes of Canada does not exist in the Arctic. The "Arctic summer," when the mean temperature averages above 32°, is only about two months in length. At Resolute the warmest month on record was July 1958 when the average surface temperature was 43°. In contrast, the mainland coastal areas usually have warmest month temperatures averaging near or over 50°.

Inspection of the July mean temperature map reveals that temperature averages are considerably more uniform throughout the Arctic in summer than in winter, and average usually between 40 and 50°. There will be, however, local differences on any island, depending on location and wind direction. Windward coasts in particular will be much colder than the insolation-heated island interiors.

Although pronounced in spring, there are no marked diurnal variations of temperature in the summer and fall months in the Arctic. Coudiness is at a minimum in spring, and with no open water temperatures tend to follow the diurnal insolation variations. In summer, however, a wind shift from land to sea or vice versa will bring a greater change in temperature than an ordinary diurnal change. Near the coasts of the islands the average daily range in July is near 10°.

Extremely high temperatures have never been experienced in the Arctic. At Resolute the highest official temperature on record is 60° and the maxima reported from most far northern stations are less than 70°. In contrast, mainland locations, even on the coast, have had temperatures above 80°.

HEATING DEGREE DAYS

We have already seen how areas of continental "temperate" Canada occasionally have winter months that average as cold or colder than anything as yet reported from the Canadian Arctic. Record low temperatures in the Arctic are likewise not as low as those reported further south. But the winter season is long, and over the years as a whole the accumulation of heating degree days below 65° greatly exceeds anything in southern Canada. Inspection of Figure 5 reveals that these values in the Arctic are between two and three times as large as similar values in southern Canada. With normal wind and sunshine conditions, heating degree days are usually proportional to the amount of fuel required to heat a properly insulated building to a temperature of 70° and so are useful in calculating fuel requirements. Although the total values vary from year to year, such variations in the Arctic are usually considerably less than 10 per cent of the annual average total.

WIND CHILL

During the past two decades the term "wind chill" has been devised and used to illustrate the "coldness" of the atmosphere in terms of temperature, wind, and other elements. Wind chill is an approximation of the rate of loss of heat from exposed flesh and is assumed to be proportional to the measured rate of cooling of suitable objects. A formula has been devised by research workers using standard air temperatures and wind speeds which gives values of dry shade atmospheric cooling in kilogram calories per square metre per hour. This has been popularly used in recent years to illustrate the wind chill in various Arctic and sub-Arctic climates.

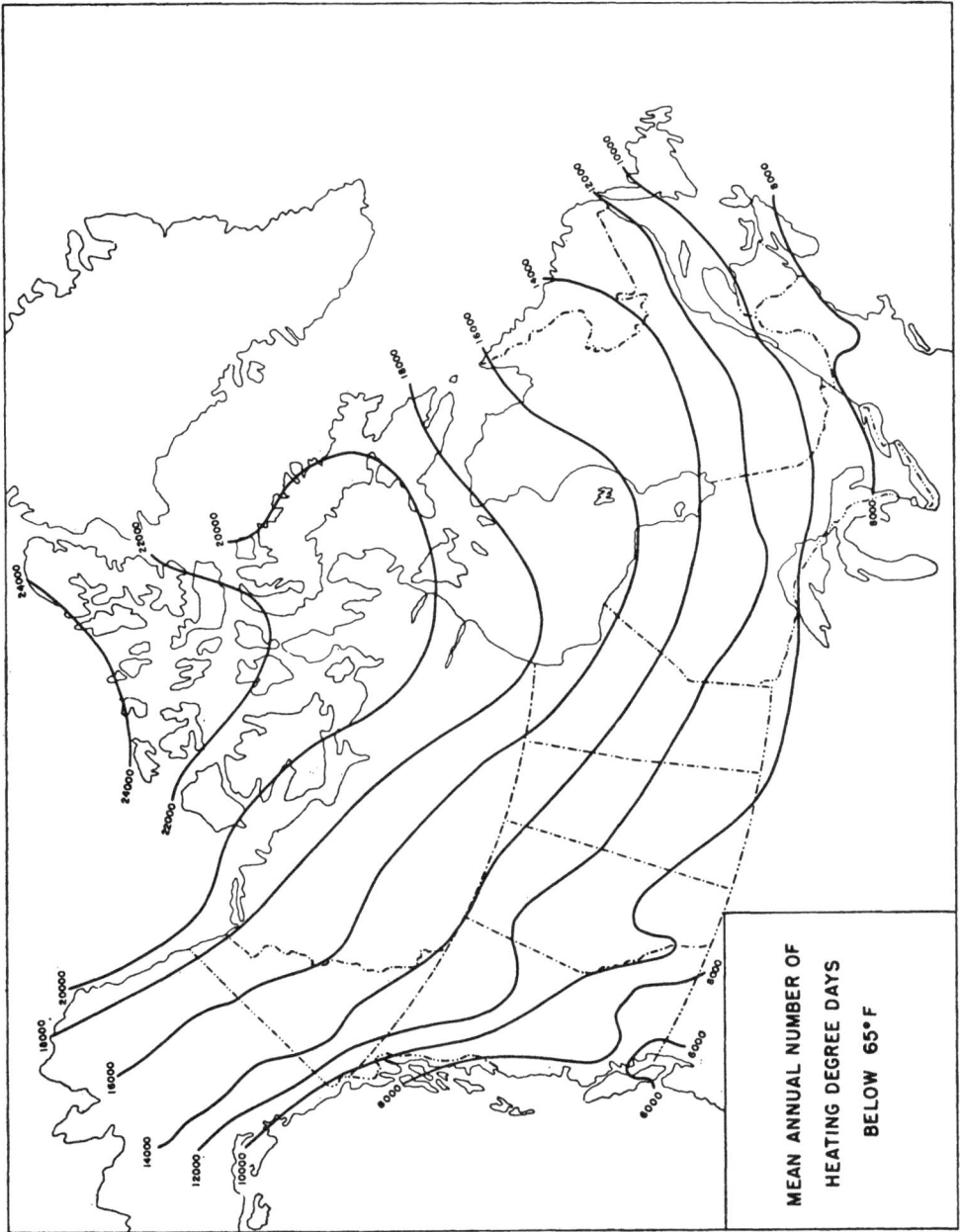

MEAN ANNUAL NUMBER OF
HEATING DEGREE DAYS
BELOW 65° F

FIGURE 5. Mean annual number of heating degree days below 65° F.

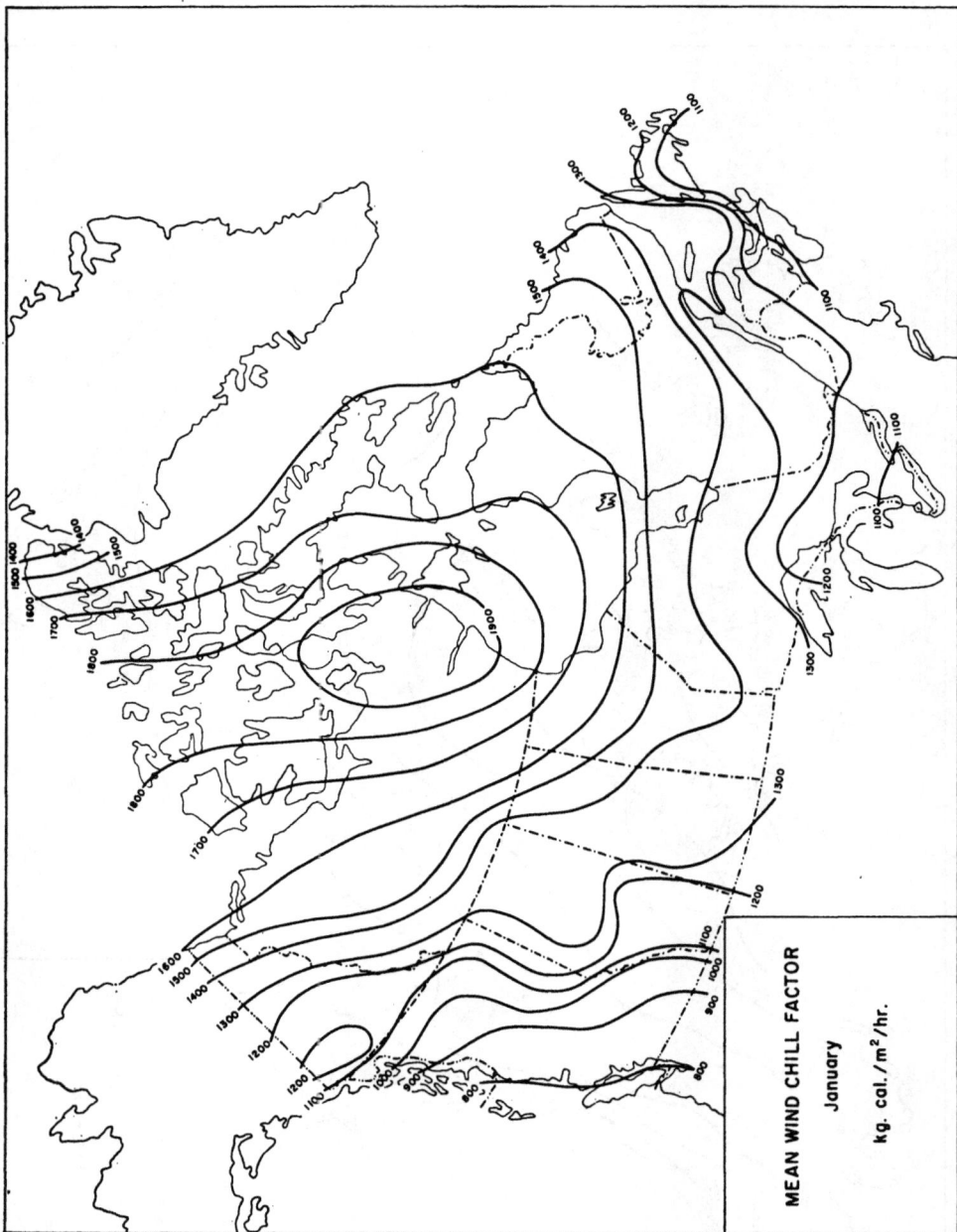

MEAN WIND CHILL FACTOR

January

kg. cal. / m² / hr.

FIGURE 6. Mean wind chill factor, January, in kg cal/m²/hr.

The values described below have been obtained from such an equation using monthly mean temperatures and wind speeds.

Most Canadian locations have their highest wind chill values in January. Figure 6 illustrates these values and shows that they are greatest in the Keewatin district west of Hudson Bay with Baker Lake and Chesterfield (average 1980 and 1950 kg cal/m²/hr respectively) having the distinction of being the coldest stations in Canada on the basis of wind chill. As average wind speeds in the Arctic islands are normally less than those around Hudson Bay, wind chill values, although high, are noticeably less. Values from the southeastern prairies and the Gulf of St. Lawrence area are as high as those in the northeastern Arctic and even greater than those from the Yukon.

CLIMATIC FLUCTUATIONS

For the Canadian sub-Arctic, homogeneous meteorological records are available from the turn of the century onwards, but for the Arctic only from the early 1920's on. January temperatures are usually representative of winter values, and thus ten-year moving means of these data have been studied. Figure 7 shows the resulting trend lines from a number of Canadian stations as well as representative stations in Alaska, Greenland, Norway, and the U.S.S.R. In each case the ten-year moving mean has been credited to the final year in each decade.

In the sub-Arctic records from Dawson and Fairbanks, a warming period from the 1910's to the early 1930's is revealed, and a further warming in the 1940's. In this continental area, the trend during the 1950's has been towards lower temperatures. An entirely different trend has been observed on the western Arctic coast and is illustrated by data from Barrow and Aklavik. Along the coast there have not been any marked climatic fluctuations since the beginning of observations in the 1920's. In the eastern Arctic, records from Arctic Bay and nearby Pond Inlet date back to the twenties, and this area has shown a warm-up similar to the one at Dawson in the forties with a subsequent cooling-off period during the fifties.

In Eurasia, the counterpart of Dawson data are shown for Verkhoyansk in the sub-Arctic. The Arctic coast of Eurasia is represented by data from Dickson Island and comparable graphs are also shown for Spitsbergen and Jacobshaven, Greenland. Study of these graphs indicates that fluctuations in January temperatures have not been uniform throughout either North America or Eurasia.

SUMMARY

By our standards in southern Canada, the Arctic winter is extremely cold, and summer, as we know it, barely occurs each year. However, on occasion, temperatures have been much lower in continental Canada than any reported from the Arctic islands. Wind chill averages are higher to the west of Hudson Bay than in the far north but the accumulation of heating degree days is greatest on the northern islands. Midwinter temperatures have fluctuated during the past several decades, generally reaching a maximum in the forties. Monthy values of mean temperature, record low temperatures, heating degree days, and wind chill factors are given in Tables I and II for a selection of Arctic stations in comparison to similar values from sub-Arctic and southern Canada.

JANUARY TEMPERATURE FLUCTUATIONS

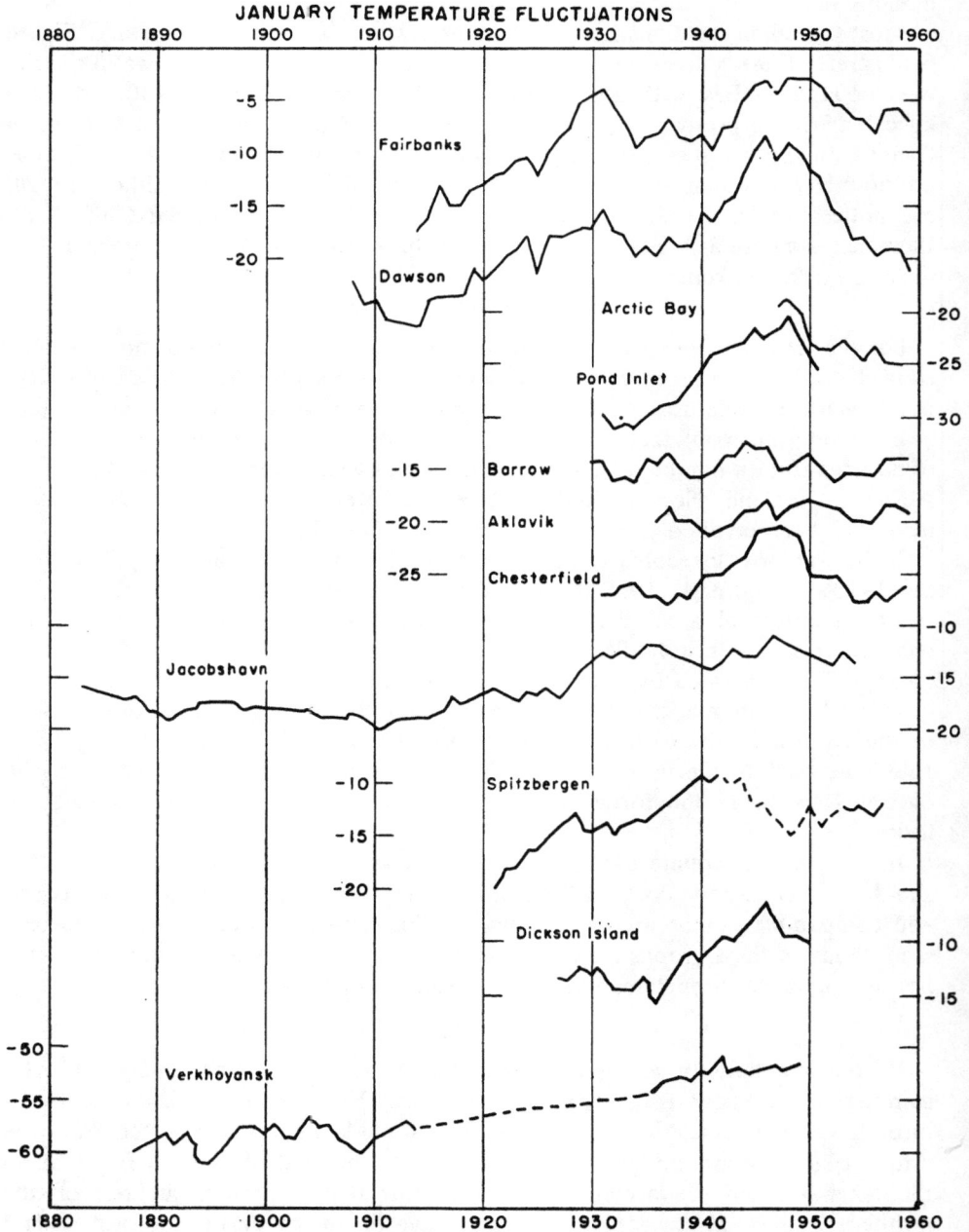

FIGURE 7.   Ten-year moving means of January temperature from representative stations in
North America and Eurasia.

TABLE I
CANADIAN ARCTIC TEMPERATURE DATA*

Station	Location Long.	Location Lat.	Location El.	Years of record	Summer Record highest	Summer Average highest	Winter Record lowest	Winter Average lowest	Winter Average Jan. wind chill†	Average annual heating degree days**
Alert	82°30'	62°20'	205'	9	68	59	−54	−51	1220	23,750
Eureka	80°00'	85°56'	8'	12	67	61	−63	−59	1700	24,740
Isachsen	78°47'	103°32'	83'	11	64	57	−65	−58	1830	24,460
Mould Bay	76°14'	119°20'	50'	11	59	55	−63	−55	1720	23,630
Resolute	74°43'	94°59'	209'	11	60	57	−61	−54	1810	22,600
Arctic Bay	73°00'	85°18'	36'	21	75	59	−57	−49	1350	20,860
Pond Inlet	72°43'	77°30'	10'	31	77	60	−64	−53	—	21,260
Sachs Harbour	72°00'	124°30'	277'	3	64	63	−54	−53	1850	21,580
Holman Island	70°30'	117°38'	30'	17	78	70	−50	−43	1700	19,610
Clyde	70°27'	68°33'	10'	16	71	64	−49	−45	1700	19,830
Cambridge Bay	69°07'	105°01'	47'	26	76	67	−63	−53	1800	21,490
Hall Lake	68°47'	81°15'	34'	4	63	62	−54	−51	1800	20,910
Coppermine	67°49'	115°05'	28'	28	87	79	−58	−48	1590	19,410
Padloping Island	67°06'	62°21'	130'	13	73	79	−49	−42	1580	18,700
Coral Harbour	64°12'	83°22'	193'	16	79	72	−61	−52	1810	19,670
Frobisher Bay	63°45'	68°33'	68'	18	76	67	−49	−44	1540	17,920
Chesterfield	63°20'	90°43'	13'	37	86	76	−60	−50	1940	19,710
Nottingham Island	63°07'	77°56'	54'	29	73	64	−42	−36	1610	17,870
Lake Harbour	62°50'	69°55'	54'	21	80	74	−49	−38	—	16,930
Resolution Island	61°18'	64°53'	127'	27	61	55	−36	−25	1580	16,110
Cape Hopes Advance	61°05'	69°33'	240'	22	81	70	−42	−34	1710	16,610
Port Harrison	58°27'	78°08'	66'	33	86	73	−57	−41	1700	16,880
For comparison Sub-Arctic										
Aklavik	68°14'	135°00'	30'	32	93	83	−62	−51	1460	17,910
Baker Lake	64°18'	96°00'	30'	13	82	78	−58	−53	1870	19,990
Fort Chimo	58°06'	68°26'	117'	20	90	84	−51	−42	1600	15,600
Dawson	64°04'	139°26'	1062'	38	95	86	−73	−54	1210	15,020
Snag	62°22'	140°24'	1925'	14	89	85	−81	−61	1120	15,600
Yellowknife	62°28'	114°27'	682'	18	86	83	−60	−51	1550	15,640
Churchill	58°45'	94°04'	115'	71	96	86	−57	−42	1800	16,910
Knob Lake	54°49'	66°49'	1681'	11	88	83	−59	−48	1610	14,890
Southern										
Vancouver	49°11'	123°10'	16'	61	92	87	0	14	760	5,520
Calgary	51°06'	114°01'	3540'	75	97	92	−49	−31	1140	9,520
Winnipeg	49°54'	97°14'	786'	87	108	96	−54	−35	1400	10,658
Toronto	43°40'	79°24'	379'	117	105	94	−26	−8	1110	7,008

*Temperatures in degrees Fahrenheit
†Wind chill factor in kg. cal./sq.m./hr.
**Heating degree days below 65°F.

TABLE II

CANADIAN ARCTIC TEMPERATURE DATA*

Station	Jan.	Feb.	March	April	May	June	July	Aug.	Sept.	Oct.	Nov.	Dec.	Annual
							Mean temperature						
Alert	−28	−28	−27	−10	11	32	39	34	15	−4	−15	−22	0
Eureka	−36	−37	−34	−16	14	37	42	38	19	−8	−22	−33	−3
Isachsen	−33	−34	−29	−11	12	31	37	33	16	−3	−18	−27	−2
Mould Bay	−29	−33	−25	−7	12	31	38	34	20	1	−15	−26	0
Resolute	−28	−30	−24	−7	14	33	40	37	23	6	−10	−21	3
Arctic Bay	−23	−25	−16	−5	19	36	43	41	29	14	−4	−15	8
Pond Inlet	−26	−29	−20	−3	19	35	42	41	31	15	−5	−20	7
Sachs Harbour	−23	−26	−18	−2	17	35	41	37	29	10	−9	−20	6
Holman Island	−17	−21	−13	1	22	38	46	43	32	17	−2	−12	11
Clyde	−15	−19	−13	−3	19	33	40	39	32	21	3	−10	11
Cambridge Bay	−29	−32	−23	−6	15	34	46	45	32	11	−11	−24	5
Hall Lake	−19	−22	−15	−8	18	32	43	40	32	17	−10	−18	8
Coppermine	−19	−21	−14	1	22	39	49	47	36	19	−4	−15	12
Padloping Island	−14	−18	−11	4	23	36	42	41	34	24	9	−6	14
Coral Harbour	−23	−23	−11	1	19	35	46	45	32	17	3	−11	11
Frobisher Bay	−16	−15	−6	6	24	38	46	45	35	24	12	−4	16
Chesterfield	−26	−26	−15	1	21	37	48	47	37	22	0	−16	11
Nottingham Island	−14	−14	−4	9	25	35	42	42	35	26	12	−3	16
Lake Harbour	−10	−10	−1	11	27	39	46	45	37	26	12	−3	18
Resolution Island	−1	−1	7	15	27	33	37	38	35	29	21	9	21
Cape Hopes Advance	−7	−8	0	11	25	35	42	42	36	29	20	7	19
Port Harrison	−15	−16	−6	11	28	39	47	47	41	31	17	−1	19
For comparison—Sub-Arctic													
Aklavik	−18	−17	−9	9	31	49	56	50	38	20	−3	−17	16
Baker Lake	−30	−27	−15	2	21	37	51	50	37	18	−3	−19	10
Fort Chimo	−13	−11	3	15	32	44	53	51	38	20	−3	−17	16
Dawson	−16	−11	6	28	47	57	60	55	43	27	−2	−13	24
Snag	−13	−7	12	26	46	54	57	52	42	25	−4	−18	23
Yellowknife	−15	−14	1	17	39	53	61	57	45	31	7	−13	23
Churchill	−17	−17	−4	11	28	42	55	53	43	30	10	−9	19
Knob Lake	−14	−9	1	18	34	47	55	51	43	32	16	0	23
Southern													
Vancouver	36	39	43	49	55	60	64	63	58	51	43	39	50
Calgary	16	17	26	39	50	56	62	60	52	42	28	19	39
Winnipeg	1	4	19	38	52	62	68	66	55	43	23	8	37
Toronto	25	24	32	44	55	66	71	69	62	50	39	28	47

*Temperatures in degrees Fahrenheit.

REFERENCES

CANADA, DEPT. OF MINES AND TECHNICAL SURVEYS ATLAS OF CANADA.   1957.   (Contains 51 climatic maps prepared by the Meteorological Branch.)

KIMBLE, G. H. T., and GOOD, D.   1955.   Geography of the northlands; Amer. Geog. Soc. Spec. Pub. no. 32, Wiley.

LANGE, R.   1959.   Zur Erwarmung Gronlands und der Atlantischen Arktis; Annalen der Meteorologie, vol. 8, pp. 9-10, 265-303.

LONGLEY, R. W.   1958.   Temperature variations at Resolute, Northwest Territories; Roy. Meteorol. Soc. Quart. J., vol. 84, no. 362, pp. 459-63.

PUTNINS, P., and STEPANOVA, N. A.   1956.   Climate of the Eurasian northlands, in Dynamic north; U.S. Navy.

RAE, R. W.   1951.   Climate of the Canadian Arctic Archipelago; Canada, Dept. of Transport, Toronto.

RUBINSHTEIN, E. S.   1959.   On the changes of climate in the U.S.S.R. during recent decades (trans. A. NURKLIK); Meteorological Translations no. 3, Meteorological Branch, Department of Transport.

THOMAS, M. K., and BOYD, D. W.   1957.   Wind chill in northern Canada; Canadian Geographer, vol. 10, pp. 29-39.

THOMAS, M. K.   1953.   Climatological atlas of Canada; National Research Council, Division of Building Research, Canada, and Dept. of Transport, Met. Div. National Research Council no. 3151, Ottawa.

UNITED STATES ARMY.   1957.   Climatic analogs of Fort Greely, Alaska, and Fort Churchill, Canada, in Eurasia; Quartermaster Research and Engineering Centre, Environmental Protection Research Division, Technical Report no. EP-77.

———   1959.   Climatic analogs of Fort Greely, Alaska, and Fort Churchill, Canada, in North America; Quartermaster Research and Engineering Centre, Environmental Protection Research Division, Technical Report no. EP-111.

WILSON, H. P., MARKHAM, W. E., DEWAR, S. W., and THOMPSON, H. A.   A series of circulars on Arctic weather and climate issued by the Meteorological Branch; Cir.-1954 (1951), Cir.-1955 (1951), Cir.-2198 (1952), Cir.-2094 (1952), Cir.-2366 (1953), Cir.-2387 (1953), Cir.-2923 (1957).

Permafrost Investigations in Canada

R. F. LEGGET, H. B. DICKENS,

AND R. J. E. BROWN

ABSTRACT

A knowledge of permafrost distribution is of vital concern to the practising engineer and geologist. The Division of Building Research is engaged in mapping permafrost in Canada on the basis of continuous and discontinuous zones. The information obtained to date has been compiled mainly from direct field observations and the mapping of permafrost by such methods is a long-term project. Recently the Division has initiated a study of climatic and terrain factors affecting the existence of permafrost in specific locations as a means of improving the ability to predict permafrost conditions. The basic energy factors affecting permafrost distribution cannot be measured with sufficient accuracy to determine the thermal condition causing permafrost which is the integrated effect of climatic and terrain factors over a long period. The objective of this study, therefore, is to investigate the thermal characteristics of the terrain components and to seek correlations between these, the climatic factors, and the distribution of perennially frozen ground.

ALMOST ONE-HALF OF THE LAND AREA of Canada is underlain by permafrost. This is the term now commonly used to describe that part of the earth's crust where the temperature is perennially below 0° C. (Bryan [1946] suggested terms that are more precise semantically but they have not been generally accepted in North America.) The existence of permafrost is of interest to both the earth scientist and the civil engineer. To the former, this phenomenon raises a number of puzzling geothermal questions; to the latter, it becomes a matter of vital concern wherever the frozen material is a water-bearing silt or clay because its presence can be fraught with serious consequences for engineering work. Figure 1 shows a typical result of neglect of necessary and well-recognized precautions when engineering construction disturbs permafrost conditions.

Appreciating the widespread existence of permafrost, the Division of Building Research of the National Research Council has been conducting investigations into permafrost since 1950 as part of its contribution to the progress of building in Canada. From both a practical and a scientific point of view, there arise two main questions: where may permafrost be encountered in Canada and why does the condition exist?

In its continuing search for answers to these questions, the Division has gathered information from a variety of sources including the technical literature, reports from others operating in permafrost areas, and from direct field observations. Emphasis has been placed on determining the southern limit of permafrost but observations from the entire permafrost region are also being recorded. Observations have been classified as indicating continuous or discontinuous permafrost defined on an areal basis as follows: *continuous* where permafrost is found everywhere in the region, and *discontinuous* where permafrost exists in combination with some areas of unfrozen material. The discontinuous zone is one of broad transition between continuous permafrost and ground with a mean temperature which is above 0° C. In this zone, permafrost may vary from a widespread distribution with isolated patches of unfrozen ground to predominantly thawed material containing islands of material that remain frozen.

In Alaska (Benninghoff, 1952) and Russia (Baranov, 1955; Sumgin *et al.*, 1940; Tsytovich, 1958; Tumel', 1946) permafrost classifications have been made on the basis of temperature and thickness of the permafrost. Classification on a temperature basis has been greatly aided in the U.S.S.R. by the large number of ground temperature installations that have been made in permafrost regions and which have been used in preparing maps of permafrost distribution. In Canada, ground temperature observations in northern regions are still insufficient to delineate permafrost zones in this way. Enough information has been collected, however, to permit an approximate classification of permafrost on an areal basis (Figure 2). This collecting of information is continuing, facilitated by a permafrost questionnaire that has been prepared by the Division and which is now being circulated to appropriate groups throughout northern Canada.

Mapping the distribution of permafrost by such means is inevitably a long-term project. To improve accuracy in the prediction of the occurrence of permafrost, it is necessary to study the basic factors affecting its formation and continued existence. Permafrost exists as a result of a thermal condition which is reflected in a mean ground temperature that never rises above 0° C. The ground temperature pattern is complicated, varying with depth from the surface and with time. Because temperature is an index of the level of heat storage, it fluctuates in response to the

FIGURE 1. Soil flow resulting from thawing of perennially frozen silt under a road.

heat losses and heat gains at the ground surface. These follow a cyclical pattern on a daily as well as a yearly basis. The mean ground temperature will change over a given period of years only to the extent that the heat losses and heat gains for that period are unequal, the difference being reflected in a change in the mean level of heat stored in the ground. It follows that any extensive permafrost condition may develop over a very long period of time owing to a very small imbalance in annual heat losses and heat gains. A local permafrost condition may, however, be created or destroyed in a short period only if a relatively large imbalance between annual heat losses and heat gains has been created. To learn more about the occurrence of permafrost, attention must therefore be directed to the factors that determine the annual heat losses and heat gains at the ground surface. These factors may be grouped broadly under climate and terrain, the latter including ground thermal properties. They may not be readily separated in the over-all picture because of the interactions between them.

Across the upper boundary of the earth's atmosphere the energy gains by solar radiation may be regarded as constant, being balanced by a corresponding loss by radiation from the earth to outer space. The distribution of solar radiation at the earth's surface varies greatly, however, not only on a regular pattern—with time of day, time of year, and latitude—but also on a much more variable basis owing to absorption and scattering produced by the earth's atmosphere. Clouds, for example, may at times provide complete interception of direct solar radiation. Climate is produced by the radiation exchange that takes place between outer space, the

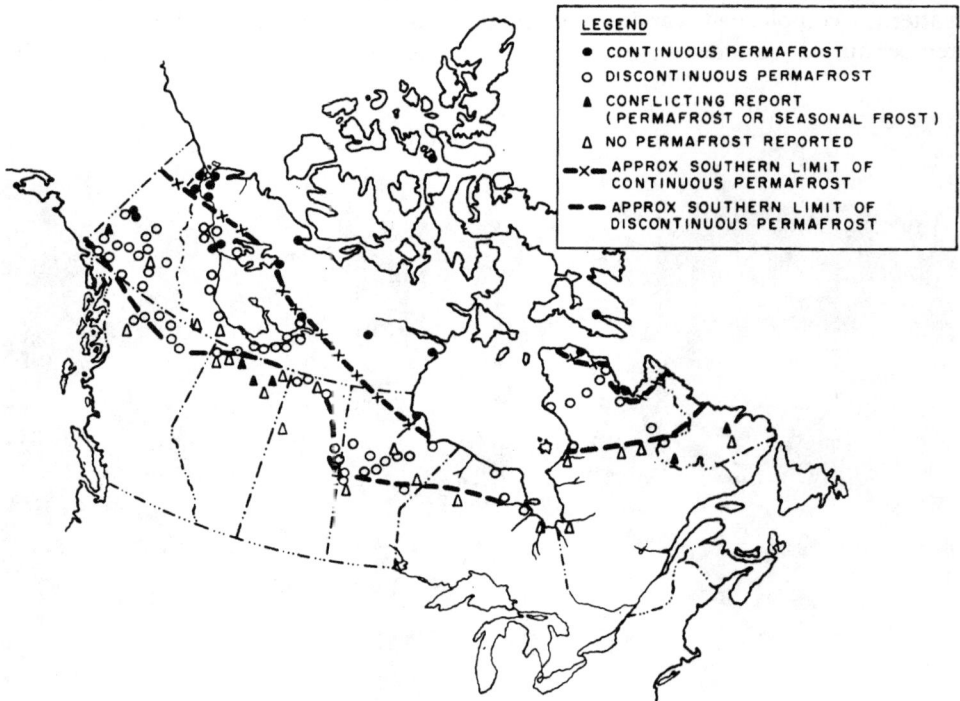

FIGURE 2.   Distribution of permafrost in Canada.

atmosphere, and the earth's surface; this in turn both affects and is affected by the energy exchanges occurring at the surface of the earth itself.

The heat exchange at the ground surface is thus highly variable in respect of climatic factors, but it is also affected by terrain, or surface, conditions including the thermal properties of the ground itself, often referred to as the geothermal conditions. It is highly variable not only with time at any one point but may also be variable from point to point on the ground. It is necessary to examine this situation carefully in order to determine just what can usefully be done in predicting permafrost occurrences and possible changes in permafrost distribution.

PERMAFROST AND CLIMATE

Air Temperature

The climatic factor that is most readily measured and most directly related to ground heat loss and heat gain is the temperature of the air. Many investigators have estimated the mean annual air temperature required to produce and maintain a perennially frozen condition in the ground, but there is much disagreement on this matter. Terzaghi (1952) reported that the permafrost layer coincides very roughly with the 0° C mean annual temperature isotherm. In Canada, this is certainly not the case because this isotherm lies a considerable distance south of known areas of discontinuous permafrost. Black (1950) reported that the mean annual air temperature required to produce permafrost varies many degrees because of local conditions and suggested that it is generally between 24° F and 30° F. Nikiforoff (1928) in his hypothesis of the origin of permafrost suggested that the southern boundary coincides approximately with the −2° C isotherm.

In parts of Canada there is some similarity between the position of the 25° F mean annual isotherm and the southern limit of permafrost (Thomas, 1953). In the Yukon Territory, however, the 30° F isotherm lies much nearer the approximate southern limit. In Manitoba, the known limit of permafrost cuts diagonally from the 25° F to the 30° F isotherm. In Ungava the permafrost limit lies far north of the 25° F, except in Labrador where there appears to be some similarity between this isotherm and the permafrost limit.

Attempts have also been made to relate permafrost distribution in Canada with freezing (Wilkins and Dujay, 1954) and thawing indices but without much success. These indices reflect the annual fluctuation of air temperature about the freezing point and indicate the amount of heat added to and withdrawn from the ground. A station with a mean annual air temperature of 0° C will therefore have equal freezing and thawing indices. It can be shown that there is only a very broad relation between permafrost distribution and air temperature, and freezing and thawing indices. Accurate prediction of the occurrence of permafrost cannot be based solely on this climatic factor. It is therefore desirable to examine the ground heat exchange picture in more detail.

Heat Exchange Components

The short-wave solar radiation received at the earth's surface is broken up into a number of components, each causing a process in which heat transfer occurs (Figure 3) (Geiger, 1950; Thornthwaite and Hare, 1955). Some of the radiation is reflected from the ground surface, the amount dependent on the reflectivity of the surface; some creates convection in the air resulting in a turbulent heat-exchange régime, some is used for evapotranspiration, and the remainder heats the ground. Heat is also carried into the ground by percolating rain and surface water.

In summer, the earth's surface gains heat from solar radiation by day and loses heat by outgoing radiation at night, the amount varying with the type of surface. In winter, the incoming radiation is reduced by day owing to the reduced angle of incidence of the sun's rays, their longer path through the atmosphere, and the shorter daily period of sunlight. North of the Arctic Circle there is a period in winter, varying directly in duration with latitude, when no radiation is received because the sun is below the horizon. The amount of incoming radiation reaching the earth may be reduced to a small quantity through reflection by clouds. The actual amount absorbed may be reduced substantially by snow cover. On a winter night the outgoing long-wave radiation is lowered owing to the reduced temperature of the earth's surface from which the heat must be drawn. Under conditions leading to a net loss outwards, the temperature of the ground surface may be reduced below that of the adjacent air and soil layers.

Incoming solar radiation and permafrost distribution are related only in a very general way. Permafrost areal extent and thickness increases generally with latitude as solar radiation decreases; examination of average daily incoming solar radiation on a horizontal surface over a two-year period (October 1957 to September 1959) bears this out.

At the earth's surface, the components of the heat exchange are affected by several factors (Swenson, 1956). The conduction-convection flux is affected by air movement and temperature. Radiation from the surface varies according to the

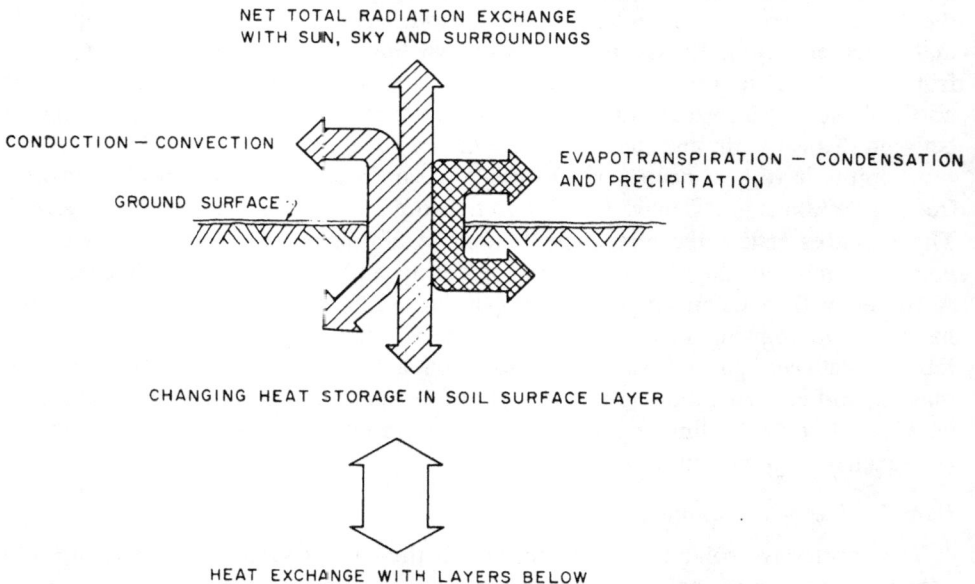

FIGURE 3.   Diagrammatic representation of components of surface energy exchange.

TABLE I

Location	Latitude	Permafrost zone	Average daily radiation gm cal/cm^2
Ottawa	46° N	None	324
Winnipeg	50° N	None	319
Edmonton	54° N	None	296
Knob Lake	55° N	Sporadic	233
Fort Simpson	62° N	Sporadic	245
Aklavik	68° N	Continuous	223
Resolute	75° N	Continuous	210

reflectivity of the surface and weather conditions. When the sun is shining, there is a heat gain to the soil regardless of soil temperature. During overcast conditions, there can be a heat gain or loss and on clear nights there is a heat loss to the atmosphere from the ground. The evaporation-condensation flux is affected by the vapour pressure difference between the soil and the atmosphere and by air movement.

The annual heat-exchange equation at the earth's surface can be written as

$$R = LE + P,$$

where R is the annual radiation balance, LE the heat involved in evaporation-condensation, and P the heat involved in conduction-convection (turbulent heat-exchange) (Grigor'ev, 1958; Budyko, 1958). In any locality where the net radiation balance is known, it follows that an increase of heat transfer by evaporation decreases the amount in turbulent exchange and vice versa. Investigations in the U.S.S.R. show that the ratio of $P:LE$ at the tree-line is about 1:6 or 1:7 increasing to about 1:3 in the south-central part of the taiga where the permafrost boundary occurs.

For a period of one year's observations at Point Barrow, Alaska, where permafrost is continuous, and conditions approximating potential evapotranspiration were assumed to exist most of the time, Mathers and Thornthwaite (1956; 1958) found that the evaporative heat flux was about 40 per cent of the net radiation. This contrasts markedly with mid-latitude areas having moist soils where about 80 to 90 per cent of the net radiation is used in the evaporation of water. These investigators attributed the lower percentage of evaporation in the net exchange at Point Barrow to the presence of permafrost. It is interesting to note that the $P:LE$ ratio at Point Barrow was found to be 1:0.7 in contrast to the Russian figures of 1:6 or 1:7 at the tree-line.

Because of the short period of observation, Mathers and Thornthwaite could not say whether there were any significant variations in evapotranspiration from one site to another. Different vegetative covers do show variations in net radiation and it is possible, on the basis of the annual heat exchange equation, that different amounts of evapotranspiration might be expected with variations in turbulent heat exchange at these various surfaces. The energy exchange régime at the earth's surface consists of a number of different components. Each of these components can be measured, although instrumentation is often complicated and costly. It is clear, however, that climate factors alone do not completely determine the heat exchange; surface and ground thermal characteristics must also be taken into account.

PERMAFROST AND TERRAIN

The effect of climate on the ground thermal régime shows marked variation with differences in ground surface conditions, either natural or man-made. These varying conditions generally arise from major differences in surface configuration, surface cover, and subsurface material. The thermal anomalies that they create have been examined to varying degrees by several workers. Only a brief review of these factors can be presented.

Configuration

The configuration of the ground surface, or the ground relief, determines the angle the ground surface makes with the horizontal and also the direction of its exposure with reference to the compass. Both these factors of slope and aspect affect the amount of solar radiation received by the surface. In zones of sporadic permafrost, this may result in permafrost occurring on north-facing slopes but not on adjacent slopes facing south. In continuous zones, the permafrost tends to be thicker and the active layer thinner on north-facing slopes. In hummocky tundra areas, the irregular permafrost table often appears as a mirror image of the ground configuration, owing to variation in solar radiation resulting from differences in microrelief (Tikhomirov, 1959). Slope and aspect may also have an influence on other factors such as air temperature, wind, and relative humidity. Another aspect of terrain is altitude. In itself, this can cause the occurrence of permafrost, even in temperate regions, owing to the reduction of air temperature with altitude.

Surface

The main types of surfaces that affect the ground thermal régime are vegetation, snow cover, water such as in lakes or rivers, bare soil, and rock.

Vegetation. The entire vegetative complex exerts an influence but it is the effect of plant cover such as mosses and lichen that appears to be paramount. The influence of a vegetative cover is not clearly understood but the suggestion that it provides resistance to heat flow by conduction only is almost certainly too simple a view. Vegetation contributes to evapotranspiration and thus a very considerable proportion of the solar radiation received can be rejected in this way without the soil beneath being affected. Mosses are strongly hygroscopic and are capable of losing moisture rapidly and in large quantities. Lichens, however, have been found to have extremely dry surfaces at times, even when lower layers near the soil are very wet. It may be that they are able to protect the soil against heat gain more by an insulating rather than an evaporative cooling effect, although some investigators have suggested that low soil temperatures and a high permafrost table are maintained by rapid evaporation at the wet basal layers.

Snow cover. Although snow is basically a climatic factor, it is more readily considered as a terrain factor for the purpose of this study, as snow cover is another type of surface that influences heat transfer between the air and the ground. As with other terrain factors, exact knowledge of the effect of snow cover on the ground thermal régime is not yet available and is difficult to determine. Observations taken in the Ottawa area (Gold, 1958) indicate that the rate of heat flow from the ground in winter is inversely proportional to the snow depth but the general validity of this relation has not yet been verified. It is known that the presence of snow will reduce the depth of seasonal frost penetration in non-permafrost areas. It has been suggested that, other things being equal, it will do this by an amount equal to twice the thickness of snow cover. Conversely, the thawing of frozen soil in spring is inhibited by the snow cover.

Tikhomirov (1952) has shown that a thin snow cover does not prevent some solar radiation from affecting the underlying soil surface. He states that a snow cover 2 to 3 cm thick will pass 50 per cent of the solar radiation and that 10 cm and 50 cm of snow will pass 2 per cent and 1 per cent of solar radiation respectively. Lachenbruch (1959) has reported that at Point Barrow, a one-foot layer of packed, drifted snow over sandy gravel has raised the mean annual temperature of the ground surface by 3° C and at Umiat, 170 miles southeast of Point Barrow, the increase has been 4.5° C.

Water surfaces. The heat storage capacity of a body of water is much higher than that of land; the reflectivity of the surface to radiation is much lower. The reflectivity is determined by a number of factors, two of which are the roughness of the water and the altitude of the sun. Geiger (1950) gives average values of reflection from a water surface of 9 per cent of the visible spectrum and 5 per cent of the ultraviolet. These values are much lower than those cited for vegetation, snow cover, bare soil, and rock. The evaporation characteristics will also differ from other types of cover. As a result of these and other factors, a water surface results in an increased temperature in the ground beneath.

There is almost always a *talik,* or unfrozen zone, beneath bodies of water in permafrost areas. The extent of this thawed zone will vary with a large number of factors — the area and depth of the water body, the thickness of ice and snow cover in winter, the flow rate in the case of a river, and the general hydrology of a lake, as well as the composition of the bottom sediments and their history. An understanding of the thawing effect of water on permafrost is therefore of unusual importance to the engineer engaged in the design of reservoirs, dykes, and dams in northern regions.

The effects of rivers are the most difficult to assess because the energy exchange relationships are complicated by water movement. Brewer (1958a) reports a warming of 3° C to a depth of 135 feet in the soil underlying the Shaviovik River in Alaska, 200 miles east of Point Barrow. Grigor'ev (1959) showed that the moderating influence of rivers on the permafrost extended a distance of 100 m away from the river bank and produced thawed layers beneath the river up to 30 m thick. He concluded that these thawed zones generally take the form of an extended trough, the width of which is proportional to the width of the river.

In the case of lakes, it has generally been assumed that, for those that do not freeze to the bottom, the thickness of the unfrozen layer beneath is proportional to the lake diameter, and that there will be no permafrost beneath, for lakes the widths of which exceed the normal thickness of permafrost in the area. Recent

Russian observations in Siberia have suggested, however, the importance of the geologic structure of the lake bottom upon heat flow and hence on the extent of the *talik* (Grigor'ev, 1959). Impermeable sediments restrict convective heat exchange at the lake bottom. In such cases it has been suggested that the thickness of taliks beneath lakes that do not freeze to the bottom will be proportional to the yearly temperature amplitude of the region regardless of the areal extent of the lake.

An important factor in the thermal effect of lakes on the permafrost below is the tendency of water at 4° C (the temperature at which it has its highest density) to accumulate at the lake bottom, and the corresponding elimination of convective action as the lake cools below this temperature. It may also be postulated that a lake, which is replenished by run-off from a drainage basin in summer, benefits from heat collected from that area by the water that will normally flow at temperatures well above freezing.

Although lakes deep enough to maintain some unfrozen water during all seasons have the greatest thawing effect on permafrost, even shallow lakes which freeze completely in winter modify the thermal régime of the permafrost below. Brewer's (1958b) studies of lakes in the Point Barrow region of Alaska showed the mean permafrost temperatures beneath these lakes to be up to 3° C higher than in the surrounding undisturbed terrain. This study showed also that lakes having a diameter of half a mile or more and a depth of at least seven feet could be expected to have basins unfrozen to a depth of 200 feet.

The influence of a very large body of water such as the ocean has been examined by Lachenbruch (1957) who cites the increase in geothermal gradient evidenced in the studies at Resolute Bay, Northwest Territories, as an example of the ocean's effect. He suggests that this change is an indication of the length and rate of emergence of the land.

Soil and rock surfaces. As surfaces, soil and rock have considerable influence upon reflectivity to solar radiation. Reflectivity values in the range of twelve to fifteen for rock (Ångström, 1925) and fifteen to thirty for tilled soil (Geiger, 1950) have been given. There will also be different evaporation rates and intakes of precipitation. The latter effects border on ground thermal problems and are difficult to differentiate from the properties discussed in the following sections.

Ground Thermal Properties

The thermal properties of the ground itself have a marked influence on the net heat exchange and resulting temperatures, supplementing the influences of climatic and terrain factors as already discussed. In the simplest cases, involving only transient heat flow by conduction, the thermal diffusivity of the soil becomes important in determining the rates at which heat is transferred from the surface into the soil itself. This affects the surface temperatures, which influence, in turn, the surface heat exchange rate. For example, soils that warm up quickly under solar radiation will achieve high-surface temperatures that can result in a substantial loss of heat by conduction-convection, thus affecting the over-all heat gain (Landsberg and Blanch, 1958). Thus the thermal conductivity, density, specific heat, and moisture content of soil all have an influence on its thermal régime.

Unfortunately, the transfer of energy through actual soils is often greatly complicated by their variability, and by the presence of moisture, particularly when a change to the vapour state is involved. When permafrost is present there are further complications in such thermal calculations because of the differences in properties between frozen and unfrozen soils as well as of the latent heat of fusion released when water is frozen. This is not readily taken into account by the heat-conduction theory.

RELATED EXPERIMENTAL WORK

Many of these factors affecting the occurrence of permafrost are also of importance in the study of the ground thermal régime in non-permafrost areas. Much of the work of the Division of Building Research on frost action, ground temperatures, and heat flow in southern regions is consequently of direct relevance to the permafrost problem.

One of the Division's earliest projects in this general field was the study of ground temperatures. This involved the direct field recording of soil temperatures at many locations in both permafrost and non-permafrost areas and under a wide variety of natural conditions (Crawford and Legget, 1957). These studies have provided information on the effect on ground temperatures of air temperature, soil type, and snow cover. An analysis of some of the observations has suggested that simple heat-conduction theory can be used to describe certain long-period temperature variations that occur within the soil (Pearce, 1958b) in the range of annual temperature variation.

Closely paralleling these studies of the natural ground thermal régime are investigations of the effect of man-made structures on ground temperatures. A cold storage plant, in which the ground beneath was subjected to freezing during operation of the building, presented the reverse situation to the thawing problem that arises in permafrost regions under heated structures (Pearce and Hutcheon, 1958; Pearce, 1958a). Other soil temperature studies have dealt with the problem of freezing water-mains and have included observations on the water and sewage installations in the permafrost area of Yellowknife (Legget and Crawford, 1952; Copp et al., 1956). Variations in soil temperatures under small, heated test structures have also been examined (Solvason and Handegord, 1959). Observations are being made beneath dykes under construction at the site of a new hydroelectric plant in an area of discontinuous permafrost.

Much emphasis has been placed on the study of frost action in soils, chiefly in relation to the basic mechanism itself (Penner, 1956; Penner, 1957) but also with respect to the correlation of frost action damage with the main climatic factors (Crawford and Boyd, 1956). It seems probable that this same frost action mechanism is involved in the formation of many of the ice structures encountered in permanently frozen soil. It may have been a determining factor in the formation, on a long-term basis, of some of the more prominent surface manifestations of permafrost such as pingos, frost medallions, and some types of patterned ground.

The Division has also studied the effect of moisture upon heat transfer in porous materials to try to improve understanding of the heat transfer mechanism in moist soils (Woodside and Cliffe, 1959; Woodside and DeBruyn, 1959). The role of moisture in heat transfer at the ground surface has been examined in terms of evaporation from both snow (Williams, 1959a) and water surfaces (Williams, 1959b). The latter study was concerned with the improvement of empirical methods of estimating heat losses from open-water surfaces and has shown that the total heat loss or gain can be correlated with the difference between mean sol-air and mean surface-water temperature.

Since 1953, the Division has conducted evapotranspiration observations at Norman Wells for another agency interested in verifying the Thornthwaite formula for potential evapotranspiration in relation to climate. These observations, as well as measurement of radiation factors, have recently been extended by the Division to local vegetative cover as a first step in the study of the effect of such cover on the basic energy components at the ground surface.

This brief summary of reports on related work carried out or in progress by the Division of Building Research is included to show that "permafrost problems" are not isolated scientific curiosities. Permafrost is a natural phenomenon, the study of which demonstrates that the solution even to the single problem of determining its extent is allied to many more general geotechnical problems.

DISCUSSION

The thermal properties of the earth's interior are still not completely understood. Enough is known, however, in this branch of geophysical study to be reasonably certain that the condition of permafrost is due to the cumulative effect of the thermal régime at the surface of the polar regions of the earth during the Recent geological period. It is known, for example, that the rate of heat-flow from the central core of the earth is so slow that its effect at the surface is of minor significance. Careful study of long-term records of (relatively) deep ground temperatures, now fortunately being obtained at a number of locations, should indicate eventually the approximate rate at which permafrost is degrading. The complications caused by the several climatic and terrain factors herein discussed make it impossible to investigate theoretically this aspect of permafrost with any degree of accuracy.

The mapping of permafrost, and especially of its southern boundary, must, however, be pressed forward, in Canada as elsewhere. Accurate determination of mean ground temperature is clearly the most certain method of determining the existence of permafrost. Unfortunately, the direct determination of mean ground temperatures is difficult to carry out under the conditions existing over most of northern Canada, so that coverage of the area with a pattern of temperature determinations would be extremely slow. The taking of observations is complicated by the annual temperature cycle so that temperatures must either be taken at some depth at which the annual variation is suitably damped, or a large number of shallow observations must be made over suitably spaced time intervals so as to provide a good average.

The making of direct observations of permafrost occurrences can be somewhat less involved, particularly because many opportunities for observation occur incidental to other work, such as the carrying out of construction projects. The distinction between permafrost and seasonal frost may sometimes be difficult to make. As in the case of the ground temperature approach, it may not be easy to determine whether the observation made is typical of the area, without some considerable extension of the pattern of observations.

It becomes important to know as much as possible about the factors that influence the ground thermal régime, many of which have been discussed here. Only with some knowledge of them will it be possible to begin to assess critically the limited observations on permafrost occurrence, and to use these with confidence in making predictions about adjacent areas. Some of this knowledge may also be valuable in assessing the effects that occur as man-made structures create changes in the ground thermal pattern.

There is no justification for assuming that natural ground temperatures, and particularly the occurrence of permafrost, can be predicted directly from estimated ground heat losses and heat gains. The relatively poor accuracy with which such estimates can be made, even under the best of conditions, precludes the possibility that any summation of them over a long period is of specific use in this way. Prediction will be possible only by using, as a datum, some known or measureable factor that is closely related to and correlates with the perennially frozen condition. It should then be possible to make corrections that could be applied to the ground temperature condition inferred from the measured or known base quantity and arrive at improved prediction.

The choice of reference factors is limited. Only a few climatic factors are regularly measured over the area of interest. Of those factors that are measured, mean air temperature would appear to be by far the most suitable. It shows some general correlation with mean ground temperature, as indicated by the correspondence between certain isotherms and the permafrost boundary, as well as by other evidence. The differences between mean air and mean ground temperature are not usually great. The identification of the factors responsible for these differences and the establishment of quantitative relations is a formidable but not impossible task.

Recognition of the complexity of permafrost and identification of the major factors affecting its occurrence provide at least a starting point for further progress. Discussion of these factors has constituted the main part of this paper rather than a recital of field studies of permafrost in Canada already made by the Division of Building Research and other Canadian agencies. It has been by the critical study of the results of such field work, and of the select bibliography available (in which contributions from the U.S.S.R. are outstanding) that the discussion herein presented has been developed. Field work at the sites of engineering undertakings in the Canadian north, and in special test areas with typical permafrost conditions, is continuing. Results thus obtained will be assessed in the light of the factors herein outlined, and in relation to other geotechnical investigations such as those noted earlier, in a continuing endeavour to delineate, understand, and explain the occurrence of permafrost in Canada.

Discussions of the subject matter of this paper with many of the authors' colleagues, with associates in other agencies of the Canadian Government, and in particular with Dr. N. B. Hutcheon, Assistant Director of the Division of Building Research, are gratefully acknowledged.

REFERENCES

ÅNGSTRÖM, A.   1925.   The albedo of various surfaces of ground; Geog. Annaler, vol. 7, pp. 323-42.

BARANOV, I. YA.   1955.   Yuzhnaia granitsa oblasti rasprostraneniia mnogoletnemerzlykh porod Materialy K Osnovam Ucheniia O Merzlykh Zonakh Zemnoi, vol. II, pp. 38-44.

BENNINGHOFF, W. S.   1952.   Interaction of vegetation and soil frost phenomena; Arctic, vol. 5, no. 1, pp. 34-44.

BLACK, R. F.   1950.   Permafrost; Smithsonian Report, pp. 273-301.

BREWER, M. C.   1958a.   Some results of geothermal investigations of permafrost in northern Alaska; Trans. Amer. Geophys. Un., vol. 39, no. 1, pp. 19-26.

———   1958b.   The thermal régime of an arctic lake; Trans. Amer. Geophys. Un., vol. 39, no. 2, pp. 278-84.

BRYAN, K.   1946.   Cryopedology—the study of frozen ground and intensive frost action with suggestions on nomenclature; Amer. J. Sci., vol. 244, pp. 622-42.

BUDYKO, M. I.   1958.   The heat balance of the earth's surface; Office of Climatology, U.S. Department of Commerce.

COPP, S. C., CRAWFORD, C. B., and GRAINGE, J. W.   1956.   Protection of utilities against permafrost in northern Canada; J. Amer. Water Works Assoc., vol. 48, no. 9, pp. 1155-68 (NRC 4056).

CRAWFORD, C. B., and BOYD, D. W.   1956.   Climate in relation to frost action, Highway Research Board, Bull. no. 111, pp. 63-75 (NRC 3746).

CRAWFORD, C. B., and LEGGET, R. F.   1957.   Ground temperature investigations in Canada; Engineering J., vol. 40, no. 3, pp. 1-8 (NRC 4302).

GEIGER, R.   1950.   The climate near the ground; Harvard University Press.

GOLD, L. W.   1958.   Influence of snow cover on heat flow from the ground—some observations made in the Ottawa area; Extrait des Comptes Rendus et Rapports, Assoc. Internationale Hydrol. Sci., U.G.G.I., Toronto, 1957, IV, pp. 13-21 (NRC 4827).

GRIGOR'EV, A. A.   1958.   O nekotorykh geograficheskikh zakonomernostiakh teploobmena i vodoobmena na poverkhnosti sushi i o putiakh dal'neishego izucheniia obmena veshchestv i energi v geograficheskom srede; Izvestiia Akad. Nauk SSSR, Seriia Geograficheskaia, no. 3, pp. 17-21.

GRIGOR'EV, N. F.   1959.   O vliianii vodoemov na geokriologicheskie usloviia primorskoi nizmennosti Ust'—IAnskogo raiona Yakutskoi ASSR; Materialy Po Obshchemu Merzlotovedeniiu, VII, Mezhduvedomstvennoe Soveshchanie Po Merzlotovedeniiu, pp. 202-6.

LACHENBRUCH, A. H.   1959.   Periodic heat flow in a stratified medium with application to permafrost problems; U.S. Geol. Surv., Bull. 1083-A.

———   1957.   Thermal effect of the oceans on permafrost; Bull. Geol. Soc. Amer., vol. 68, no. 11, pp. 1515-30.

LANDSBERG, H. E., and BLANCH, M. L.   1958.   Interaction of soil and weather; Soil Sci. Soc. Amer., vol. 22, pp. 491-5.

LEGGET, R. F., and CRAWFORD, C. B.   1952.   Soil temperatures in water works practice; J. Amer. Water Works Assoc., vol. 44, no. 10, pp. 923-39 (NRC 2910).

MATHER, J. R., and THORNTHWAITE, C. W.   1956.   Microclimatic investigations at Point Barrow, Alaska, 1956; Publications in Climatology, Drexel Institute of Technology, Laboratory of Climatology, Centerton, N.J., vol. IX, no. 1.

———   1958.   Microclimatic investigations at Point Barrow, Alaska, 1957-1958; Publications in Climatology, Drexel Institute of Technology, Laboratory of Climatology, Centerton, N.J., vol. IX, no. 2.

NIKIFOROFF, C.   1928.   The perpetually frozen subsoil of Siberia; Soil Sci., vol. 26, pp. 61-78.

PEARCE, D. C.   1958a.   An analysis of frost action beneath cold storage warehouses; Trans. Engineering Inst. Canada, vol. 2, no. 4, pp. 153-6 (NRC 5095).

———   1958b.   Ground temperature studies at Saskatoon and Ottawa; Extrait des Comptes Rendus et Rapports, Assoc. Internationale Hydrol. Sci., U.G.G.I., Toronto, 1957, IV, pp. 279-90 (NRC 4796).

PEARCE, D. C., and HUTCHEON, N. B.   1958.   Frost action under cold storage plants; Refrigerating Engineering, vol. 66, no. 10, pp. 3-7 (NRC 4939).

PENNER, E.   1956.   Soil moisture movement during ice segregation; Highway Research Board, Bull. 135, pp. 109-18 (NRC 4278).

———   1957.   The nature of frost action; Proc. Thirty-Eighth Convention of the Canadian Good Roads Assoc., pp. 234-43 (NRC 4627).

SOLVASON, K. R., and HANDEGORD, G. O. 1959. Soil temperature measurements at Saskatoon; Trans. Engineering Inst. Canada, vol. 3, no. 2, pp. 67-73 (NRC 5344).

SUMGIN, M. I., KACHURIN, S. P., TOLSTIKHIN, N. I., and TUMEL', V. F. 1940. Obshchee merzlotovedenie; Akad. Nauk SSSR.

SWENSON, E. G. 1956. Weather in relation to winter concreting; RILEM Symposium, session A, general report no. 8575 (NRC 3830).

TERZAGHI, K. 1952. Permafrost; J. Boston Soc. Civil Engineers, vol. 39, pp. 1-50.

THOMAS, M. K. 1953. Climatological atlas of Canada; Division of Building Research, National Research Council, and Meteorological Division, Department of Transport (NRC 3151).

THORNTHWAITE, C. W., and HARE, F. K. 1955. Climatic classification in forestry; Unasylva, vol. 9, no. 2, pp. 50-9.

TIKHOMIROV, B. A. 1959. Vliianie rastitel'nosti na letnee protaivanie pochvy v oblasti mnogoletnemerzlykh porod; Materialy Po Obshchemu Merzlotovedeniiu, VII, Mezhduvedomstvennoe Soveshchanie Po Merzlotovedeniiu, pp. 207-9.

———— 1952. Znachenie mokhovogo pokrova v zhizni rastenii krainego severa; Botanicheskii Zhurnal, vol. 37, pp. 629-38.

TSYTOVICH, N. A. 1958. Osnovaniia i fundamenty na merzlykh gruntakh; Akad. Nauk SSSR.

TUMEL', V. F. 1946. Karta rasprostraneniia vechnoi merzloty v SSSR; Merzlotovedenie, t. 1, no. 1, pp. 5-11.

WILKINS, E. B., and DUJAY, W. C. 1954. Freezing index data influencing frost action; Proc. Seventh Canadian Soil Mechanics Conference, December 10-11, 1953, Associate Committee on Soil and Snow Mechanics, National Research Council, Canada, Tech. Mem. no. 33, pp. 36-9.

WILLIAMS, G. P. 1959a. Evaporation from snow covers in eastern Canada; Division of Building Research, National Research Council, research paper no. 73 (NRC 5003).

———— 1959b. An empirical method of estimating total heat sources from open water surfaces; Presented to Seminar on Ice Problems, Eighth Congress of the International Association of Hydraulic Research, Montreal, August 1959.

WOODSIDE, W., and CLIFFE, J. B. 1959. Heat and moisture transfer in closed systems of two granular materials; Soil Sci., vol. 87, no. 2, pp. 75-82 (NRC 5030).

WOODSIDE, W., and DE BRUYN, C. M. A. 1959. Heat transfer in a moist clay; Soil Sci., vol. 87, no. 3, pp. 166-73 (NRC 5098).

Seismic Refraction Soundings in Permafrost near Thule, Greenland

HANS ROETHLISBERGER

ABSTRACT

The applicability of various seismic methods for engineering purposes has been investigated in the Thule area. Special attention has been given to the refraction method in the cases where shallow ice (up to 200 feet) occurs overlying frozen ground (till), or where frozen ground (till, outwash) up to a few hundred feet thick overlies bedrock. Seismic velocities have been measured in different types of sediments of the Thule formation and in the crystalline basement rock. Very high velocities were found for all types of rock; the temperature was about $-10°$ C, and most pores and cavities were probably filled with ice. It was discovered that for shallow soundings of a few hundred feet, the seismic methods can probably be used more elaborately in permafrost than in unfrozen material, as later pulses can be identified on the records shortly after first breaks. A negative velocity gradient in frozen ground is believed to be responsible for this.

THE SEISMIC REFLECTION METHOD has been used extensively for soundings on glaciers in Greenland and elsewhere, but the use of the refraction method is rarely mentioned. However, the reflection method has its limitations when the ice becomes shallow. In a 1957 survey a lower limit of ice thickness of about 200 feet was found on the TUTO ramp, which is the access route to the Greenland ice-sheet from Thule (Roethlisberger, 1959). In the two following years, refraction studies were carried out on ice less than 200 feet thick, and on frozen till to establish its thickness down to bedrock. Velocities in a variety of rock types were also measured in order to obtain more general conclusions about the applicability of the refraction method in the formations of the Thule area and in similar formations throughout the Arctic.[1] The seismic equipment used for reflection and refraction soundings was the high-frequency system, type P-15, manufactured by the Southwestern Industrial Electronics Company.

VELOCITY MEASUREMENTS

The refraction technique can be applied successfully only when a lower velocity material occurs over a higher velocity refractor. Thus, the knowledge of velocities is essential for proper planning of a refraction survey. Again the velocities in different rock types must be known if identification of an unknown refractor is to be attempted from refraction soundings. As the Thule area is quite complex geologically, a special effort to determine velocities in different formations was justified.

Geology

A thorough geological survey of the area was carried out recently by Davies, Krinsley, and Nicol (in preparation). They report various metamorphic rocks,

[1]For the complete report see Roethlisberger (1961).

mainly gneiss, from the Precambrian basement complex, three distinct formations of sedimentary rock, and diabase dykes and sills. Large areas are covered by glacial deposits. The sedimentary formations, believed to be of late Precambrian age, and formerly named the Thule formation in the literature, are the Wolstenholme formation, consisting of quartzite and sandstone; the Dundas formation consisting of black shale with some dolomite and grey sandstone; and the Narssârssuk formation. It is subdivided into the lower red member, the Arferfik dolomite member, and the upper red member and consists of red siltstone, grey dolomite, and sandstone. A cyclic repetition of sediments is typical for the lower and upper red members. Diabase dykes occur throughout the area both in the crystalline basement and in the sediments, while the sills are intruded into the Dundas formation only.

Faults usually delineate the boundary between the sediments and the basement rock, except in a few places where the base of the Wolstenholme formation rests unconformably on metamorphic rock. The main occurrence of the sediments is in a belt not more than ten to fifteen miles wide which extends in an east-west direction. Only small outcrops of the Wolstenholme formation are located south of the southern limit of the sedimentary belt. Although the faults follow an east-west trend, they may not be extrapolated very far under glacial deposits or the ice-cap because they are not straight and form a complex system.

Technique

It is not as simple as it might seem to measure the velocity of elastic waves (seismic velocity) accurately on rock outcrops owing to their weathering. Not only is the velocity changed in the weathered zone, but also the frequency of the first break becomes so low that accurate measurements over a short path are impossible. This is generally true in temperate climates. However, in cold regions where the permafrost is well developed, it is much easier to make reliable velocity measurements at the surface. It was found that the following method furnished the best results with the least amount of effort.

A flat surface was chosen close to an outcrop, where there was no doubt about the exact nature of the bedrock, but where the rock was covered with a thin and most probably uniform layer of top soil, and the surface showed frost pattern features. The top soil consisted of heterogeneous glacial drift or decomposed debris of the underlying bedrock. A row of geophones was laid out so that they were close to a straight line and approximately equally spaced, and also that each geophone could be placed on a patch of fine material (centre of fines) of the frost pattern. The distance between geophones was usually about 50 or 100 feet. The explosive charges were placed close to both ends of the line also in centres of fines. If conditions permitted, additional charges were set off at greater distances up to several hundred feet from either end of the geophone line. Either a hole was dug to the frost-table or a crater was blasted with a preliminary charge so that the explosives were always in contact with the frozen ground. For 500-foot lines, about two pounds of 60 per cent dynamite were used.

In this method neither geophones nor shot-point are placed on the bedrock to be investigated, but on a low-velocity top layer, and so use is made of refracted waves which are affected by the thickness of the low-velocity layers and by the velocities of these layers. Nevertheless, the scatter of individual points around a straight time-distance curve was less than when the geophones were placed on exposed bedrock and direct waves were measured. The exposed rock is probably slabbed off and invisibly shattered by frost which could account for scatter in the direct velocity measurements. The small scatter in the refraction method indicates amazingly homogeneous conditions in the low-velocity layers as well as in the undisturbed rock underneath. In an attempt to give an explanation for this, some features of the permanently frozen ground and the seasonally thawing and freezing active layer on top will be discussed as far as they are pertinent to the propagation of elastic waves.

Permafrost is impermeable to water and may consist of debris or jointed rock cemented by ice. Physically speaking, most types of permafrost may be considered as massive rock. Owing to thermal contraction, this massive rock will crack in a polygonal pattern in winter, from the surface down to a limited depth but these cracks will heal when they fill with melt-water in early summer or when the thermal expansion closes them again as the seasonal heat wave moves downwards. Only the ice wedges developing in the cracks and growing over the years need to be considered in velocity studies. This is the only weathering process to occur in the permafrost zone.

Similarly in the active layer where extreme weathering takes place, the conditions are rather simple. The active layer is of very limited thickness, in the Thule area usually between three and seven feet, but cannot be neglected because of the small velocities. One of its outstanding features is the sorting of particles of differ-

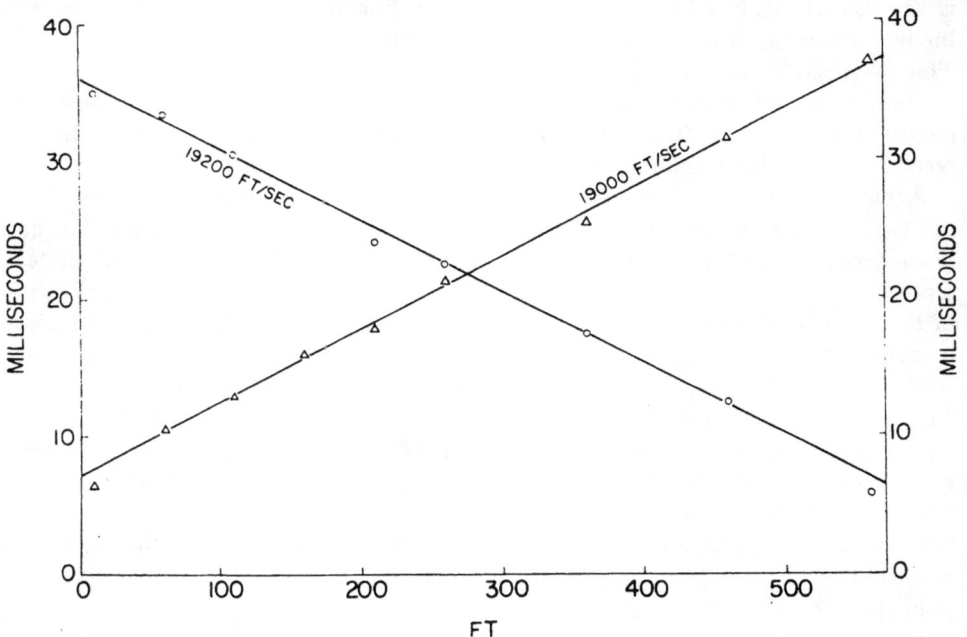

FIGURE 1.   Time-distance curves of a velocity profile on a dolomite plateau, V 4.

ent size. A vertical sorting in which the coarse particles move to the top can be distinguished from a horizontal sorting, in which they move away from centres of fines and accumulate in patches and rows forming a sorted net pattern. On a slope, lines of coarse material alternate with stripes of fines. As stated above, the fines were chosen as bases for the geophones and the charges were placed in the fines as well. The findings of Corte (personal communication) indicated the fines to be better locations than accumulations of coarse debris or areas of little sorting. The important features of the centres of fines and stripes are: (1) greater homogeneity over the full thickness of the active layer; (2) lesser penetration of the thawing, that is, a thin active layer; and (3) retention of moisture assuring good transmission of the elastic waves.

No detailed study of the low-velocity layers was carried out. However, it can be stated that the total effect, that is, the vertical shift of the time-distance curve (intercept time), varied between two and ten milliseconds. Individual points were seldom more than one millisecond in error. A typical example of a time-distance graph is given in Figure 1. The frequencies of the first few cycles on the records varied between 200 and 500 cps. The frequency bands of the equipment were set for a range of either 70 to 425 cps or 220 to 425 cps.

Results

The velocity data are presented in Table I and the first column refers to the locations marked on the map in Figure 2. The velocity values have been determined from the slope of time-distance graphs where all available information from a number of records was considered. Both first breaks and first peaks were usually plotted. The peaks sometimes lined up better in a straight line, but probably the velocities were slightly too low because of decreasing frequency with distance. The plus or minus ranges given with the velocity values are not defined errors, but indications of the accuracy assumed to have been achieved in the graphical evaluation. They also take into account the differences which have been encountered

TABLE I

P-WAVE VELOCITIES

Place on map	Formation	Rock type	Velocity ft/sec		Velocity m/sec	
V 1	Basement Rock	Gneiss	20400	± 400	6220	± 120
V 2	Wolstenholme	Quartzite	18700	± 300	5700	± 90
V 3	Dundas	Black shale	13000–14000	± 500	3960–4260	± 150
V 3	Dundas	Sandy shale	14500	± 500	4420	± 150
V 4	Narssârssuk	Dolomite	19100	± 200	5820	± 60
V 5	Narssârssuk	Sandstone	17200	± 200	5240	± 60
V 6	Dike	Diabase	(18500 ?	± 500)	(5640 ?	± 150)
S 1, 2, 3	Glacial deposit	Till	15700	± 300	4780	± 90
S 5	Glacial deposit	Outwash	15200	± 300	4630	± 90
S 2	Ice-cap	Ice	12200	+ 500	3720	+ 150

between the two directions of shooting in a profile, between close-up and more distant shots, and between different profiles, if more than one was located in the same material.

No attempt has been made to correlate the velocity values with petrological data. However, since representative outcrops for the several formations were chosen, general information on the nature of the rocks may be obtained from the geological literature (Davies, *et al.*). The relatively low velocity in diabase cannot be accepted as representative. Because the measurements were carried out close to the edge of a cliff along a low ridge formed by the dyke, it is likely that the cracks of the generally shattered diabase have not been filled with ice owing to excessive drainage. An alternative explanation would be that the cracks are filled with ice, but that the shattering is extreme. Also in this case the low velocity would probably be limited to a shallow zone at the surface.

The rest of the rocks show high velocities for the type of material under consideration, cementation by ice probably being the reason. That ice is a good rock cement is illustrated by the high velocities found in glacial deposits. The till consists of boulders, cobbles, and pebbles in a matrix of sand and silt. The outwash shows only cobbles and pebbles at the surface, but sand and silt are also present in deeper layers according to a drill log. It is noteworthy that the velocity in these glacial deposits is clearly higher than in the shale of the Dundas formation.

As a hypothesis, one can assume that the velocity in frozen ground and shale

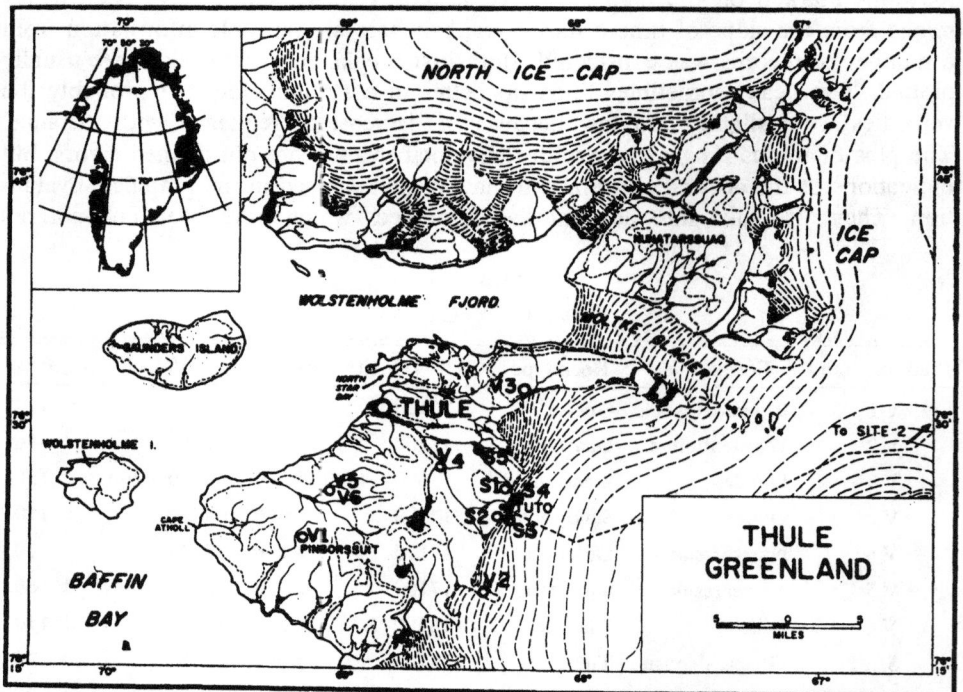

FIGURE 2.   Thule area where velocity measurements and soundings (V and S respectively) were carried out.

will depend primarily on the amount of air, ice, and clay minerals present. When ice fills an air space and acts as a cement, the velocity increases. When it becomes abundant enough to be an important fraction of the ground, then the velocity is bound to decrease with additional ice, because the velocity in ice is smaller than in most rock fragments. Clay minerals cause a low velocity since the water absorbed by them will not be completely frozen at temperatures below 0° C. The amount of unfrozen water will depend on the temperature and a considerable velocity variation with temperature will result for rocks containing clay minerals. Stationary ground temperatures of −10° C to −12° C have been reported around Thule and TUTO.

SOUNDINGS

The refraction soundings were primarily experimental, to prove if and where the method could be applied successfully. Only the cases of ice over glacial deposits, glacial deposits over bedrock, and the three-layer case with ice, till, and high-velocity bedrock were examined. Some of the results were of morphological significance and some had an engineering application. The locations where soundings have been carried out are marked with an "S" followed by a number on Figure 2.

Technique

On the glacial deposit, the same arrangement was used as for the velocity measurements, that is, the geophones and charges were placed on centres of fines with the charges at the frost table. The geophones, spaced 100 to 250 feet apart, were left in place and the shot-points were moved out along a straight line until the high velocity showed up in the first breaks. The geophones were then moved to the opposite end of the profile and the shots were fired from the side of the profile where the geophones had been before. The procedure was similar on the ice-cap, except that no selected places for geophones or shots were needed. Both were placed at the surface of the ice. The geophones were buried slightly in the ice and covered with a pile of ice chips to prevent them from thawing into the ice in which case they would be supported by the rubber cap instead of the base.

The longest profile on frozen ground was 2800 feet long and twelve-and-a-half pounds of high explosive were used at the end point shots. On the ice, the same amount of explosive was used for 3200 feet. Later in the survey, it was learned that low amplitude records furnished information on later arrivals and from then on every shot was fired twice, once with a large charge for first breaks, once with a lesser charge for later arrivals.

Results

The best information on the applicability of the refraction method can be obtained from the velocity table. There are a number of formations with distinctly different velocities, which means that the refraction method will work if the formation with the lower velocity is on top of the other. This will generally be the case, because the succession of ice, glacial deposit, sedimentary rock, and basement rock shows successively larger velocities, with the exception of shale where it underlies glacial deposits. The differences in velocities are not large. This requires long profiles relative to the depth to be measured and results in a lower accuracy.

On the terrace where TUTO is located (S1, S2), a high-velocity layer was found at a depth of 300 to 400 feet, the velocity being 21,500 ft/sec. The difficulty here is that no bedrock outcrops occur in the vicinity and the trend of the main fault limiting the sediments to the south does not allow a prediction of the rock type

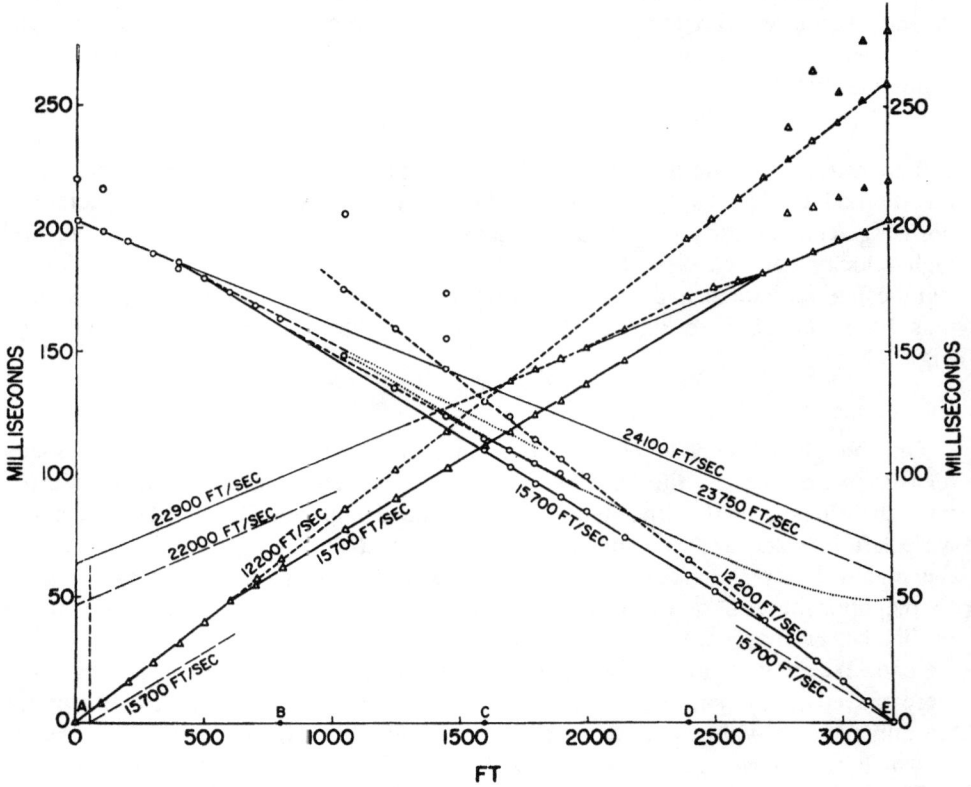

FIGURE 3.   Time-distance curves from soundings through ice and frozen ground to bedrock at S3. Light, solid and dashed lines show first approximation with topographic and top-layer corrections. Heavy, dashed and dotted lines show alternative and more elaborate interpretations.

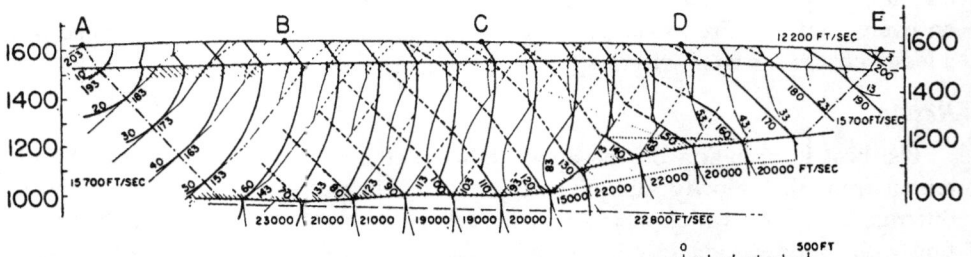

FIGURE 4.   Pairs of wave fronts originating at A and E and results of soundings at S3. The solid, dotted, and light-dashed interfaces represent different interpretations of the time-distance curves in Figure 3.

underlying the till. The velocity found in the refraction profiles would suit gneiss best, but a diabase sill or even dolomite have also to be considered. However, if the presence of sedimentary rock is not ruled out, shale might be present at a lesser depth not revealed by the refraction method. Shale is not likely to be found on top of a hill, and no reflection signals of the type to be discussed later were found on the records; hence, bedrock at depths less than 300 to 400 feet is not believed to be present. Later excavation to 50 feet below the surface of the terrace did not reveal bedrock.

Close to the edge of the ice-cap (S3), a refraction profile 3200 feet long was established approximately parallel to the edge. An almost uniform ice thickness of 75 to 100 feet was determined over the whole length of the profile. There is no doubt about the underlying material being till, and accordingly the velocity measured at the ice surface was about 15,700 ft/sec. A high-velocity layer was found at a depth of roughly 600 feet below the top of the till, the apparent velocity being 22,800 ft/sec which is an unlikely high value, but which can be explained, at least partially, by dipping. The time-distance graph for the shot-points A and E is given in Figure 3. There are later arrivals which can be associated with the first-break refraction signals, which lead to more detailed information on the high-velocity refractor. The results of different assumptions are given in Figure 4, where solid lines correspond to the solid lines of Figure 3 and so forth. The dotted lines pertain

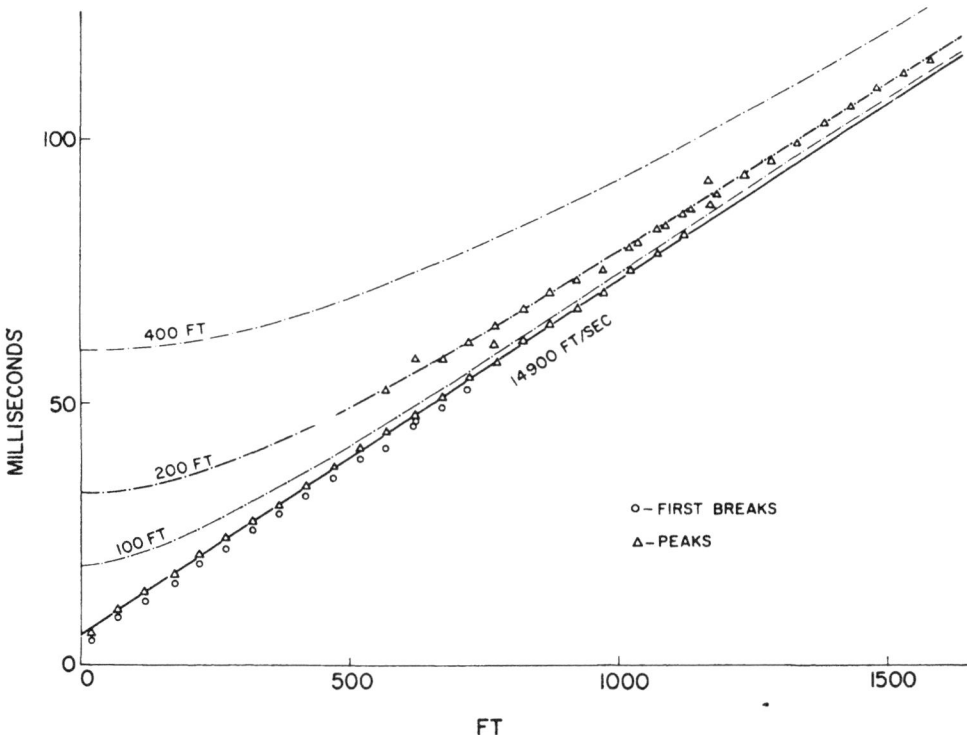

FIGURE 5.   Time-distance curves from a profile in flood-plain at S5. Theoretical curves for the first peaks of reflections from 100, 200, and 400 feet are compared with the measured values. The 200 feet curve fits the observed travel times closely.

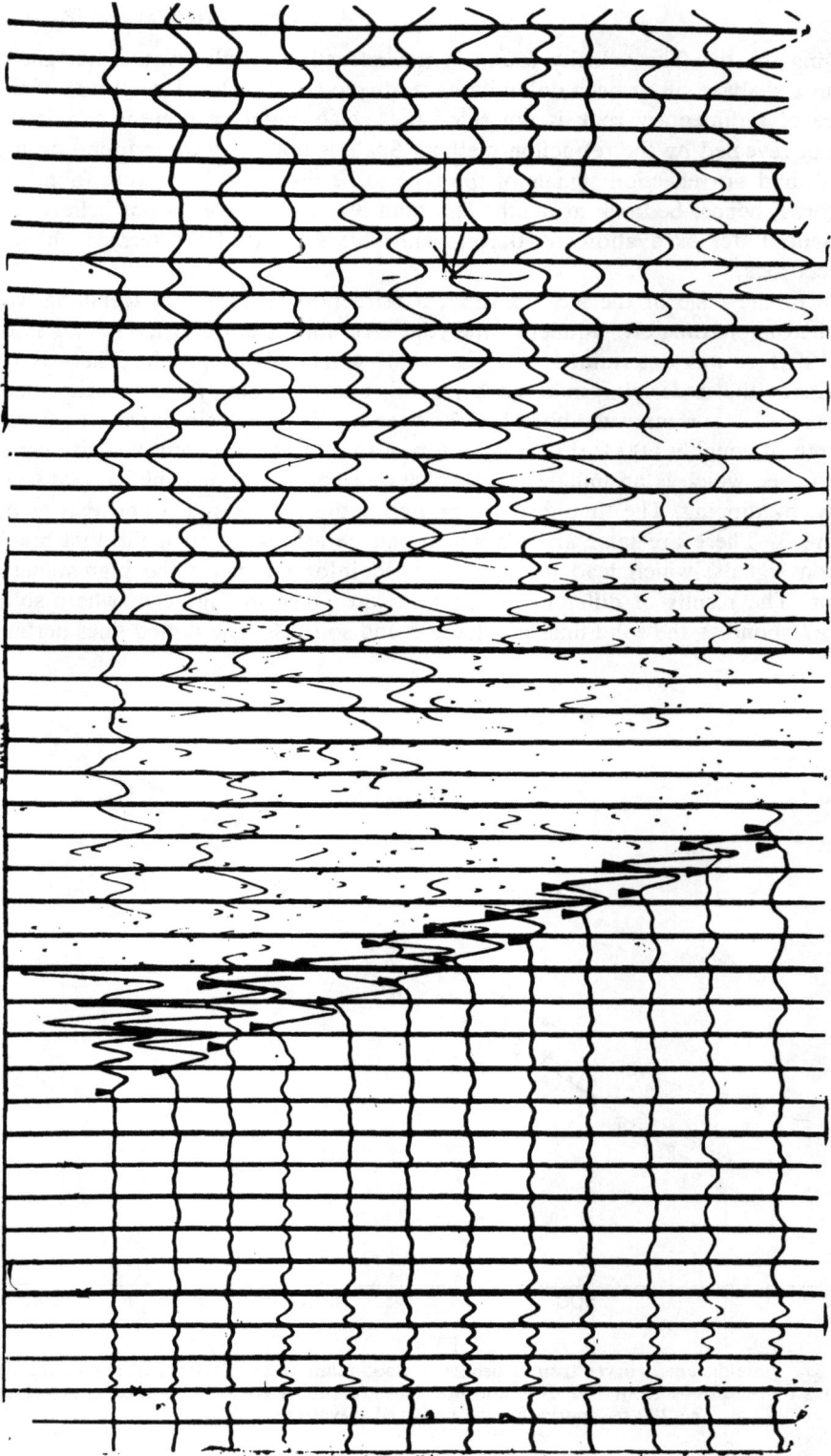

FIGURE 6. One of the records from which the travel times plotted in Figure 5 were derived. Time lines in 5 millisecond intervals.

to the assumption that one set of later arrivals belongs to reflected waves from an internal reflection in the frozen ground. The wave-front method (Hagedoorn, 1959; Thornburgh, 1930) has been applied and proved to be very useful.

Shallow Reflections with Refraction Techniques

Although it was not expected that a refraction survey in the flood-plain halfway between Thule and TUTO (S5) could be obtained successfully, a 1600-foot profile was investigated. Information could be obtained on a drill-hole sunk down to 253 feet in the flood-plain in 1951 and the drill log reported 203 feet of gravel and sand over shale. It was hoped to find perhaps later arrivals due to PS or SP or SPS waves. Figure 5 gives the results of the survey. The first arriving signals rapidly became weak and of low frequency with distance while a much stronger later arrival with the original high frequencies occurs on the record. Figure 6 is a record of this type. The later signals can be attributed to the reflected wave, the reflection occurring at the top of the low-velocity shale. The drill log indicated the presence of a clay layer at the top of the shale, making the reflection boundary probably more pronounced. The rapid disappearance of the direct wave can readily be explained by an increasing temperature with depth, causing a negative velocity gradient. Such a negative velocity gradient with depth is probably unique for the permafrost and is the reason for the applicability of special methods on a small scale, as in this case of reflection soundings. It is the same mechanism that made the use of the later arrivals possible in the examples on refraction soundings.

CONCLUSIONS

Although the principles of the seismic methods are the same in unfrozen areas as in permafrost, the latter has special properties which are important for shallow soundings. The special properties are in particular extremely high velocities and relatively small velocity discrimination, but at the same time the velocities are probably more homogeneous in most frozen formations than in similar unfrozen ones. Debris cemented by ice may eventually have higher velocities than con-solidated rock.

In material containing clay minerals, the velocity is believed to decrease with depth because of the increasing temperature, causing the energy of the direct wave along the surface to diminish rapidly with distance. Against the weak, direct wave the later pulses of refracted and reflected waves stand out spectacularly, as rarely experienced in a temperate climate.

In the Thule area, where the mean ground temperature is slightly below $-10°$ C, the refraction technique proved successful in most cases where shallow ice occurred over frozen ground, or where frozen ground overlay bedrock with or without ice on top. Where the frozen ground occurred on lower-velocity shale, a flat incidence reflection technique could be used. In order to draw conclusions on the applicability of the method in other areas, more should be known of the effect of temperature on the velocity in frozen ground, and generally of velocities in different types of permafrost.

ACKNOWLEDGMENTS

The help in the field by T. Fohl, K. C. Thomson, R. M. Van Noy, D. White, A. A. Wickham, B. M. Hamil, and other United States Army Snow, Ice and Permafrost Research Establishment and United States Army Engineer Research and Development Detachment personnel assisting the project, is gratefully acknowledged. Moreover, R. M. Van Noy and A. A. Wickham have carried out most of the computations.

REFERENCES

CORTE, A. E.   In preparation.   Relationship between four ground patterns, structure of the active layer, and type and distribution of ice in the permafrost.

DAVIES, W. E., KRINSLEY, D. B., and NICOL, A. H.   In preparation.   Geology of the North Star Bugt Area, Northwest Greenland.

HAGEDOORN, J. G.   1959.   The plus-minus method of interpreting seismic refraction sections; Geophysical Prospecting, vol. 7, no. 2, pp. 158-82.

ROETHLISBERGER, H.   1959.   Seismic survey 1957, Thule Area, Greenland; Snow Ice and Permafrost Research Establishment, Corps of Engineers, U.S. Army, Technical Report 64.

————   1961.   The applicability of seismic refraction soundings in permafrost near Thule, Greenland; Cold Regions Research and Engineering Laboratory, Corps of Engineers, U.S. Army, Technical Report 81.

THORNBURGH, H. R.   1930.   Wave-front diagrams in seismic interpretation; Bull. Amer. Assoc. Pet. Geol., vol. 14, no. 2.

Surface Features of Permafrost
in Arid Areas[1]

WILLIAM E. DAVIES

ABSTRACT

Most permafrost studies in America have been in wet soils. In North Greenland, at Brønlunds Fjord and Polaris Promontory, cursory examination of permafrost in extremely arid areas shows that permafrost conditions contrast greatly with those in areas of wet soils such as on the Arctic slope of Alaska.

Soils in the areas studied are silty clay, sand, and gravel. The physical characteristics of the active zone are similar to those in other areas of permafrost. The active zone extends twenty to forty inches at the maximum in summer. Soil temperature at the surface is 40 to 46° F. Temperatures in the active zone in gravel grades from freezing at the permafrost layer to 37 to 42° F six inches below the surface. Soil moisture is 1.2 to 7.8 per cent; the higher content is close to the permafrost boundary. Temperature in silty clay soil is similar to that in gravel; moisture content is 1 to 11 per cent near the surface and 8 to 22 per cent just above permafrost. The top of the permafrost is uniform with flat or gently sloping surface. Ice wedges and related features are not extensive.

The surface shows few features characteristic of permafrost or severe frost action. In the silty clay, heave and slump are lacking. The only surface markings are many small, polygonal, desiccation cracks. Flats underlain by gravel have irregular polygons with broad, shallow depressed edges; shrinkage cracks filled with fine sand are in the sand and gravel. Permafrost in arid areas produces an increase in soil compaction, uniform continuing moisture content, and a high-shear strength.

FROM 1956 THROUGH 1958 members of the United States Geological Survey had the opportunity of making incidental observations of permafrost conditions in North Greenland. These investigations were made as part of the Arctic programme of the Air Force Cambridge Research Center. Permission to visit North Greenland was extended through the courtesy of the Danish government. Dr. Axel Nørvang, of the Zoological Museum at Copenhagen, scientific representative of the Danish government, co-operated in the field work.

Northern Greenland is a high arctic desert (Fristrup, 1953). Precipitation is very low; wind velocity is high. This desert extends from Thule on the west to Germania Land on the east, but excludes the coastal areas along the northern part of Peary Land and in the vicinity of Nord where the precipitation exceeds thirteen inches (Figure 1). Most of North Greenland is a plateau 2000 to 3000 feet in elevation. Broad, steep-walled valleys cut the plateau into a series of large mesas. The flat floors of the valleys contain large deposits of alluvial material which mask most of the glacially-carved bedrock features. Along the coast are extensive areas of raised beaches and flats formed on marine silt and clay (Troelsen, 1952).

Climatic data in the high arctic are scarce. Regular meteorological observations have been made at only three stations, Alert on Ellesmere Island, Nord in north-

[1]Publication authorized by the Director, United States Geological Survey.

east Greenland, and Thule in northwest Greenland. A significant series of observations were made at Brønlunds Fjord, 1948-50, by the Danish Peary Land Expedition (Fristrup, 1952). Supplemental data, covering a shorter period of time, were obtained by the United States North Polar Expedition (U.S.S. "Polaris") 1871-2 (Bessels, 1879).

The gross features of the climate as defined from these sparse data show an average annual precipitation in the area of about 2.4 inches, of which 80 per cent or more is in the form of snow. Thule averages 2.5 inches per year, Brønlunds Fjord 2.3 inches, and Polaris Promontory 0.7 inches per year. The latter figure is based on only a year of observation and probably is too low. The mean annual temperature for Thule is 11° F, Polaris Promontory 4° F, Brønlunds Fjord 4° F. Maximum temperatures at all the stations are in the order of 53° to 64° F; minimum temperatures range from −43° to −45° F. Average relative humidity in the area is 66 to 78 per cent.

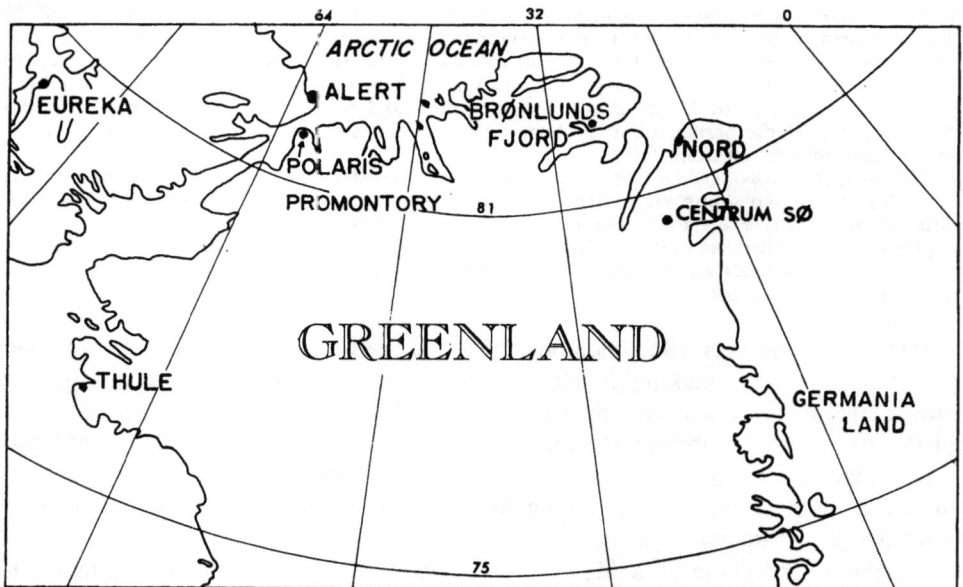

FIGURE 1.   Map of northern Greenland.

Perennially frozen ground apparently is present a few feet below the surface throughout the area. Surface features commonly associated with permafrost in areas of wet or saturated soils generally are lacking. In the area around Brønlunds Fjord, permanently frozen marine silt and clay form terrace benches up to 220 feet above sea-level. Shell remains in this marine formation have been dated 5390 years old.[2] Above 220 feet a series of kame terrace deposits overlap the marine silt and clay. The silt and clay also are present as alluvium that fills a former lagoon (Figure 2). It is perennially frozen eighteen to twenty-four inches below the ground surface during periods of maximum thaw (Table I). Moisture content in

[2]C14 date by Meyer Rubin, United States Geological Survey, Washington, D.C.

the active zone in August 1957 was 1 per cent a few inches below the surface and 21 to 45 per cent at twenty-four inches just above the permafrost.

TABLE I

CHARACTERISTICS OF ACTIVE ZONES

Soil	Thickness	Depth	Temp. °F	Per cent moisture
Brønlunds Fjord (composite of 4 pit studies)				
Firm brown clay	6–24 in.	surface	46	1–14
		6 in.	40–42	2–13
		12	37–40	5–19
		18	34–40	7–22
Permafrost	—	24	32	21–45
Polaris Promontory (composite of 5 pit studies)				
Fine sandy silt	6–15 in.	1 in.	40–46	1.9–5
		12	36–39	1.2–5.9
Sandy gravel, very few fines	16–31 in.	18	34–36	1.2–4.3
		24	33–37	1.2–4.5
		32	32–36	2.5–7.8
Permafrost	—	32–40	32	—

Both the alluvial and marine silt and clay show little surface expression of the permafrost. Slump and heave are absent. The only surface feature is a broad expanse of desiccation cracks developed on surfaces underlain by moist soils. These cracks form four- to six-sided polygons, two to eight inches on a side.

FIGURE 2. Alluviated bed of former lagoon, Brønlunds Fjord. Desiccation cracks are in silt-clay deposited from erosion of adjacent marine formation (photo by George Stoertz).

Individual polygons have slightly raised centres and are separated from one another by cracks up to a quarter-inch wide and as much as a foot in depth. The strength of the silt-clay soil is very high, ranging from three to more than ten expressed in terms of California Bearing Ratio. Bulk density was from 113 to 125 pounds per cubic foot. In marine silts and clays adjacent to the lagoon desiccation cracks occur, the moisture content is lower, and the strength is over ten (CBR). In kame deposits adjacent to the silt and clay beds, surface features commonly associated with permafrost are present. Polygons 1000 feet on a side are conspicuous. These polygons are bounded by trenches a foot wide and up to 2 feet deep. Similar features on gravel terraces and flats are discernible on aerial photographs throughout much of the arid part of northern Greenland. In limestone on the lower slopes of Kølen, eight miles east of Brønlunds Fjord, large geometrically-arranged frost patterns on bedrock are common. These disruptions are in the form of ridges of broken rock two feet high and four feet wide forming a series of rectangles, sixty feet on a side (Figure 3).

In northwestern Greenland, at Polaris Promontory, permafrost observations also were made in gravel outwash that lies along a major river valley (Figure 4). Polaris Promontory consists of a plain, ten to fifteen miles wide, trending southwest-northeast. The plain rises to 265 feet in elevation and is bordered by mountains 2000 to 3000 feet high. On the southwest side (Polaris Bay) and the northeast side (Newman Bay) of the plain there are moraines as much as 200 feet high. Below these moraines and outwash plains the Polaris lowland is formed of marine silt and clay similar to that at Brønlunds Fjord. Radiocarbon dating of shells in the marine clay by Meyer Rubin, United States Geological Survey, Washington, D.C., gave an age of 6100 ±300 years. Shells collected from a raised beach cut into one of the moraines yielded an age of 3780 ±300 years. No surface features indicative of permafrost are present in the silt and clay or moraine deposits although perman-

FIGURE 3.   Frost heave in limestone bedrock, south slope of Kølen, eight miles east of Brønlunds Fjord.

ently frozen ground is two feet below the surface. From the air, ancient polygon patterns, not easily discernible on the ground, show as slightly different shades of colour in the soil (Figure 5). The frost features are no longer active and are being eroded by wind and stream action.

Soil on the Polaris outwash plain is tightly packed sandy gravel with six to fifteen inches of fine sandy silt at the surface. Moisture content ranges from 2 per cent at the surface to 4.1 per cent and 7.8 per cent at the boundary with permafrost thirty-one to forty inches below the surface.

The surface of the outwash plain is marked by numerous large but indistinct polygons (Figure 4). The polygons are 60 to 100 feet on a side and are bounded by shallow depressions a foot or two wide and up to six inches deep; slopes are gentle. The depressions are filled with dark brown silt. In addition to the polygons the surface is cut by a series of small ancient drainage channels six to twenty feet wide and up to a foot deep. These channels are apparently relics from a time of greater precipitation and soil moisture content. Plant growth and deflation by wind are slowly erasing or hiding them.

Centrum Sø in northeastern Greenland is bordered on the west by a large triangular-shaped delta terrace that lies between two large rivers. The terrace is nine to twenty feet above lake level and is composed of gravelly sand. The gravel is pea-sized and constitutes about 20 per cent of the deposit. The permafrost surface rises from river level on the flanks of the terrace to seven or ten feet below the ground surface in the centre of the terrace. The terrace surface is free of most features related to permafrost except in its central part. Here polygons from seven to forty feet on a side exist. Originally these polygons had gently-raised centres bounded by depressed edges. The depressions, six to eight inches deep and as

FIGURE 4.   Outwash plain, Polaris Promontory. Shallow depressions bound large polygons. Smaller desiccation cracks are developed in silt soil overlying gravel.

much as a foot wide, are filled with compact dark brown silt which supports a luxuriant growth of plants. Some of the polygons retain raised centres and depressed borders. Others have been subjected to severe wind erosion and the sand from the raised central areas had been removed by deflation. The silt with plant cover in the original depressed borders has resisted deflation and now remains as ridges up to a foot high and wide. The deflated polygons are mainly on the south side of the terrace facing a broad valley that extends to the ice-cap. In areas to the leeward of the deflated polygons sand has been deposited masking many of the depressed boundaries. The deflated polygons grade laterally into the ones with depressed boundaries. At present the rate of deflation and burial greatly exceeds that of polygonal development.

The short reconnaissance of the surficial geology of northern Greenland has given data which indicate that the formation of surface features of permafrost in this arid area is arrested or retarded in spite of the existing extremes in temperatures. The active zone in this arid area is one of strength and stability in contrast to that in areas of higher precipitation and moister soils. The absence of permafrost features in silt and clay and their presence in coarse-grained soils is a striking anomaly. In most permafrost areas maximum development of such features is in the fine soils rendering them weak and unstable during periods of thaw. Studies have not progressed far enough to permit a factual explanation of the anomaly. However, it is apparently related to arid climatic conditions. The low rate of precipitation is greatly exceeded by evaporation and the moisture in the active zone is low except close to the base of the zone. In addition, it is believed that the soil moisture content, either as ice or free water, is uniform throughout the year at given points in the active zone. Observations made during the melt of wet snow in summer indicate that surface moisture penetrates only a few inches of the surface soil. The two feet of soil with low moisture separating the surface and the

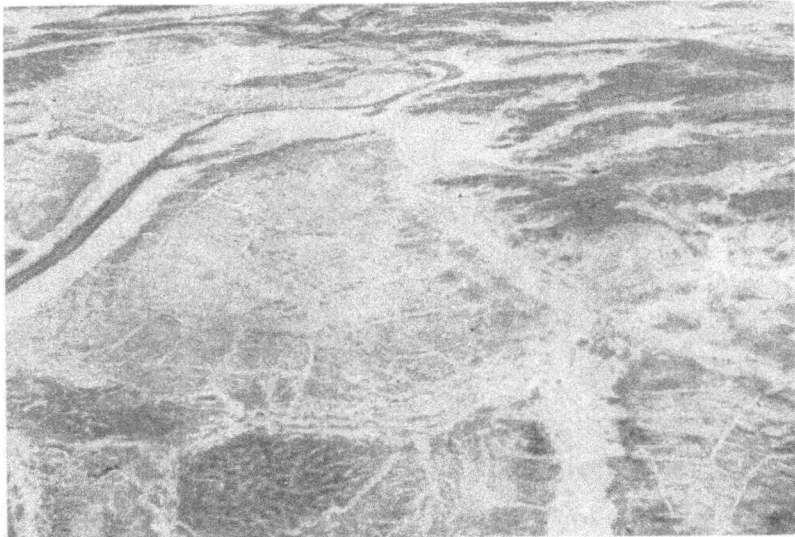

FIGURE 5.   Traces of ancient polygons in marine silt-clay, Polaris Promontory.

high-moisture band at the base of the active zone serves as a buffer to frost action. The frost action that effects surface configuration is probably confined to a thin band at the surface which receives moisture occasionally from rain and wet snow. The frost features in this thin band would be of the magnitude that develop in temperate zones. Such features disappear when the ground thaws. Because of this condition no surface configuration is present that is normally thought of as characteristic of permafrost.

Areas lacking sharp permafrost features similar to the one just described occur in the Copper River Basin, Alaska, and in parts of northern Canada. Such areas appear to be characterized by less than six inches of precipitation per year, where soil conditions are homogeneous.

ACKNOWLEDGMENTS

The author wishes to express his gratitude to George Stoertz of the United States Geological Survey and Captain Donald Klick, U.S.A.F., Cambridge Research Center for use of data collected by them during field work. The author also wishes to thank Stanley M. Needleman, Air Force Cambridge Research Center for facilitating field investigations in northern Greenland.

REFERENCES

BESSILS, EMIL. 1879. Die amerikanische Nordpol-Expedition; Wilhelm Engelmann, 643 p.
FRISTRUP, BØRGE. 1952. Physical geography of Peary Land, I. Meteorological observations for Jørgen Brønlunds Fjord; Medd. om Grønland, bd. 127, nr. 4, 143 p.
——— 1953. Wind erosion within the Arctic deserts; Geog. Tidsskr., bd. 52, pp. 51-65.
TROELSEN, J. C. 1952. Notes on the Pleistocene geology of Peary Land, North Greenland; Medd. fra Dansk Geol. Forening, bd. 12, hefte 2, pp. 211-20.

A System for the Refrigeration of Drilling Fluids for Rotary Drilling and Coring in Frozen Earth Materials[1]

G. ROBERT LANGE

The United States Army Snow, Ice, and Permafrost Research Establishment has developed a system for rotary drilling and coring in glacier ice and permafrost. A portable refrigerator of 5-ton capacity is used to chill the drilling fluid so that the system may be used to take frozen cores and maintain hole walls in the frozen state with ambient air temperatures of up to 70° F Two heat exchangers are available for use with the refrigerator, one for chilling compressed air and the other for chilling diesel fuel. More success has been experienced with the diesel fuel. It is suggested that such a system might be used to secure cores of certain poorly consolidated unfrozen materials by inducing the freezing of the material ahead of the bit and thereby increasing its strength to the point where it might be cored.

[1]Abstract only.

An Investigation into

Methods of Accelerating the

Melting of Ice and Snow

by Artificial Dusting[1]

K. C. ARNOLD

ABSTRACT

During the summer of 1959 the author conducted experiments on snow, sea ice, and lake ice surfaces in an attempt to accelerate their rate of melting by artificial dusting. Seven different types of materials were used, differing in particle size and in the presence or absence of salt or sand. Each of these materials was spread in amounts of 100, 300, 500, 700, and 1000 grams on squares one square metre in area. This gave thirty-five squares to study on each of the snow, sea ice, and lake ice surfaces.

During the summer, repeated measurements were taken in each of these squares, using a level and rod to measure the departure of each square from a system of reference stakes. A comparison was also made with untreated areas, which served as a control. Cores and stereoscopic photographs were taken in each square of the sea and lake ice surfaces, illustrating the penetration of the dust into the ice surface. Meteorological readings are available at three-hour intervals in the areas studied.

The results are discussed and some comparisons made between the different surfaces and materials. Former practical applications of this method are referred to, and possible future applications are suggested.

IN NATURE IT OFTEN HAPPENS that a snow or ice surface becomes covered, by wind or other action, with a layer of dark material. Depending on the thickness, porosity, and heat-absorbing qualities of this material, the rate of melting of the underlying snow or ice surface can be accelerated or retarded. These contrasting processes have been commented upon frequently. Troll (1949) discusses these ablation forms, drawing examples from many latitudes, and gives many references. Sharp (1949) discusses many different ablation forms occurring on one glacier, the Wolf Creek glacier in the Canadian Yukon, and draws attention to the role played by direct and indirect solar radiation.

The protective role of a thick layer of dark material is well illustrated in dirt cones. Lewis (1940) discussed the well-developed dirt cones that are found on the northern margins of Vatnajökull. A later paper on Icelandic dirt cones by Swithinbank (1950) was written after the eruption of Mount Hekla in 1947, a source of much wind-blown material. A layer of dust as thin as half a centimetre can keep the underlying snow from melting. G. Warren Wilson (1953) who describes dirt cones found in Jan Mayen Island discusses both the protective and the melting effects of wind-blown volcanic dust.

[1]Published with the permission of the Director, Geographical Branch, Department of Mines and Technical Surveys, Ottawa.

Cryoconite holes or dust wells are excellent natural examples of the role played by a thin layer of dust in accelerating the melting of ice. This alone has a considerable literature. Recent work in high latitudes has been published by Gajda (1958) and by Gerdel and Drouet (1958). They describe cryoconites and other ablation phenomena found near the edge of the Greenland ice-sheet near Thule.

Attempts to reproduce these effects artificially are not new. Trushin (1957), in a paper on recent Russian experiments, states that research has been carried on in Russia since 1885. Landsberg (1940) experimented with coal dust spread on snow, and on small blocks of ice that were periodically weighed. An unusual aspect of research in this field appeared in a recent magazine article[2] on the life and letters of Benjamin Franklin. A letter written in 1761 to a British lady gave advice on suitable choices of colour for summer hats; after he had laid out various coloured squares of cloth on snow, Franklin had observed the depth to which they melted. It is interesting that the reconstruction of the experiment for the magazine photograph shows that the behaviour of the pieces of cloth faithfully reproduced another experiment by Hand and Lundquist (1942) nearly two hundred years after Franklin's letter.

The experiments at Isachsen in 1959 were designed to make a quantitative study of the relative effectiveness of cinders and ash which differed in particle size, and in the presence or absence of an addition of salt or fine sand. The relative effectiveness of these materials on snow, sea ice, and lake ice surfaces was also of interest.

METHOD AND LIMITATIONS OF EXPERIMENTS

The material used was supplied by a Toronto firm, and consisted of the following seven types: coarse cinder, coarse cinder with 75 per cent rock salt added, coarse cinder with 25 per cent rock salt added, fine cinder, fine cinder with 25 per cent rock salt added, fly ash, and fly ash with 25 per cent fine sand added. The proportions stated are by weight. The material as received was not entirely suitable for the experiments, as it had not been sufficiently graded with regard to particle size. Further grading was done in the field, giving particles from 10 to 2 mm for coarse cinder, 2 to 0.2 mm for fine cinder, and less than 0.2 mm for fly ash. Unfortunately this cut down the amount of material available and set the limit of 1000 gm/sq m for the densest concentration of dust spread.

The snow and sea ice test areas were on winter ice about one km from the Canadian–United States Joint Weather Station at Isachsen, Northwest Territories (78°47′N, 103°30′W). They were 50 m apart and about 250 m from the shore. The nearest suitable body of fresh water was a small lake about five km northwest of the weather station and sixty m above sea-level.

On May 28 the material was spread on the snow and sea ice test areas, after all snow had been removed from the sea ice test area. It was spread in seven strips, each containing five-m squares, with concentrations of 100, 300, 500, 700, and 1,000 gm/sq m. These strips were separated by lanes half a metre in width to allow access to the individual squares (Figure 1). On May 30 the material was spread on the lake ice in an identical manner to the sea ice test area. Ablation stakes were set in untreated control areas about ten m away from each test area.

[2]*Life*, vol. 48, no. 8, February 29, 1960, p. 60.

In the snow test area the depth of snow lying on the sea ice was measured with a steel ruler by tapping lightly with a small hammer until the resistance of the underlying sea ice was felt. Care was taken to disturb the snow as little as possible. Five readings were taken in each square, and a wooden frame was made as a guide to ensure that the five readings were taken in approximately the same place. These readings were taken to 0.5 cm. The standard error of the mean of the five readings in each square was calculated. The mean for all thirty-five squares in the snow test area was about one cm throughout the experiment, and the results have been tabulated in one cm intervals.

Measuring the depth of snow was relatively easy, as the sea ice underneath formed a convenient reference base. There was no convenient base for the ice test areas, which were measured by periodic spirit levelling (Figure 2). A Wild reversible level was used to measure the departure of the different squares from the control area. Five readings, with the bubble in direct and reversed position, were taken in each square. The accuracy of the measurements was not the same throughout the experiments. The mean of the standard errors of the means of individual squares ranged from 0.1 cm for the sea ice area on May 29 to 2.2 cm for the lake ice test area on July 5. The results have been tabulated in roughly comparable intervals. The standard error may be taken as a rough index of the unevenness of dusted surfaces.

Three-hourly surface weather observations are available from the Canadian–United States Joint Weather Station at Isachsen. This station is about 25 m above sea level. Selected observations of cloudiness, temperature, humidity, wind, and precipitation have been plotted, and these serve for a general record of the weather during the experiments, and for comparisons with similar experiments elsewhere.

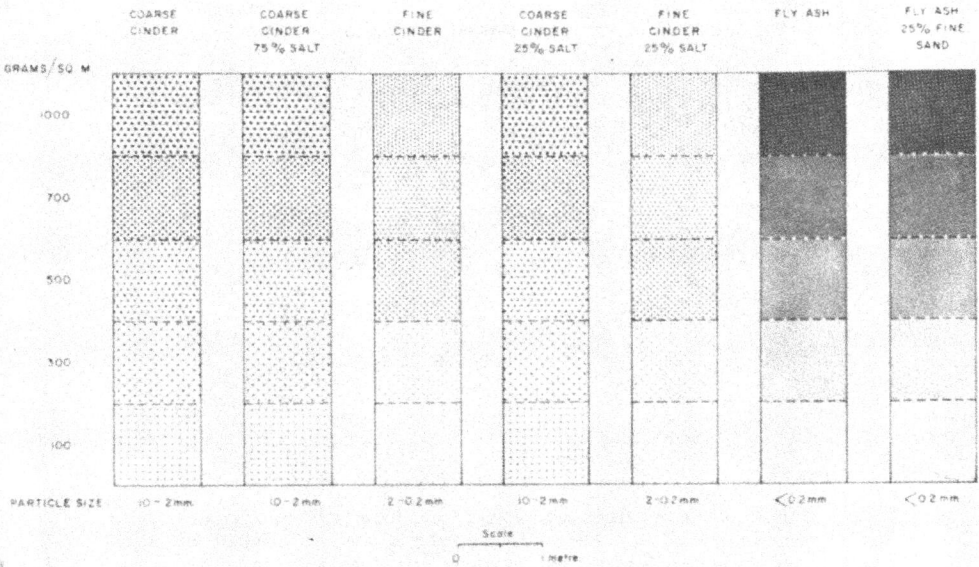

FIGURE 1. Arrangement of dust showing material, concentration and particle size.

For a more refined study detailed micrometeorological observations would be necessary, particularly of the radiation balance in the various areas. Exact measurements of the absorption of radiation by particles under pools of water, or of individual particles that have melted into the ice, would need delicate instruments and methods. Thermographs were set up about 25 cm above the snow surface in improvised screens, between the snow and sea ice test areas and at the lake ice test area. Their records shows that conditions at these two sites were similar as far as temperature is concerned.

Cores three inches in diameter were taken from each square in the sea ice test area on June 14, and from the lake ice test area squares on June 15. These illustrate the penetration of the dust into the ice (Figures 3-8). In other studies this could be done at frequent intervals but care should be taken to use larger test

FIGURE 2. Sea ice test area, June 14. The strip containing fly ash and 25 per cent fine sand is nearest to the camera.

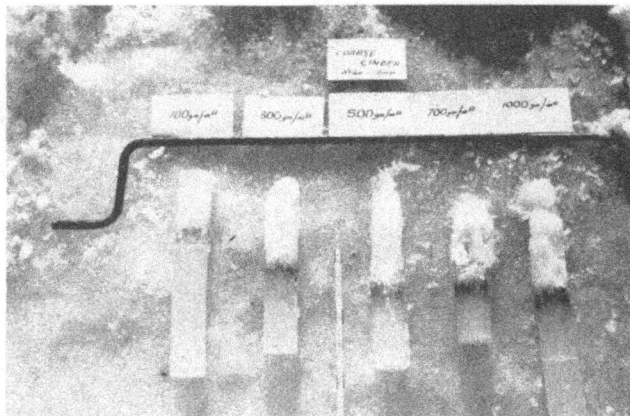

FIGURE 3. Core from sea ice test area, June 14. Coarse cinder. The under side of the metal bar represents the original ice surface before dusting, and the cores are placed at their correct depth. Note the penetration of material into the solid part of the core.

areas if this is done, as the removal of a number of cores would be a disturbing factor.

Stereophotographs were taken in each square of the sea ice test area on June 29, and this was repeated on the lake ice on July 5. The camera used was a Kodak Revere 35 mm stereo camera. These pictures are interesting as a record of conditions at the end of the measurements, and illustrate the nature of the penetration of particles into the ice, and the unevenness of the bottom of the pools that had then formed. Some rough measurements are possible, but there are limitations. The inner orientation of the camera was not known, and a small fixed focus camera would be more useful, taking photographs from both ends of a rigid base. A suitable target for control of the pictures would be an accurately constructed open square marked with graduated divisions on the sides, and with some levelling arrangement. Accurate measurements from photographs of an object under water, when the camera is in the air above it, poses some interesting problems, as the rays of light from different parts of the object do not appear to converge in a single point, owing to different amounts of refraction. However, this technique

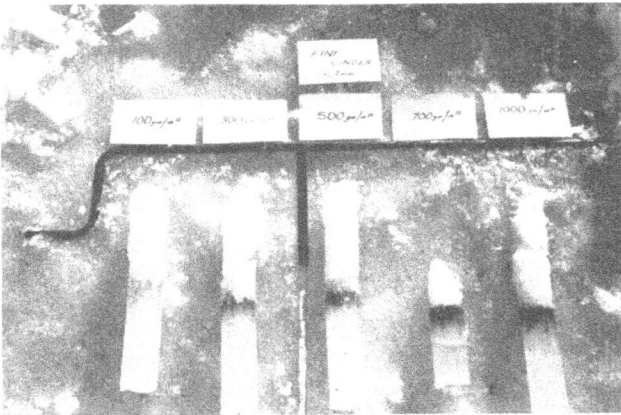

FIGURE 4.   Core from sea ice test area, June 14. Fine cinder.

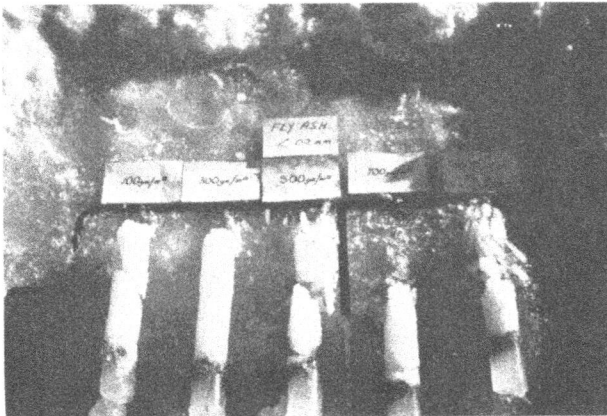

FIGURE 5.   Core from sea ice test area, June 14. Fly ash.

might, with suitable refinements, prove very useful if a refined micrometeorological programme called for accurate measurements of the amount of ice lost.

One limitation that affected the early stage of the experiments was the amount of snow falling in early June. This was a problem until about June 15. Had the fall been heavier there would not have been sufficient material to spread more. The snow test area was not interfered with. In the sea ice and lake ice test areas the snow was kept clear as far as possible, to give each of the areas equal exposure to solar radiation. Most of the snow in these areas was built up by wind drifting in the lee of the piles that had been shovelled clear before. It was not possible to give the same attention to the lake ice area as to the sea ice area, owing to the distance involved, and this may be a limitation in comparing the early differences.

Sweeping the snow off the areas in this manner necessarily made conditions artificial, but fortunately, the snow test area, lying on sea ice, later formed a good

FIGURE 6. Core from lake ice test area, June 15. Coarse cinder. Note the contrast in penetration of material compared to the sea ice cores, Figures 3-5.

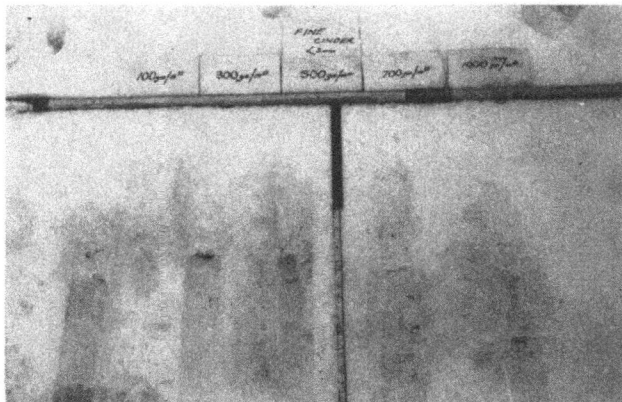

FIGURE 7. Core from sea ice test area, June 15. Fine cinder.

example of sea ice that had been dusted and not further interfered with (Figure 9). This was watched, but not levelled, throughout the summer. Although in the early stages this area seemed to lag about ten days behind the neighbouring sea ice test area, at the end of the summer both areas melted through to the sea water at about the same time. The sweeping of snow from the two ice test areas, therefore, would not appear to affect the validity of the experiments taken over a whole season of melting.

EFFECT ON SNOW

The snow test area was chosen to be as even in depth as possible, and close to the sea ice test area. The snow in Station Bay was well wind-packed, and had a density of 0.35. The sastrugi made it impossible to find a perfectly even area, and a site was chosen 50 m from the sea ice test area. The initial surface (Figure 10) varied from 28 to 12 cm in depth, but only in the 100- and 300-gm concentrations of fly ash and 25 per cent fine sand were slopes steep enough seriously to influence the depth measurements and amount of direct solar radiation received. The area

FIGURE 8. Core from sea ice test area, June 15. Fly ash.

FIGURE 9. The former snow test area, June 29. Isachsen weather station in background.

was dusted on May 28, and the depth of snow was remeasured on the following three days (Figures 11-13; Table I).

Larger amounts of the various types of coarse cinder uniformly resulted in an increased loss of snow. When coarse cinder was used without salt, this increase began to decline with amounts greater than 500 gm/sq m. Coarse cinder with 75 per cent salt added acted rapidly, but its final effectiveness was about 20 per cent less than coarse cinder alone. The darkening of the snow surface was perceptibly less than with the other materials. Coarse cinder with 25 per cent salt gave results that were very closely related to the degree of concentration of the material. The 1000-gm concentration was little more effective than coarse cinder alone, and the smaller concentrations were less effective. The value of an added amount of

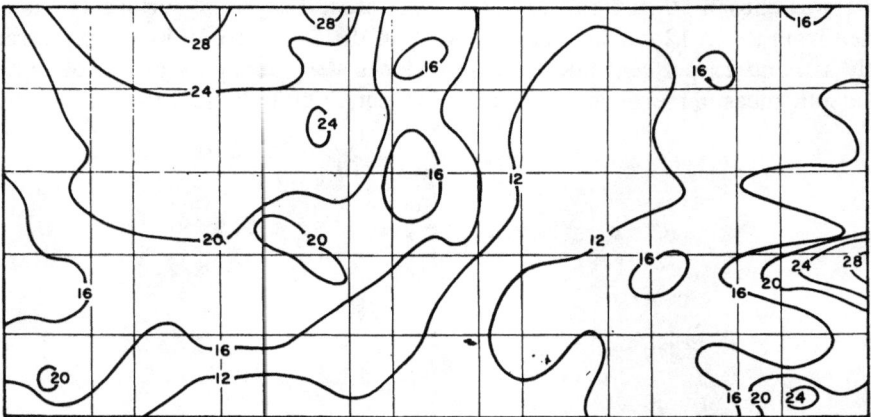

FIGURE 10.   Snow. Snow depth before dusting, May 28.

GRAMS/SQ.M	COARSE CINDER	COARSE CINDER 75% SALT	FINE CINDER	COARSE CINDER 25% SALT	FINE CINDER 25% SALT	FLY ASH	FLY ASH 25% FINE SAND
1000	8	9	8	7	10	10	8
700	8	8	8	7	10	9	7
500	8	6	7	4	9	8	7
300	6	6	6	2	8	7	9
100	4	3	7	1	7	8	7
PARTICLE SIZE:	10-2 mm.	10-2 mm.	2-0.2 mm.	10-2 mm.	2-0.2 mm.	<0.2 mm.	<0.2 mm.

control area gained 1 cm. from snowfall during this period
m.s.e.m. 0·9 cms.

FIGURE 11.   Snow. Loss, in cm, May 28-29 (1 day).

TABLE I

MEASUREMENTS IN SNOW TEST AREA, MAY 28 TO MAY 31

Snow	Coarse cinder (1 2 3 4 5)	Coarse cinder, 75 per cent salt (6 7 8 9 10)	Fine cinder (11 12 13 14 15)	Coarse cinder, 25 per cent salt (16 17 18 19 20)	Fine cinder, 25 per cent salt (21 22 23 24 25)	Fly ash (26 27 28 29 30)	Fly ash, 25 per cent fine sand (31 32 33 34 35)
Loss, cm May 28 May 29 m.s.e.m. 0.9 cm	8, 8, 6, 6, 4	9, 8, 6, 6, 3	8, 8, 7, 6, 7	7, 7, 4, 2, 1	10, 10, 9, 8, 7	10, 9, 8, 7, 8	8, 7, 7, 9, 7	*Control area gained 1 cm from snowfall*
Loss, cm May 28 May 30 m.s.e.m. 0.9 cm	10, 8, 7, 6, 3	8, 7, 7, 6, 2	12, 10, 8, 7, 5	8, 6, 3, 4, 0	12, 11, 9, 9, 6	12, 10, 10, 7, 7	11, 10, 11, 12, 8	*Control area gained 2 cm from snowfall*
Loss, cm May 28 May 31 m.s.e.m. 0.8 cm	11, 10, 10, 8, 5	9, 8, 6, 6, 3	14, 13, 12, 11, 6	12, 8, 4, 2, 0	12, 11, 9, 12, 7	13, 12, 12, 12, 8	12, 13, 14, 17, 10	*Control area gained 3 cm from snowfall*

rock salt appears small, if a continuing solar radiation absorption effect from a single application is desired. Even at the end of the first day, the coarse cinder with 75 per cent salt had been almost overtaken by plain coarse cinder.

Fine cinder was the most effective, by a small margin, of all seven materials at the end of the experiment. An addition of 25 per cent salt again produced an initial advantage of about 25 per cent on the first day. This became a 15 per cent disadvantage by the end of the experiment. Increased concentrations of both types

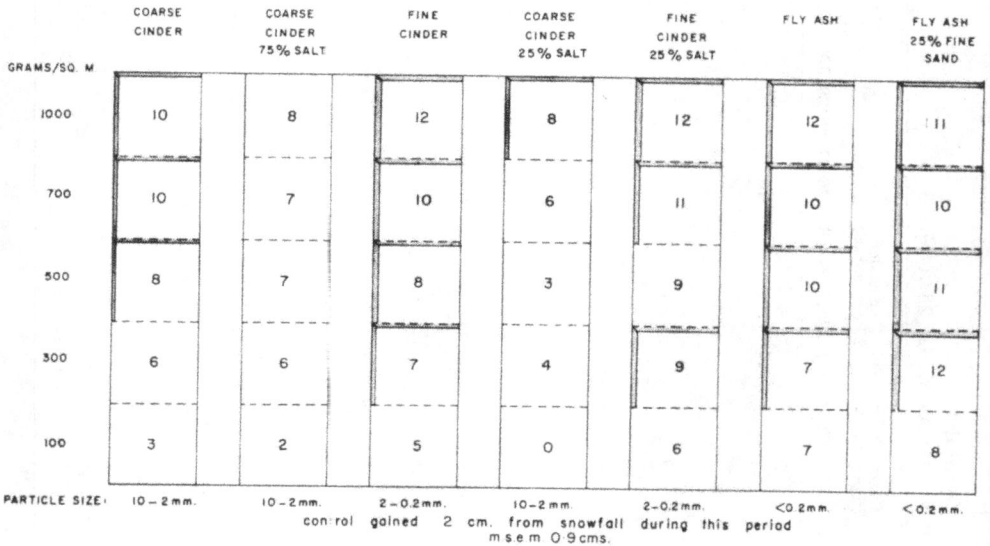

GRAMS/SQ. M	COARSE CINDER	COARSE CINDER 75% SALT	FINE CINDER	COARSE CINDER 25% SALT	FINE CINDER 25% SALT	FLY ASH	FLY ASH 25% FINE SAND
1000	10	8	12	8	12	12	11
700	10	7	10	6	11	10	10
500	8	7	8	3	9	10	11
300	6	6	7	4	9	7	12
100	3	2	5	0	6	7	8
PARTICLE SIZE	10 – 2 mm.	10 – 2 mm.	2 – 0.2 mm.	10 – 2 mm.	2 – 0.2 mm.	<0.2 mm.	<0.2 mm.

control gained 2 cm. from snowfall during this period m s e m 0.9 cms.

FIGURE 12.   Snow. Loss, in cm, May 28-30 (2 days).

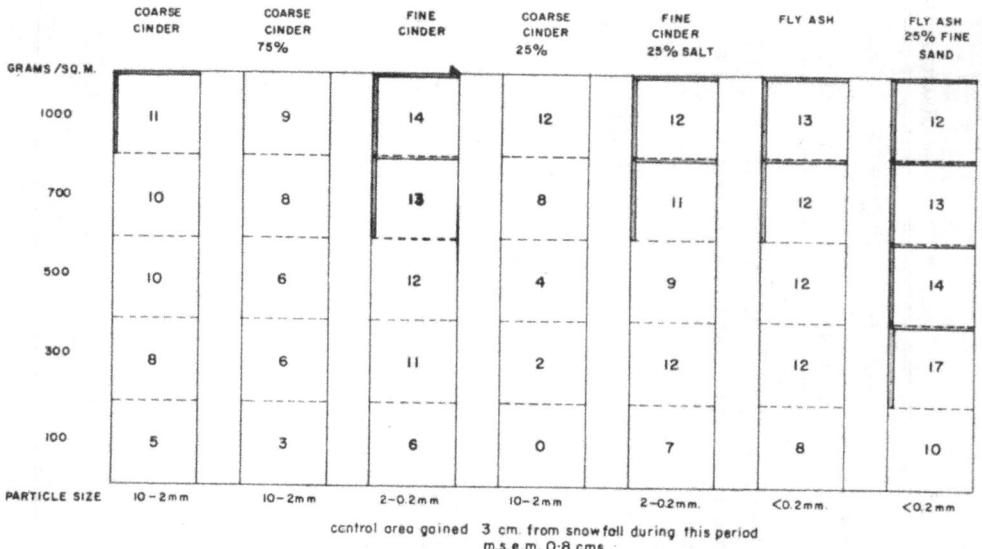

GRAMS /SQ. M.	COARSE CINDER	COARSE CINDER 75%	FINE CINDER	COARSE CINDER 25%	FINE CINDER 25% SALT	FLY ASH	FLY ASH 25% FINE SAND
1000	11	9	14	12	12	13	12
700	10	8	13	8	11	12	13
500	10	6	12	4	9	12	14
300	8	6	11	2	12	12	17
100	5	3	6	0	7	8	10
PARTICLE SIZE	10 – 2 mm	10 – 2mm	2 – 0.2mm	10 – 2mm	2 – 02mm.	<0.2mm.	<0.2 mm

control area gained 3 cm. from snowfall during this period m.s.e.m. 0.8 cms.

FIGURE 13.   Snow. Loss, in cm, May 28-31 (3 days).

of fine cinder generally gave increased effectiveness, but this was not as well marked as with coarse cinder.

Fly ash, and fly ash mixed with 25 per cent fine sand, were very similar in effectiveness. The unevenness of the snow in the two finest concentrations of fly ash mixed with fine sand made these measurements less reliable. The greatest standard error in these squares was 3.5 cm. The differences between the 100- and 1000-gm concentrations are much less than for the other materials. Both types of fly ash were markedly more effective than fine and coarse cinder in equal 100-gm concentrations, and retained this advantage up to the 700-gm concentrations. The 1000-gm concentrations were only a little less effective than the same amount of fine cinder.

In summary, the effects can generally be explained in terms of albedo. With large concentrations of materials, particle size was of little importance as a factor; with small concentrations, its importance increased. For a given weight of material, fine particles will darken a given area more than large ones. When the concentration is increased, the albedos tend to become more equal. This is a simplification, for other factors are involved. A large particle can absorb more heat than a small one, and retain this longer while melting into the snow. But this later becomes a disadvantage, for individual large particles become isolated in this way and are less accessible to solar radiation. Paulsen (1955), quoting work done in Austria, emphasizes the importance of albedo in its relation to the melting of different types of snow. New snow had an average albedo of 0.8. This became 0.5 for old snow. Dirty ice had an albedo of 0.09 to 0.2.

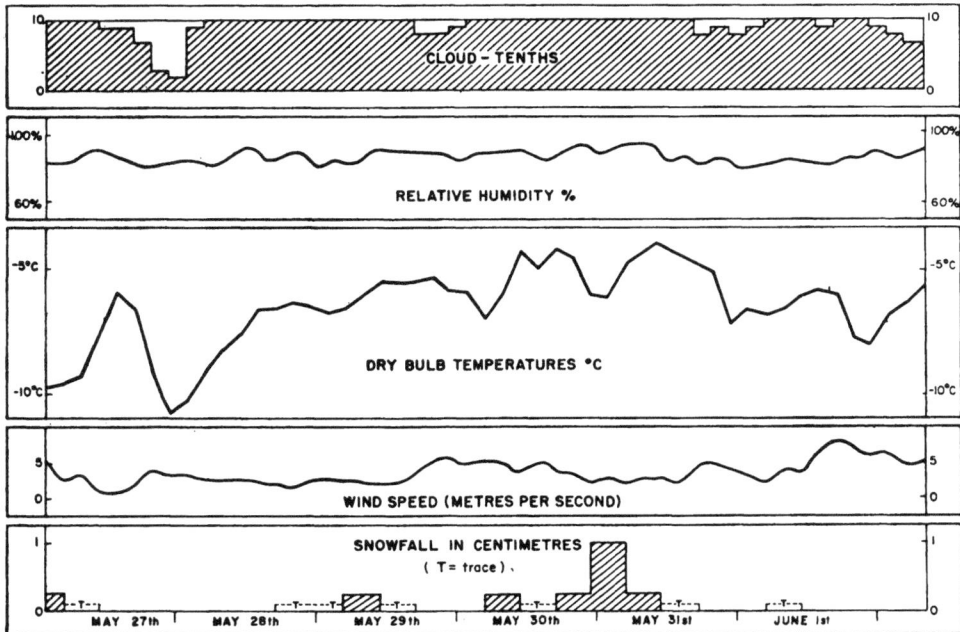

FIGURE 14. Selected meteorological observations, Isachsen weather station, during snow-dusting experiment.

The weather observations during this period (Figure 14) show that air temperatures were about —5° C. This agrees well with the work of Trushin (1957) who comments on the effectiveness of this method below freezing temperatures. The level of cloudiness was high throughout, illustrating that direct sunshine was not necessary and that diffuse solar radiation played a large part in melting snow. Although three cm of new snow fell on the area during the experiment, no square showed a gain in depth. This was not caused by the dusted areas keeping abreast of the falling snow. The test area was covered with new snow immediately after the fall on the night of May 30-31. Subsequent melting of this snow suggests that solar radiation can penetrate a cover of new snow. Hand and Lundquist (1942) stated that 1 per cent of the incoming radiation penetrated 15 cm of uniform fine granular snow. Six cm of snow and slush cut the incoming radiation down to the same amount. A layer of new ice also acts as a barrier (Paulsen, 1955). This should be considered if a decision must be made to spread more dust after a fresh fall of snow.

EFFECT ON SEA ICE

The sea ice test area was laid out on level winter ice that had formed in a sheltered part of Station Bay. On May 28, when the dust was spread, this ice was 220 cm thick. On June 11 the salinity of the sea water was twenty-one parts per thousand. A profile through the sea ice was taken on July 4, when the ice was 160 cm thick. At this date the salinity of the ice varied from 0.5 parts per thousand at 10 cm below the surface of the ice to 1.5 parts per thousand at 10 cm above the sea water. It is probable that these values would have been from two to three times greater at the time that the dust was spread.

The sea ice strips were measured six times between May 29 and June 29 (Figs. 15-20; Table II). Until June 14, when air temperatures rose above 0° C, the

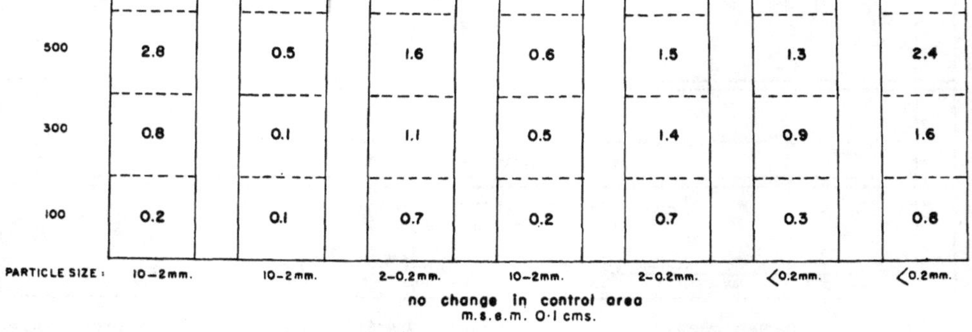

GRAMS/SQ. M	COARSE CINDER	COARSE CINDER 75% SALT	FINE CINDER	COARSE CINDER 25% SALT	FINE CINDER 25% SALT	FLY ASH	FLY ASH 25% FINE SAND
1000	2.6	2.0	2.9	1.8	2.5	2.2	2.5
700	1.8	0.7	1.7	0.9	2.8	1.6	2.3
500	2.8	0.5	1.6	0.6	1.5	1.3	2.4
300	0.8	0.1	1.1	0.5	1.4	0.9	1.6
100	0.2	0.1	0.7	0.2	0.7	0.3	0.8
PARTICLE SIZE	10–2mm.	10–2mm.	2–0.2mm.	10–2mm.	2–0.2mm.	<0.2mm.	<0.2mm.

no change in control area
m.s.e.m. 0·1 cms.

FIGURE 15.   Sea ice. Loss, in cm, May 28-29 (1 day).

TABLE II
MEASUREMENTS IN SEA ICE TEST AREA, MAY 28 TO JUNE 29

Sea ice	Coarse cinder					Coarse cinder, 75 per cent salt					Fine cinder					Coarse cinder, 25 per cent salt					Fine cinder, 25 per cent salt					Fly ash					Fly ash, 25 per cent fine sand					Control
	1	2	3	4	5	6	7	8	9	10	11	12	13	14	15	16	17	18	19	20	21	22	23	24	25	26	27	28	29	30	31	32	33	34	35	
Loss, cm May 28 May 29 m.s.e.m. 0.1 cm	2.6	1.8	2.8	0.8	0.2	2.0	0.7	0.5	0.1	0.1	2.9	1.7	1.6	1.1	0.7	1.8	0.9	0.6	0.5	0.2	2.5	2.8	1.5	1.4	0.7	2.2	1.6	1.3	0.9	0.3	2.5	2.3	2.4	1.6	0.8	area lost 0.0 cm
Loss, cm May 28 June 8 m.s.e.m. 0.6 cm	13.5	10.5	11.5	1.0	2.0	5.0	5.0	4.5	1.5	1.0	12.5	10.5	9.5	6.5	5.0	7.5	8.0	4.0	3.0	2.5	11.5	13.0	8.0	7.0	6.0	3.0	10.5	9.0	9.0	8.5	10.5	10.5	10.0	8.5	8.5	area lost 0.0 cm
Loss, cm May 28 June 12 m.s.e.m. 0.7 cm	18.5	13.0	12.5	8.0	3.0	8.5	7.0	6.5	6.0	1.5	17.5	15.5	14.0	10.5	7.0	14.5	9.5	8.0	7.5	3.5	18.0	19.0	13.0	12.5	9.5	14.5	13.0	12.5	10.5	10.5	16.5	15.0	12.5	10.5	11.5	area lost 0.1 cm
Loss, cm May 28 June 14 m.s.e.m. 0.8 cm	23	19	18	8	6	10	8	7	6	5	25	16	15	11	8	15	13	9	7	6	24	21	13	11	9	14	12	12	10	10	20	14	12	10	10	area lost 0.2 cm
Loss, cm May 28 June 22 m.s.e.m. 2.0 cm	60	42	40	26	22	32	26	20	18	16	62	48	46	30	24	36	22	22	22	22	64	46	40	28	22	54	38	34	26	24	64	54	44	30	24	area lost 8 cm
Loss, cm May 28 June 29 m.s.e.m. 2.0 cm	98	100	100	72	46	64	62	68	42	34	100	104	84	64	52	84	72	50	48	42	100	98	68	64	50	100	66	58	58	46	92	80	70	72	72	area lost 20 cm

relative effectiveness of the different materials on ice was similar to the results on snow. On May 29, measurements were made of both the sea ice and the snow test areas twenty-four hours after the dust had been applied. The depth of snow lost was much greater than that of ice, but when the difference in density is considered (0.35 for snow, about 0.9 for ice), the amount melted in both areas appears to be roughly the same. This is true for the most effective, 1000-gm concentrations. It does not apply as well to the smaller concentrations.

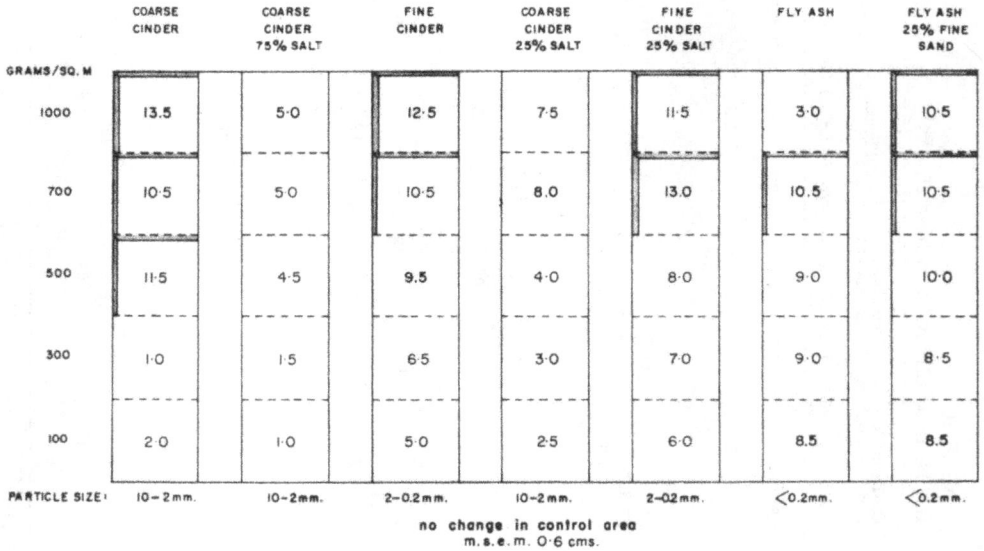

GRAMS/SQ. M	COARSE CINDER	COARSE CINDER 75% SALT	FINE CINDER	COARSE CINDER 25% SALT	FINE CINDER 25% SALT	FLY ASH	FLY ASH 25% FINE SAND
1000	13.5	5·0	12·5	7·5	11·5	3·0	10·5
700	10·5	5·0	10·5	8.0	13.0	10.5	10·5
500	11·5	4·5	9.5	4·0	8·0	9·0	10·0
300	1·0	1·5	6·5	3·0	7·0	9·0	8·5
100	2·0	1·0	5·0	2·5	6·0	8.5	8.5
PARTICLE SIZE:	10-2mm.	10-2mm.	2-0.2mm.	10-2mm.	2-02mm.	<0.2mm.	<0.2mm.

no change in control area
m. s. e. m. 0·6 cms.

FIGURE 16.　Sea ice. Loss, in cm, May 28-June 8 (11 days).

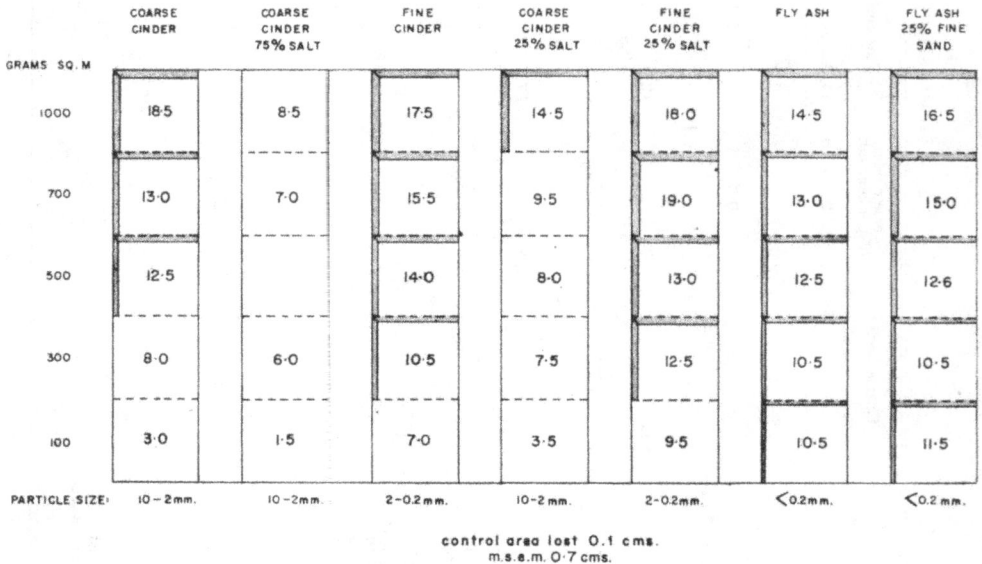

GRAMS SQ. M	COARSE CINDER	COARSE CINDER 75% SALT	FINE CINDER	COARSE CINDER 25% SALT	FINE CINDER 25% SALT	FLY ASH	FLY ASH 25% FINE SAND
1000	18·5	8·5	17·5	14·5	18·0	14·5	16·5
700	13·0	7·0	15·5	9·5	19·0	13·0	15·0
500	12·5		14·0	8·0	13·0	12·5	12·6
300	8·0	6·0	10·5	7·5	12·5	10·5	10·5
100	3·0	1·5	7·0	3·5	9·5	10·5	11·5
PARTICLE SIZE:	10-2mm.	10-2mm.	2-0.2mm.	10-2mm.	2-0.2mm.	<0.2mm.	<0.2 mm.

control area lost 0.1 cms.
m.s.e.m. 0·7 cms.

FIGURE 17.　Sea ice. Loss, in cm, May 28-June 12 (15 days).

By June 14, all seven materials showed that an increase in concentration gave an increased loss of ice. At this date the untreated control area had lost only 0.2 cm of ice. A 100-gm concentration of coarse cinder with 75 per cent salt, which was the least effective, lost 5 cm of ice. A 1000-gm concentration of fine cinder was the most effective combination at this date. This gave a loss of 25 cm ice. The addition of 25 per cent rock salt made little difference. Coarse cinder without any added salt was almost as effective as fine cinder. An addition of

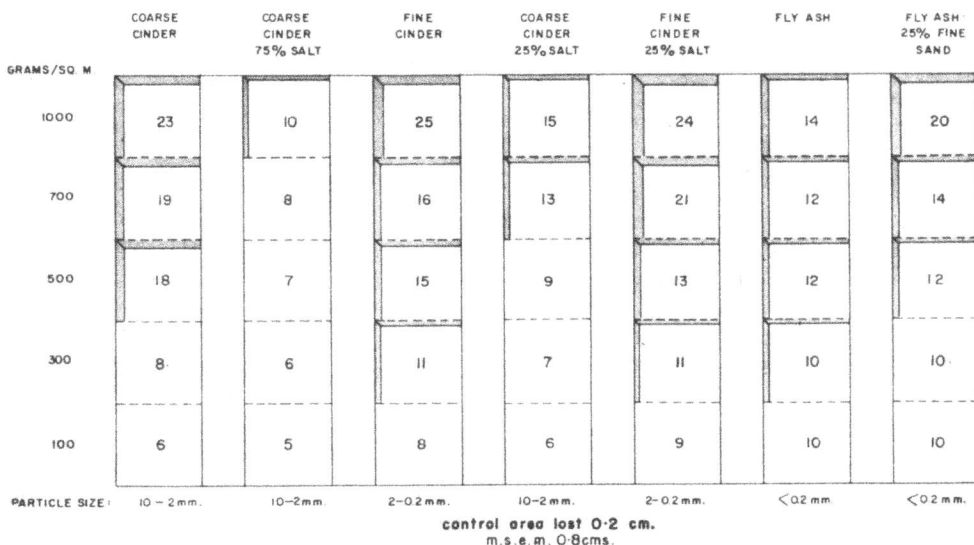

GRAMS/SQ. M	COARSE CINDER	COARSE CINDER 75% SALT	FINE CINDER	COARSE CINDER 25% SALT	FINE CINDER 25% SALT	FLY ASH	FLY ASH 25% FINE SAND
1000	23	10	25	15	24	14	20
700	19	8	16	13	21	12	14
500	18	7	15	9	13	12	12
300	8	6	11	7	11	10	10
100	6	5	8	6	9	10	10
PARTICLE SIZE	10 – 2 mm.	10–2mm.	2–0.2mm.	10–2mm.	2–0.2mm.	<0.2 mm	<0.2 mm.

control area lost 0·2 cm.
m.s.e.m. 0·8cms.

FIGURE 18.   Sea ice. Loss, in cm, May 28-June 14 (17 days).

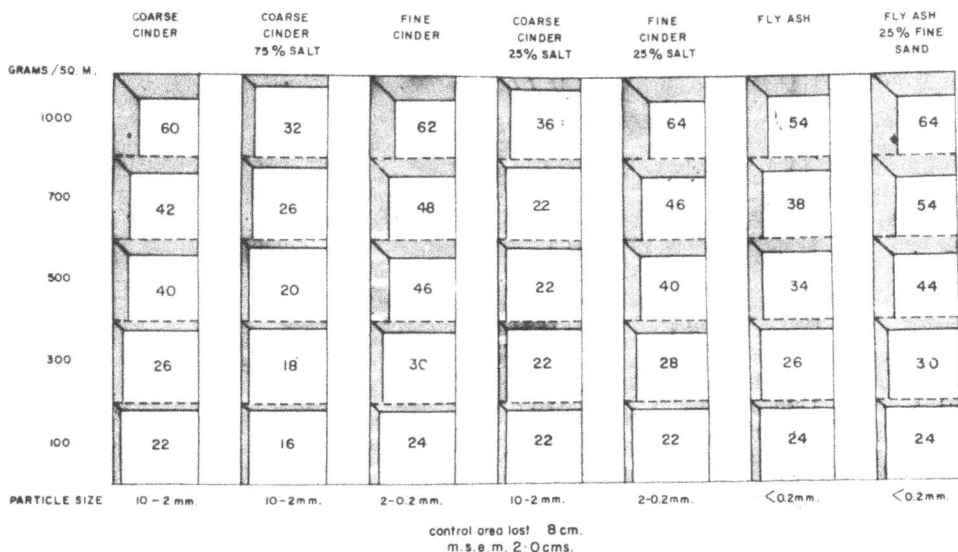

GRAMS/SQ. M	COARSE CINDER	COARSE CINDER 75% SALT	FINE CINDER	COARSE CINDER 25% SALT	FINE CINDER 25% SALT	FLY ASH	FLY ASH 25% FINE SAND
1000	60	32	62	36	64	54	64
700	42	26	48	22	46	38	54
500	40	20	46	22	40	34	44
300	26	18	3C	22	28	26	30
100	22	16	24	22	22	24	24
PARTICLE SIZE	10 – 2 mm	10–2mm.	2–0.2 mm.	10–2 mm.	2–0.2mm.	<0.2mm.	<0.2 mm.

control area lost 8 cm.
m.s.e.m. 2·0cms.

FIGURE 19.   Sea ice. Loss, in cm, May 28-June 22 (25 days).

25 per cent rock salt cut the effectiveness of a 1000-gm concentration by about 35 per cent; a 75 per cent addition of rock salt cut the effectiveness by about 55 per cent. When used in a 100-gm concentration the effectiveness was about equal for the three types of coarse cinder. Fly ash, in a 1000-gm concentration, was less effective than coarse or fine cinder. At the early stage of the experiment the 1000-gm concentration of fly ash was probably retarded by a layer of new ice some 4 cm thick that had formed in a pool of water above this material. By June 14, fly ash was probably about 25 per cent less effective than fine or coarse cinder in equal 1000-gm concentrations. In 100-gm concentrations it was more effective, and in 300-gm concentrations it was equal in effectiveness. An addition of 25 per cent fine sand appeared to make little difference.

The effect of particle size was again greater for coarse material than for fine. The ratio of effectiveness of the 1000-gm concentration to the 100-gm concentration was about 4-1 for coarse cinder, 3-1 for fine cinder, and 2-1 for fly ash. The addition of salt was of no value in accelerating the melting over a long-term period, although it was evident that it had an immediate effect when the dust was spread. The added salt was still noticeable on June 12, when salinity samples were taken in the pools of water above each of the 1000-gm concentrations. The results were:

Material	Salinity 0/00
Coarse cinder	2.9
Coarse cinder, 25 per cent salt	4.7
Coarse cinder, 75 per cent salt	7.4
Fine cinder	1.6
Fine cinder, 25 per cent salt	4.7
Fly ash	2.1
Fly ash, 25 per cent fine sand	1.9

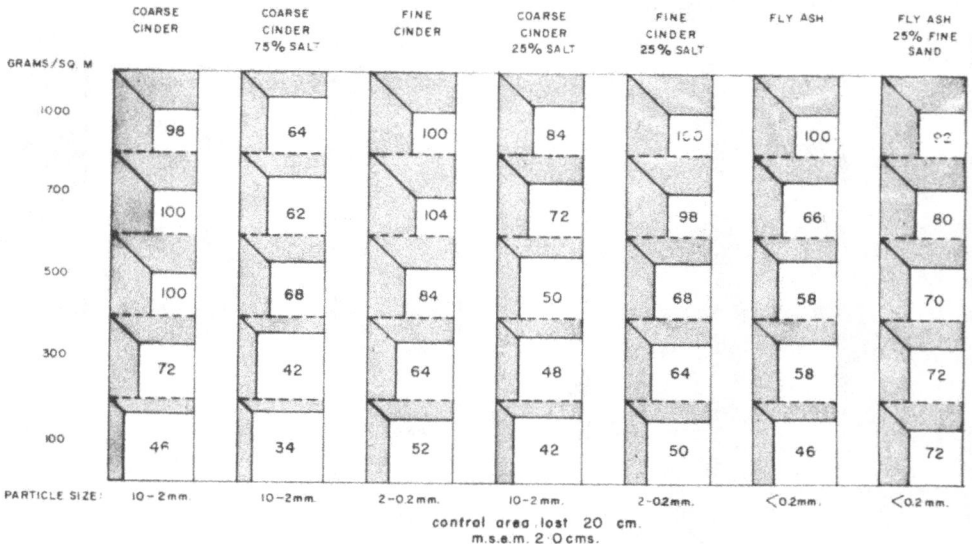

FIGURE 20. Sea ice. Loss, in cm, May 28-June 29 (32 days).

After June 14, the results began to be masked by other factors. Air temperatures were generally above 0° C, and the pools of water that covered the greater concentrations began to grow and coalesce, first between different concentrations within a single material, and later between different materials. This caused heat exchange between the different samples. Another disturbing factor was that, with increasing depth of the pools, the 1000-gm concentrations were less accessible to the direct rays of the midnight sun. At Isachsen in mid-June the sun is 35 degrees above the horizon at midday, but only 12 degrees above the horizon at midnight. In addition, the smaller concentrations of fly ash with 25 per cent fine sand were affected by a stream flowing into that area.

By June 29, the differences between materials and the 1000-, 700-, and 500-gm concentrations had become less apparent. The more effective combinations had lost about 100 cm of ice. In comparison, untreated ice had lost only 20 cm. A 700-gm concentration of fine cinder was the most effective combination, by a small margin. Fly ash appeared to be 10 to 20 per cent less effective in larger concentrations, and about equal in effectiveness in smaller concentrations.

The whole area melted through to sea water in the second week of July. This was about twenty days in advance of the general break up of the sea ice in Station Bay. Small ledges of ice remained where the 100-gm concentrations had been spread. These slowly wasted away by melting and some wave action.

Selected meteorological observations during both ice-dusting experiments are shown in Figure 26.

EFFECT ON LAKE ICE

The lake ice was dusted on May 30. On this date the ice was 235 cm thick. This test area was measured on five occasions between June 1 and July 5 (Figures 21-25, Table III). Measurements on June 9 and June 13 appear to show that

GRAMS/SQ M	COARSE CINDER	COARSE CINDER 75% SALT	FINE CINDER	COARSE CINDER 25% SALT	FINE CINDER 25% SALT	FLY ASH	FLY ASH 25% FINE SAND
1000	2.0	0.4	3.0	2·0	3.0	3.2	3·6
700	0.2	0.0	3.0	0.2	2.4	2·2	3.0
500	0.2	0.0	1.2	0.2	0·6	2·0	1·0
300	0·2	0·0	1·8	0·0	0.2	0·4	1·0
100	0.0	0·0	0.4	0.0	0.0	0.0	0·8
PARTICLE SIZE:	10–2mm.	10–2mm.	2–0.2mm.	10–2mm.	2–0.2mm.	<0.2mm.	<0.2mm.

control area lost 0.0 cm.
m.s.e.m. 0·2 cms.

FIGURE 21.   Lake ice. Loss, in cm, May 30-June 1 (2 days).

TABLE III

Measurements in Lake Ice Test Area, May 30 to July 5

	Lake ice	Coarse cinder					Coarse cinder, 75 per cent salt					Fine cinder					Coarse cinder, 25 per cent salt					Fine cinder, 25 per cent salt					Fly ash					Fly ash, 25 per cent fine sand					Control area lost
		1	2	3	4	5	6	7	8	9	10	11	12	13	14	15	16	17	18	19	20	21	22	23	24	25	26	27	28	29	30	31	32	33	34	35	
Loss, cm; May 30–June 1; m.s.e.m. 0.2 cm	2.0	0.2	0.2	0.0			0.4	0.0	0.0	0.0		3.0	3.0	1.2	1.8	0.4	2.0	0.2	0.2	0.0	0.0	3.0	2.4	0.6	0.2	0.0	3.2	2.2	2.0	0.4	0.0	3.6	3.0	1.0	1.0	0.8	0.0 cm
Loss, cm; May 30–June 9; m.s.e.m. 0.4 cm	8.0	5.0	5.0	6.5	0.0		5.5	5.0	5.0	0.0	0.0	9.0	8.5	7.0	1.0	0.5	6.5	3.0	0.5	0.5	0.0	6.0	7.5	4.5	2.0	0.5	3.0	2.5	2.0	1.0	0.5	3.0	4.0	3.0	2.0	1.5	0.0 cm
Loss, cm; May 30–June 13; m.s.e.m. 0.6 cm	9.0	7.0	5.0	6.5	3.0		7.0	7.5	6.0	6.0	2.0	9.5	14.0	14.0	4.0	2.0	6.0	7.5	5.5	5.5	2.0	8.0	10.0	10.0	14.5	2.0	5.0	3.5	4.0	2.5	1.5	6.5	5.0	5.5	5.0	2.0	0.0 cm
Loss, cm; May 30–June 15; m.s.e.m. 0.9 cm	19	15	11	9	5		10	11	10	9	6	23	24	19	14	6	12	14	12	10	6	18	18	15	18	7	14	11	11	9	5	15	13	14	15	9	0.0 cm
Loss, cm; May 30–July 5; m.s.e.m. 2.2 cm	48	42	42	42	40		38	40	42	42	42	54	72	72	60	42	52	58	50	48	42	74	86	78	76	46	46	46	64	72	40	58	46	56	60	40	16 cm

melting in the lake ice test area was much less than in the sea ice test area, for comparable intervals of time. A probable explanation was that the lake ice test area was not swept clear of drifting snow as regularly as the sea ice test area, due to its greater distance from Isachsen. By June 15 results in the lake ice test area were generally about 15 per cent less effective than on the sea ice.

The relative effectiveness of the different materials and concentrations was very similar, fine cinder being again the most effective. The 1000- and 700-gm concen-

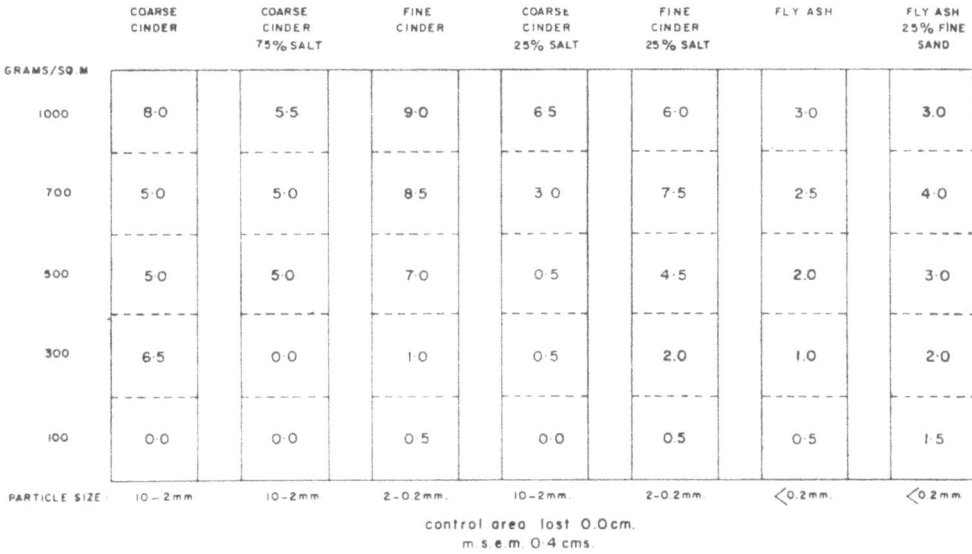

GRAMS/SQ.M	COARSE CINDER	COARSE CINDER 75% SALT	FINE CINDER	COARSE CINDER 25% SALT	FINE CINDER 25% SALT	FLY ASH	FLY ASH 25% FINE SAND
1000	8·0	5·5	9·0	6 5	6·0	3·0	3.0
700	5·0	5·0	8 5	3 0	7·5	2·5	4·0
500	5·0	5·0	7·0	0·5	4·5	2.0	3·0
300	6·5	0·0	1·0	0·5	2.0	1.0	2·0
100	0·0	0·0	0 5	0·0	0.5	0·5	1·5
PARTICLE SIZE	10–2mm	10–2mm	2–0.2mm	10–2mm	2–0.2mm	<0.2mm	<0.2mm

control area lost 0.0cm.
m.s.e.m. 0·4 cms

FIGURE 22.   Lake ice. Loss, in cm, May 30-June 9 (10 days).

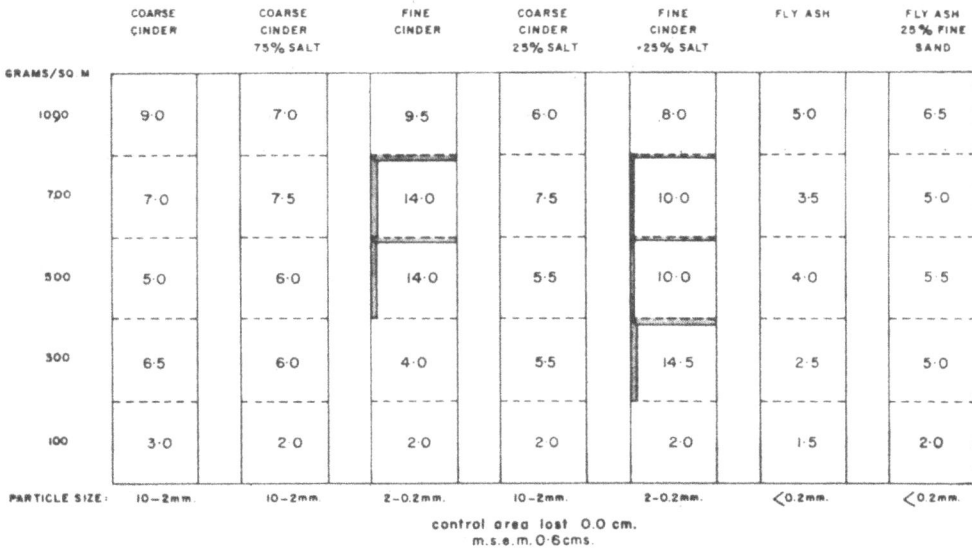

GRAMS/SQ M	COARSE CINDER	COARSE CINDER 75% SALT	FINE CINDER	COARSE CINDER 25% SALT	FINE CINDER +25% SALT	FLY ASH	FLY ASH 25% FINE SAND
1090	9·0	7·0	9·5	6·0	8·0	5·0	6 5
700	7·0	7·5	14·0	7·5	10·0	3·5	5·0
500	5·0	6·0	14·0	5·5	10·0	4·0	5 5
300	6·5	6·0	4·0	5·5	14·5	2·5	5·0
100	3·0	2 0	2 0	2 0	2·0	1·5	2·0
PARTICLE SIZE	10–2mm	10–2mm	2–0.2mm	10–2mm	2–0.2mm	<0.2mm	<0.2mm

control area lost 0.0 cm.
m.s.e.m. 0·6cms

FIGURE 23.   Lake ice. Loss, in cm, May 30-June 13 (14 days).

trations gave almost equal results. An addition of 25 per cent rock salt decreased the effectiveness of fine cinder by about 25 per cent, when used in 1000-, 700-, and 500-gm concentrations. Coarse cinder was from 20 to 40 per cent less effective than fine cinder when used in the heavier concentrations. The addition of 75 and 25 per cent rock salt decreased the effectiveness of a 100-gm concentration of coarse cinder by 45 and 35 per cent respectively. However, the addition of salt to both coarse and fine cinder appeared to lessen the differences between the 300-,

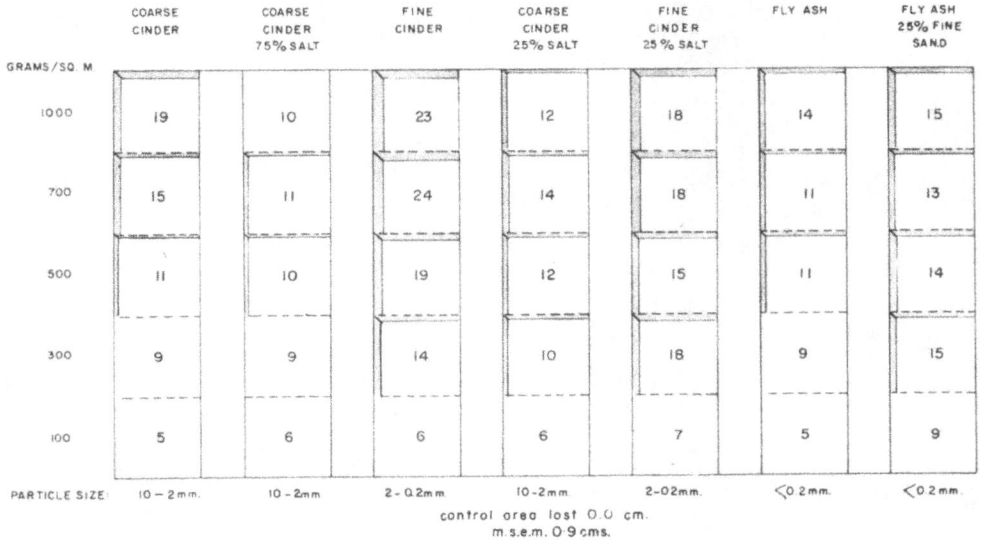

GRAMS/SQ. M	COARSE CINDER	COARSE CINDER 75% SALT	FINE CINDER	COARSE CINDER 25% SALT	FINE CINDER 25% SALT	FLY ASH	FLY ASH 25% FINE SAND
1000	19	10	23	12	18	14	15
700	15	11	24	14	18	11	13
500	11	10	19	12	15	11	14
300	9	9	14	10	18	9	15
100	5	6	6	6	7	5	9
PARTICLE SIZE	10 – 2 mm.	10 – 2mm.	2 – 0.2mm.	10 – 2mm.	2 – 02mm.	< 0.2mm.	< 0.2 mm.

control area lost 0.0 cm.
m.s.e.m. 0.9 cms.

FIGURE 24.   Lake ice. Loss, in cm, May 30-June 15 (16 days).

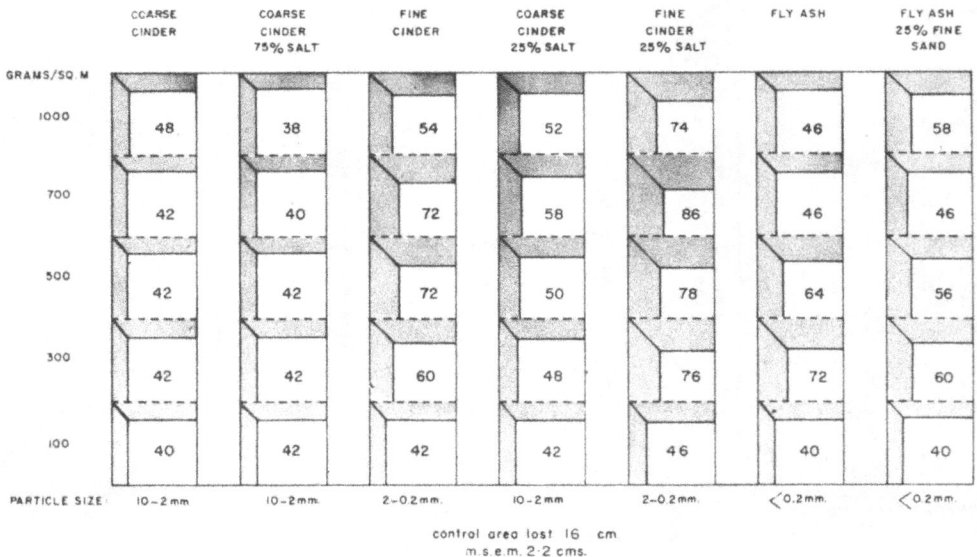

GRAMS/SQ. M	COARSE CINDER	COARSE CINDER 75% SALT	FINE CINDER	COARSE CINDER 25% SALT	FINE CINDER 25% SALT	FLY ASH	FLY ASH 25% FINE SAND
1000	48	38	54	52	74	46	58
700	42	40	72	58	86	46	46
500	42	42	72	50	78	64	56
300	42	42	60	48	76	72	60
100	40	42	42	42	46	40	40
PARTICLE SIZE	10 – 2 mm	10 – 2mm.	2 – 0.2mm.	10 – 2mm.	2 – 0.2mm.	< 0.2mm.	< 0.2 mm.

control area lost 16 cm
m.s.e.m. 2.2 cms.

FIGURE 25.   Lake ice. Loss, in cm, May 30-July 5 (36 days).

500-, 700-, and 1000-gm concentrations. A 1000-gm concentration of fly ash was only 60 per cent as effective as a similar quantity of fine cinder. The effect of concentration was less marked for fly ash than for the other materials. In a 100-gm concentration the effectiveness of all materials was very similar. An addition of 25 per cent fine sand to fly ash made little difference to the over-all effect of this material, although effects of concentration were least pronounced in the strip containing this mixture.

The last measurement of the lake ice test area was made on July 5, when the relation between materials, concentration, and effectiveness was much more obscure, probably because of the same disturbing factors that affected the last measurement in the sea ice test area. However, differences were much less clear in the lake ice area, probably because the last measurement was taken seven days after that in the sea ice test area. Any differences at this time should be viewed with caution, but it would appear that the untreated lake ice control area lost only about 75 per cent as much ice as the sea ice control area, in a comparable period of time. An average of all seven materials suggests that a 1000-gm concentration on lake ice was about 60 per cent as effective as on sea ice; a 100-gm concentration was about 90 per cent as effective. Within the lake ice test area, an average of all concentrations suggests that the 2.0- to 0.2-mm particle size was the most effective. The average loss for this particle size was 65 cm. The very fine and coarse particles were about 80 and 70 per cent as effective, respectively.

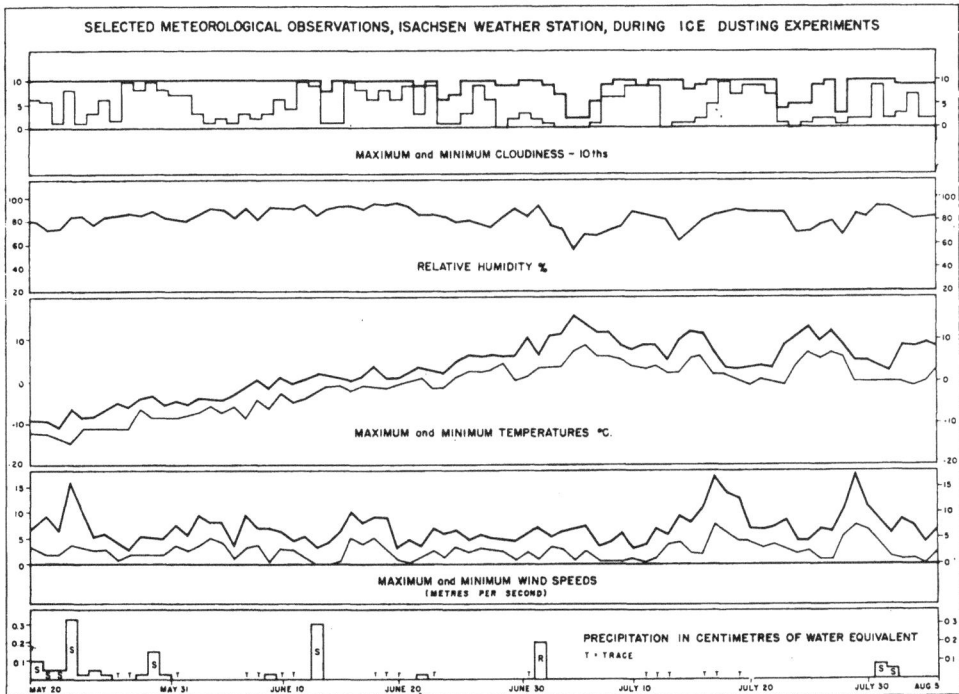

FIGURE 26.   Selected meteorological observations, Isachsen weather station, during ice-dusting experiments.

After July 5, a widening shore lead and the candled state of the ice made it impossible to inspect the lake ice test area again. It is not known if this area melted through to the bottom of the lake ice. The lake was still more than half covered by ice in the second week of August.

Selected meteorological observations during the ice dusting experiments are shown in Figure 26. The level of cloudiness was generally quite high except for a period in late June and early July and another during the third week of July. The relative humidity was generally about 80 per cent. Maximum temperatures rose above 0° C about June 14, and minimum temperatures followed about a week later. The rate of melting increased quite sharply after June 14. Until June 12 drifting snow occurred when wind speeds were greater than 5 m/sec. The gale at the end of July hastened the break-up of the sea ice in Station Bay. After the end of May, measurable precipitation fell on six days and traces occurred on sixteen days. Snow began to fall again in early August.

DISCUSSION

When used in concentrations of up to 1000 gm/sq m, all materials accelerated the rate of melting. During the month of June the 1000-gm concentrations increased the rate of melting about five times; 100-gm concentrations gave an increase of about two-and-a-half times.

The amount of material would have to be increased considerably before protection of the underlying ice would occur, and any future experiments in this direction might use amounts in a logarithmic ratio as 0.5, 1, 2, 4, and 8 kg/sq m.

These experiments were made in a high latitude, 78½°N, where the sun is above the horizon from April 21 to August 24. Similar studies in lower latitudes would be interesting to determine how applicable these results would be in planning practical use of this method in the more heavily populated areas of southern Canada. Kimball (1931) stated that in Svalbard, in a similar latitude to Isachsen, the amount of solar radiation received during the three weeks centring on the summer solstice was greater than that of any station in the continental United States. At the edge of the earth's atmosphere the increasing length of day in high latitudes would make this true. The situation at the surface of the earth is more complicated, for the obliquity of the sun's rays in high latitudes would lessen the amount of solar radiation received at the earth's surface because of greater atmospheric absorption. It is possible that the greater clarity of the polar atmosphere may reduce the effect of this adverse factor.

Temperatures during the snow-melting experiments were between —10° C and —5° C. Up to 25 cm of sea ice was lost while temperatures were between —10° C and 0° C. It would be interesting to repeat the experiment at temperatures below —10° C in an attempt to discover the lowest temperatures at which dusting would give a worth-while effect.

The differences in effectiveness of the same quantities of material on sea and lake ice is difficult to explain. It is possible, but doubtful, that the difference was due to local climatic factors. The thermograph records showed that temperature conditions were similar. It is possible that cloudiness may have been higher at the lake, which was partially enclosed by hills about 800 feet high. These hills may have also affected the wind-speeds at the lake. It is unlikely that relative humidity or precipitation would have varied greatly between the two sites. It is unlikely that differences in orientation of the test areas could account for the differences observed. The long side of the sea ice test area was oriented 109° true; that of the lake ice was 154° true. It is possible that differences in structure between lake and sea ice may have caused the difference in effectiveness. Cores taken in the sea ice (Figures 3-5) showed that the material penetrated the ice in depth, and may have followed the brine pores in the sea ice. The lake ice cores (Figures 6-8) did not show this action to the same extent. The material was mainly confined to the broken ice above the solid core, and only a few particles penetrated more deeply.

For practical applications of this method, fine cinder with particles from 2 to 0.2 mm in size would appear to be most advantageous for ice or snow. For snow alone, the use of finer material such as fly ash might be considered, if evenness of spreading is a problem, with only a slight reduction in effectiveness. A concentration of 1000 gm/sq m gave the best results, but further experiments might show a greater concentration to be still more effective. It is interesting that the concentration of 1000 gm/sq m is within 3 per cent of the amount suggested by Bonin and Teichmann (1949).

In remote areas it would be more practical to use unconsolidated materials available locally, and experiments with local materials of different degrees of darkness would be a useful prelude to any large-scale use of this method.

PRACTICAL APPLICATIONS

This short discussion of practical applications in different latitudes is not intended to be in any sense exhaustive, but it may suggest some future applications in Canada.

Iakobsen (1934) describes accelerated melting of snow on an air field in Saratov province on the middle Volga (52°N, 47°E). A 16-hectare area of snow, averaging 40.5 cm in depth, was covered with ashes on March 18. The snow had completely melted by April 7, and the air field was used on April 12. In contrast, untreated snow was still 10.5 cm deep on April 19.

Georgievskii (1937) described some experiments in melting snow and ice near Schmidt Cape (69°N, 179°W) on the north coast of Siberia south of Wrangel Island. He found that snow in dusted areas melted at an average rate of 4.3 cm a day, eight times faster than untreated snow. Laktionoff (1957) in a recent paper on the effects of ice on navigation, suggested the spreading of dark material on the surface of the ice as one of the means to combat it.

Bensin (1952) described the use of coal dust to accelerate the warming up of the soil in spring at Fairbanks, Alaska (65°N, 148°W). He used coal dust in a concentration of about 1000 gm/sq m. He found that soil temperatures were increased to a depth of 10 cm, and that this gave a better root system to plants. He suggested the use of aeroplanes to spread dust in amounts of 1100 kg to the hectare (1000 pounds to the acre).

Avsiuk (1953) studied the melting-rate of the Tien Shan glaciers southeast of

Lake Baikal (42°N, 80°E). He estimated that a covering of five tons to the sq km of coal dust and loess would accelerate the rate of melting of snow by three to six-and-one-half times. Glacier ice melted 1.35 times faster, and the run-off in dry seasons could be increased by 54 per cent. Protection of the underlying ice occurred if the covering was more than 1 cm thick.

Konovalov and Miasnikov (1956) describe the use of this method to fight ice jams on the Irtish (about 55°N, 75°E) and North Dvina (64°N, 42°E) rivers. The ice jams usually occurred at the same place on the rivers from year to year, often at shallow shoals. An aircraft similar in size to a Norseman was used to spread dust from an altitude of about five m. The load carried was about 1000 kg, and the cost was estimated at ten to fifteen roubles per hectare. Antrushin (1956) describes use of this method to free approaches to the port of Archangelsk (64°N, 40°E). Three hundred to five hundred kg of dust mixed with 50 to 100 kg of rock salt was spread on each hectare. Waste oil was then spread in an amount of about 10 litres to the hectare. The cost was estimated at thirty-five roubles per hectare. Ice treated in this way broke up about fifteen to twenty days before untreated ice. In 1954, 60,000 kg was spread, and in 1955 this amount was increased by ten times. The costs involved in 1955 were estimated at 2 per cent of the costs of using ice-breakers to open the channel.

Taketa and Marukami (1956) describe the use of this method to increase power resources. Carbon black was spread on Miura reservoir (37°N, 137°E) in central Honshu Island. It was spread both by hand and by a specially constructed blower. The method was effective under a cover of new snow 20 cm deep.

In Canada artificial dusting was used in 1933 on Lake Laberge, Yukon Territory (61°N, 135°W) (Bostock, personal communication). Lake Laberge formed part of the water route from Whitehorse to Dawson. The break-up of the lake was about three weeks later than usual that summer, and a channel had been formed by dusting that allowed freight canoes to reach the headwaters of the Yukon River.

Possibilities for similar applications exist in Canada where problems of spring flooding and extension of the navigation season have considerable economic interest. Brochu (1958) studied the annual flooding of the Chaudière River (46°N, 71°W), and suggested that dusting the ice would reduce the amount of dynamiting of ice jams necessary, which has been the method of control favoured so far.

In remote northern areas, the use of local materials should be considered, where possible. Tracked vehicles might compete economically with aircraft in spreading material. At Isachsen (79°N, 103°W) some small experiments were made with a small rotary spreader towed behind a light cargo carrier, type CL-70. A larger trailer with a built-in spreader might be designed to distribute material in measured quantities. Controlled excavations in ice, to serve as temporary docks or fuel tanks, might be made in this way.

One fortuitous example of a possible practical use occurred at Isachsen. Sea ice that had been crossed by vehicles with muddy tracks later became crossed by long drainage channels. The upper surface of the sea ice is low in salinity during the summer, and the water in these tracks had a salinity of only 0.8 parts per thousand. A suitable pattern of tracks laid out before the thaw commences may provide a supply of water suitable for many purposes at remote northern stations, where an adequate supply of water from other sources is often lacking.

ACKNOWLEDGMENTS

This study was part of the contribution of the Geographical Branch to the programme of the Polar Continental Shelf Project of the Department of Mines and Technical Surveys for the 1959 season. I am indebted to Dr. R. T. Gajda for the suggestion that led to this study, and to my fellow geographers, D. St-Onge and R. M. Moskal for a great amount of invaluable assistance in the field, and for their suggestions. My wife gave me much help in checking calculations and making tables, without the compensation of interesting work in the field.

REFERENCES

ANTRUSHIN, N.   1956.   Aircraft are destroying ice; Grazdanskaya Aviatsiya, vol. 4, p. 33.

AVSIUK, G. A.   1953.   The problem of artificially increasing the melting rate of the Tien Shan glaciers; Institutt Geografii Nauk SSSR, Moscow.

BENSIN, B. M.   1952.   Coal dust absorbs solar heat; Heating and Ventilating, vol. 49, no. 9, p. 86.

BONIN, J. H., and TEICHMANN, O. E.   1949.   Investigation of solar energy for ice melting; Armour Research Foundation, final report.

BOSTOCK, H. S.   Personal communication.

BROCHU, M.   1958.   Dynamiting the ice of the Chaudière in the spring of 1958; Rev. Can. de Geogr., vol. 12, no. 3-4, pp. 159-62.

GADJA, R. T.   1958.   Cryoconite phenomena on the Greenland ice cap in the Thule area; Can. Geogr., vol. 12, pp. 35-44.

GEORGIEVSKII, N.   1939.   Experiments to accelerate snow and ice melting on the Schmidt Cape in 1937; Severnyi Morskoi Put, vol. 13, pp. 29-35.

GERDEL, R. W., and DROUET, F.   1958.   The cryoconite of the Thule area; Sipre research report no. 50.

HAND, I. F., and LUNDQUIST, R. E.   1942.   Observations of radiation penetration through snow; Monthly Weather Review (U.S.), vol. 70, pp. 23-5.

IAKOBSEN, N.   1934.   The acceleration of snow melting; Grazdanskaya Aviatsiya, vol. 4, no. 6, pp. 26-7.

KIMBALL, H. H.   1931.   Solar radiation intensities within the Arctic circle; Monthly Weather Review (U.S.), vol. 59, pp. 154-7

KONOVALOV, I. M., and MIASNIKOV, M. V.   1956.   Utilizing solar radiation for increasing the navigation period; Rechnoi Transport, vol. 15, no. 1, pp. 13-18.

LANDSBERG, H. E.   1940.   The use of solar energy for the melting of ice; Amer. Meteor. Soc. Bull., vol. 21, pp. 102-7

LAKTIONOFF, A. F.   1957.   The effects of ice upon shipping routes, sea and rivers ports, and the means to combat it; 19th Intern. Navig. Congr. (Brussels), sec. I, vol. 3, pp. 177-217.

LEWIS, W. V.   1940.   Dirt cones on the northern margins of Vatnajokull; J. Geomor., vol. 3, pp. 16-26.

PAULSEN, H. S.   1955.   The role of solar radiation in the melting of snow and ice; Naturen, vol. 79, pp. 66-73

SHARP, R. F.   1949.   Studies of superficial debris on valley glaciers; Amer. J. Sci., vol. 247, pp. 289-315

SWITHINBANK, C. W. M.   1950.   The origin of dirt cones on glaciers; J. Glac., vol. 1, no. 8, pp. 461-5

TAKETA, M., and MURAKAMI, M.   1956.   On artificial snow melting; Seppyo, vol. 17, no. 3, pp. 21-7

TROLL, C.   1949.   Melting and evaporation of snow and ice in their relation to the geographical distribution of ablation forms; Erdkunde, bd. 3, heft 1, pp. 18-29.

TRUSHIN, V. F.   1957.   The effects of the continuous covering of snow with black powder on snow melting and melt water runoff; Meteorologiia i Gidrologiia, 1957, no. 3, pp. 44-5.

WILSON, J. WARREN.   1953.   The initiation of dirt cones on snow; J. Glac., vol. 2, no. 14, pp. 281-7.

Ground-Water Hydrology in Alaska[1]

D. J. CEDERSTROM

ABSTRACT

Systematic studies of ground-water hydrology in Alaska began in 1947. Intensive studies were carried out in Fairbanks, Anchorage, and the Matanuska valley. Localized studies have been made elsewhere—in Kotzebue, the Pribilof Islands, and Bethel, in particular. Work in this field continues today.

The Tanana valley, in which Fairbanks is located, with its gravelly fill was found to be one of the most prolific sources of ground water in the world. In this area permafrost is a minor problem in developing ground water, as is high iron content in the water. In Anchorage and the Matanuska Valley ground water occurs sporadically in glacial deposits but many wells have yielded water copiously. Many failures are known, however, in part because of poor construction. Municipal supplies of ground water have been developed at Fairbanks, Anchorage, and Palmer.

Alaska north of southeastern Alaska may be characterized as an area where great alluvial valleys contain much ground water. Northward, however, permafrost becomes more and more of a problem. Possibilities of development of water from hard rock are little known.

The negative results of drilling at Kotzebue and the more favourable results of investigations in the Pribilof Islands and at Bethel are briefly discussed.

IN 1947 THE GEOLOGICAL SURVEY began systematic studies of the ground-water resources of Alaska, at which time the writer was assigned to the project and began studies at Fairbanks. Although this paper deals with the ground-water investigations only, it may be mentioned that work on surface water and quality of the water began at much the same time and all three phases of hydrologic study have continued in Alaska to the present. The ground-water programme is currently under the direction of Mr. Roger Waller, to whom I am indebted for data on hydrologic activities in Alaska since the fall of 1954.

Very little was known of the ground-water hydrology of Alaska in 1947. A rather large number of small-diameter driven wells were then supplying domestic needs in Fairbanks, some relatively inefficient drilled wells supplied nearby military installations, and a number of shallow drilled wells supplied domestic requirements in the Matanuska valley and Anchorage. But, elsewhere, the wells in Alaska, as stated in U.S.G.S. Circular 169 (1952), were few and far between. In fact, well failures, as a result of lack of knowledge of geology, hydrology, and well construction, loomed larger in the public eye than successful wells.

In the ten years or so after World War II there was an awakening of interest in ground-water supplies on the part of individuals, business firms, and territorial and federal agencies. Our work happened to coincide with this awakened interest and could not have been started at a more opportune time. We, on our part, did everything possible to foster that interest, to assist with information at all times and, wherever appropriate, to suggest the desirability of bigger and better studies of the

[1]Publication authorized by the Director, United States Geological Survey.

ground-water resources. Our studies and test-drilling activities were integrated with the then current efforts to develop water.

The total of accumulated data on ground-water hydrology in Alaska is such that we now have much knowledge of, or can judge rather well, what conditions will be in a fairly large part of the state. In general, Alaska north of the sixtieth parallel may be characterized as a region of rugged mountain chains with broad, alluvium-filled valleys, most of which produce water in large quantity. Precipitation is rather low, but this is no drawback where the storage capacity of sediments is tremendous, as in the alluvial areas, and where conditions of recharge are good to excellent. Real problems exist in the development of ground-water supplies but most of these are not markedly different from problems in similar areas of the United States.

As noted above, systematic studies were begun in 1947 in Fairbanks, where many small-diameter wells existed, some reaching a depth of 150 feet. The Tanana valley fill consists of alternating sands and gravels which yield water easily. The permeability of the sediments obviously was very high; we drilled several test holes adjacent to pumping wells, and with these as observation wells made aquifer tests. Such tests were not simply tests to determine the amount of water discharged at a certain pumping level. These were controlled operations in which water level and discharge data were used in a very complex series of calculations to determine permeability and storage capacities of the sediments, mutual interference, sources and time of recharge, hydrologic barriers, and other information.

One test showed a transmissibility of 800,000 gpd (gallons per day per foot). This means that each mile-wide vertical strip of the aquifer will transmit 800,000 gallons of water per day under a gradient of one foot per mile. This value seems reasonable to me at the present time. The value obtained is probably not characteristic of all the water-bearing gravels in the Fairbanks area, but it is certainly of the right order of magnitude for the particular gravel stratum tested. Without further calculations of any kind, the value for transmissibility suggests that it should be possible to develop wells of extremely high yield from similar gravels.

No properly developed wells existed in the area in 1947; a bountiful supply of water for domestic use, up to 40 gpm (gallons per minute), was easily obtained by advancing a two-inch pipe down to the first gravel lens encountered (in some wells as little as twenty or thirty feet, in other wells as much as 150 feet in order to get beneath the permafrost). Drilled wells, six-, eight-, or ten-inches in diameter, obtained several hundred gallons a minute with ten or fifteen feet of draw-down. Most drilled wells were finished with the lower end of the casing open and with the lower several feet of casing slotted. The first real attempt to develop the gravels properly was made around 1950 at Ladd Air Force Base. The results were spectacular. An eighteen-inch diameter well was drilled and in it, at first, eight feet of screen having slots 0.060 inches wide was placed opposite the water-bearing gravel as an extension of the casing. In the subsequent development procedure, much of the finer-grained fraction of the formation near the well was drawn through the screen by pumping or surging, leaving an envelope of the coarser element around the screen. The "60-slot" screen held back the largest grains of medium and coarser sand grains and permitted removal of finer material. The well then yielded 1500 gpm with thirty-seven feet of draw-down which was reduced to nine feet with the substitution of twenty feet of 80-slot screen for the eight feet of 60-slot screen. The 80-slot screen permitted removal of medium sand and the smallest fraction of coarse sand. The specific capacity of 66 gpm per foot of draw-down was three times greater than that of any well drilled in the area prior to that time and was, without question, the most efficient well in Alaska.

Iron is troublesome in many Fairbanks well waters — and, as far as is known, high iron content is characteristic of the Yukon valley system and of much of the Kuskokwim valley. It is thought that the high iron content is a function of the reducing environment existing in the sediments having a high organic content. In such an environment, carbon dioxide is generated and the water becomes highly corrosive and attacks iron-bearing minerals.

The entire Tanana valley appears to be filled with saturated gravelly and sandy materials from which tremendous quantities of water might be obtained with relative ease. The amount of water in storage is so great as to stagger the imagination. In addition, were there ever any appreciable removal of water from this vast reservoir, immediate recharge would occur from the Tanana River, as well as from precipitation and underflow. Even with the present scant specific knowledge of the Yukon hydrologic system as a whole, I venture to say that it is one of the world's great water-bearing provinces.

The type of knowledge discussed above was gained from the collection and study of existing data and the observation of current operations. A more direct attack was made on the hydrology of the silt-covered hills just north of Fairbanks, where there has been extensive agricultural activity for many years. Here the Geological Survey, beginning in the fall of 1948, contracted for the drilling of a series of test holes in the hill area by the cable-tool method. Test holes were drilled in a number of localities; it was established that water in small quantity should be commonly available in the upper portion of the ordinarily schistose bedrock but that fairly good yields might be expected in the shallow creases on the flanks of the hills marking old stream courses filled with gravelly materials mantled with silt. There are now perhaps a dozen or so wells scattered around on the hills north of Fairbanks, but I do not know of any located so as to develop the fairly thick sands lying beneath the topographically insignificant stream valleys. Test drilling here by the cable-tool method was discontinued in 1949.

With the expense of cable-tool drilling in mind, an experimental jetting rig was assembled a few years later (1954) and demonstrated in two months' work that in many localities on the hill slopes north of Fairbanks inexpensive wells might be easily constructed. One of the interesting sidelights on this operation is that late in August 1954, "open house," previously announced in the local newspapers, was held at a drilling site on the Farmers Loop Road. Nearly 100 people attended the demonstration of the operation and efficiency of the jet-drilling machine. In 1955 a Fairbanks farmer built a simpler but similar jet-drilling rig and has successfully completed a number of wells.

As noted in U.S.G.S. Circular 275, drilling in permafrost, *per se*, is easier than drilling in thawed ground and ordinarily poses no important mechanical difficulties in well construction. Hydrologically, permafrost may be considered simply as an impermeable formation. It acts as a cap rock over the deeper thawed formations in places and is directly responsible for the occurrence of flowing artesian wells on the low slopes north of Fairbanks. On the lower elevations in the Yukon valley system, water commonly can be obtained above the permafrost, beneath it, or in places where permafrost is absent. Wells drilled through permafrost have a tendency to freeze if not pumped regularly, and special measures have to be taken to keep them open. Permafrost is also a problem in the distribution of water. In recent years the municipality of Fairbanks installed a system in which water circulates continuously through mains laid in frozen ground. In this system, water passes through the power-generator condensers, from which appreciable heat is gained. The prevention of freezing in the small take-offs from the mains is ingeniously accomplished by causing water to move continuously in the service connection loop.

The Matanuska valley is another area that was considered unproductive of water, or nearly so. In the early days of the colony several deep wells had been drilled in the town site of Palmer. These encountered bedrock a short distance below the surface and very low yields were obtained. Further, in some of these wells brackish water was encountered. We saw no reason to think that these conditions were representative of the valley as a whole, which is filled with glacial sediments. Encouraged by the Alaska Department of Health, among others, the Geological Survey assigned Frank Trainer to the area at the earliest practical moment. The area was mapped geologically and the records of existing domestic wells were assembled and studied. It early became apparent that it should be possible to drill wells of fairly high yield in many places, even though none existed at the time of the study. When the Town of Palmer wished to acquire a ground-water supply a site was recommended a short distance northwest of the town on the basis of Mr. Trainer's geological map, and test wells were drilled. These proved successful, and permanent production wells were installed which have been supplying the town now for several years.

The wells in the Matanuska valley obtain their water from buried glacial outwash deposits. With proper construction and development, moderate to high yields are possible in many places. Conspicuous failures have resulted where such gravels are absent and attempts have been made to obtain water from till. Till is not entirely hopeless as a reservoir for water; sandy till or till with thin stringers of sand may yield enough water for domestic needs. However, drilling or digging in the till may be most difficult and the final results very disappointing.

When the Geological Survey's ground-water studies began in the Anchorage area there were not more than half a dozen deep drilled wells. The data gained from these were confusing and, in part, conflicting. It was decided early that test drilling was needed. Accordingly, a test-drilling outfit was assembled and put into operation. The drilling machine was rented from the Alaska Department of Health. The driller and helper were Geological Survey employees. Commercial drilling for private individuals in the Anchorage area was well under way by the time our test well near the present Anchor Park development was drilled to a depth of 396 feet. Nevertheless, this well, completed in July 1952, was the first well in the area in which a succession of water-bearing formations was adequately tested. It was also the first well in the area to furnish data indicating that a million gallons of water a day might be obtained from a single well source.

Shortly thereafter, a well near the Ranger Station on Oilwell Road proved to have a comparable yield. A Geological Survey test hole was then drilled near this well and, at the appropriate depth, an aquifer evaluation test was made, using the Army well as an observation well; a transmissibility value of more than 100,000 gpd was determined. This was the first such test made in Anchorage, although several had been made previously in Fairbanks. This well was also the first to explore the full thickness, 447 feet, of the glacial sediments in the Anchorage area and to reach bedrock. In this hole Tertiary shales were encountered whose presence in the area was previously unknown. The drill was still in shale at 617 feet, where drilling was discontinued.

Early in 1953 an engineering report made by a private firm for the Corps of Engineers at Elmendorf Air Force Base, Anchorage, recommended the drilling of a large-diameter test well as the most desirable step towards establishing ground water as a possible source of significant additions to the Ship Creek water supply then in use. Early that spring, such a well was begun by the Corps of Engineers and did not reach bedrock until more than 700 feet of sediments had been penetrated. This well was actually pumped at a rate of 1340 gpm (about two million gallons a day), although its specific capacity (gallons per minute per foot of draw-down) was not as great as that of several existing wells, which, however, had been pumped at much lower rates.

When work was under way at the Corps of Engineers' test well, the Geological Survey drilled a smaller-diameter test well nearby to a depth of 336 feet. During the course of pump testing the Corps of Engineers' well, observations were made on the Geological Survey well, permitting calculations of transmissibility and storage coefficient. The values obtained proved to be almost identical with values obtained several years later when the first city of Anchorage well was pumped and the Corps of Engineers' well was used as an observation well. Subsequently in 1956 the Corps of Engineers drilled six test holes between this location and the U.S.G.S. test hole at the Ranger Station to the southwest; the results of testing these wells, all less than 150 feet deep, were most gratifying. Three of these are now in use and supplement the Ship Creek supply as needed.

Returning to consideration of our own efforts in that area in 1953, a start was made on several test wells drilled near the mountain flanks on Ski Bowl Road. Here the sequence of glacial sediments is thinner, but highly permeable beds are present in which water occurs under low artesian head. Three shallow holes were drilled in the gravels between Ski Bowl Road and Ship Creek. These gravels proved to be very highly permeable and might furnish water in quantity, particularly since tests and subsequent calculations indicated that under conditions of heavy pumping these beds would be recharged by Ship Creek and would yield naturally filtered Ship Creek water — a desirable feature at times of spring run-off when Ship Creek is heavily charged with trash and contaminants.

At this particular location the artesian water is forty feet below the surface, whereas water in the stream gravels stands about ten feet below the surface. By establishing hydraulic connection, shallow water should flow by gravity into the deeper beds. Several things might be accomplished: (1) the artesian head would build up and the formations thus recharged would subsequently yield water at less cost; (2) appreciable losses of head due to heavy pumping would be minimized; and (3) summer water from Ship Creek would raise the temperature well above the 32-degree water available from surface supplies in the winter. A recharge well was completed in 1954 which accepted water at a rate of 140 gpm for a while but eventually dwindled down to a steady 50 to 60 gpm. My own feeling is that with the proper equipment and the knowledge of geology and hydrology now available, more recharge wells might be constructed at moderate cost and contribute substantially to the water assets of the area.

In 1954 a commercial driller completed a well south of the International Airport in which fifty-five feet of gravelly material was penetrated. Ten feet of 60-slot screen and fifteen feet of 80-slot screen was placed and after six days of surging the well yielded 270 gpm with one-and-one-half feet of draw-down, truly a fine yield; one hesitates to think of the efficiency that might have resulted had the entire formation been screened.

In 1954 the city of Anchorage decided to attempt to turn to ground water for its municipal supply. An adequate supply of ground water, in addition to supplying needs which the Ship Creek system did not, would eliminate both the need for a filter plant and costs of its subsequent operation and also the problem of freezing mains, being of a higher temperature than Ship Creek water in the colder months.

Accordingly work began in 1955. City test hole no. 1, located half a mile northwest of the Anchor Park U.S.G.S. test well, was so successful that it was immediately converted into a producing well. No. 2, located nearer the mountains, was less successful and was left as an observation well, but nos. 3, 4, and 5 were quite successful. Taking into account small interference effects, these four wells will produce a total of about 4500 gpm (six-and-one-half million gallons a day) under continuous pumping conditions, according to the city engineer (Johnson National Drillers Journal, vol. 30, no. 1, Jan-Feb., 1958, p. 8). It is of interest to note that most of the water-bearing beds developed were within 336 feet of the surface (no. 4 is completed at 200 feet) and the formations were screened with 60-, or 80- and 125-slot screens (125-slot screens will permit removal of grains in the fine-gravel range). Many of the formations, seemingly having an undesirable amount of fine sand, proved better than estimated from examination of the samples.

I will mention briefly the test well drilled at Kotzebue; in addition to the usual troubles of personnel, logistics, and weather, we had to drill the well twice, and, further, found only brackish water. Of particular interest here was a water encountered between seventy-nine and eighty-six feet (within an otherwise frozen section) that was saltier than sea water. It is suggested that this water originated by a process of fractionation by freezing.

One of the most interesting spot jobs accomplished was to recommend a ground-water supply in one of the Pribilof Islands in the Bering Sea. Time for field work was limited — the work was actually done between five o'clock and seven-fifteen one morning — but in that time enough was seen to lead to the conclusion that here was a simple classical occurrence of Ghyben-Herzberg water. Accordingly, a well site was recommended. A drilling outfit was brought up from Seattle next year and work began. Two years later I was telephoned in Washington about my wells. Almost fearfully I asked how they had come out, and sighed with relief upon learning that the wells had been highly successful. "But we are in difficulty in that the Health Department complains that the chloride content has gone up to 150 parts per million." Would that all problems of Alaskan hydrology were as minor!

Mechanical Aspects of the Contraction Theory of Ice-Wedge Polygons[1]

ARTHUR H. LACHENBRUCH

ICE-WEDGE POLYGONS in permafrost play an important role in many natural processes and cause serious engineering problems, but their origin is not completely understood. Much evidence points to the general qualitative view that they result from the freezing of water in seasonally recurring thermal contraction cracks.

Theoretical methods are used to test certain implications of this contraction theory and to gain insight into the controlling mechanical and thermal processes. The general problem is separated into four more or less distinct parts: (1) the relation between temperature distribution and the induced-stress distribution in the ground prior to cracking; (2) the relation between ground-stress distribution at the time of cracking and the depth of crack propagation; (3) the modification of the ground-stress distribution owing to the formation of a crack; (4) mechanical factors controlling the interrelation of cracks in the polygonal fracture system.

Linear relations between stress, strain, and their first time-derivatives seem to lead to contradictions with the loose restrictions imposed on the contraction theory by qualitative field observation. It is possible, however, to find a reasonable non-linear deformation law consistent with thermal requirements and leading to polygon dimensions compatible with those observed.

Such an analysis can also be applied to the related problems of polygonal jointing in cooling basalt and in drying mud.

[1]Abstract only.

Hydrodynamic Analysis of

Circulation and Orientation

of Lakes in Northern Alaska

R. W. REX

ABSTRACT

The systematic orientation of elliptical lakes occurring in the unconsolidated sediments of the Arctic Coastal Plain of Alaska has been attributed by some investigators to winds blowing along their major axes (Black and Barksdale, 1948) and by others to winds blowing along their minor axes (Livingston, 1954; Mackey, 1956a, b, 1957, 1958; Zenkovitch, 1959). The application of hydrodynamic principles resolves this question in favour of orientation of the lake minor axis parallel to the dominant wind direction.

Details of predicting lake waves and computing longshore currents, orbital velocities associated with shoaling waves, orbital velocities required to move sediments, and equilibrium shoreline configurations are presented with an example for the Barrow area of Alaska. Data are provided so that the reader may repeat these computations for other cases. Although the area studied here is in the Arctic, the same hydrodynamic principles apply to oriented lakes, bays, and lagoons elsewhere in the world.

One may reverse this line of reasoning and utilize geomorphological evidence to make meteorological deductions. The details of Black and Barksdale's (1948) descriptions of the Alaskan lakes suggest that the summer intensity of the polar northeasterly winds is decreasing and that there has been a recent increase of southerly winds.

THE ARCTIC COASTAL PLAIN of northern Alaska is dotted with thousands of lakes which often show a striking orientation, with their major axes trending N9°W to N21°W. All degrees of gradation exist between simple low-centre polygon pools and large oriented lakes up to nine miles long and three or more miles wide. In 1949, Black and Barksdale published their classic paper describing these oriented lakes, attributing the orientation of the lakes to wind-generated waves. Black and Barksdale pointed out that the present prevailing winds blow from the east-northeast and that there is a secondary maxima from the west-southwest. They then *assumed* that the direction of maximum lake elongation should be parallel to the prevailing wind direction and that therefore the lake orientation was an indicator of a prior Pleistocene dominant northwesterly wind.

On the basis of field work on the Arctic Coastal Plain, a number of observers reached somewhat different conclusions on the time of lake orientation. Kirk Bryan, Jr., in 1951 (personal communication) suggested that the lake orientation was produced by the present wind system with the prevailing wind parallel to the minor axes of the lakes. D. J. Livingston, E. J. Taylor, and the author reached similar conclusions as the result of field work in the area in 1952. Livingston reported his observations in 1954, utilizing the equation of continuity and an analogy to rip currents to show that wind-generated waves should establish a two-cell circulation that corresponds to the present shape of the lake. Hutchinson

(1957) comments that Livingston is undoubtedly correct in his interpretation. Mackay (1956a, 1956b, 1957, 1958) in his papers on the Mackenzie River delta and Liverpool lakes region described large numbers of oriented lakes very similar in appearance to those in the Barrow area. Mackay points out that the minor lake axes of the Mackenzie lakes are parallel to the prevailing wind direction, and that the lake orientation is being affected by present lake waves. He further argues that the Barrow lakes described by Black and Barksdale (1949) and Livingston (1954) and the Mackenzie delta lakes are formed by similar mechanisms. Odum, on the other hand (1952, p. 264), utilized Black and Barksdale's concept of lake orientation parallel to the prevailing winds as partial justification for the reconstruction of Pleistocene wind conditions in the area of the Carolina bays. The principal weakness of Odum's treatment is in forecasting surface winds from the 10,000-foot level. The forecasting techniques employed by Odum assume the same tropospheric stability and structure, the same zonal and meridional components in the Pleistocene as exist at present. The papers of Arrhenius (1950) and Nyberg (1956) on the equatorial upwelling in the Pacific show that this assumption is seriously in error. For this reason a more direct method is needed of correlating wind direction and lake orientation when wind-generated waves are the lake-orienting mechanism.

In 1959 Zenkovitch summarized his concepts of cuspate spit genesis along the shores of the Chukotsky Peninsula and of the orientation of lakes in areas of unconsolidated deposits. Zenkovitch attributes a circular lake outline to waves generated by a wind from one direction and an elliptical outline to waves generated by winds prevailing from one or two opposite directions. In the latter case, the minor axis of the ellipse lies in the direction of the more active wind. This is in agreement with Livingston (1954) and others and in contradiction to Black and Barksdale (1948). The obvious differences between these two concepts of lake orientation suggest reference to the hydrodynamic principles governing wave and near-shore processes to see which hypothesis is most reasonable.

DESCRIPTION OF THE LAKES ON THE ARCTIC COASTAL PLAIN

As Black and Barksdale (1949, p. 111) have reported, there are two principal types of lakes: those with simple saucer-shaped basins often about six feet deep, and deeper lakes with shelves, which are more common in the higher southern regions of the upper coastal plain. The shelves are generally strongly developed along the lake sides which parallel the major axis of the lake and are not present, or are only slightly developed, along the other shores.

Investigations in the Barrow area by the author during the field seasons of 1952, 1953, and 1954 indicate that thermal, ice, animal, and wave erosion cause collapse of ice-wedge polygon rims, permitting many lakes to form and grow initially by the coalescing of flooded areas of polygons. Thermal erosion includes the processes associated with an increase of heat content of the ground, such as the thawing of ground ice and solifluction. In addition, the melting of interpolygon ice-wedges and polygon ground ice in turn causes ground subsidence, permitting lakes to flood adjacent areas of polygons. The sod covering the ground persists under water and prevents wave erosion until it is destroyed. This frequently happens when the lake ice breaks up, is blown back and forth across the lakes, and plows up the shore. The ice sometimes pushes up muck ridges two or three feet high which are later washed away by storm waves. Once established, the lakes gradually grow through the thawing of ground ice and the destruction of the shores by all types of erosion. However, not all lakes are formed in this manner. Some lake basins originate as segmented lagoons (Black and Barksdale, 1948); some originate as ox-bow lakes, and many go through various cycles of lake formation, filling in with vegetation, being converted to polygonal ground, and then forming new lakes. The lakes on the Arctic Coastal Plain of northern Alaska are certainly polygenetic, but systematic orientation is a general property and any valid orientation mechanism must operate within the framework of polygenetic lakes.

WIND AND WAVES

The Arctic wind circulation is dominated by the polar anticyclone which in all seasons except winter is centred over the Arctic Ocean. So far as is known, the directions of the prevailing wind correspond to the mean pressure distribution. This gives rise to a prevailing east-northeast wind in the Barrow area (Weather Bureau, 1953) and general northeast winds in interior Alaska (Petterssen et al., 1956, p. 45). The prevailing east-northeast winds, and a secondary component blowing in the opposite direction, usually blow with speeds of 10 to 30 mph, and frequently last for a week or more without interruption. The reversal of the prevailing wind is usually associated with the incursion of an air mass from the region of the prevailing westerlies of the temperate circulation. The mean wind speed for Barrow during the period of open water is 13 mph (Petterssen et al., 1956, p. 97).

Wind blowing over the water surface subjects the surface to a stress. In the case of a rough-water surface, there are additional pressure gradient forces associated with turbulence at the air-water interface. The first waves to be generated are small ripples whose restoring force is dominantly surface tension; hence they are called capillary waves. In the case of increased wind stress, capillary waves grow in size to where the acceleration of gravity is the dominant restoring force; they are called gravitational waves. The energy of the wave propagates along the air-water interface. The intensities of the components of wave motion in the water away from the interface (down) decrease according to hyperbolic functions. The water surface may be considered a wave guide. If the depth of water is greater than one-fourth of the wave-length, the medium may be considered to be semi-infinite or deep water. On the other hand, in shallow water there is a rigid boundary to the wave guide which changes the character of the wave propagation. Shallow-water conditions are defined as water depths less than one-twentieth of the wave-length. Intermediate depth conditions occur between the deep- and shallow-water conditions and give rise to what we shall call the shoaling transformation of deep-water waves. Details of the mathematical treatment of the shoaling transformation are presented in a number of references such as Sverdrup et al. (1942); Eckart (1952); Beach Erosion Board (1954); Wiegel and Fuchs (1955); and Eagleson (1956).

When waves move into shallow water from deep water, their velocity decreases as the depth decreases until the point where the wave breaks. This response of wave velocity to bottom depth in a constricting wave guide can result in refraction of the wave crests. The general effect of bottom topography is to modify the wave crests to conform to the bottom contours. Detailed discussions of wave refraction and associated phenomena are given in Munk and Traylor (1947); Beach Erosion Board (1954); Arthur, Munk, and Isaacs (1952); Chien (1953); and Pierson, Neumann, and James (1955).

CURRENTS PREDICTED BY
HYDRODYNAMIC THEORY FOR AN OBSERVED WIND

In the fall of 1952, the author observed the orientation of the ice on the lakes and rivers of the Arctic Coastal Plain. The period of freeze-up occurred during a storm blowing from the east-northeast and the axis of ice orientation was perpendicular to the wind direction. Continuous anemometer records were available for Barrow for this storm of late September 1952. The wind speed increased gradually to twenty to twenty-seven knots from the east-northeast and east, blew at this speed for four days, and then gradually diminished to a gentle breeze. Figure 1 shows the relative radial distribution of wind kinetic energy for the eleven-day period including the entire storm. On the basis of Weather Bureau data (Figure 2), this storm was thought to be typical for the Barrow region. Wind stress on the lakes generated waves and currents which in turn might be expected to move sediment in the lakes.

In order better to understand the relation of the oriented lakes to present and past winds, the author, utilizing wave forecasting, wave refraction, orbital velocity, longshore current, and sediment transport computations, will review semi-quantitatively with this example the relation between wind direction and force and lake shape and circulation. The terms and symbols used in this discussion are given in Tables I and II and Figures 3 and 4.

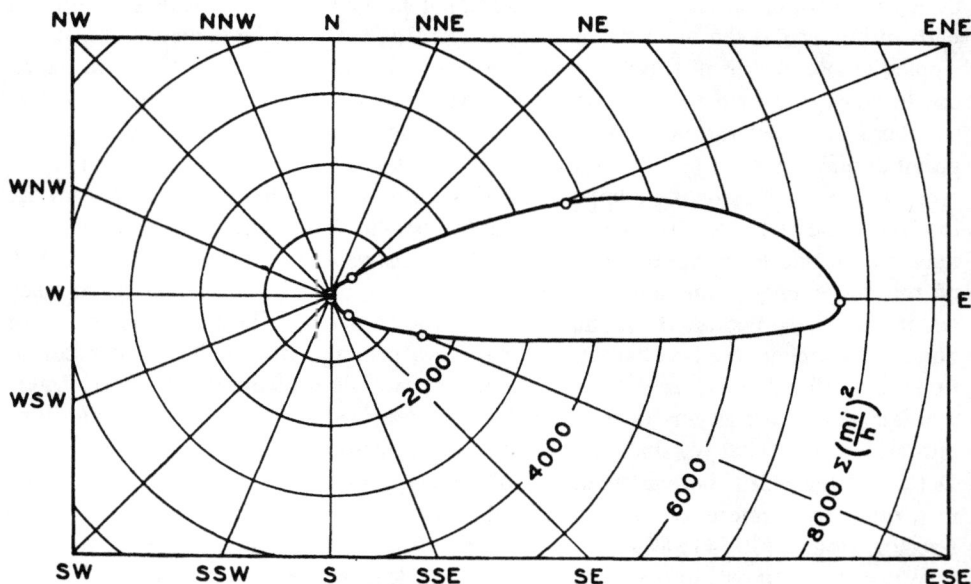

FIGURE 1.   Directional index of wind energy during the period of lake freeze-up summed in kinetic energy units of (miles/hour)^2 over an eleven-day period, September 20 to October 1, 1952.

TABLE I

T	=	wave period (sec)
σ	=	$2\pi/T$
L	=	wave-length (ft)
k	=	$2\pi/L$
C	=	wave velocity
U	=	wind velocity (ft/sec)
F	=	fetch or distance which wind blows over open water (ft)
g	=	acceleration of gravity (32.2 ft/sec²)
H	=	height of wave (trough to crest elevation difference) (ft)
H_0	=	significant wave height or the average height of the highest one-third of a selected number of waves, this number being determined by dividing the time of record by the significant period (deep-water conditions)
T_0	=	significant period which is usually taken as the average period of the highest one-third waves in interval under study (T_0 will hereafter be given simply as T)
h	=	water depth (ft)
z	=	distance below average water surface (ft)
deep water	=	$h > 1/4\,L$
intermediate depth water	=	$1/4\,L > h > 1/20L$
shallow water	=	$1/20\,L > h$
u	=	horizontal component of wave orbital motion
$u_{m,0}$	=	maximum horizontal component of wave orbital motion at bottom derived by evaluating equation for u given in Table II and used in Figure 8.

Non-dimensional functions for predicting wind-generated waves have been prepared by Johnson (1950, p. 390) using data collected at Abbotts Lagoon, California, and elsewhere (Figures 5 and 6). An example will help explain their use. Consider our example of the September 1952 storm with its mean peak wind speed of twenty knots (34 ft/sec) and a typical lake width (i.e., fetch) of 5000 feet. Using Johnson's two graphs (Figures 5 and 6), we take U as 34 ft/sec and

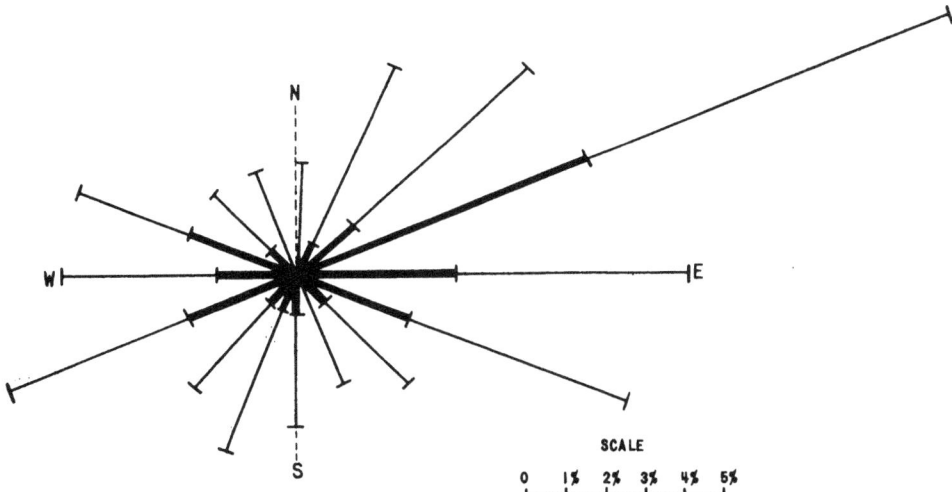

FIGURE 2. Direction, velocity, and frequency of winds at Barrow, Alaska. Length of vector indicates percentage of time that wind blows in indicated direction. Wind blow towards centre. Heavy line indicates winds of velocity greater than 15 mph (31 per cent of total). Light line indicates winds of velocity from 1 to 15 mph (69 per cent of total). Calms exist less than 1 per cent of the time (Black and Barksdale, 1949).

replot wave period T and wave height H as a function of fetch F (Figure 7). A family of curves similar to those of Figure 7 for different wind speeds would give the picture of wave period and height for wind conditions through the entire storm. Changes of wind speed and direction over lakes are followed by rapid readjustment of the equilibrium wave conditions in a matter of minutes. Figure 7 gives an equilibrium significant wave period $T = 2.1$ seconds, and a significant wave height $H_0 = 1.1$ feet for our example of 5000-feet fetch. The wave array grows in height and period as the wave front progresses from initial formation on the upwind side of the lake towards the downwind shore. From Table II we see that a period T of 2.1 seconds corresponds to a deep-water wave-length of 22.6 feet. Remember that deep-water conditions occur for depths greater than one-quarter the wave-length; intermediate depths between $1/4\ L$ and $1/20\ L$; and shallow water at less than $1/20\ L$. The typical oriented lake near Barrow is often six to nine feet deep and saucer-shaped, so the wave starts out as a deep-water wave and usually remains a deep-water wave until the downwind shore is approached. Here the wave first becomes an intermediate-water wave and then becomes a shallow-water wave. The wave has the *same period* through this transition and, because of the short wave period, for all practical purposes the same height (Figure 8) (Beach Erosion Board, 1954, p. D-24). The shoaling transformation for the waves in question becomes important at a water depth of about five feet and from here on into the breaker zone where the predicted breaker would be about 1.1 to 1.5 feet high,

TABLE II

PROGRESSIVE WAVE EQUATIONS

Term	Deep water $h/L_d > 1/4$	Intermediate	Shallow water $h/L_d < 1/20$
C^2	g/k	$g/k\ \tanh\ kh$	gh
L	$\dfrac{gT}{\sigma}$	$\dfrac{gT}{\sigma}\ \tanh\ kh$	$T\sqrt{gh}$
u	$\dfrac{H}{2}\sigma e^{-k(h-z)}\cos Q$	$\dfrac{H}{2}\dfrac{\cosh\ kz}{\sinh\ kh}\cos Q$	$\dfrac{H\sigma}{2hk}\cos Q$

Abbreviations: $k = 2\pi/L$; $\sigma = 2\pi/T$; $Q = kx\text{-}\sigma t$ where $x =$ distance of displacement and $t =$ time in seconds.

In order better to understand the difference between the equations, note the information given in Table III.

TABLE III

Hyperbolic function	Large A	Small A
$\sinh\ A \approx$	$1/2e^A$	A
$\cosh\ A \approx$	$1/2e^A$	1
$\tanh\ A \approx$	1	A

depending on the slope of the shore (Figure 9) (Beach Erosion Board, 1954, p. 48). The bottom profile of an artificially drained lake near Barrow was observed to be saucer-shaped, and it is concluded that the use of the shoaling transformation is reasonable. A highly irregular bottom might introduce complexities into the details of the shoaling transformation of waves.

Computation of the maximum orbital velocity of the wave near the lake bottom (Table I, Figures 3 and 10) shows that bottom surge in this storm becomes strong enough at depths less than eight feet to form sand ripples, and that sand can be readily moved within the zone lying between eight feet and the breakers. For further discussion of this subject, the reader is referred to the Beach Erosion

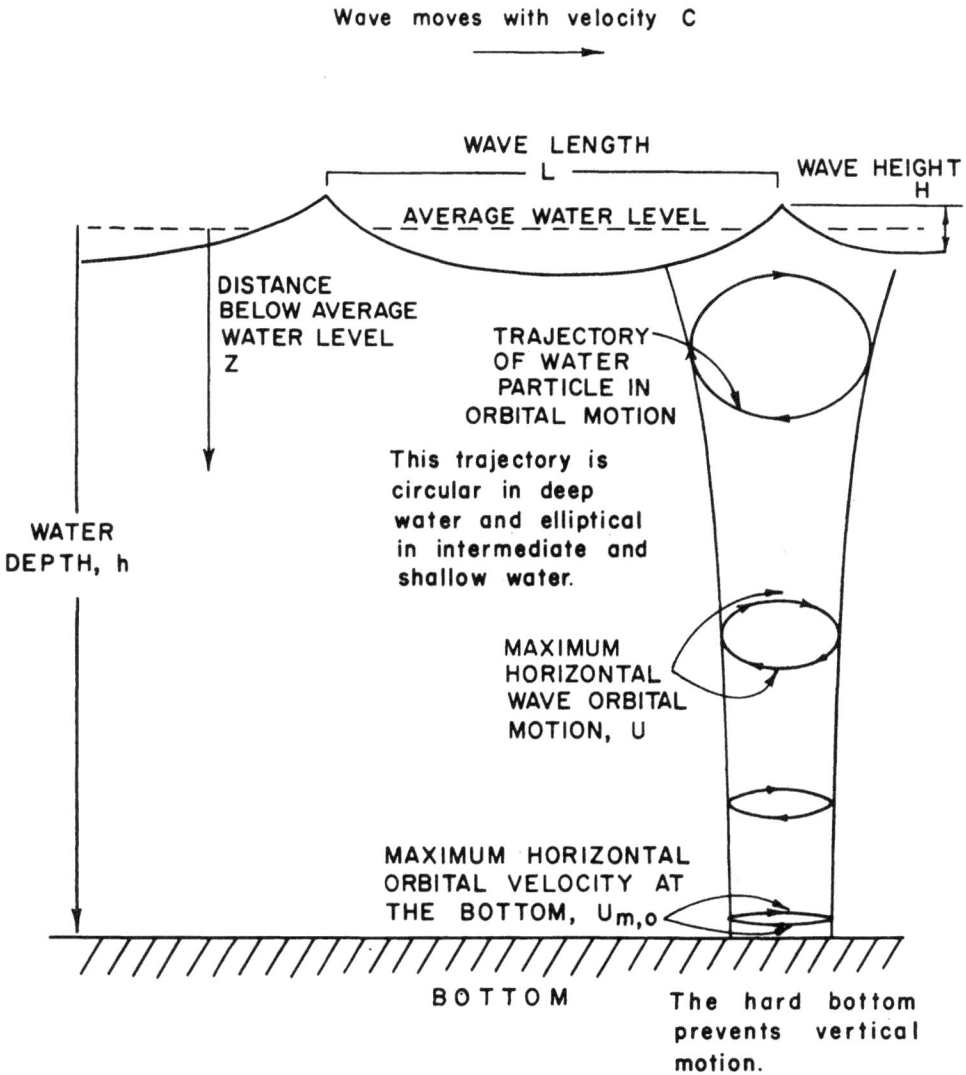

FIGURE 3.   Terminology used in describing lake waves.

Board Report no. 4 (1954) which gives a review of wave effects on near-shore features. Sand movement velocity thresholds are from Inman (1957, p. 31). The effect of grain size is given by Figure 11 (Inman, 1957, p. 16).

The average longshore current in the surf zone along a straight beach generated by the significant waves of 2.1 sec period and 1.1 feet deep-water height as predicted by the technique of Inman and Quinn (1952, Figures 12, 13, 14) is given in Table IV. Extension of the formula (Figures 12, 13 and 14) to higher angles is not felt justified because of the increasing importance of the differences between curved and straight beaches, although Inman (personal communication) indicates

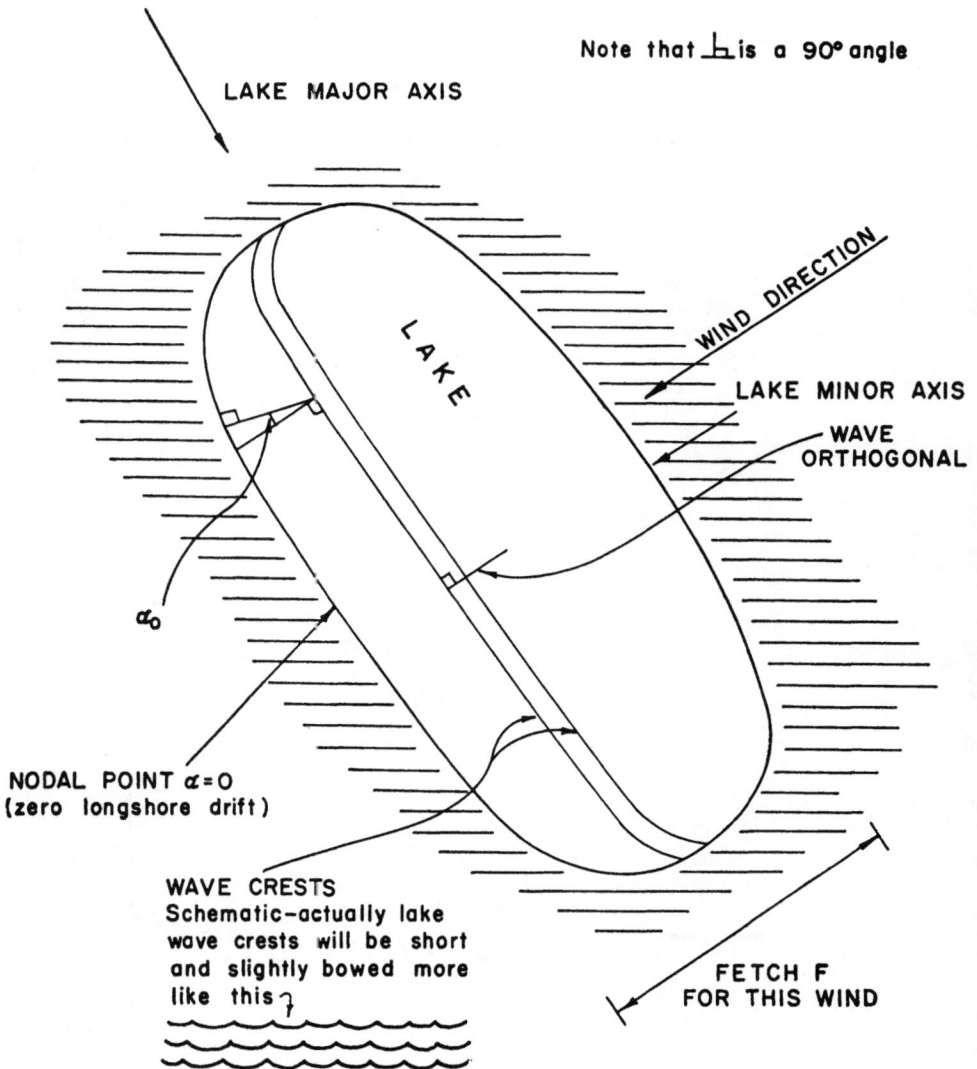

FIGURE 4.   Terminology used in describing lake orientation.

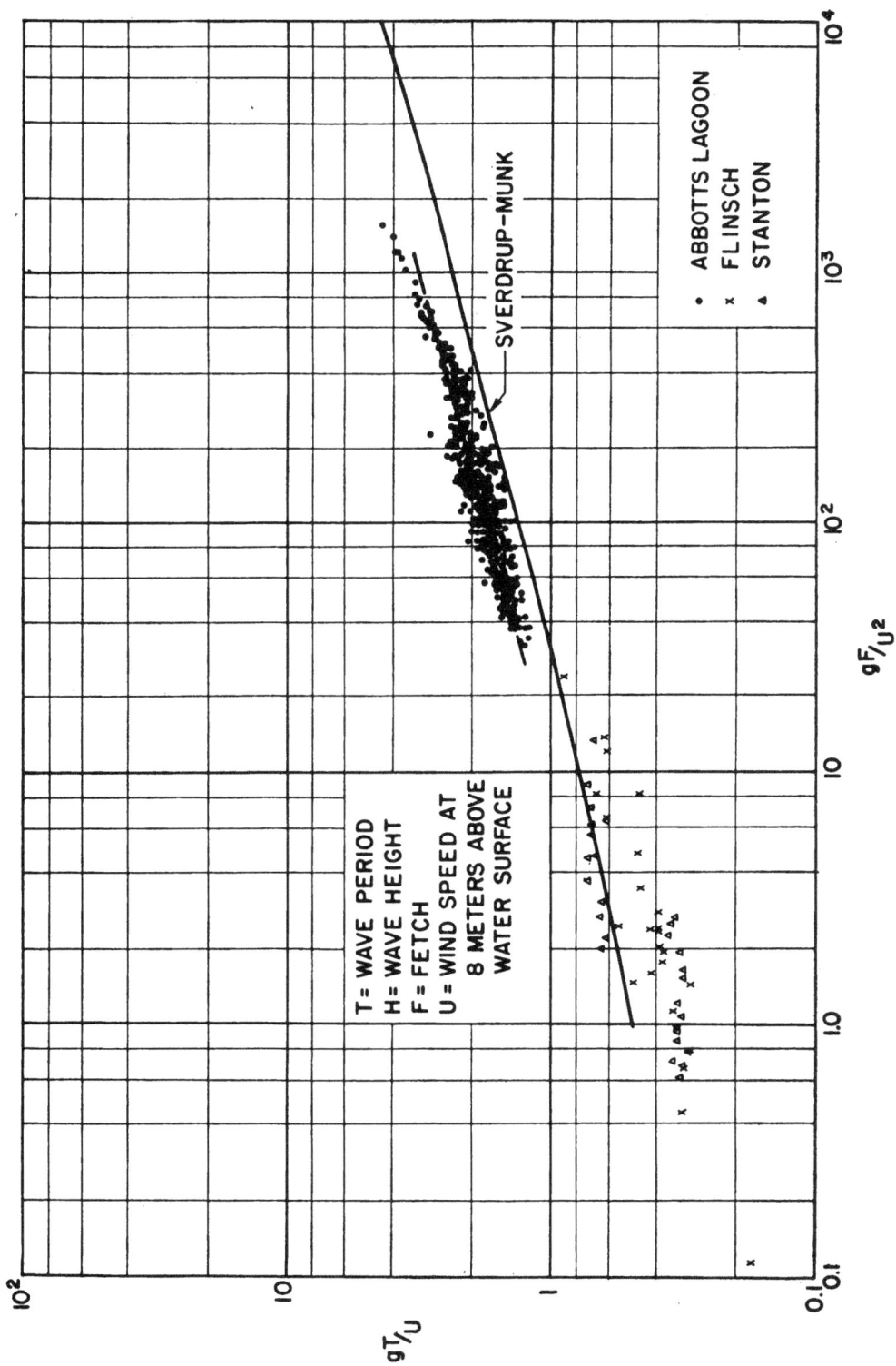

FIGURE 5. Non-dimensional plot of wind speed, period, and fetch (Johnson, 1950). See Figure 6 for height graph.

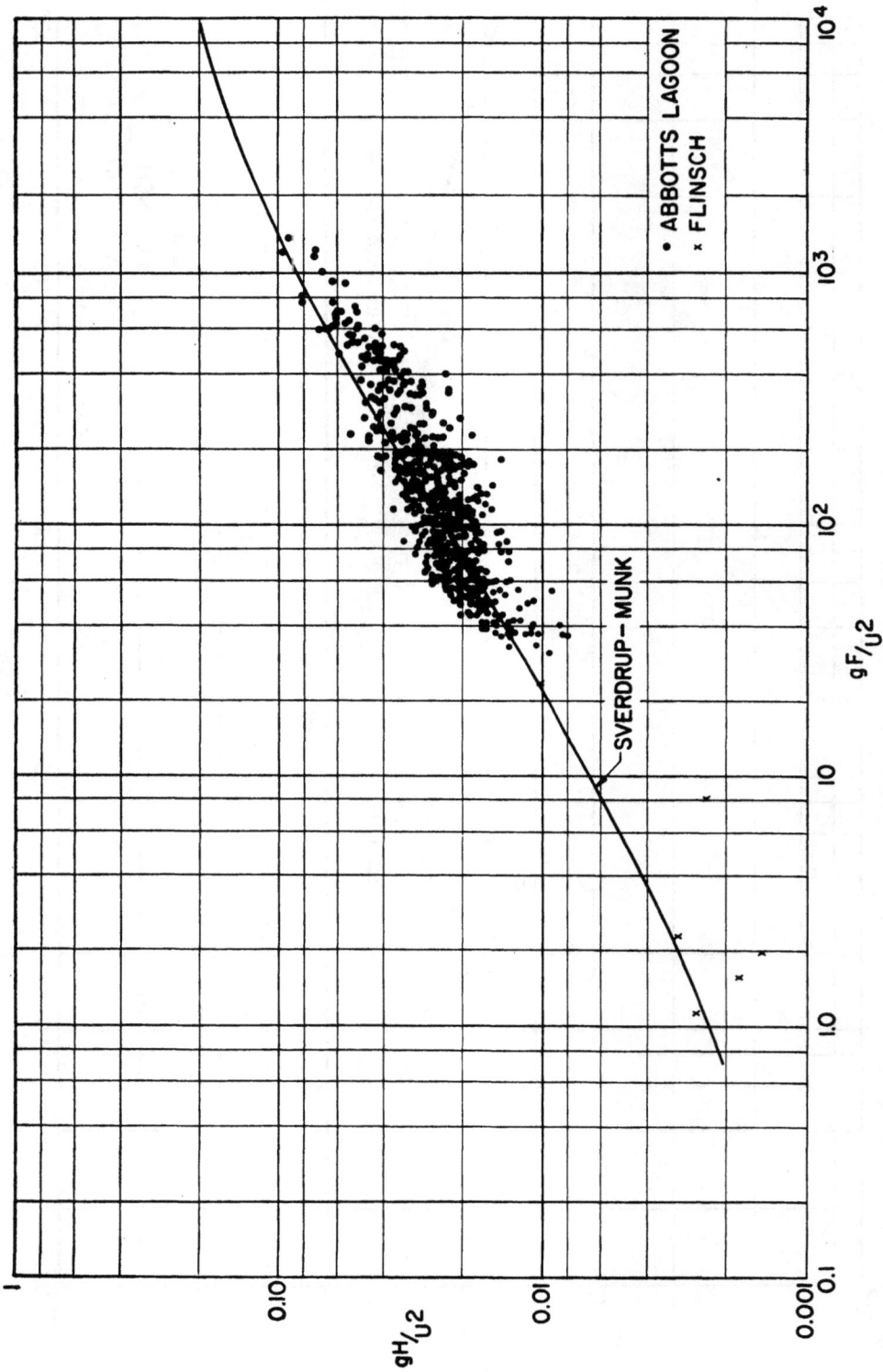

FIGURE 6. Non-dimensional plot of wind speed, height, and fetch (Johnson, 1950). The symbol notation is the same as for Figure 5.

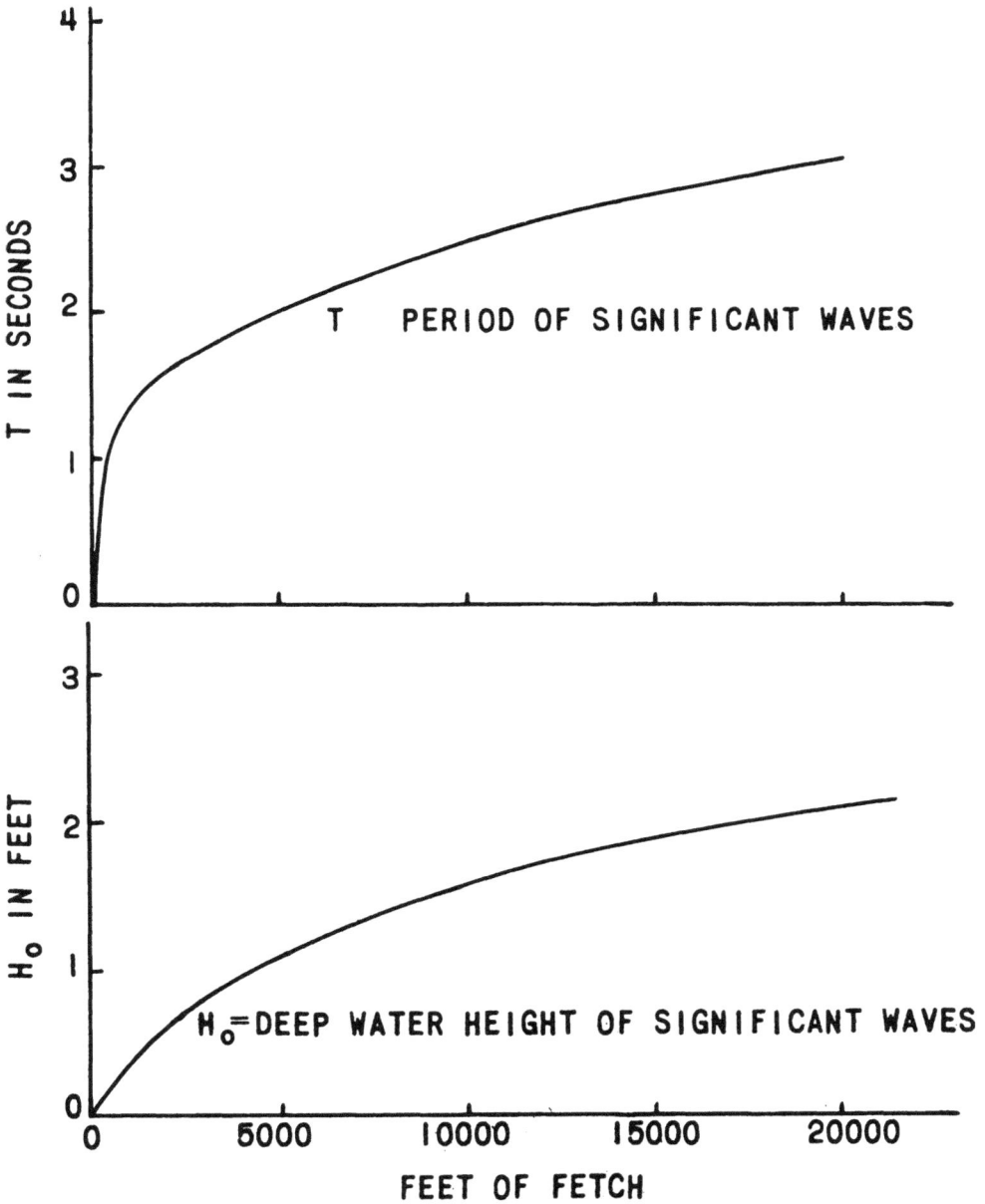

FIGURE 7.   Replot of data from Figures 5 and 6 for our example of a twenty-knot (34 ft/sec) wind showing the increase of wave height and period with increasing fetch and deep water conditions.

TABLE IV

Longshore Current Velocities in the Surf Zone

Slope of beach	3 per cent	5 per cent	10 per cent
$\alpha^1 =$ 0°	0.0 ft/sec	0.0 ft/sec	0.0 ft/sec
5°	0.4 ft/sec	0.5 ft/sec	0.6 ft/sec
10°	1.0 ft/sec	1.2 ft/sec	1.4 ft/sec
15°	1.6 ft/sec	1.8 ft/sec	2.0 ft/sec
20°	2.2 ft/sec	2.5 ft/sec	2.7 ft/sec
25°	2.8 ft/sec	3.1 ft/sec	3.5 ft/sec
30°	3.3 ft/sec	3.7 ft/sec	4.2 ft/sec
35°	3.9 ft/sec	4.4 ft/sec	4.7 ft/sec

α^1 is the breaker angle or the angle between the shore and the wave crests or equally the angle between the wave orthogonals and the normal to the shoreline. α_0 is the angle the waves make with the shoreline while they are still in deep water.

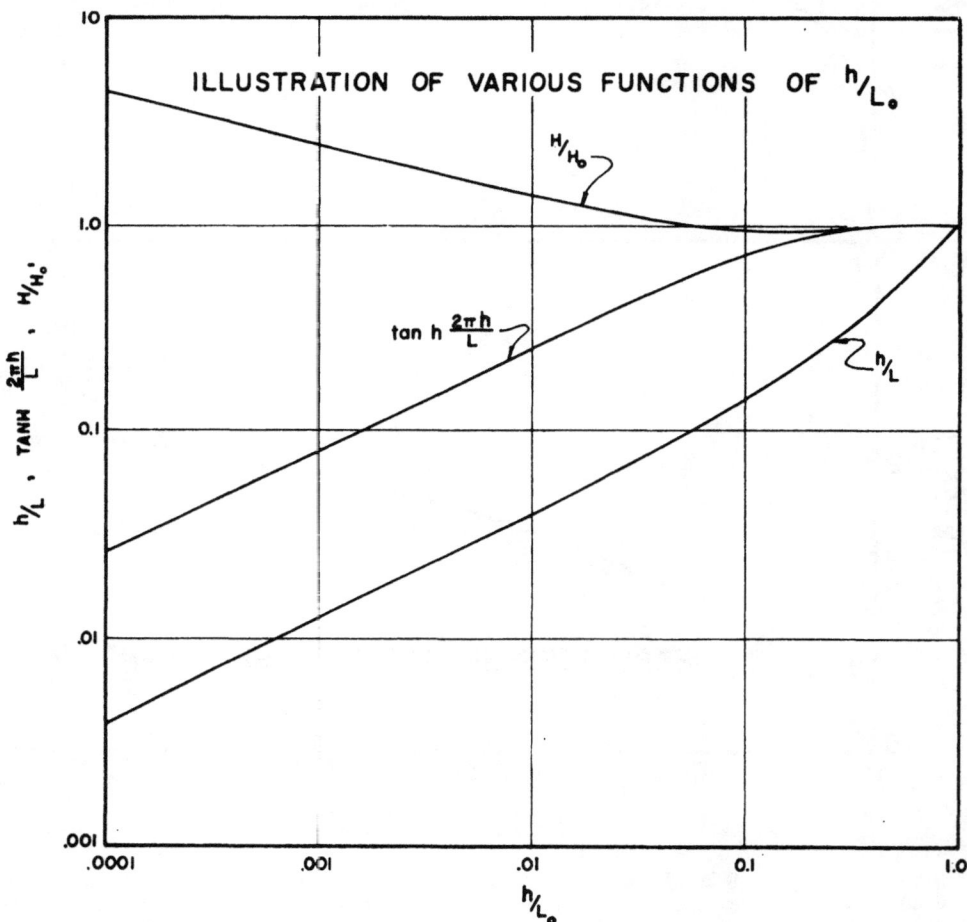

FIGURE 8.   Various functions of h/L_0. In the text we use H/H_0 to show the change of wave height through the shoaling tranformation and H_0 is the deep-water wave height (Beach Erosion Board, 1954).

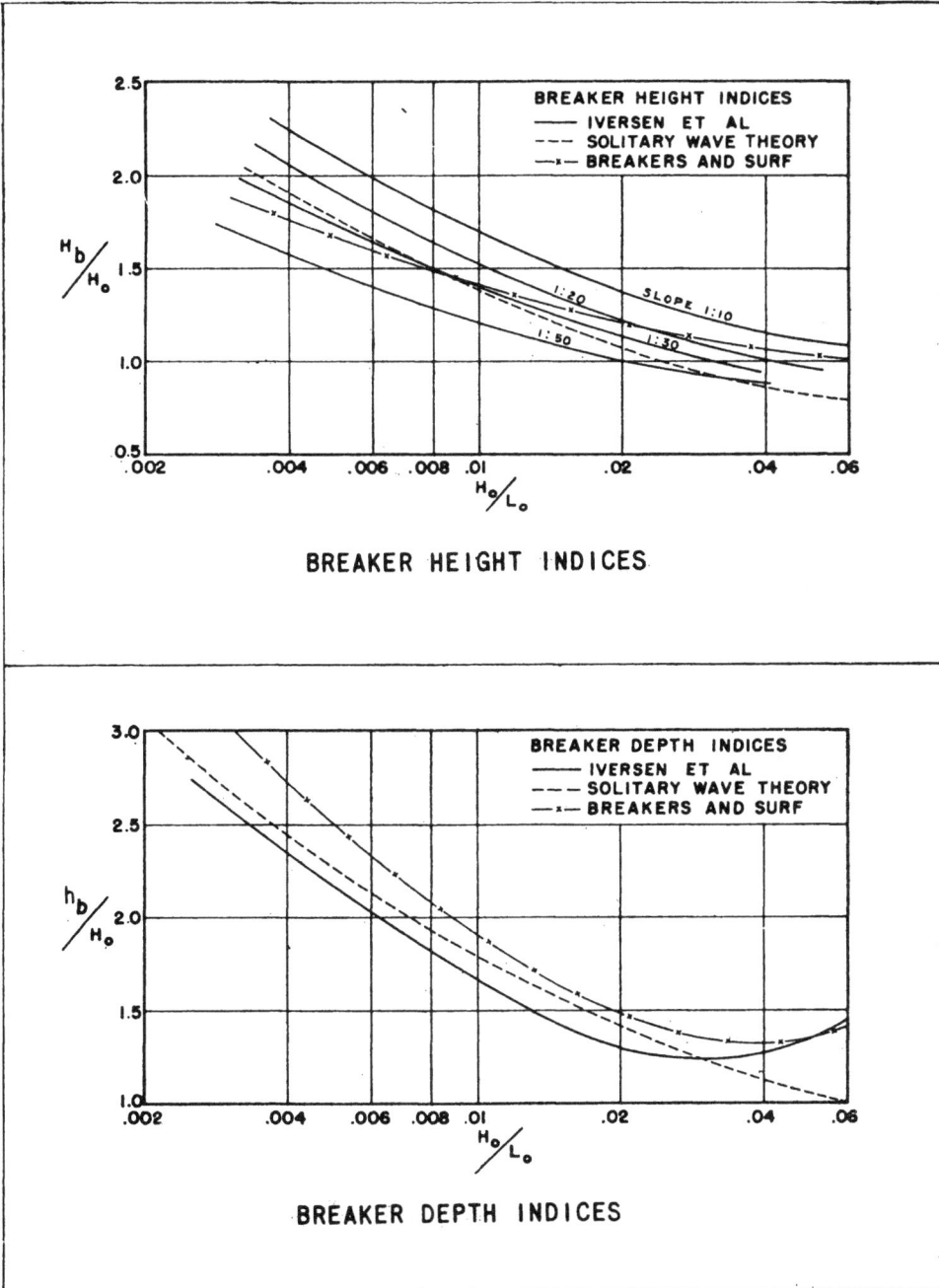

FIGURE 9.   Breaker height and depth indices where H_b is the wave height at the point of first wave breaking (Beach Erosion Board, 1954).

that the formula proved valid to angle of $\alpha = 50°$ in Mission Bay, San Diego, California.

Approximately 0.3 ft/sec water current is sufficient to move sand along the bottom (Inman, 1957, p. 31), so we see that regardless of which value of beach slope is assumed here, the longshore currents in the surf zone should be sufficient to move sand except in the zone where $\alpha = 0$ which is the nodal point of littoral drift. The addition of the traction effect of the orbital velocity of the waves (Figure 10) makes it possible to transport sand with still lower longshore current velocities. We see that for depths less than eight feet, wave orbital motion and surf zone currents are predicted to move sand in response to waves generated by our twenty-knot wind.

Application of littoral transport concepts is thought to be justified in this situation because the entire Arctic Coastal Plain is mantled with sands, silts, gravels, and peat of the Gubik formation consolidated only by freezing of interstitial water. Furthermore, we see that sand should be moved by waves and current which form on these lakes under the influence of observed winds. Wind-generated waves on an irregularly-shaped lake will move in an array with their orthogonals roughly parallel to the wind. The littoral drift of sediment along the shore redistributes

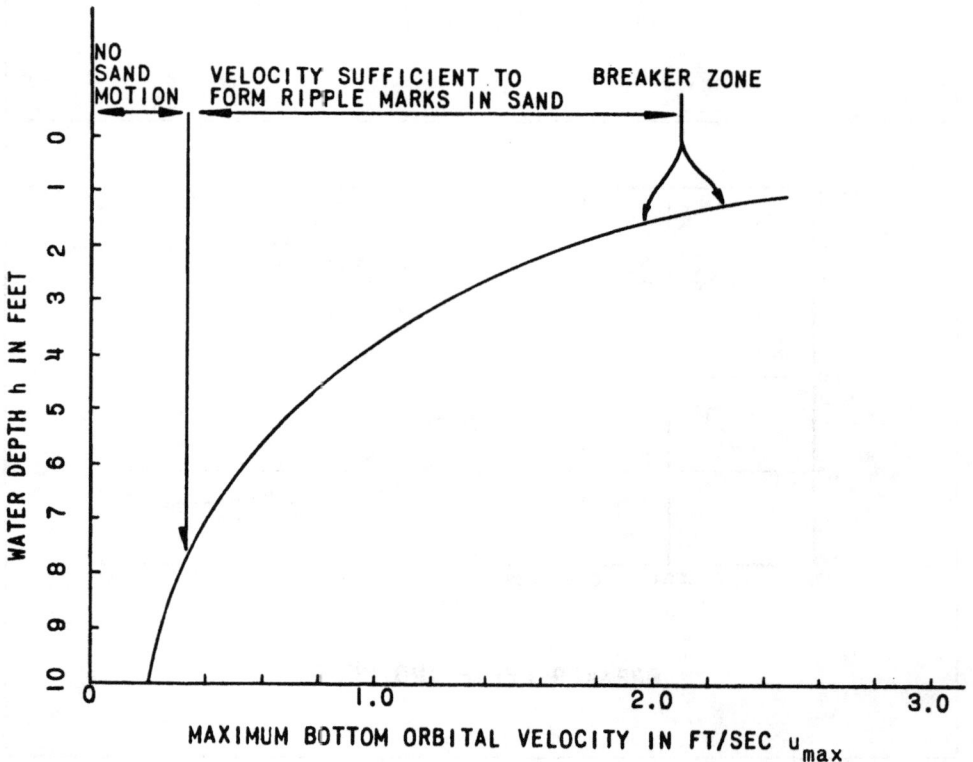

FIGURE 10.   Maximum wave orbital velocity in ft/sec on the bottom for the predicted significant waves of $T = 2.1$ sec and $H_0 = 1.1$ ft is $u_{max} = H\pi/T \sinh (2\pi h/L)$. Derived from equations for u, Table II.

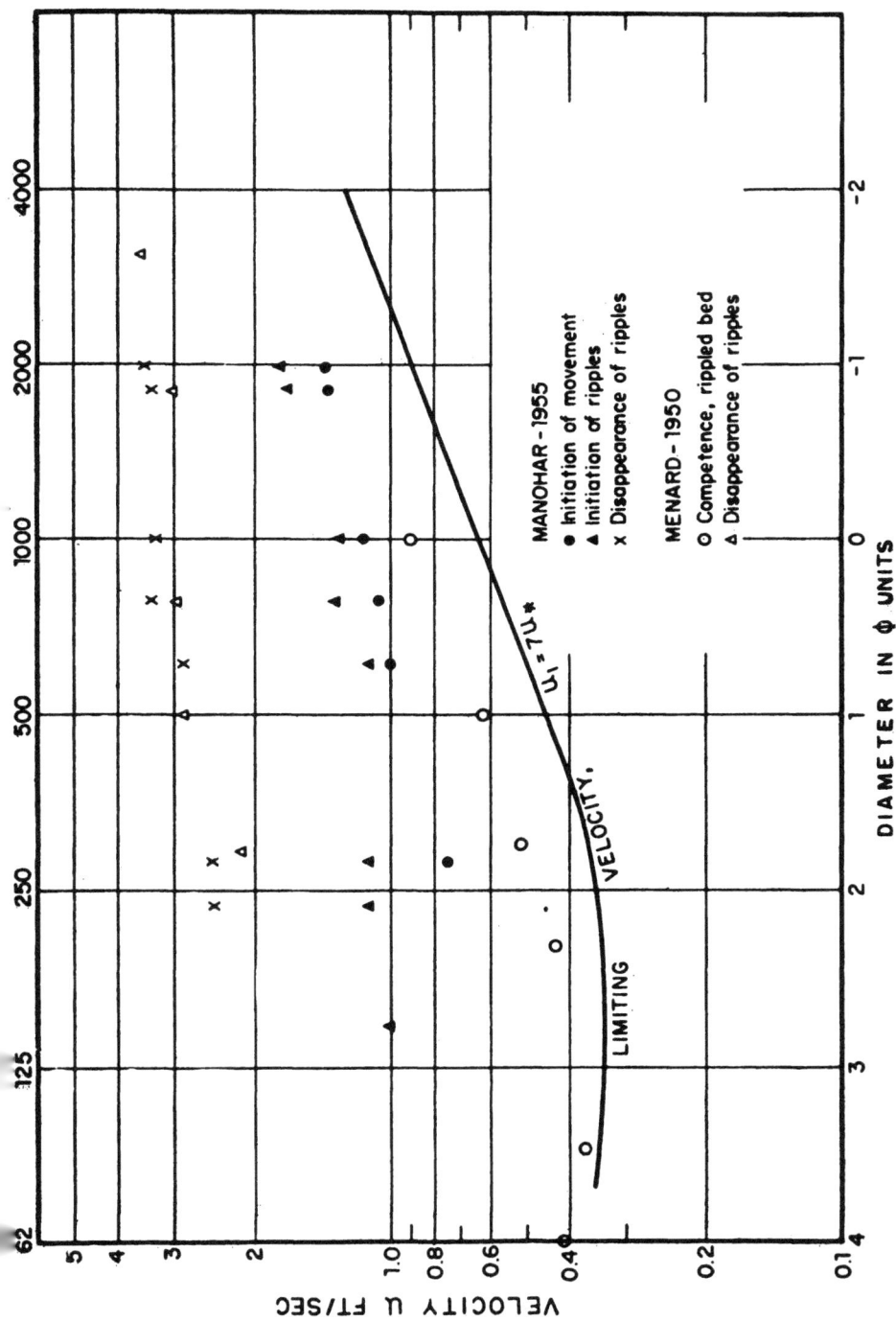

FIGURE 11. Limiting or minimum velocity for the initiation of motion of sand of a given size. Limiting velocity, arbitrarily defined as equal to $7u_*$, where u_* is the threshold or critical friction velocity. For unidirectional flow this relation would give a limiting velocity equivalent, for example, to the mean velocity measured 1 ft above a bottom which has a roughness length of 2 cm. Field observations near the surf zone indicate that planation and disappearance of ripples does not occur unless the maximum velocity associated with the wave crest somewhat exceeds that listed by Menard and Manohar (Inman and Quinn, 1952). For a more detailed discussion of this subject, see Inman and Quinn (1952).

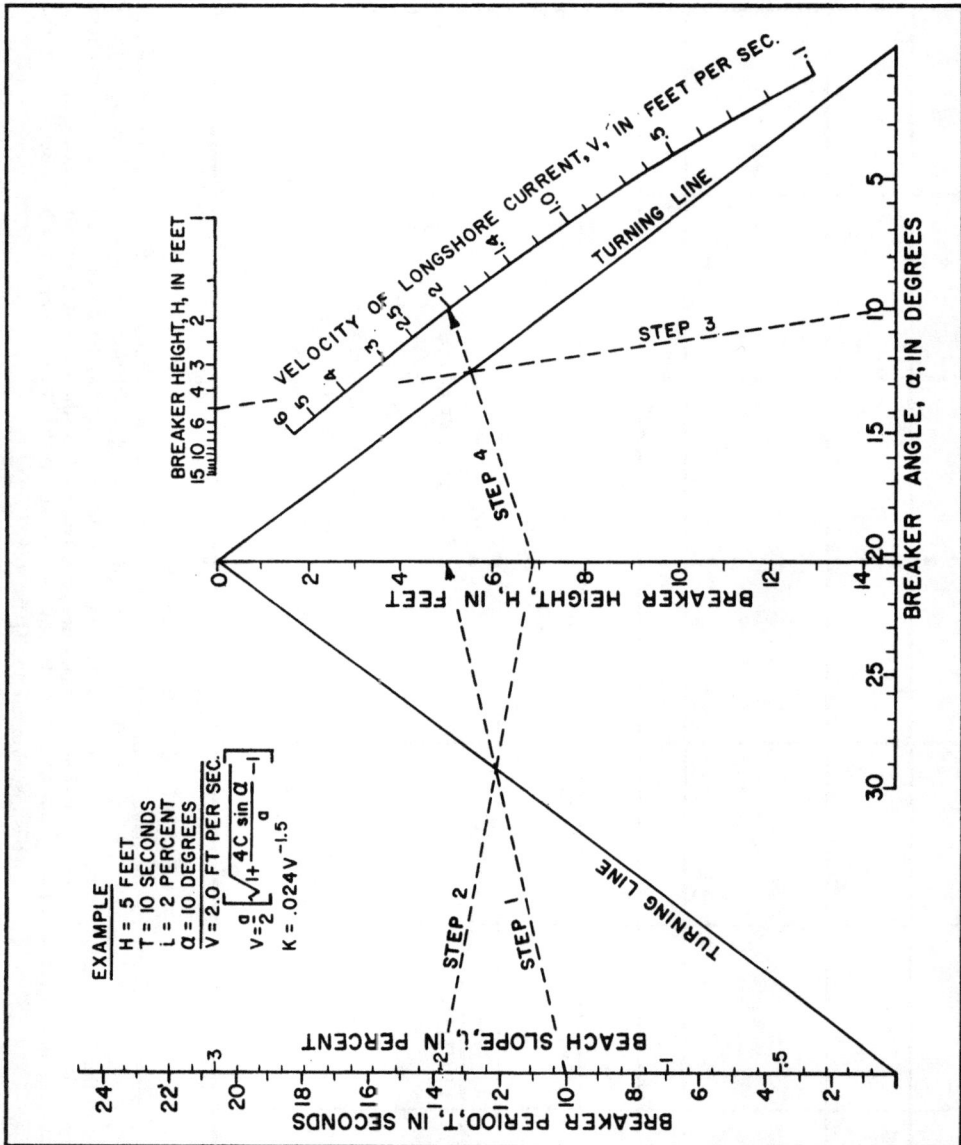

FIGURE 12. Alignment chart for the computation of longshore current. This figure is for use on a natural beach with a slope ranging up to 3 per cent. Procedure: (1) place straight edge from appropriate T to H and determine intersection with turning line; (2) turn straight edge about intersection to i, and determine intersection with H scale; (3) determine intersection on second turning line between H and α scale; (4) align intersections of H-scale and

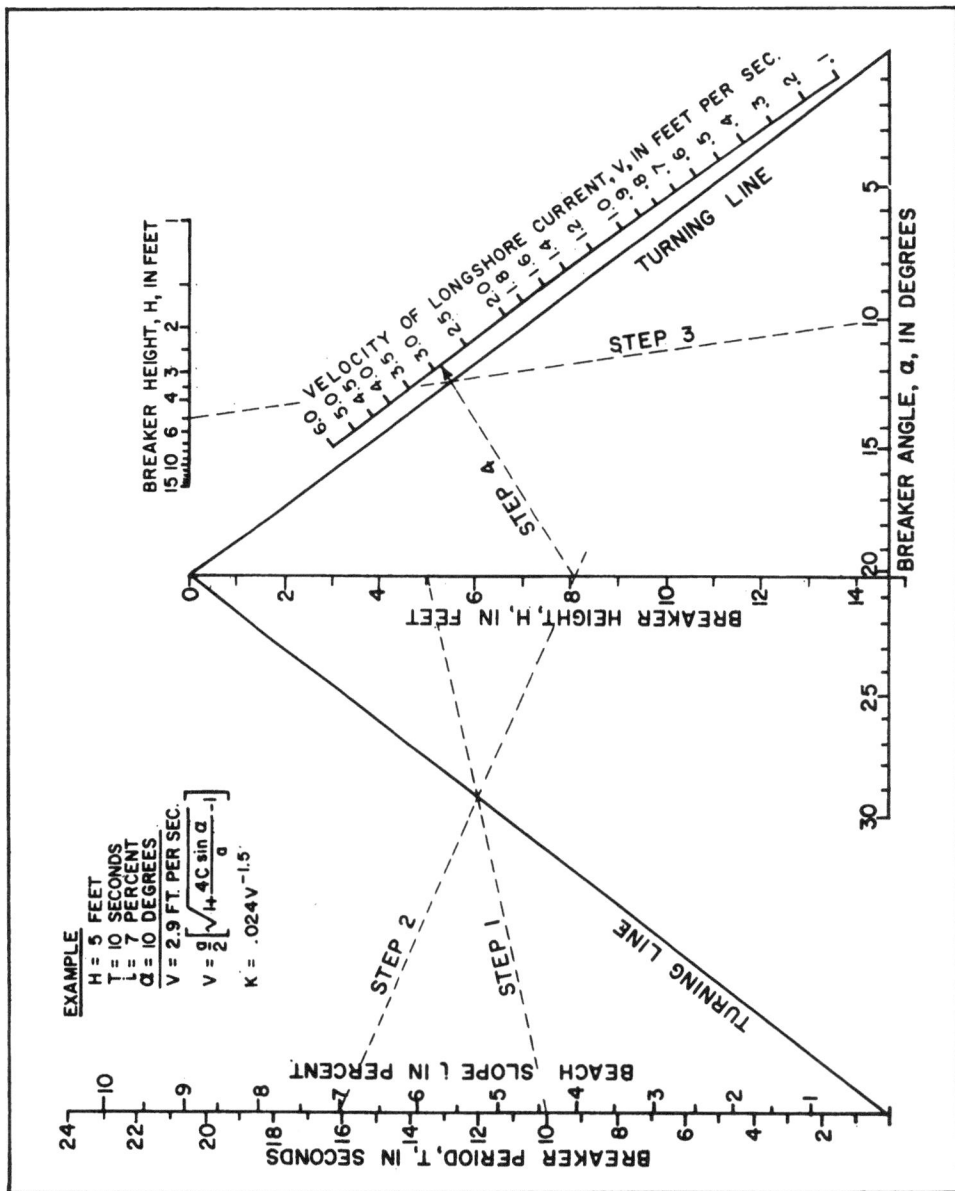

FIGURE 13. Alignment chart for the computation of longshore current. This figure is for use on a natural beach with a slope ranging up to 10½ per cent. The procedure is the same as for Figure 12.

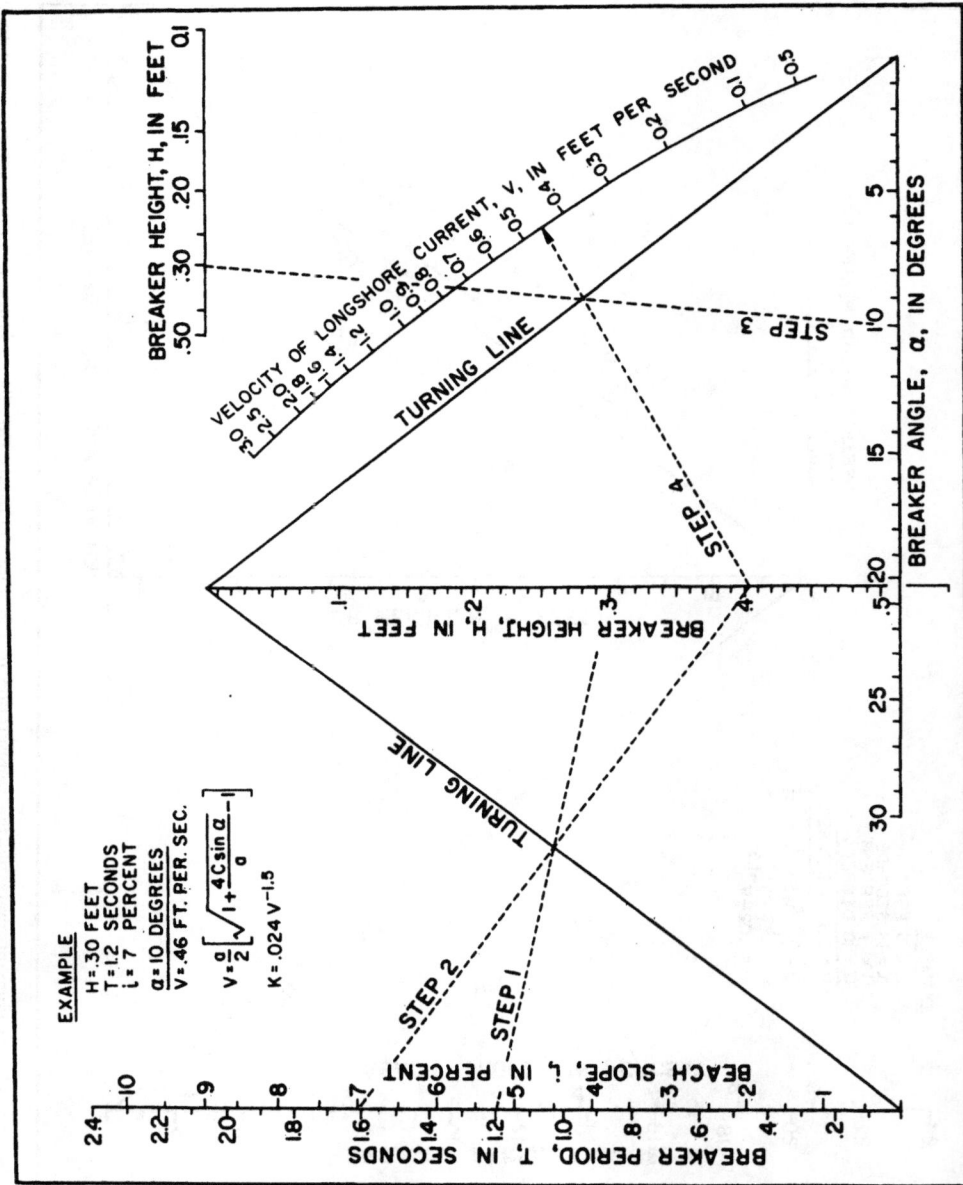

Figure 14 Alignment chart for the computation of longshore current. This figure is for use on model beaches with ...

material to positions of least displacement and thereby gives rise to a smoothed shoreline. The *nodal point* of a lake shore is the point where the littoral drift is zero. The terminology used in describing lake orientation was illustrated in Figure 4.

The littoral drift has been described by a sine function where a maximum of drift occurs at $\alpha_0 = 30°$ for a straight beach (Saville, 1950) and $\alpha_0 = 50°$ for a curved beach (Bruun, 1954) where α_0 is the angle between the deep-water wave orthogonals and the normal to the shoreline. The nodal point lies at the point where $\sin \alpha_0 = 0$, that is, $\alpha_0 = 0°$, which is the centre of the downwind shore (Bruun, 1954). The equilibrium shape for a shoreline of finite length composed of sand and other easily moved material is a cycloid (Bruun 1954, p. 6) where:

$x = $ const. $((\sin 2\alpha_0)/4) + (\alpha_0/2))$; $y = $ const. $((\sin^2 \alpha_0)/2))$; $0 \leq \alpha_0 \leq \pi/2$, const. $=$ twice diameter of circle describing cycloid; α_0 in radians.

Inspection of actual bays and lagoons where waves have dominant directions shows a minor departure from the predicted equilibrium form for angles of $\alpha_0 = 50°$ with a sharper shoreline curvature in this range (Bruun 1954, p. 9). This suggests that maximum littoral drift in nature occurs at an angle of $\alpha_0 = 50°$ where the fraction of maximum littoral drift $= 0.57$ ($\sin \alpha_0 + \sin 2\alpha_0$) (Figure 15).

This minor departure of the equilibrium form of a downwind shore from the cycloid can most easily be understood if one considers the wave orthogonals not as unidirectional but rather as an array from one sector. Figure 16 shows the great shift of the nodal point with a shift of wind direction of $45°$ while the zone of maximum littoral drift shows practically no change for this variation of wind direction. The zone which includes the nodal point is essentially one of minimum sediment transport and therefore one of littoral deposition or very little erosion. On the other hand, the waves and currents in the zone of maximum transport pass littoral drift through this zone faster than it is supplied from up-shore. This portion of the shore therefore has a deficit of sand and gravel and the banks of the lake

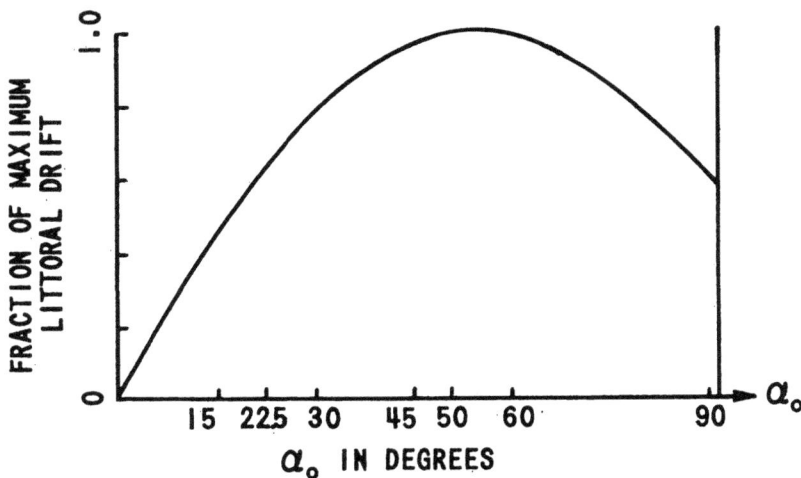

FIGURE 15.   Littoral drift as a function of the deep-water wave orthogonal angle α. (Bruun, 1954).

are exposed to direct wave erosion. Because of the very small shift of the zone of maximum erosion with moderate shifting of the wind direction, there is a concentration of erosive action in small zones at $\alpha_0 = 50°$ and the wave equilibrium form changes from a cycloid to a more rectangular shape. This effect is evident in the oriented lakes of the Arctic Coastal Plain which, while they approximate outlines of wave equilibrium, nevertheless do show the partial rectangular modification resulting from the effect of a wave array. In lakes with abundant sand, there may be actual deposition in the nodal areas, while in lakes enlarging in soil with abundant ground ice, there may be erosion along the entire shoreline; nevertheless most active erosion occurs at values of $\alpha_0 = 50°$.

Any unidirectional wind will generate waves which propagate downwind and naturally will leave the upwind shore unaffected. In order to modify the upwind shore, it is necessary for the wind to shift to approximately the opposite direction; however, here again an array of wind directions will suffice to bring about the development of an oriented shoreline. The equilibrium form of a lake in unconsolidated and easily transported materials, where the winds blow with equal force for equal periods of time from all directions, would be circular. Any prevailing direction of the wind then results in lake orientation with minimum erosion in the zone occupied by the nodal point, maximum erosion at an angle of $\alpha_0 = 50°$ between the dominant wave orthogonals and the normal to the shoreline, and intermediate erosion at values for α_0 from 60° to 90° and 22° to 45° (Figure 16).

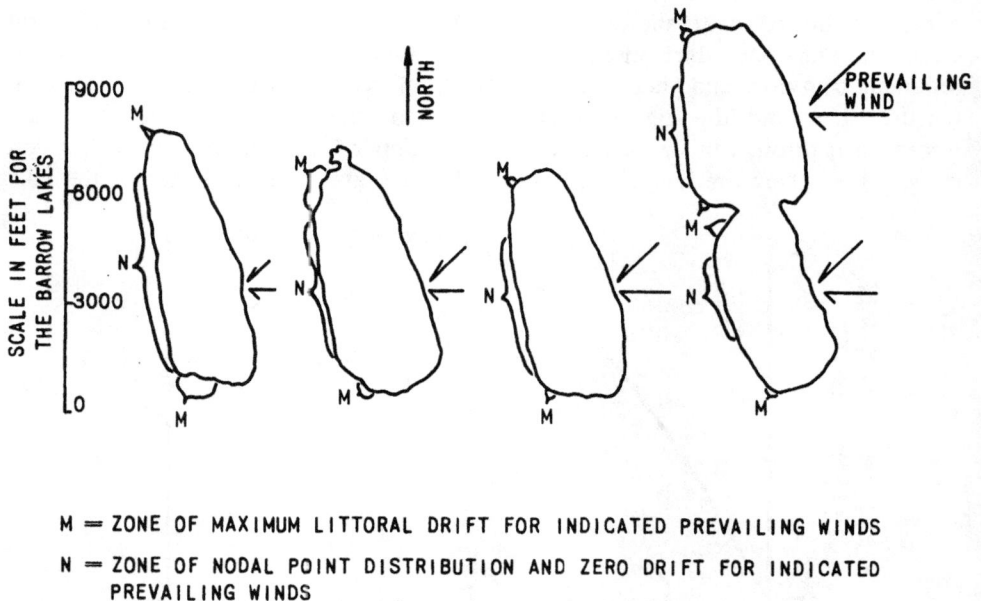

M = ZONE OF MAXIMUM LITTORAL DRIFT FOR INDICATED PREVAILING WINDS

N = ZONE OF NODAL POINT DISTRIBUTION AND ZERO DRIFT FOR INDICATED PREVAILING WINDS

FIGURE 16.   Oriented lakes near Barrow, Alaska showing the small variation of zone of maximum drift and large shift of zone of zero sediment drift as a result of a 45-degree shift in wind direction. The two dominant wind directions in the Barrow area are opposed so the prevailing wind is shown by the arrows and the second strongest wind corresponds to a reversal of the same arrow directions.

This clearly gives rise to an orientation of the lake with the major axis normal to the dominant wave direction and therefore the dominant prevailing wind. Any resistant features along the shoreline such as logs, rocks, sod, or sediments harder than usual tend to divide the shoreline into a series of finite segments each of which undergoes orientation producing an embayed shoreline.

Inspection of the wind directions at Barrow, Alaska (Figure 2) shows an easterly maximum with the dominant wind coming from the east-northeast and a second westerly maximum with the west-southwest component slightly dominant over the west-northwest component. The wind distribution at Barrow therefore would generate two opposed wave arrays with orthogonals normal to the major axes of the oriented lakes. These wave arrays would, according to littoral transport concepts, move sediment along the lake shores to establish an equilibrium form similar to that observed. In addition, thermal erosion, antecedent features, and varying resistance to erosion of the shoreline can give rise to the irregular scalloped shorelines sometimes observed. It is concluded therefore, that there is no justification for calling on a shift of wind during Pleistocene time to produce the oriented lakes because the lakes at Barrow and elsewhere on the Arctic Coastal Plain show a close agreement with the predicted wave equilibrium form for the present wind pattern.

The wave and littoral drift concepts utilized here are equally applicable to other water bodies of similar size and it seems probable that the shapes of many oriented lakes, lagoons, and bays found on other coastal plains may be explained either as wave equilibrium forms or as intermediate between wave equilibrium and antecedent forms.

It should be pointed out that exactly opposed wind directions are not required. Actually the more frequent occurrence may be two generally opposed wind directions but with an obtuse angle between the two wind directions. This could give rise to "clam-shell lakes" such as those observed in the Vidauri district, Texas (Johnson, 1942, p. 322), the "triangular lakes" of the Mackenzie River delta (Mackay, 1956a, 1956b), and similar heart-shaped lakes from east of the Andean foothills in Bolivia.

We have seen, then, that the degree of development of oriented segments of shoreline in a lake in unconsolidated sediments reflects the meteorological conditions in the area during the period of open water. Black and Barksdale (1949, p. 115) observe that the present tendency in the Alaskan lakes is towards destruction of the northerly orientation of the lakes and towards an easterly orientation in those lakes with sufficient fetch for wave action to be effective. This observation indicates that either a southerly or a northerly component of summer winds is increasing in importance. As it is usually the northern lake shore (in those similar cases observed by the author) which shows a more pronounced evidence of reorientation, we conclude that summer southerly winds are increasing in importance. This agrees in a general way with evidence for regional warming of the Arctic and suggests increased incursions of frontal storms from the south at Barrow.

The larger and deeper lakes of the upper coastal plain are more complicated in their internal structure than the Barrow lakes. Some show shelves, large sand waves, and other internal features suggesting a more complicated circulation pattern than discussed in this paper.

The treatment of the subject in this paper pertains primarily to currents in the surf zone and portion of the lake basin where waves affect the bottom sediments. The complexities of single cell versus double cell circulation out in the main body of the lake are not discussed and not necessarily pertinent to the question of lake orientation. The main lake currents are usually much slower than those in the surf zone and are very poorly understood. For this reason, Livingston's (1954) treatment is incomplete, although it does yield the correct conclusion concerning the directions of lake orientation.

ACKNOWLEDGMENTS

It is a pleasure to acknowledge the helpful interest and advice of Professor D. L. Inman and Professor R. S. Arthur of the Scripps Institution of Oceanography in the writing of this paper, and of Mrs. J. R. Rex for the meteorological statistical work and assistance in the field. This study was aided by a contract between the United States Office of Naval Research and the Arctic Institute of North America. Reproduction in whole or in part is permitted for any purpose of the United States government. Aid was also received from the President's Travel Fund, Stanford University. The assistance in the field of the directors, Professor Ira L. Wiggins and Mr. T. Mathews in 1952-3 and 1954 respectively, and staff of the Arctic Research Laboratory, and Dr. W. E. Lyons and his group at the United States Navy Electronics Laboratory is gratefully acknowledged. The constructive criticism of Dr. Z. V. Jizba, and Mr. P. Bruggeman has been most helpful in the preparation of the manuscript.

REFERENCES

ARRHENIUS, G.   1950.   Late Cenozoic climatic changes as recorded by the equatorial current system; Tellus, vol. 2, pp. 83-8, 236.
ARTHUR, R. S., MUNK, W. H., and ISSACS, J. D.   1952.   The direct construction of wave rays; Trans. Amer. Geophys. Un., vol. 33, pp. 855-65.
BEACH EROSION BOARD.   1954.   Shore protection, planning, and design; Dept. Army, Corps Engineers, B.E.B. Rept. No. 4, 242 p. plus appendices and revisions of 1957 (loose-leaf).
BLACK, R. F., and BARKSDALE, W. L.   1949.   Oriented lakes of northern Alaska; J. Geol., vol. 57, pp. 105-18.
BRETSCHNEIDER, C. L.   1952.   The generation and decay of wind waves in deep water; Trans. Amer. Geophys. Un., vol. 33, pp. 381-9.
BRUUN, P.   1954.   Forms of equilibrium of coasts with a littoral drift in Coast stability; Atelier Elektra, Copenhagen. (Also published as Univ. Calif. Inst. Eng. Res., Ser. 3, Issue 347, (1953), Berkeley.)
CHIEN, N.   1953.   Some ripple tank studies of wave refraction; Inst. Eng. Res., Waves Res. Lab. Tech. Rept., ser. 3, Issue 358, Univ. Calif.
EAGLESON, P. S.   1956.   Properties of shoaling waves by theory and experiment; Trans. Amer. Geophys. Un., vol. 37, pp. 565-72.
ECKART, C.   1952.   Propagation of gravity waves from deep to shallow water, in Gravity Waves, U.S. Nat. Bur. Standards, Circular 521, pp. 165-73.
HUTCHINSON, G. E.   1957.   A treatise on limnology; vol. 1; J. Wiley.
INMAN, D. L.   1957.   Wave-generated ripples in nearshore sands; Dept. Army, Corps Engineers, Beach Erosion Board Tech. Memo. No. 100, 42 p. plus append.
INMAN, D. L., and QUINN, W. H.   1952.   Currents in the surf zone; Submarine geology report 23, Scripps Inst. Oceanography, Univ. Calif. at La Jolla, Ref. 52-10.
JOHNSON, D. W.   1942.   Origin of the Carolina bays; Columbia Univ. Press.

JOHNSON, J. W.   1950.   Relationship between wind and waves, Abbotts Lagoon, California; Trans. Amer. Geophys. Un., vol. 31, pp. 386-92.

JOHNSON, J. W., and RICE, E. K.   1952.   A laboratory investigation of wind-generated waves; Trans. Amer. Geophys. Un., vol. 33, pp. 845-54.

LIVINGSTON, D. A.   1954.   On the orientation of lake basins; Amer. J. Sci., vol. 252, pp. 547-54.

MACKAY, J. R.   1956a.   Notes on oriented lakes of the Liverpool Bay area, Northwest Territories; Rev. Canad. de Géog., vol. 10, pp. 169-73.

———   1956b.   Oriented lakes of the Liverpool Bay area, N.W.T.; Ann. Assoc. Amer. Geograph., vol. 46, p. 261.

———   1957.   Les lacs orientes de la région de la Baie de Liverpool: Discussion; Rev. Canad. de Géog., vol. 11, pp. 175-8.

———   1958.   Anderson River map-area, N.W.T.; Geographical Branch, Mem. 5, pp. 74-81.

MUNK, W. H., and TRAYLOR, M. A.   1947.   Refraction of ocean waves: a process linking underwater topography to beach erosion; J. Geol., vol. 55, pp. 1-26.

NYBERG, A.   1956.   On the variation of the general circulation of the atmosphere during past ages; Swedish Deep Sea Expedition 1947-8, Repts., vol. 5, part 5:2, pp. 237-44.

ODUM, H. T.   1952.   The Carolina bays and a Pleistocene weather map; Amer. J. Sci., vol. 250, pp. 263-70.

PETTERSSEN, S., JACOBS, W. C., and HAYNES, B. C.   1956.   Meteorology of the arctic; OPNAV-PO3-3, March, U.S. Navy Technical Assistant to Chief of Naval Operations for Polar Projects (Op-O3A3), Washington, D.C.

PIERSON, JR., W., NEUMANN, G., and JAMES, R. W.   1955.   Practical methods for observing and forecasting ocean waves by means of wave spectra and statistics; U.S. Hydrographic Office Publication no. 603.

SAVILLE, JR., T.   1950.   Model study of sand transport along an infinitely long, straight beach; Trans. Amer. Geophys. Un., vol. 31, pp. 555-65.

SVERDRUP, H. U., JOHNSON, M. W., and FLEMING, R. H.   1942.   The oceans; Prentice-Hall.

WEATHER BUREAU.   1953.   Local climatological data 1952, Barrow, Alaska; U.S. Dept. Comm.

WIEGEL, R. L., and FUCHS, R. A.   1955.   Wave transformation in shoaling water; Trans. Amer. Geophys. Union, vol. 36, pp. 975-84.

ZENKOVITCH, V. P.   1959.   On the genesis of cuspate spits along lagoon shores; J. Geol., vol. 67, pp. 269-77.

Dating Rock Surfaces by Lichen Growth and its Application to Glaciology and Physiography (Lichenometry)

ROLAND E. BESCHEL

ABSTRACT

Most of the arctic and alpine crust lichens, especially the genera *Rhizocarpon* and *Lecidea* grow very slowly. This can be concluded indirectly from maximum diameters on rock surfaces of known age or repeated measurements. The over all constant increment after an initially sigmoidal growth allows dating of rock surfaces exposed up to 1000–4500 years B.P., depending on the climatic conditions. Lichenometry permits relative dating of events which led to the exposure of bare rock surfaces within the age limit of the lichens in similar macroclimates. This can be converted to an absolute scale if one event is dated by other means, e.g. historical information, or if the growth rate is measured directly. From lichen measurements obtained in West Greenland, the Alps, and the Ruwenzori Mountains, the synchronism of glacier behaviour within the advance period of modern times (400–40 B.P.) appears very high. Early hypothermal moraines and boulder streams can be separated clearly from early modern ones, even if other morphological criteria fail.

Lichen growth rates are inversely proportional to the hygrocontinentality of the area. This permits calculation of this or similar combined climatic factors through lichenometry, or the prediction of lichen growth rates from the known climate.

Lichenometry is especially useful where dendrochronology is impossible.

LICHEN GROWTH

General statements about Lichen growth are based on studies in the Alps (Beschel, 1950, 1957, 1958a, b). A visit to the Sukkertoppen and Holsteinsborg districts of West Greenland in the summer of 1958, subvented by the Arctic Institute of North America, has permitted comparison of conditions in the Alps with those in the Arctic.

Whenever a new rock surface becomes exposed to the atmosphere it is sprayed with the wind- and water-transported reproductive bodies (diaspores) of plants. They will lodge in capillary cracks or other small pockets and be trapped. Only a very few, glass-like surfaces offer no such places. Small diaspores are more easily retained by a surface, and the smaller bodies have also a greater chance of long distance dispersal (Geiger 1957, p. 45).

Whether a diaspore will be able to germinate and to grow into a new plant depends on many circumstances. Granting a certain stability of the substratum there must also be sufficient amounts of water, warmth, light, and nutrients. Any of these factors may become limiting for the growth of even the best-adapted plant. Liquid water vanishes quickly from rock surfaces after the cessation of rain, fog, dew, or melting of snow and only organisms resisting repeated and prolonged desiccation may grow. Mineral nutrients may become available either through weathering or through active chelation by the plants, or may be blown upon or washed over the surfaces. Nitrogen from the atmosphere may be fixed by many micro-organisms. The time of active life during which favourable temperature, light, and liquid water are all present is so short during each year that the scarce mineral nutrients may be sufficient for such interrupted growth. Admittedly, the vicinity of bird perches and other spots where animal excreta and organic refuses accumulate develop, especially in polar regions, a surprisingly luxuriant vegetation. This is especially true for plants rooted in the soil, where entirely different water relations exist. Rock surfaces are covered with plants restricted to such localities and adapted to excessive fertilization. Plants common outside such environments barely enter them. Even the plants adapted to intensive manuring become scarce under adverse climatic conditions. Generally the rule applies that edaphic conditions have a greater importance in selecting the organisms that may develop on a rock surface, but the climate is more likely to determine their growth.

When the first plants germinate and grow, other diaspores arrive. In the resulting mixture only the largest plants will be of an age equal to that of the exposed substratum. As different plant species grow with different speed, some will be visible while others are still microscopic, even if they all arrived at the same time. Contacts between plants and the effects of chemical antagonisms increase. Parasites and plants favoured by the presence of others may invade and competition accelerates. Within an area of equal exposure time, the micro-environment varies, permitting faster growth on some surfaces, delaying it on others. As any smaller plant may be either younger or stunted by its micro-environment, a consideration of the largest plants of one species as age indicators will simultaneously select the individua growing under optimal local conditions. Only the most common and competitive species will show a close correlation between their maximum sizes and the age of the exposure of the substratum, if it lies within the age limit of the plant. The individua or their aggregates must be clearly bordered as any confluence with similar units would invalidate an age determination based on size. Many lichens — individual crusts, foliaceous thalli, and disk-like aggregates of fruticose types — as well as cushions of mosses and flowering plants may be used for dating, if the entities have a more or less circular shape. Although many lichens show either a radial arrangement of their lobes or areoles, or are surrounded by a hypothallus of different colour and shape that remains when identical species meet, any oblong patch should only be considered in its shorter diameter.

If climatic changes are disregarded for the moment, it will be seen that aggregates of individua increase their diameters with equal speed. The organic matter produced by their members is not shared between them if the asexually-produced offspring is soon separated from the parent. In an individuum with free transport of synthesized matter, however, the older parts may contribute to the marginal growth. This will become accelerated until aging reverses the trend. Crustaceous and many foliaceous lichens stand between these extremes. They do not grow with constant velocity during their whole life. The diameter of a thallus increases at first very slowly and considerable time elapses before the lichen becomes visible.

Then growth gathers speed and many thalli pass through a "great period" until the increase in a given time period drops to a constant value that is maintained for a long time. When environmentally possible, maximum diameters are approached, and the rate of growth slows down. Frey (1959; p. 269) photographed lichens that did not change their appearance in twenty years. It is not known how long such a thallus may persist in this stationary phase, but weathering and competition will finally remove it. The great period may be absent in thalli growing close to the extremes of their climatic amplitude. Fluctuations of the climate vary the growth velocity, but the slower a lichen grows, the smoother will be the size-time curve. If one considers growth in constant time intervals expressed as percentages of the previous size, the fastest relative increase appears in the first years of life. The relative increase-time graph falls in a hyperbolic curve. Thompson (1942; p. 141) doubted the validity of this consideration, which was used first by Minot (1890) but it shows, at least, that a lichen does not grow with constant acceleration in its youth. Perhaps a reduction of this acceleration starts at the time of germination. It is not yet possible to express with a formula the actual dimensions of the great period for a given situation, or to predict the changes of this part of the growth curve in different environments. The constant speed after the great period can be correlated better with the climate. As the maximum diameters of a given lichen species on substrata with exposure times exceeding the species' life span vary in different climates, the climates may be indirectly effective through the speed of weathering they cause. An analysis of this complex of factors is impossible at present.

The growth curve of a disk-shaped lichen may be interpreted in the following way. As long as the organic substances produced by a thallus are shared by all its parts, a constant production rate per area will lead to an exponential increase of thallus area and diameter. Yet the surface of exchange with the surroundings, allowing diffusion of waste products outwards and diffusion of mineral nutrients inwards, will decrease per unit area of the thallus and a deceleration dependent on size will act from the onset of growth. As soon as a radius is approached past which transport becomes ineffective, the central parts will cease contributing to the marginal growth. Only an outer ring of constant thickness will be responsible for further increase in diameter, and such increase will become linear. The central areas may channel their remaining productivity into reproduction and maintenance. They may remain on the substratum, as is the case with many crustaceous lichens, or may die and break away. As shown by many foliaceous lichens, the remaining rings do not depend on the centre and continue to grow at a uniform speed.

A thallus that is fixed to the substratum only at the centre, as rock tripes (Umbilicaria) are, presents a special case. Persistence of the oldest parts is necessary for individual existence, and a higher integration of all parts seems to be indicated by the lack of a period of constant growth rate. The growth curve of Umbilicariae is sigmoidal but asymmetrical, for inflection occurs closer to the onset of growth. The point of inflection shows the highest speed of growth. If the growth velocity is plotted against time, which is the first differentiation of the growth curve, it corresponds to a maximum after an eighth to a sixth of the life span. The bell-shaped velocity curve is very skewed. The growth pattern cannot be expressed by the Verhulst-Pearl curve, as explained by Buchanan (1949, p. 108), for example.

Depending on the environment, a rock surface may be completely covered with lichen thalli after 100 to 300 years. The individua are then still far below their maximal sizes. A comparison of dated moraines of different ages showed that the closing of the lichen cover on boulders did not change the linear increase in diameter of the largest thalli. These plaques must have replaced other individua of both different and similar species with constant speed, as if the smaller competitors had not been present. This cannot be explained satisfactorily. Observations of identical surfaces over long time intervals may substantiate and clarify the process. An expansion of the studies of Bitter (1898) is sorely needed.

The time-range for which dating with lichens (lichenometry) can be applied depends on the life-span of the growing plants. For periods of many decades, and even of a few centuries members of the genera *Alectoria, Parmelia, Stereocaulon, Umbilicaria, Rinodina, Lecanora*, as well as cushion plants, can be used for the dating of the exposure time. Woody plants, especially willows and dwarf birches in treeless regions, within the same time periods permit a direct count of their age through the annual growth rings. Several species of the crustaceous lichen genera *Rhizocarpon* and *Lecidea,* however, continue to grow for much longer time periods. Their largest thalli seem to be about 600 to 1300 years old in the Alps and 1000 to 4500 years in West Greenland. Perhaps these values are exceeded in high arctic or antarctic regions.

METHODS OF LICHENOMETRY

Direct measurement of the same plants at sufficiently long time intervals must remain the basis for any growth analysis. The longevity of many lichens excludes observation by a single worker. At present Frey (1959) photographed and measured lichens over the longest time interval, thirty-seven years. The interpretation of the results is very difficult, for individual plants of the same species grow with very different speeds. As outlined before, this may be due to microenvironmental differences, accentuated to a varying degree during shorter climatic fluctuations. The growth velocities will further vary inherently in the great period, which itself is altered in amplitude and duration by differences of the environment in space and time. No averaging can simplify the results.

Indirect measurement through the comparison of circular thalli with the largest diameters on substrata of known ages of exposure gives a simplified picture that is very useful. Only the oldest, and at the same time the optimally-developed individua, are considered. Nevertheless, a number of uncertainties arise. The plants being compared must belong to the same species or at least to closely related ones with nearly identical growth curves. As the determination of most crustaceous lichens can only be made using the microscope, any rapid field method introduces errors. As the first thalli grow in a very scattered manner on a new rock surface, often with distributions of one plant to a few square metres, and as only few of these will have found close to optimal microenvironmental conditions, the rock surface to be studied should at least be of the order of 100 square metres. By measuring as many common species as possible the accuracy of the method may be increased. The size relations of different species remain rather constant as long as the growth velocities stay constant. If, on a given surface area, a number of species show this relation, the possible size of a further species that may be absent or by chance smaller can be inferred. An indirect measurement treats the old and large thalli as if they had had the same environment, especially the same climate in their youth many centuries ago, as the small thalli on a much more recently exposed substratum experienced in the most recent past. Indirect measurement of the growth velocity is only possible where it can be based on the known age of exposure of at least one and preferably more substrata. If no such information is available the time differences between the exposure of several surfaces can, however, be expressed relatively. But it must be emphasized again that no standard growth velocity can be expected in a large region. The effect of the climate is of paramount importance and the difference in the constant speed of diameter increase between a humid and an arid region may be a factor of twenty (Figure 1).

When the maximum diameters of several lichen species have been measured on a few dated surfaces, growth curves can be plotted and the constant growth velocity after the great period can be derived from the graph. The age of other surfaces in the vicinity can be found by the insertion of their largest lichen diameters onto the

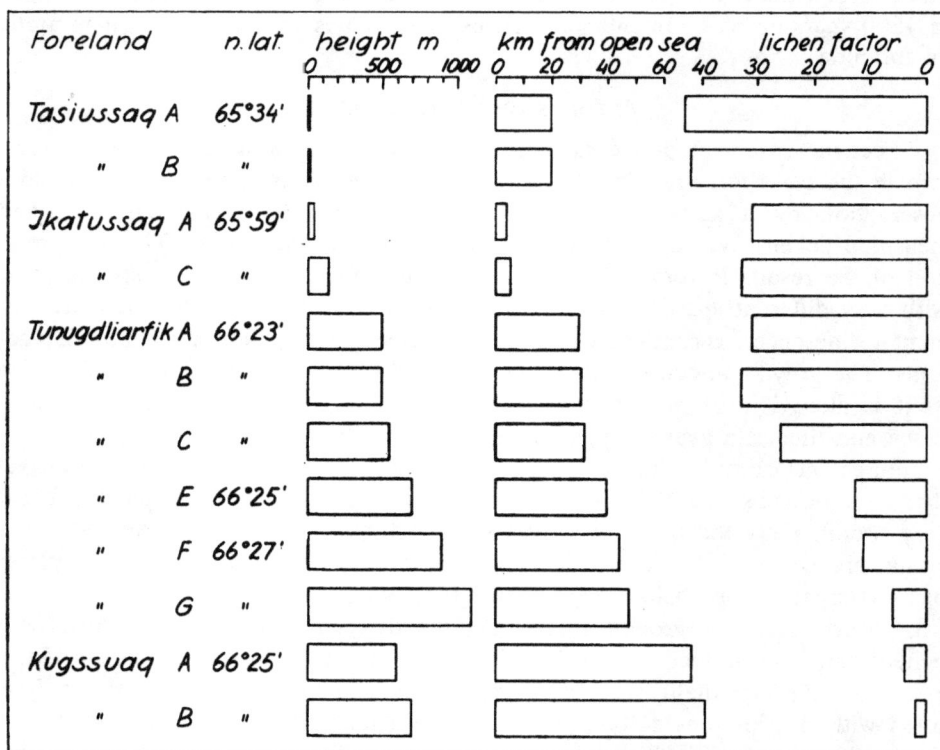

FIGURE 1.   The lichen factor (maximum diameter of century-old thalli of *Rhizocarpon tinei* in mm) in relation to continentality in glacier forelands of West Greenland.

graph and reading the corresponding time. When simultaneous events exposed surfaces farther away, the differing growth curves permit a comparison of the climate. By repeated interpolation and extrapolation along one or both of these lines many situations arise where one conclusion can be checked against another, thus increasing the accuracy of the method.

APPLICATIONS IN ALPINE AND POLAR REGIONS

A rock surface may become exposed to the atmosphere by a large variety of agencies. The rock may form from lava. It may emerge from the sea in a drained lake basin or in the bed of a river that has changed its course. A retreating glacier may expose it, or a perennial snow-patch may decrease in size during a climatic amelioration. Frost shattering, rock falls, and avalanches create new surfaces on the cliffs of the bedrock as well as in the resulting talus. Human interference constantly increases. In addition to these rather drastic creations of new surfaces, wind, water, and ice, as well as chemical and other mechanical erosion, modify already existing surfaces. Solifluction and slumping of soil over buried and melting ice turns stones and exposes formerly buried surfaces.

Sea-Level Changes

On the sea shore conditions for lichen growth vary to an extreme degree. The same levels on emerging bedrock will experience during the ice-free season rather different degrees of wetting by waves depending on the prevailing wind directions, the slope of the shore, and the outline of the cost. These conditions may change during an uplift. The degree of salt spray will vary within even greater height differences. During the winter an ice-foot forms and prevents lichen growth. As Paterson (1951, p. 6) has shown, snowbanks accumulate in the niche of ice-foot and bedrock which prevent lichen growth at still higher levels either owing to the short vegetation periods on surfaces covered so long by snow, or owing to the fast weathering caused by these conditions. The flatter the shore, the smaller is this influence. Lichen-free zones on bedrock above the present high-water level cannot be taken as an indication of uplift. In sheltered bays with bedrock sloping at a low angle, or on beaches with numerous ice-rafted boulders, favourable conditions may coincide. The uplift could be fast enough to be reflected by smaller lichen diameters at lower levels. Wherever maximum-sized lichens are found close to high-water mark it can at least be concluded that sea-level was not higher than at present within the lifetime of these plants.

Lake Levels

There are numerous cases in mountain regions of glaciers damming lakes which emptied when the ice retreated. Whenever the inundation has lasted longer than a year the original lichen cover has been killed and sharp trimlines result at the former water level. Large surfaces are exposed within a short time and the conditions are ideal to apply lichenometry. If the lake existed for only a few years, the lower thallus layers of crustaceous lichens may remain for at least a century on the rocks. Such lichen corpses cannot be determined any more taxonomically but are very useful indicators of drastic events that lasted only a few years.

Outwash Fans and River Beds

In active deltas and valleys with braided streams, for instance in front of glaciers far below the snowline, boulders will be moved or buried under fresh material at a fast rate. Their surfaces are mostly bare and the actual age of deposition cannot be deduced from the plant cover. Some channels will have been dry for a longer time than others and will have correspondingly larger lichens. Whenever the disturbance ceases in one sector colonization has already progressed to unpredictable amounts in the sector's various parts. Lichen diameters in an outwash zone between a terminal moraine and the glacier front do not indicate the time which elapsed since the ice retreated from these spots but give a much shorter one.

Within the zone of spraying or occasional wave-splashing on the borders of little brooks lichens may grow much faster than those without this irrigation, especially in arid regions. Such an optimal water supply will make the growth primarily temperature-dependent. A further analysis of the complex influence of climate can then be attempted if the age of the surface is known.

Glacier Forelands

Kinzl (1929) has chosen the term Gletschervorfeld or simply Vorfeld (plural Vorfelder) for the area that was ice-covered during the later centuries and has since been exposed by the receding glaciers. This may be rendered best in English as "foreland," as suggested by Miss A. Herling. The term outwash, connected with fan, delta, plain, or apron, refers obviously to a slightly inclined area influenced by melt-water streams. Clearly it extends beyond the terminal moraines. It should not be applied to moraines or bedrock which were recently ice-covered. Outwash is not precise in its restricted application either for wind action in these areas is of high importance and the areas could be called outblow also. It seems, moreover, to be a neologism for the Icelandic sandur (plural sandr), a term not restricted to a certain particle size of the glaciofluvial deposits. Bergström (1955) speaks of lichen zones surrounding a glacier snout; they may in some cases but not if the ice advanced below the tree-line. Weidick (1959) uses the term melting zone synonymously with foreland. It may better be applied with a wider meaning that includes melting zones of perennial snow accumulations which have decreased in the recent past.

When ice has moved over rock surfaces they are scoured clean of lichens. Boulders, which fell on the glacier in its accumulation zone, will become embedded in the moving ice and will also lose their lichen cover. Boulders deposited on the ice surface in the ablation zone will easily fall into crevasses but, in exceptional circumstances, may remain on the surface until they are deposited at the snout. It can hardly be expected, however, that, after a ride on a glacier, an originally lichen-covered rock will be redeposited with an orientation permitting the continued growth of the plants. Examining some one hundred glaciers the writer could never find any lichens on boulders or bedrock in front of receding ice. The more surprising is the finding of a boulder, with undamaged crustaceous and even foliaceous lichens, in a shaft "to the bottom of the glacier 30 m from the front of the ice cliff and covered by 42.5 m of ice" in the Red Rock area near Thule (Wolfe ms.;

Goldthwait in litt.). It is assumed that the plants were covered by static ice in a pocket that protected them. Their C^{14}-dated age was 200 years (Blake in litt.). As far as known to the writer, whether they were still alive was not investigated. It remains doubtful if such plants could survive their emergence, had the ice freed them during a retreat. In the Arctic more than in the Alps, melt-water streams accompany the margins of a glacier snout. Lateral moraines may be partially washed away or the vegetation may be damaged outside of them and outwash deposited there. The limits of such damage are always ill-defined.

The most surprising difference from the Alps, however, is a sharp trimline outside many moraines if their flanks face east or north. These lines separate zones of strikingly different lichen cover and cut straight across both bedrock and boulders. Boulders farther away have only their basal parts bare, and have large lichen thalli at a higher level, while only the tips of the largest boulders closer to the moraines are covered with large lichens. No moving ice could cause such a selective killing of the plant cover. In more arid regions the phenomenon is especially pronounced. The explanation can be found at higher altitudes. There steep outer flanks of moraines bear extensive snowbanks if they face into a direction of more shadow. In extreme cases such snow accumulations outside the foreland have even developed into thin glacierets as a "Flankenvereisung." The trimlines caused by marginal snowbanks are sometimes higher than the moraines. In these situations, snow accumulated in the lee and shadow of the ice tongue itself and the lateral moraine thawed out only after the glacier thinned. Thus pale bands seen to accompany the margin of many ice bodies on aerial photographs of the Arctic should not be considered as forelands, but as melting zones of former marginal snowbanks. The maximum lichen diameters within these trimlines may be used nevertheless as age indicators of the last glacier advance; the dimension of the glacier, however, is better derived from the frontal moraines. Their degree of colonization corresponds very well with that of substrates in the former snowbanks. The largest of the lichens may be even smaller on the moraines. Whenever massive moraines are deposited they contain some ice under the insulating mantle of till. During delayed thawing, especially in permafrost regions, the material slumps, and boulders are turned. In such cases the lichen cover is rather irregular. Boulders, already colonized, will lie beside bare ones, and buried or crypt surfaces may bear lichens or lichen corpses which never could have grown in their present orientations. Sections of moraine arcs with lesser height, especially where deposited on bedrock, give a much better chance of accurate lichenometrical dating. This is one of the advantages of the forelands of smaller glaciers. Large bodies of ice take more time to respond to climatic fluctuations and to come into an equilibrium with new conditions. Small glaciers oscillate their fronts faster and accumulate less moraine material in one advance phase, but form more moraines during a longer time. Depending on the subglacial relief, some outlet glaciers of large ice-caps may also respond very fast. When the ice has receded from terminal moraines, over bedrock or ground moraine, these areas can be studied with lichenometry and useful information about the speed of recession is obtained. In many cases outwash eliminates such evidence and may even destroy or bury terminal moraines. This is less likely in forelands of smaller glaciers.

Snow Patches

In addition to snowbanks on moraines, many perennial snow accumulations have decreased substantially in rather recent time. Nearly bare rock surfaces in the vicinity of present snow-patches have often been interpreted as an effect of present microclimatic conditions. This cannot be doubted for areas very close to the present summer snow limit, but the outer zones of these areas, with very slight vegetation cover, present an identical picture to the younger parts of glacier forelands. These surfaces bear young thalli, including even chionophobic lichens. Whereas lichen corpses are common in such localities in the Alps, the area visited in West Greenland has very few. It can be concluded that in both regions snow-patches have shrunk considerably in the last one hundred years, but excessive snow cover lasted much longer before that time in West Greenland than in the Alps. Any study of vegetation zonation and weathering around present snow-patches must consider that the plant cover is far from in equilibrium with the present climate, and that rock surfaces especially reflect conditions that prevailed up to a century ago.

Rock Falls and Talus

Where single rocks have broken out the resulting new surfaces on the bedrock are mostly too small to allow lichenometrical dating of the event. Mechanical damage increases in the gullies into which the rocks are channelled. There, thalli are restricted to protected corners. Where a glacier ends on steep bedrock colonization below is delayed owing to the same effect. The extent of lichen cover on the resulting talus cones is a good indication of the intensity of rock fall. A uniform immature colonization permits dating of the time of formation. If the steepness of a slope prevented the formation of terminal moraines, an advance maximum of a glacier can also be dated from the talus that accumulated during that time. With proper care, more than one advance with similar extent may be discerned in the talus if lichen diameters fall within distinct size classes on a high percentage of the boulders. Extensive rock falls and rock avalanches, when recorded historically, help to find the growth velocities of lichens on bedrock exposed and talus formed at those times.

Solifluction

The extent of vegetation cover on patterned ground has long been used to estimate its speed of formation. A higher accuracy can be obtained when lichen diameters are measured. As the surface areas within one stone polygon or stone strip are mostly too small, homologous zones in a number of similar formations have to be considered together. The extent and speed of the turning of stones may be found from the size of lichens or lichen corpses on surfaces that would not have been colonized in their present orientations.

Motion within rock glaciers and boulder streams presents an intriguing problem for lichenometry. In the multitude of moraine-like ridges thrust up within the moving mass of boulders, microclimatic conditions — especially the duration of snow cover — vary to a high degree, even if the surface pattern did not change. With the continued alteration of surface forms, however, the microenvironment of a rock lichen varies enormously in space and time. Maximum-sized lichens are extremely rare unless motion within the boulder stream decreases substantially. Even then the whole mass may continue to glide. There are boulder streams with a nearly continuous lichen cover of maximum-sized thalli on their surfaces, while the lower edges consist of bare boulders mixed with fine-grained material. If a boulder stream borders on surfaces with very old lichens, the whole mass must have moved at least the distance within which formerly present flanking snowbanks had killed the lichens and prevented recolonization until the end of the last larger glacier advance. Within old boulder streams younger phases are common in higher parts that push tongue-shaped over the older material. Whenever distinct ridges traverse a boulder stream the marginal parts have a much better lichen cover with greater individual thallus diameters than the areas near the centre of these arcs. These size differences are inversely proportional to the rate at which boulders are turned over. If a few phases of movement have each been initiated in a pronounced advance period of glaciers, only the lateral parts of the ridges formed at these times will have lichen diameters corresponding vaguely to the age of the advance period, whereas their central parts will have much younger surfaces.

All degrees of combination may occur among the various agents discussed. Nivation may form a cirque and talus cones under its steep hind walls. During a period of more extensive perennial snow cover a small glacier originates in the cirque and heaps much of the loose material into moraines. These may or may not contain ice when they start moving as rock glaciers or boulder streams if the slope permits. This sequence would occur during one climatic fluctuation from warmer to colder and back to the original conditions. If such fluctuations are repeated the resulting processes overlap to a different extent. For instance, in the Stubai Alps moraines of old advances have been found which, while they were nearly covered by perennial snow, have barely been started to move as boulder streams by a younger advance. A measurement of lichen diameters allows not only the analysis of such a sequence but also permits, in many cases, dating of the events.

Weathering

The intensity of denudation processes depends on the form of the landscape, the climate, and the texture and composition of the rock. As Paterson (1951, p. 21) has shown convincingly, weathering is especially rapid where rocks are soaked with water and the temperature often oscillates through the freezing point. Such conditions will be found around melting snowbanks and elsewhere. Frost shattering and scaling due to nivation seem to be more intensive in sub-Arctic and low Arctic regions than in the high Arctic. Heat shattering was not observed by the writer in West Greenland, although day and night temperatures may also differ greatly in the Arctic. For example, the temperature of south-facing gneiss near the mouth of the Söndre Strömfjord changed on August 6, 1958, from over 50° C (the end of the instrument scale) to 1.4° C in ten hours. Yet on north-facing slopes or in their shadow, weathering has removed very thoroughly minor marks of ancient glaciation such as striae and chattermarks, while they may be still visible on south-facing rock.

Despite the slow weathering on sunny surfaces lichens may be very scarce there, owing to the arid microclimate. This was especially noticeable on a gently southwards-sloping ancient moraine that consisted of boulders of very light gneiss, and was located in the eastern part of the Tunugdliarfik valley at 1000 msm at the Polar circle. Viewed from the north, the rock surfaces appeared black owing to the nearly continuous cover of *Alectoria pubescens, A. minuscula, Buellia atrata, Umbilicaria havaasii,* and the black hypothalli of several *Rhizocarpon* species. Viewed from the south, however, the landscape was strikingly white and no lichen was visible. The lichen cover was sharply limited on each boulder. Similar conditions are common at greater altitude in the continental areas of the Söndre Strömfjord, but are rather rare in the humid zone of the outer coast. These features have to be considered in studies of weathering. The sparse lichen cover on a south-facing slope and the south-facing stones of an old fox trap in the vicinity of a snowbank on South Ryders Ø in Melville Bugt, described by Paterson (1951, p. 9 ff.), also seems due to the effects of irradiation and not to the direct and indirect influences of prolonged snow cover, as Paterson thinks.

Not much is known about how individual lichen species contribute to weathering. The basic studies of Bachmann (1904, 1911, 1917) have not been expanded for a long time. Mechanical aspects were stressed by Fry (1927). Schatz *et al.* (1954, 1956) added a new aspect by pointing out the effect of lichens as chelators. Crustaceous lichens attack a variety of minerals and "excrete indigestible mineral grains," especially quartz, on their thallus surface, thus digging gradually deeper. When the surface has roughened, with quartz veins or large feldspar crystals protruding more than 2 cm, and is covered by maximum-sized lichen thalli, many centuries and possibly millennia have passed since the creation of the surface. In such places muriform and spheroidal nivation can be excluded as denudation processes. Other surfaces may acquire a rather continuous lichen cover but will ring hollow when struck. Recent muriform scaling shows in a few places. Large plates of 5 to 10 mm thickness can easily be wedged loose with a knife from their surroundings. These places are convenient for lichen collecting, for it takes little time, and adds less weight than usual to the rucksack. From the maximum diameters of slow-growing species one may conclude how quickly this weathering proceeds which leads to an accumulation of iron stains below the rock surface in the layer of loosened texture. Still other rocks become coated with an intensive iron stain, the *Gletscherlack,* which prevents lichen growth in most instances. *Gletscherlack* is less common in the visited area of West Greenland than in the Alps and is especially rare in dry regions. Generally it appears that lichen cover cannot protect a rock from weathering and only retards it to a very slow rate in the Arctic. It is very doubtful if rock lichens contribute anything towards soil formation. Whenever crustaceous lichens have covered a rock they resist the further invasion of other plants to a high degree. Some of them are the oldest beings alive in polar regions, if not in the whole earth. It may happen that shrubby lichens (*Stereocaulon, Alectoria*) and mosses (*Grimmia*) appear first on a rock surface and are in turn replaced by foliaceous lichens (*Umbilicaria, Parmelia*) until crustaceous species (*Rhizocarpon, Lecidea*) have achieved a final hold on inclined surfaces. The latter are very slow invaders and seem to have gained their fame as "pioneers of the xerosere" without justification. The premature generalizations of Clements (1916) may even have delayed an earlier consideration of their importance as age indicators in geomorphological and ecological studies.

Human Action

New surfaces exposed through quarrying or road construction are still very rare in the Arctic, but may be locally important to obtain a dated base for indirect measurement of growth velocities. The same applies for the slag heaps of mines or the gravel ridges piled up by large-scale gold-washing. Ideally suited are gravestones in cemeteries if they indicate the date of the burial and have been made of lichen-free material, e.g. imported rocks. The gravestones of Sukkertoppen gave the key to lichenometrical studies in this area. Cairns or buildings and their foundations, when of known age, are less reliable because stones at least partially covered with lichen were used for their construction and only the least covered surfaces can be assumed contemporaneous. Obviously the dating of such structures if of unknown age becomes very problematical. They would have to be much larger, like the statues on Easter Island for instance, to give sufficient accuracy for lichenometry. Shelters for hunters, traps, or prepared nesting-places for eider ducks, were also constructed with camouflage in mind and the lichen-covered surfaces turned to the outside whenever possible.

Climatic Variation in Space

When dated substrata are present in sufficient amount, the local optimum growth velocities may be used as indicators of the combined influence of the macroclimate. The speed of lichen growth appears to give sensitive and accurate information, which is better than the climatological conclusions that may be obtained with chorological and phytosociological methods. In the Alps linear relationships have been found between the growth velocities of lichens and the hygrocontinentality of an area. The latter is formulated by Gams (1931-2) as the angle κ between the abscissa and the straight line connecting all points on a graph, where the annual precipitation (x) in mm plotted against the altitude (y) in m reveals a constant relation; thus cot κ = x:y.

Owing to the lack of a dense net of temperature recordings in mountain ranges, the altitude is chosen instead, for temperature decreases rather constantly with increasing altitude and is thus less favourable to growth the higher the elevation. Precipitation sums are known from numerous places by means of totalizers, although the results are not very accurate. Gams' formula is a rough but useful approximation of climatic conditions in limited regions, usable until the number and standardization of meteorological instruments increases. A great improvement can be expected with the use of automatic recorders in which radioactive substances write on films (Dahl, 1949, 1959). The formula of the hygrocontinentality cannot be applied for a comparison of different latitudes, or of regions close to sea-level. As it is based on the average annual precipitation, differences in the seasonal distribution of moisture are not considered. Linear relations of lichen growth velocities and hygrocontinentality exist only within areas of similar distribution patterns of precipitation during an average year. Lichen growth is slower in those areas of equal hygrocontinentality which have less precipitation during the actual vegetation period.

Combined climatic formulae tend to apply only for limited regions if they are simple, and will be inaccurate, or very complicated, when used for areas far apart. The already-integrated climate, as it expresses itself in lichen growth, may be used to solve this dilemma. As the "lichen factor" the writer suggested (1957, p. 180) the local optimum diameters in mm of century-old thalli of *Rhizocarpon geographicum* (L.)DC. This species has been divided into a number of smaller taxonomic entities, of which *Rhizocarpon tinei* (Tornab.) Runem., the most common, occurs on siliceous rock in all cool temperate and cold climates. To date, none of the others has been found growing faster than *R. tinei*, but several grow with the same velocity in the same places. Even if one considers the division into several species valid, no harm seems done if they are retained as *R. geographicum* coll. for lichenometrical studies. The lichen factor can be measured easily in the glacier forelands of known history, especially if century-old moraines are present. Even without a detailed climatological basis, the lichen factor seems well suited for a comparison of different polar and alpine regions. Precipitation and temperature differ more at the same altitude between the outer coast and the inner fjord regions in West Greenland, than between the humid northern margin of the Alps and their continental central massifs. Growth velocities of lichens also differ to a larger extent in West Greenland (Figure 1). They do not decrease linearly with the distance from the coast, but drop only slightly in the outermost 20 to 30 km, decrease to about a tenth within the next 20 km, and farther inland diminish very little. The steepest gradient of growth velocities coincides with the zone in which *Cassiope tetragona* is the dominant heath plant in mesic conditions. *C. tetragona* is restricted to exposed habitats in the coast region, and to the vicinity of snow beds in the continental area. The growth velocity gradient reflects a climatic gradient and it is possible that its steepest part coincides with the limits of the coastal fog region. This vague generalization, the one possible at present, is based on only a small number of glacier forelands where analogous moraines could be assumed to be contemporaneous. Confirmation and a higher' accuracy are expected when many lichen photographs will be repeated after about ten years and growth velocities will be measured directly. How well this scheme applies in other Arctic regions remains to be seen.

The microclimatic conditions seem to vary with the relief to a higher degree in West Greenland than in the Alps. Irradiation and duration of snow cover are very likely more variable over short distances. This necessitates greater care and larger minimum surfaces for lichenometry. Towards high Arctic regions the irradiation differences between north- and south-facing substrata should decrease again, while the distribution of the snow cover will be of at least the same magnitude of importance as it is in the continental regions of central West Greenland. Where only exceptional microclimatic circumstances permit lichen growth, a correlation with these different factors must be sought and the rules obtained for the dependence of the growth velocity on the macroclimate will not necessarily hold. Llano (1959 and personal communication), for instance, has shown that in Antarctica lichens are often limited to the basal parts of rock surfaces where the scant snow cover is sufficient to melt and refreeze into thin ice-sheets, which in turn cause the formation of miniature greenhouses enclosing higher temperatures and humidities. Geiger (1957, p. 173 ff.) describes the effects of a similar phenomenon.

The relation of the growth velocities between pairs of identical or related species, or even types of a similar growth form, varies less in a climatic gradient than the growth velocities. Certain differences, however, occur with a change of the macroclimate, which may depend to a higher degree on single factors. *Sporastatia testudinea* grows up to twice as fast as *Rhizocarpon tinei* in very humid areas of the Alps, yet it is the slower of the two in continental areas. *Rhizocarpon jemtlandicum* grows only with two-thirds of the speed of *R. tinei* on the West Greenland coast, but with equal velocity farther inland. The amount of precipitation seems to be the primary factor to develop inherent differences in both cases. The fastest-growing species of *Lecanora* subgen. *Aspicilia* increase outside the spray and inundation zone of brooks, with about twice the speed of *R. tinei* in most of the visited glacier forelands of the Alps, as well as in the coastal region of West Greenland. The velocities become equal, or *Aspicilia* grows even slower than *R. tinei*, in the higher altitudes of continental areas of the Alps and West Greenland. This could reflect a stronger limiting tendency of low winter temperatures for *Aspicilia*. As the growth velocity of *Umbilicaria* varies continuously it has to be compared with the rather constant increment of *Rhizocarpon* at a specified point. The diameter at the time when *Umbilicaria* measures two-thirds of its final size may be divided by the maximum diameters of *R. tinei* at the same time. The resulting quotient varies between three and eight in the visited areas of the Alps and Greenland, depending on the species, but according to measurements by De Heinzelin (1956 and *in litt.*) it rises to twenty-eight in the foreland of the Stanley glacier in the Ruwenzori Mountains, Central Africa. There the surfaces of rocks seem to be wet most of the year owing to excessive fog and drizzle. The temperature fluctuates only slightly around the freezing point throughout the year. Although the species common there (*U. haumaniana* and *U. africana*) may be inherently fast growing, the microclimatic difference between the actual rock surface inhabited by *R. tinei* and the atmospheric layer of the adjoining 2 cm in which the rock-tripes live, must be very important.

Climatic Variation in Time

As stated previously, the maximum diameters of common crustaceous lichens on dated moraines of different age fall on a straight line in a diagram of size versus time after the great period, which is limited to a few decades at the most. This permits the conclusion that climatic fluctuations within the dated period are not reflected in the measured lichen growth. The extrapolation of this principle permits the dating of other moraines. However, climatic fluctuations have caused a variation of the accumulation-ablation balance of glaciers which is greatly responsible for the movement of their snouts. Through the dating of the correlated deposits with lichenometry and other geomorphological and biological methods, climatic fluctuations may be elucidated indirectly. From the evidence so far obtained a similar pattern appears for various regions of the Alps (Beschel, 1950, 1957,

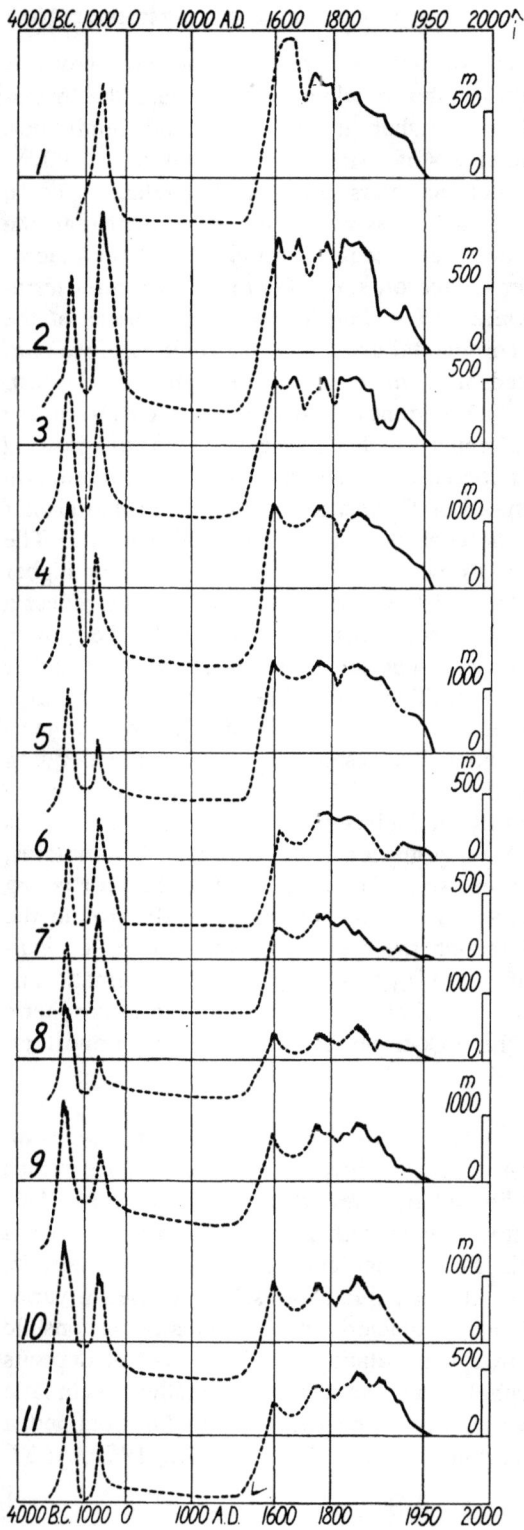

FIGURE 2. Glacier advance ↑ and retreat ↓ in hypothermal time, from lichenometrical, geomorphological, historical, and dendrochronological evidence.

(1) Glacier Stanley occidental, Ruwenzori, Central Africa

(2) Ghiacciaio di Gran Neiron, Gran Paradiso, Italy

(3) Grünauferner, Stubai Alps, Tyrol

West Greenland (Locations of the glaciers are indicated on Figure 1.)

(4) Tasiussaq A

(5) Tasiussaq B

(6) Ikatussaq A

(7) Ikatussaq C

(8) Tunugdliarfik A

(9) Tunugdliarfik B

(10) Tunugdliarfik C

(11) Kugssuaq A

1958b; Heuberger and Beschel, 1958), West Greenland (Beschel, 1958c, 1959, and ms.), and the Stanley glacier in Central Africa (De Heinzelin, 1953, 1956), confirming Ahlmann's conclusions (1948, p. 72) (Figure 2).

The studied glaciers have all passed through a well-circumscribed advance period, which must have started about four to five centuries ago. It lasted until the end of the last century and culminated in a number of smaller peaks, which appeared rather simultaneously — as far as lichen dating is exact — but vary in their extent regionally and even more so locally. This seems mainly due to differences in the landscape. Whenever a younger advance was more extensive it destroyed the frontal moraines of older ones, but as older moraines presented an obstacle, they have not always been removed completely and diversion structures (*apparati di diversione* according to Capello, 1952) may have resulted. Only a few glacier forelands retained moraines of all the minor advances, yet it seems very probable that the majority of fronts of smaller glaciers moved synchronously. How far a glacier retreated between the advances is less certain and has been indicated with broken lines in Figure 2. The whole advance period has been called *neuzeitlich* by Heuberger and Beschel (1958, p. 95) which may be translated as modern, as it took place in modern times. The term Recent has been used too ambiguously. The journalist's "little ice age" (cf. Lawrence, 1958) refers to a longer time period that is more aptly called hypothermal by Cooper (1958).

After a probably rapid advance culminating around 1600 A.D., the glaciers remained for the most part in a rather advanced position, reaching another maximum around 1680. When the advances of the seventeenth century climaxed on the Ruwenzori cannot yet be ascertained. Moraines of this first subphase have been called Fernau moraines by Kinzl (1929, 1932) in the Alps. The next advance reached its maximum extensions around 1740-50 and 1770-80, with perhaps little change in between. Corresponding moraines of the earlier time are more pronounced in West Greenland and on the Stanley foreland. Only moraines of the later time have been found in the Alps so far, although advances around 1740 are historically known (Richter, 1891). A rather fast retreat must have followed after 1780 until the next advance started at the beginning of the nineteenth century. The culminations of about 1820 and 1850 are not separated by an extensive shrinking, and many glaciers must have been rather stationary during this period. Afterwards the tendency to recede gathered momentum, although slowed down by minor advances around 1870-80, 1890-5, and 1920-5.

In the Alps, and in coastal West Greenland it matters little which of the subphases was the most extensive, as all have been of rather similar intensity. On the Stanley glacier the advances were succeedingly less accentuated after the seventeenth century, but it remains to be seen whether this applies also for other African glaciers. The trend of the visited glaciers in continental West Greenland lies in the opposite direction. The farther inland a foreland is situated, the more strongly developed are the moraines of 1850 as compared with those of the eighteenth century, and even the moraines of 1870 and 1890 gain in importance. Weidick (1958a,b, 1959) summarizes historical evidence of glacier behaviour on Greenland's west coast and finds indications of a successively later maximum advance within the modern phases, with increasing latitude. Much more evidence will be needed to test whether the mentioned subphases occurred synchronously on the whole earth, and if the maximum subphase occurred later with higher latitudes and continentalities.

No moraines have been found which would date from the centuries just preceding the modern advance period. Lichenometry can give only a minimum age of the moraines which lie closest to the forelands. Geomorphological studies in the Alps suggest (Heuberger, 1954, Heuberger and Beschel, 1958) that one or two earlier hypothermal glacier advance periods occurred, of which the Larstig advance was short and intensive. It may have taken place between 800 and 500 B.C. Larstig moraines are present in front of small glaciers or in localities where no glaciers exist today. Contrarily, an older advance, which may have been hypothermal as well, is indicated by massive moraines in front of larger glaciers, the Hochmoos-ferner in the Stubai Alps, Tyrol, for example. The retreat after this Hochmoos advance must have been substantial, otherwise the larger glaciers would have increased to a larger extent in the Larstig phase as well and the Hochmoos moraines would not have been partially overridden by Larstig moraines of a neighbouring small glacier. Although Bergström (1955, p. 472) presents evidence from lake sediments for early hypothermal glacier advances in the Ruwenzori, their intensity and time is not known. Early hypothermal moraines have been found in West Greenland by the author. Two phases may be distinguished. One is pronounced in front of glacierets, while neighbouring glaciers of somewhat larger size have destroyed such moraines nearly or completely (if they formed them at all) during the modern advance period (Ikatussaq C, Tunugdliarfik C). Another, older phase is well established in front of large glaciers. The age of both phases exceeds the maximum age of lichens in the coastal region, while a penetration of old shore-lines at about 30 msm suggests hypothermal origin (Tasiussaq A). Probably synchronous moraines of the older phase in front of the foreland of a large outlet glacier (Kugssuaq B) have about 4000-year-old lichens in the continental region. There are slightly larger lichens present outside, but the moraine is too small to give a high probability of accurate dating, even if the estimate of the growth velocity is correct. Near the small outlet glacier, Kugssuaq A, lies a moraine arc of a former glacieret, in part overridden by a boulder stream. This moraine has lichens suggesting Larstig age, while the foreland of Kugssuaq A ends with its outermost moraines of the modern advance period directly on bedrock with maximum-sized lichens. The conditions for the preservation of older moraines would be favourable there. An absence may be due to a larger extent of the modern advances. Had the phase forming the old moraine at Kugssuaq B also caused the moraine of the former glacieret near Kugssuaq A, Kugssuaq A would also show traces of it. The possibility of two separate advances is more likely: the older causing the larger outlet glacier to increase beyond the limits of the modern advance; the younger of short duration but high intensity, exceeding the dimensions of the modern advance only in very small glaciers. The situation seems analogous to the Hochmoos and Larstig phase of the Stubai Alps. In any case, further evidence is needed. The glacier behaviour before the modern advance period as indicated in Figure 2 is highly tentative and should be considered only as the most likely possibility derived from the indications known at present.

REFERENCES

AHLMANN, H. W. 1948. Glaciological research on the North Atlantic coasts; Roy. Geogr. Soc., Res. ser., no. 1.

BACHMANN, E. 1904. Die Beziehungen der Kieselflechten zu ihrem Substrat: Deutsch. Bot. Ges., Ber., vol. 22, pp. 101-94.

———— 1911. Die Beziehungen der Kieselflechten zu ihrer Unterlage. II. Granat und Quartz; Deutsch. Bot. Ges., Ber., vol. 29, pp. 261-73.

———— 1917. Die Beziehungen der Kieselflechten zu ihrer Unterlage. III. Bergkristall und Flint; Deutsch. Bot. Ges., Ber., vol. 35, pp. 464-76.

BERGSTRÖM, E. 1955. British Ruwenzori Expedition, 1952: Glaciological observations — preliminary report; J. Glaciol., vol. 2, pp. 468-76.

BESCHEL, R. E. 1950. Flechten als Altersmasstab rezenter Moränen; Zschr. Gletscherk. Glazialgeol., vol. 1, pp. 152-61.

———— 1957. Lichenometrie im Gletschervorfeld; Ver. Schutz. Alpenpfl., Jb., vol. 22, pp. 164-85.

———— 1958a. Flechtenvereine der Städte, Stadtflechten und ihr Wachstum; Naturw. Med. Ver. Innsbruck, Ber., vol. 52.

———— 1958b. Ricerche lichenometriche sulle morene del Gruppo del Gran Paradiso, N. Giorn. Bot. Ital., n.s., vol. 65, pp. 538-91.

———— 1958c. Lichenometrical studies in West Greenland; Arctic, vol. 11, p. 254.

———— 1959. Glacier foreland succession in West Greenland; Proc. IXth Int. Bot. Congr. Montreal, vol. 2, pp. 29-30.

BITTER, G. 1898. Ueber das Verhalten der Krustenflechten beim Zusammentreffen ihrer Ränder; Jb. wiss. Bot., vol. 33, pp. 47-127.

BUCHANAN, R. E. 1953. Some elementary mathematics of plant growth, in LOOMIS, W. E., Growth and differentiation in plants; Iowa State College Press, pp. 101-11.

CAPELLO, C. 1952. Gli apparati morenici di diversione; Com. Glaciol. Ital., Boll., n.s., vol. 3, pp. 25-44.

CLEMENTS, F. E. 1916. Plant succession; Carnegie Inst. Washington Publ. 242.

COOPER, W. S. 1958. Terminology of post-Valders time; Bull. Geol. Soc. Amer., vol. 69, pp. 941-5.

DAHL, E. 1949. A new apparatus for recording ecological and climatological factors, especially temperatures over long periods; Physiol. Plant., vol. 2, pp. 272-86.

———— 1959. Scandinavian studies with special reference to arctic-alpine vegetation; Proc. IXth Int. Bot. Congr. Montreal, vol. 2a, p. 8.

DE HEINZELIN, J. DE B. 1953. Les Stades de recession du Glacier Stanley occidental (Ruwenzori, Congo Belge); Expl. Parc National Albert, ser. 2, fasc. 1.

DE HEINZELIN, J. DE B., and MOLLARET, H. 1956. Biotopes de haute altitude Ruwenzori, I; Expl. Parc National Albert, ser. 2, fasc. 3.

FREY, E. 1959. Die Flechtenflora und -vegetation des Nationalparks im Unterengadin. II. Die Entwicklung der Flechtenvegetation auf photogrammetrisch kontrollierten Dauerflächen; Schweiz. Natlpk., Erg. wiss. Unters., N.F., vol. 6, pp. 239-319.

FRY, E. JENNIE. 1927. The mechanical action of crustaceous lichens on substrata of shale, schist, gneiss, limestone, and obsidian; Ann. Bot., vol. 41, pp. 437-60.

GAMS, H. 1931. Die klimatische Begrenzung von Pflanzenarealen und die Verteilung der hygrischen Kontinentalität in den Alpen; Ges. Erdkde. Berlin, Zschr., no. 9-10.

———— 1932. Die klimatische Begrenzung von Pflanzenarealen und die Verteilung der hygrischen Koninentalität in den Alpen; Ges. Erdkde. Berlin, Zschr., no. 1-2, 5-6.

GEIGER, R. 1957. The climate near the ground; rev. 2nd printing of enlarged translation of 2nd German edition, Harvard University Press.

HEUBERGER, H. 1954. Gletschervorstösse zwischen Daun- und Fernau-Stadium in den nördlichen Stubaier Alpen (Tirol); Zschr. Gletscherk. Glazialgeol., vol. 3, pp. 91-8.

HEUBERGER, H., and BESCHEL, R. E. 1958. Beiträge zur Datierung alter Gletscherstände im Hochstubai (Tirol); Schlern Schr. (Innsbruck), vol. 190, pp. 73-100.

KINZL, H. 1929. Beiträge zur Geschichte der Gletscherschwankungen in den Ostalpen; Zschr. Gletscherk., vol. 17, pp. 66-121.

———— 1932. Die grössten nacheiszeitlichen Gletschervorstösse in den Schweizer Alpen und in der Montblanc-Gruppe; Zschr. Gletscherk., vol. 20, pp. 270-397.

LAWRENCE, D. B. 1958. Glaciers and vegetation in southeastern Alaska; Amer. Scientist, vol. 46, pp. 88-122.

LLANO, G. A. 1959. Lichens of the Antarctic; Proc. IXth Inst. Bot. Congr. Montreal, vol. 2, p. 232.

PATERSON, T. T.   1951.   Physiographic studies in north west Greenland; Medd. om Grønland, vol. 151, no. 4.
RICHTER, E.   1891.   Die Geschichte der Schwankungen der Alpengletscher; Deutsch. Öst.
Alpenver., Zschr., vol. 22, pp. 1-74.
SCHATZ, A., CHERONIS, N. D., SCHATZ, VIVIAN, and TRELAWNY, G. S.   1954.   Chelation
(sequestration) as a biological weathering factor in pedogenesis; Proc. Penns. Acad. Sc.,
vol. 28, pp. 44-51.
SCHATZ, VIVIAN, SCHATZ, A., TRELAWNY, G. S., and BARTH, K.   1956.   Significance of lichens
as pedogenic (soil-forming) agents; Proc. Penns. Acad. Sc., vol. 30, pp. 62-9.
THOMPSON, D'ARCY W.   1942.   On growth and form, 2nd edition, Macmillan.
WEIDICK, A.   1958.   Frontal variations at Upernaviks Isstrøm in the last 100 years; Dansk
Geol. Fören., Medd., vol. 14, no. 1, pp. 52-60.
————   1958.   Gletscheraendringer i Grønland og Europa i historisk tid; Tidskr. Grønland,
April, 1958, pp. 137-45.
————   1959.   Glacial variations in west Greenland in historical time. Part I. south west
Greenland, Medd. om Grønland, vol. 158, no. 4.
WOLFE, J. N.   Mss.   9. Ecological study of the vegetation patterns (in the Red Rock area
of northern Nunatarssuaq, Greenland).

Mud Volcanoes in the
Copper River Basin, Alaska[1]

DONALD R. NICHOLS AND

LYNN A. YEHLE

ABSTRACT

Two groups of mud volcanoes, consisting largely of clayey silt cones which discharge gas and highly mineralized water, occur within fifteen miles of Glennallen. The four cones of the Tolsona group range in height from twenty-five to sixty feet, and lie west of the Copper River near coal-bearing rocks of Tertiary age. Three of these cones are active. They discharge methane and nitrogen gas and water composed of chlorides of sodium and calcium. The three cones in the Drum group lie east of the Copper River near the slopes of the volcanic Wrangell Mountains. They are 150 to 310 feet high, and emit carbon dioxide gas and warm sodium chloride and bicarbonate waters.

The source of gas in the Tolsona group may be from buried marsh or coal deposits because the gas contains only a trace of carbon dioxide and lacks hydrocarbons heavier than methane. Gas from the Drum group probably emanates from volcanic sources. Water in Tolsona springs may be a mixture of meteoric, connate, and/or highly saline ground water; the Drum springs may also include small amounts of volcanic water.

Formation of the cones was by quiet intermittent accretion, and in the Drum cones, probably included eruptive phases. Most cones formed largely before or during the last major glaciation.

SEVEN MINERALIZED SPRINGS and their related cones, referred to by local residents as mud volcanoes, have been investigated in the course of an engineering and glacial geology study of the southeastern part of the Copper River Basin, Alaska (Figure 1*). In this paper, mud volcanoes are defined as cone-shaped mounds composed of clayey silt from which mud, gas, and mineralized water have been discharged. This definition, in contrast to that given in the *Glossary of Geology and Related Sciences* (Howell, 1957, p. 194), does not limit the mud, water, or gas to a particular chemical composition.

The geographic distribution and the physical and chemical characteristics of the seven mud volcanoes permit their division into two general groups. The Drum group lies east of the Copper River and consists of the Shrub, Upper Klawasi, and Lower Klawasi mud volcanoes (Figure 2*). This group has the largest cones, 150 to 310 feet high, and their springs are characterized by carbon dioxide gas and warm sodium bicarbonate and chloride waters. The Tolsona group lies west of the Copper River and consists of the much smaller (twenty-five to sixty-foot high) cones of the Shepard, Nickel Creek, Tolsona No. 1, and Tolsona No. 2 mud volcanoes (Figure 2*). Except for the Shepard mud volcano, which is inactive, the cones contain springs which discharge methane gas and cool sodium and calcium chloride water. Several other mineralized springs in the Copper River Basin have little or no surface expression.

[1]Publication authorized by the Director, United States Geological Survey.
*See separate container of figures.

The current exploration programme for petroleum in Alaska has focused interest in the mud volcanoes as possible indicators of the presence of petroleum. The increasing local need for potable water by an expanding population also raises the question of what relation the mud volcanoes bear to the saline ground water in the Copper River Basin. Mud volcanoes frequently have been considered indicative of oil-producing strata (DeGolyer, 1940, p. 23-4). Unfortunately, wells have not been drilled through the Pleistocene sediments in the immediate area of the mud volcanoes to test this theory in the Copper River Basin. A preliminary report (Nichols, 1956) describes the ground water of the Glennallen area and concludes that the principal ground water horizon lies at depths greater than 200 feet where it is confined under artesian pressure beneath fine-grained sediments and/or permafrost. This water is quite hard and has a salty taste. Even shallow wells placed in low terraces and in the flood plain of the Copper River near Glennallen encounter water that is highly saline.

Despite the prominence of several of the mud volcanoes and their proximity to old trails, earlier workers did not investigate them. As described more fully later, Theodore Chapin's 1914 field notes record small springs on the north bank of the Tazlina River but his report (Chapin, 1918) makes no mention of them or of the many more prominent mud cones. Apparently Bradford Washburn's brief description (1941, p. 227) of what is here termed the Lower Klawasi mineral spring is the first published reference to the mud volcanoes. R. E. Frost, in a symposium on frost action in soils (1952, p. 236-8), shows pictures of the Lower Klawasi cone and identifies it as a mud volcano but classifies it as a type of pingo.

This paper is part of the results of an areal mapping project sponsored by the United States Geological Survey and supported in part by the Office of the Chief of Engineers, United States Army. The studies have included four summers and one winter of field work. Appreciation is expressed to Arthur Grantz for facilitating the analyses of some samples and to him and J. R. Williams for discussion of the problems. G. W. Whetstone and his successor, F. B. Walling, District Chemists of the Branch of Quality of Water, Water Resources Division, United States Geological Survey, Palmer, Alaska, performed the water analyses. The gas analyses were obtained through the interest of W. M. Deaton of the Helium Activity Laboratory, United States Bureau of Mines. J. R. Williams, H. R. Schmoll, and D. E. White reviewed the manuscript and offered numerous helpful comments, although the authors must accept responsibility for the conclusions reached.

GEOGRAPHY

The Copper River Basin lies in south-central Alaska (Figure 1*). It is an intermontane basin ranging from 500 to 4000 feet above sea-level and rimmed by 4500- to 16,500-foot peaks of the Alaska Range and the Talkeetna, Chugach, and Wrangell Mountains. Mendenhall (1900, p. 297) and Schrader (1900, p. 284-6) first termed it the "Copper River Plateau," and later both used this term interchangeably for "Copper River Basin." As they used it originally, "Copper River Plateau" included river valleys and other physiographic forms. As this usage is not accepted generally, the term is dropped in favour of "Copper River Basin." Chapin (1918, p. 48) introduced the term "Copper-Susitna lowland" as a physiographic term for the Copper River Basin and correctly included that portion of the Susitna drainage that lies in the intermontane basin area. However, the earlier usage of the name "Copper River Basin" establishes its priority. Without reliable maps or aerial photographs it was difficult to make a subdivision of the basin, and Chapin did not distinguish between the east and west portions which are characterized by sharp differences in morphology. The descriptive term "trough" had been used informally for the eastern area below 2000 feet (Nichols, 1956, p. 2; United States Army, 1955, p. 5) and "plateau" (United States Army, 1955, p. 5) and "piedmont" (Nichols, 1956, p. 2) for areas above 2000 feet, largely in the west. The latter two terms are now deemed inappropriate and are dropped, and Chapin's term "Copper-Susitna lowland," which has priority, is adopted and restricted to areas above 2000 feet.

Thus the Copper River Basin is divided into two physiographic subunits as follows. (1) The *Copper-Susitna Lowland* has a rolling to hummocky surface and lies between 2000 and 3000 feet in altitude. It is situated largely in the northern and western parts of the basin with extensive glacial drift deposits, numerous lakes, and a few scattered bedrock hills, some of which rise to 4000 feet. (2) The *Copper River Trough*, which lies below the lowland, is a flat to gently sloping lacustrine plain extending in an arc along the north, west, and south sides of the Wrangell Mountains. It descends from an altitude of 2000 feet at the northeast end to 500 feet near Chitina at the south end.

South of Chitina, the Copper River flows through a deep bedrock canyon across the Chugach Mountains and empties into the Pacific Ocean. The Copper River and its principal tributaries have cut canyons up to three miles wide and 550 feet deep in the floor of both the lowland and the trough. The Copper River and all but its minor tributaries head in glaciers.

The Copper River Basin has a typically subarctic continental climate. At Gulkana FAA airfield, in the centre of the trough, the mean annual temperature is 26.8° F and the mean annual precipitation is 11.4 inches. The distribution of permafrost in the basin requires that it be classified as part of the discontinuous permafrost zone (Nichols, 1956, p. 7). Permafrost probably is absent only under lower terraces and flood plains of the major rivers and under the larger lakes. Elsewhere it generally is encountered from one to ten feet beneath the surface and extends to a depth of 120 to 250 feet (Nichols and Watson, 1955).

REGIONAL GEOLOGY

The Copper River Basin is bordered by mountains composed of igneous and sedimentary rocks, some of which have undergone slight to intense metamorphism. The rocks range in age from middle Palaeozoic to Tertiary. They consist largely of schist, greenstone, greywacke, slate, shale, and sandstone, locally associated with minor amounts of altered limestone, tuffaceous beds, and basalt flows, and are intruded by granular igneous rocks (Moffit, 1938, p. 19; 1954, Plate 7). Large areas of the Wrangell and Talkeetna Mountains and local areas of the Chugach Mountains are underlain by considerable thicknesses of basaltic and andesitic lava flows.

The northwestern part of the basin is underlain by volcanic rocks of late

Palaeozoic to Triassic age (Dutro and Payne, 1957). Underlying the southwestern part of the basin and perhaps the Tolsona group of mud volcanoes are Cretaceous shale and sandstone and semi-consolidated Tertiary sandstone and conglomerate with a few thin lignitic beds (Miller *in* Miller, Payne, and Gryc, 1959, Plate 3 and p. 52; Grantz, 1953, Figure 3). An aeromagnetic map of the Copper River Basin (Andreason and others, 1958) suggests that Wrangell andesitic lavas and Pleistocene unconsolidated deposits probably underlie the Drum group of mud volcanoes at relatively shallow depth and extend at least as far west as the Copper River. In the trough, bedrock exposures are few and, except for greenstone at one locality and a few scattered limestone outcrops, only Pleistocene sediments, including a few exposures of near-surface andesite flows, are exposed.

During the Pleistocene, glaciers from the surrounding mountains repeatedly invaded the basin, at times covering the entire basin surface and flowing across divides to the north, west, south, and east. In the early stages of each glaciation, glaciers in the Chugach Mountains coalesced and dammed the Copper River; the result was the formation of a large lake in the basin. Mud flows resulting from violent volcanic eruptions (Ferrians, Nichols, and Schmoll, 1958), ash deposits, and andesite lava flows are interbedded with till, outwash, and lacustrine deposits in the trough and form a complex succession of heterogeneous beds. During the last major glaciation, valley glaciers pushed into the basin and coalesced into lobes which spread over portions of the lowland and trough. One of the largest lobes

FIGURE 3.   Cone and crater of Upper Klawasi mud volcano. View to the south from the Shrub mud volcano showing barren drainageway in the foreground. The esker complex at the right appears to head in the 300-foot high drumlinoid Upper Klawasi mud volcano. September 15, 1955.

advanced in the trough and, during maximum advance, covered the area of mud volcanoes. This lobe fronted in a lake which covered large areas of the basin not occupied by ice (Nichols, 1956, p. 4). In the northwestern part of the basin lacustrine deposits suggest that lake levels stood above 2800 feet (J. R. Williams, personal communication, 1959). Local strand lines and lake deposits in the area of the upper Susitna River and Maclaren River indicate the presence of levels as high as 3200 feet (United States Army, 1956). After maximum glaciation ice and lake margins fluctuated widely and lake levels were lowered. This is evident from the complicated interfingering of lacustrine and glacial deposits in river bluffs and by numerous well- to poorly-developed strand lines at altitudes at and below 2450 feet (Ferrians and Schmoll, 1957), in the vicinity of the mud volcanoes. Mud volcanoes that are capped by till, and those that lie below these strand lines and are covered by lacustrine deposits, indicate that most of the cones existed during and perhaps prior to the last major glaciation.

After retreat of the ice and drainage of the lake, the present river valleys were carved into the Pleistocene sediments. Between river valleys, muskegs and marshes, which occupy depressions on the old lake floor, are perched, poorly drained, perennially frozen lake sediments. The smooth surface of the Pleistocene lake bottom is relieved by numerous wave-modified drumlins and by a few eskers in the trough. On gentle to moderate slopes of the lowland, glacial land forms are less modified and sharp-crested moraines, eskers, and other ice-contact features are prominent.

DESCRIPTION OF CONES

Drum Group

The Drum group consists of the Shrub, and Upper and Lower Klawasi cones, all of which lie west of the Copper River (Figure 2*). The Shrub and Upper Klawasi mud volcanoes share a number of characteristics. They lie at an altitude of about 3000 feet at the toe of outwash fans built by streams which originate in glaciers on the west side of Mount Drum. Thus they lie well above the 2450-foot strand line. Their cones are covered by dense spruce forests broken only by barren drainageways from the crests. Large, northward-sloping ice-marginal drainage channels terminate just south of these cones. The only two large esker systems on the southern and western slopes of Mount Drum commence downslope from the Shrub and Upper Klawasi cones (Figure 3), and continue up the Copper River valley to the north. The cones are elliptical in ground plan and drumlinoid in profile (Table I and Figure 3) with the north-south basal diameter being the larger. In both cones the southern slopes are the steeper and the northern the gentler, suggesting overriding by ice moving northward. Glacial drift, consisting of coarse gravelly sand with striated erratics up to five feet in diameter, caps the crest and slopes of the Shrub and Upper Klawasi mud volcanoes. The drift contains rocks of highly diverse lithology that are derived from the Wrangells and various points within the Chugach Mountains. The fact that these ice-modified cones are much larger than any other drumlins in the Copper River Basin and contain active water and gas springs and saline mud in craters near their crest suggests that they are composed largely of effusive clayey silt built up by accretion from vents.

In contrast to the Upper Klawasi and Shrub cones, the Lower Klawasi mud volcano lies much lower, at an altitude of about 1800 feet and in a small re-entrant in the general westward slope. Its cone is nearly symmetrical, and is evidently younger than others in the Drum group, for glacial drift or lacustrine deposits are not present. Clayey silt, presumed to have emanated from the cone vents, covers the surface.

Shrub mud volcano. Unlike most of the other cones, water and gas do not emanate from the crest cf the Shrub mud volcano. A vegetation-bare basin or crater about 120 feet in diameter and floored by clayey silt is present on the southwest side of the cone about thirty-five feet below the crest. In 1955 this basin had seven pools, two to eighteen inches in diameter, from which gas and silt-laden water were discharged. Only a trickle of water from the pools drained across the basin and down the steep-sided cone. The flow of water must have been considerably greater in the past as indicated by the presence of two barren drainageways on the south and west sides of the cone where non-salt-tolerant plants have been killed.

In 1956 only four pools, one to two inches in diameter, remained and both gas and water emanations were sharply reduced in volume. There was no surface drainage away from the pools. The surface of this and other cones has been much trampled by moose and, possibly, caribou which graze in the area and drink the salty water. These animals, by stepping in or near the vents may cause them to be

FIGURE 4.   Cone and crater of Upper Klawasi mud volcano. Gas and mineralized water are discharged near the centre of the Upper Klawasi mud volcano pool, which is ninety-six feet in diameter. The lower slope of Mount Drum is visible in the left background. July 21, 1954.

temporarily, or perhaps even permanently clogged. This may result either in a shift in gas outlets or, in the case of weak activity, complete cessation of discharge.

Upper Klawasi mud volcano. This is one of the few mud volcanoes that has shown a sharp increase in discharge of both gas and water during the period of observation. The United States Coast and Geodetic Survey description of the triangulation station (1941, p. 13) on the west rim of the crater, states that the crater is 150 feet in diameter and "is filled with a grayish-white volcanic mud, has 2 muddy warm springs 15 feet in diameter in its center and has an outlet to the eastward." Between 1941 and 1954 the two muddy springs had increased in size and at present they form one large pond ninety-six feet in diameter (Figure 4).

The water in the pond is gray although local areas are covered by a dark brownish scum, which probably results from the attachment of silt particles to the water film about small gas bubbles. In July 1954 the water temperature was measured by thermistor cable, and it averaged about 86.5° F with an air temperature of 42° F. No measurements were made in subsequent years, but the water felt slightly warmer. Location of gas emission areas has varied from year to year. In 1958 the principal emission appeared to be just north of the centre of the pool, with smaller emissions visible along the periphery on the west and southeast.

The clayey silt in the crater walls is crudely stratified to a depth of at least four feet, with a white caliche-like precipitate capping each layer of pinkish-brown clayey silt. Shallow depressions on the muddy crater floor also are covered by an evaporite crust, composed largely of halite and various carbonates, chiefly of sodium. A sample of the mud of the crater floor included, in addition to clay, scattered grains of mineral detritus such as quartz, feldspar, pyroxene, zircon, and a few grains of an unidentified fine-grained, high-index mineral.

Lower Klawasi mud volcano. Gas and water emanate from the crest of the Lower Klawasi mud volcano, and through a pond which is 175 feet in diameter (Figure 5). The pond surface is depressed fifteen feet below the crater rim and water drains through a narrow cut on the northwest side. Gas bubbles to the pond surface at numerous points, suggesting the presence of multiple gas vents. The number of vents appears to vary from year to year and probably depends on the degree of activity. A dark scum, probably composed of dark brown silt which clings to the gas bubbles, lies over the main centres of activity (Figure 6). Over the five-year span of observation, the rate of gas and water emanation from the pond seems to have decreased.

Dead stumps standing in the pond are coated up to two-and-one-half feet above water level with a rind of almost pure sodium chloride, one-quarter to one-half inch thick. Logs, tree stumps, branches, and twigs are imbedded within the laminated clayey silt composing the crater walls. A mechanical and hydrometer analysis of the inorganic material near the base of the crater wall showed that 98 per cent of the particles passed the no. 200 United States Standard sieve (.074 mm) but only 16 per cent of the sample was smaller than 0.005 mm. It is thus classified as a clayey silt despite its strong plasticity. Most of this material consists of clay, optically resembling montmorillonite, and contains scattered small angular grains of clear quartz, white quartz, bluish-grey chalcedony, and nodules of limonite. Grain sizes range up to three-quarters of an inch in diameter but average one-third of an

FIGURE 5. Aerial view of Lower Klawasi mud volcano. The Lower Klawasi mud volcano lies on the lower slopes of the Wrangell Mountains. It is composed of clayey silt and its base is approximately 6000 feet E-W and 8200 N-S. The cone is about 150 feet high. The pool in the crater is depressed fifteen feet below the crest and is 175 feet in

inch. Pebbles are more numerous in sediments exposed in the lower part of the crater wall, perhaps indicating decreasing activity during recent time. A depression west of the crest is joined to the main crater by a beheaded gully which appears to have once drained the large pool. Presumably springs issue from this depression, but this has not been verified. Two small springs occur at the base of the cone on the east and northeast. All these springs are believed to be associated with the construction of the Lower Klawasi cone.

A dense white spruce forest covers the cone except along the north slope, the base of the east slope, generally within fifty feet of the crater, and in small areas elsewhere (Figure 5). Flow of water from the pool is restricted at present to the north side of the cone. The lack of vegetation on that side may be caused by the high salinity of the water. It is not entirely clear that salinity of the surface material is the only factor inhibiting the presence of vegetation, because tree stumps scarred by fires dot the bare northern slope. Forest fires do not completely account for the configuration of the vegetation-free areas, however.

A dense dendritic gully system, which is particularly well exhibited on the northern vegetation-free slope (Figure 7), is incised two to four feet into the clayey silt of the cone. The crest line of intergully areas and remnant mounds are capped by small angular rock fragments one-quarter to one-half inch in diameter. Representative fragments include black porous lava, dense red lava, rhyolite, glassy lava, basalt, limestone, limonite, chert, chalcedony, quartz, quartz-tourmaline rock, quartzite, and a few pieces of a warty encrustation composed of calcite. These fragments must have been deposited either from streams draining the pond or from explosive eruptions at the crater. In any case, they are not representative of rock

FIGURE 6. Surface features of Lower Klawasi cone. The Lower Klawasi crater pond drains through an outlet cut fifteen feet in the crater rim. Crater walls are composed of clayey silt, organic matter, and in the lower half, scattered small pebbles. The black scum in the middle surrounds the principal gas emission area.
September 7, 1956.

types in glacial drift near the cone but, along with the fine-grained, clay-rich matrix, they are characteristic of eruptive products of shallow-depth hydrothermal systems (White, 1955, p. 1128). Locally, mound crests are capped by a square foot or more of travertine fragments one to two inches thick and one to four inches across. Several serpentine-shaped, travertine pipes (Figure 8) one to four inches in diameter, were seen on the surface of the cone at several points and appear to be indicative of a widespread, but probably ephemeral, vent system that developed locally. Such vents would help to explain the erratic distribution of vegetation-free areas and widespread gully systems that do not appear to be related to run-off or drainage from the crater.

Tolsona Group

In the Tolsona group (Table I) the Nickel Creek mud volcano is similar to the Shrub and Upper Klawasi cones of the Drum group, while the Shepard, Tolsona No. 1, and Tolsona No. 2 cones are miniature replicas of the Lower Klawasi cone. The relatively steep-sided Nickel Creek cone is almost twice as high as the other cones in its group, but is less than half as high as any cone of the Drum group. The diameter of the Tolscna No. 2 cone, which is the largest in its group, is only about half the size of the Shrub cone which is the smallest in the Drum group.

The Tolsona cones all lie about 2100 feet in altitude (Figure 2) and below nearby strand lines. Except for the Nickel Creek cone, they are capped by sand and massive pebbly silt identical to some of the deposits of probable lacustrine origin which cover most of the trough and extend up to 2450 feet in the lowland. A few of the pebbles are striated, most are well rounded, and are similar to rocks of the Chugach and Wrangell Mountains. The Nickel Creek cone is mantled by debris which may have been derived directly from a glacier or which may have

FIGURE 7. Surface features of Lower Klawasi cone. Drainage from the Lower Klawasi mud volcano pool is incised three to five feet in the slope of the cone and flows across a six- to ten-foot-wide flood plain. The youthful dendritic drainage is characteristic. Whitened stumps in the background are fire-scarred. July 19, 1954.

TABLE I

Comparison between Certain Physical Characteristics of Mud Volcanoes, Copper River Basin, Alaska

Mud Volcano	Diagrammatic cross-section N S	Approximate dimension * of cone		Alt. * of crest	Approx. * diam. of "crater"	Surf. water temp., °F.	Est. water disch., gpm.
		Base	Hgt.				
Drum Group							
Shrub		3600 4200	310	2950	120	54	< ¼
Upper Klawasi		4200 6700	300	3017	150	86.5	2-5
Lower Klawasi		6000 8200	150	1875	175	82	5-10
Tolsona Group							
Nickel Creek		800 1000	60	2025	150	cold	< ¼
Shepard		1300 1600	25	2172	15	—	—
Tolsona No. 1		600 900	25	2045	30	38-55	< ¼
Tolsona No. 2		2000 2300	40	2085	150	40-60	< ¼

* =in feet,   • = active spring,   o = inactive spring.

been ice-rafted. Effusive stratified clayey silt, locally iron-stained and containing organic matter, occurs within the crater of all cones. Particle sizes larger than silt are lacking.

The slopes of the Nickel Creek cone are covered by white spruce, except on barren drainageways. The gentler slopes of the Shepard, Tolsona No. 1, and Tolsona No. 2 cones have been burned over and are covered by diamond willow with scattered small white spruce and an undergrowth of sedges, grasses, and wild sweetpea. Saline-tolerant grass species grow on the crests and marginal to the active spring areas (Lloyd Spetzman, 1959, written communication).

Nickel Creek mud volcano. The Nickel Creek cone is composite, formed by the growth of a subsidiary crater fifteen to twenty feet below and south of the main crater. The rim of the main crater is two to five feet above its floor while that of the subsidiary crater is one to two feet. In June 1958 a slight hydrogen sulphide odour was detected around the spring areas and locally the mud and grass on the main crater floor had an orange oxidation stain. The minor amounts of discharged

FIGURE 8.   Surface features of Lower Klawasi cone. Sinuous, travertine-walled pipe remnants occur on the middle slopes of the Lower Klawasi mud volcano. Desiccation cracks are common in the clayey silt exposed in vegetation-free areas. The pick is eighteen inches long. September 7, 1956.

water did not flow from the main cone as surface run-off. The subsidiary cone had no active springs.

Shepard mud volcano. This is the only cone that showed no activity during the period covered by field investigations. In fact, activity probably ceased prior to 1941 when the United States Coast and Geodetic Survey established a triangulation station directly over a small depression at the crest of the cone, which may have been a spring area. The summit of the cone is a circular flat area with no bordering rim.

Tolsona No. 1 mud volcano. The Tolsona No. 1 mud volcano has gently sloping sides rising to a slightly domed crest on which several active spring areas are located (Figure 9). The number of areas of activity and of individual gas and water vents has varied from year to year. Most vents feeding the small pools are

FIGURE 9.   Springs at crest of Tolsona No. 1 cone. The slightly domed crest of the Tolsona No. 1 mud volcano has two mineral spring areas, one to the left of the figure, and the other in the foreground to the left. July 14, 1958.

FIGURE 10.   Spring at crest of Tolsona No. 1 cone. The active spring in the left foreground of Figure 9 has numerous small vents which discharge gas, water, and clayey silt. They vary from one-half to eighteen inches in diameter. Gas emission is sporadic and discharge of water is insufficient for surface run off. July 14, 1958.

funnel-shaped and are one-half to eighteen inches wide at the surface (Figure 10). Emission of gas, which, when ignited burns with a blue flame, is somewhat sporadic. The variability in the location and size of vents may be due to normal shifting of vents, to animals stepping in or near the vents, or to the freezing of water in the vents each winter with the formation of new vents the following spring. On March 1, 1955, and again on November 13, 1959 (O. J. Ferrians, Jr., written communication, November 14, 1959), this spring was completely frozen over and covered by snow with no evidence of gas or water emission after freeze-up. Thus, the gas must be contained during the winter months and released during the summer months.

Tolsona No. 2 mud volcano. This cone has a circular platform, 150 feet in diameter, which lies fifteen to twenty feet below and south of the cone crest. The present springs issue from two small ponds which lie on the north edge of this platform. A one-half to one-foot-high rim, faintly suggesting a crater wall, surrounds the platform. At the crest of the cone is a sharp-peaked mound about one-and-one-half feet high which may represent a frost boil or an earlier site of spring activity.

Springs are at present much less active than during an earlier period when the platform was constructed. In mid-July of 1958, water discharge from the two small pools was insufficient to permit surface run-off from the cone. One pool was fed by two mildly active vents. Each vent was funnel-shaped, measuring three feet or more in diameter at the surface and narrowing to about four inches at a depth of twenty-two feet. The other pond, which was three feet in surface diameter, had but one vent. This cone was not visited in the winter, but local residents reported that its springs continue to be active throughout the year.

Other Mineral Springs

Other mineral springs have been reported or were seen on aerial photographs in the southeastern Copper River Basin (Figure 2), but none of them have been visited by the authors. J. R. Williams and Arthur Grantz of the United States Geological Survey report a gas and water spring northwest of Moose Lake, which lies three-and-one-half miles west of Tolsona No. 2 spring. Analyses of gas from this spring (Miller, Payne, and Gryc, 1959, p. 53), which lacks a cone, are reproduced with other analyses in Table II. In his field notes, Chapin (1914) mentions an area of mud volcanoes three miles east of the mouth of Tolsona Creek and two-and-one-half miles south of the Shepard cone. He describes it as a "circular area 15 feet across [with] over 50 mud volcanoes. Some craters two-and-one-half feet across to small ones less than an inch. Mud and water bubbles out and have built up mounds four to five feet high." No statement was made regarding the salinity or temperature of the spring water. Both of these reported springs probably contain gas and water chemically related to springs of the Tolsona group.

A fairly large spring lies two-and-one-half miles north of Copper Center and one-and-one-half miles east of the Copper River. The spring does not appear to have built a cone although it does have a considerable discharge, killing the vegetation adjacent to its flood plain. Despite its proximity to the Copper and Klawasi rivers, saline water does not seep from the coarse materials in either bluff.

This spring, which may be an incipient mud volcano, probably is related chemically to the Drum group.

Summary

The physical characteristics of the mud volcanoes may be summarized by classifying them as to size and morphology into (1) steep-sided cones such as the Shrub, Upper Klawasi, and Nickel Creek mud volcanoes, which are capped by glacial erratics, and whose present gross form is a modification of pre-existing mud volcano cones by glacier ice, and (2) the low, broad shield-type cones as exemplified by the Lower Klawasi, Shepard, Tolsona No. 1, and Tolsona No. 2 mud volcanoes. In this latter group, the gross form has developed during or after melting

TABLE II

ANALYSES OF GAS FROM SOME MINERAL SPRINGS, COPPER RIVER BASIN ALASKA*

	Drum Group				Tolsona Group		
Component	Shrub	Upper Klawasi	Lower Klawasi†		Nickel Creek	Moose Lake	Tolsona No. 2
	per cent	per cent	per cent	per cent	per cent	per cent	per cent
Methane	0.0	0.1	Tr.	Tr.	54.4	48.6	69.4
Ethane	0.0	0.0	0.0	0.0	0.0	0.2	0.0
Propane	0.0	0.0	0.0	0.0	0.0	0.0	0.0
n-butane	0.0	0.0	0.0	0.0	0.0	0.0	0.0
i-butane	0.0	0.0	0.0	0.0	0.0	Tr.	0.0
n-pentane	0.0	0.0	0.0	0.0	0.0	0.0	0.0
i-pentane	0.0	0.0	0.0	0.0	0.0	0.0	0.0
cyclo-pentane	0.0	0.0	0.0	0.0	0.0	0.0	0.0
Hexanes plus	0.0	0.0	0.0	0.0	0.0	0.0	0.0
Nitrogen	0.6	1.1	24.8	3.4	44.9	50.6	30.0
Oxygen	Tr.	0.0	6.0	0.6	0.1	0.1	0.1
Argon	Tr.	Tr.	0.4	0.2	0.1	0.1	0.1
Helium	Tr.	Tr.	Tr.	Tr.	0.1	0.1	0.1
Hydrogen	0.0	0.0	0.0	0.0	Tr.	0.0	0.0
Hydrogen sulphide	0.0	0.0	0.0	0.0	0.0	0.0	0.0
Carbon dioxide	99.3	98.8	68.8	96.4	0.4	0.2	0.4
Total in mole per cent	99.9+	100.0+	100.0+	100.6+	100.0+	99.9+	100.1
Sulphur odour	—	None	None	None	None	—	None
Calc. total Btu.	—	1	—	—	551	496	703
Date coll.	8–11–56	6–18–58	6–17–58		6–19–58	—	7–14–58
Date anal.	—	8–22–58	8–22–58		8–22–58	12–16–51	8–22–58
Sample no.	N-6.200	N-8.45	N-8.9		N-8.62b	—	N-8.111

*Mass spectrometer analyses by Helium Activity Laboratory, United States Bureau of Mines, Amarillo, Texas.

†Sample contaminated by air during collection. Left column, analysis of air contaminated sample; right column calculated on an air-free basis.

Tr. = trace, less than 0.05 per cent.

of the glaciers in the Copper River Basin. The cones east of the Copper River are all larger than those to the west. The only two cones with a large pond in the crater also lie east of the Copper River. Two cones, the Shrub spring east of the Copper River, and Tolsona No. 2 to the west, have platforms or shallow craters below the crest and present-day activity is centred on the side of the cone.

ANALYSES

Gas

Analyses of gas emanating from the cones provide the best means of distinguishing between the mud volcanoes of the Drum and Tolsona groups. Table II tabulates

TABLE III
ANALYSES OF MUD VOLCANO AND GROUND WATERS, COPPER RIVER BASIN, ALASKA*
(Chemical concentrations in parts per million)

Constituent	Drum Group		Tolsona Group			Ground Water	
	Shrub	Lower Klawasi	Nickel Creek	Tolsona No. 1†	Tolsona No. 2	354 foot well**	Copper River seep
SiO₂	65	132	10	16	7.1	19	—
Fe	0.04	0.03	0.67	0.26	—	2.2	—
Al	—	—	—	0.16	—	—	—
Ca	94	119	2,760	787	1,580	1,900	2,080
Mg	502	130	65	111	94	520	392
Na	9,390	10,400	2,600	4,660	4,000	1,150	} 670
K	275	433	24	60	26	44	
Ba	—	—	—	14	—	—	—
Li	—	—	8.0	—	—	—	—
Mn	—	—	0.97	—	0.02	—	—
NH₄	11	—	—	5.6	—	—	—
Zn	—	—	—	0.02	—	—	—
CO₃	0.0	0.0	—	0.0	—	0.0	0.0
HCO₃	7,350	7,290	90	143	48	53	226
SO₄	0.0	666	230	0.0	5.5	0.0	60
Cl	12,000	12,500	9,100	8,870	9,450	6,470	5,680
F	0.4	—	0.4	0.3	—	—	—
Br	—	—	17	17	—	—	—
I	—	—	2.2	3.7	—	—	—
NO₃	5.5	—	—	0.7	—	—	—
B	120	—	—	35	—	—	—
PO₄	—	—	0.16	—	—	—	—
Dissolved solids	26,100	28,000	14,900	14,600	15,200	10,200	8,990
Hardness:							
Non carb.	0.0	0.0	7,080	2,310	4,280	6,840	6,620
Total	2,300	832	7,150	2,430	4,330	6,880	6,800
Spec. cond.	37,000	39,500	24,900	23,600	24,794	—	16,400
Density	1.018	1.018	—	1.008	—	—	1.00
pH	8.2	7.7	6.8	7.1	6.3	7.4	
Date coll.	8–11–56	9–7–55	6–19–58	9–21–56	7–14–58	12–28–54	5–26–55
Lab. no.	21045	3653	5106	20910	4881		2979

*Analyses by Palmer Laboratory, Water Resources Division, United States Geological Survey.
†Collected by F. Rucker.
**Collected by Alaska District, Corps of Engineers, United States Army.

mass spectrometer analyses by the United States Bureau of Mines, Helium Activity Laboratory, of gas samples collected from both groups of mud volcanoes. All samples from the Drum group can be classified readily as carbon dioxide gases with a minor amount of nitrogen as an accessory.

Gas samples from the Tolsona group, on the other hand, have a negligible carbon dioxide content and are characterized by containing 48 per cent or more methane, the remainder being largely nitrogen.

Water

Water analyses provide an interesting comparison between springs of the Drum and Tolsona groups and furnish a basis for speculation as to the origin of the mud volcanoes. Table III illustrates that, although all the springs are highly saline, there are significant differences, the most notable being the relatively high bicarbonate content in waters of the Drum group. These waters also appear to have a significantly higher silica, magnesium, sodium, potassium, and boron content. The high sulphate value obtained from an analysis of Lower Klawasi spring water suggests that this anion may be much higher in Drum waters than in Tolsona waters. In contrast, analyses of ground waters and springs in the Tolsona group have a consistently higher iron and calcium content and chloride is almost their sole anion. As will be shown later, some of these differences are most readily apparent by translating concentrations in parts per million to ratios.

SOURCES OF GAS AND WATER

Gas

The striking difference in composition between gases from the two groups of springs suggests that they are derived from entirely different sources. In the Drum group, two basic processes exist for the formation of carbon dioxide (Ovchinnikov, Ivanov, and Yarotsky, 1959, p. 51): (1) thermometamorphic, which is in part strictly magmatic, and (2) biochemical, which results from the decay of organic matter. According to Ovchinnikov, Ivanov, and Yarotsky, the latter process usually leads to lower concentrations of carbon dioxide and to admixture with heavy carbohydrates and hydrogen sulphide. These authors state further that "in all areas of active volcanism, . . . CO_2 is the basic, and often sole, component in gases produced from numerous high-temperature sources." According to W. M. Deaton of the Bureau of Mines (written communication to Arthur Grantz, November 17, 1955) the combination of an almost complete lack of hydrocarbons and the presence of almost 100 per cent carbon dioxide in gases emanating from the Shrub mud volcano indicates that the gas probably does not originate from a petroleum source. However, D. E. White (1957b, p. 1671) warns that "a high content of CO_2 is probably less diagnostic of volcanic waters than many geologists have believed." He also states that "oil field gases nearly always contain some CO_2 and a few consist largely of this gas. This CO_2 probably was derived largely from organic carbon" (1957b, p. 1670). Nevertheless, the composition of the gas and the intimate geographic association of the Drum mud volcano group to the flanks of the Wrangell Mountains suggest that the gas is related to Cenozoic volcanic activity from basic igneous bodies (Piip, 1937, p. 225, 248, 249).

Gas from the Tolsona mud volcano group, on the other hand, is less clearly associated with volcanism. It probably has a closer relation to coal beds or marsh deposits as suggested by Deaton (written communication, August 22, 1958) rather than with an oil-producing structure. If the gas originated from a petroleum source it should contain some hydrocarbons heavier than methane and other constituents should not be limited almost solely to nitrogen which actually is dominant in the Moose Lake spring sample. Tertiary or older coal beds are exposed in a hill seven-and-one-half miles west of the Tolsona mud volcanoes (Miller, Payne, and Gryc, 1959, Plate 3, pp. 35, 52), and buried Pleistocene marsh deposits are believed to be quite common in the area. The gas may also originate from slightly organic, non-petroliferous connate water associated with Cretaceous (?) deposits.

Water

The source of the water in the springs is not as readily apparent as the source of the gas. The high salinity of both groups of springs and of the ground water in the central trough area suggests that they all originate or, at least, receive contributions from a common source. Three principal sources, or combinations thereof, have been considered.

The source immediately brought to mind by comparing analyses of the spring waters is connate water associated with marine sediments, possibly an oil-field brine. Connate water, as used in this paper, refers to fossil interstitial water of marine or non-marine origin entrapped in rock at the time the material was deposited. Oil-field brine is assumed to include all highly saline water associated with petroleum and may consist of connate water or salt-enriched meteoric water.

Although the composition of water from the Tolsona group in particular is similar to connate water, analytical data from the Drum group are less suggestive of a connate source (Table IV).

A second possibility is that the waters are fed by magmatic emanations. This is supported, in part, by analyses of the warm Drum spring waters. Many of the chemical constituents occur in amounts consistent with the common range for those in sodium chloride volcanic springs as given by White (1957b). However, if volcanic waters are involved, they must be well diluted, particularly in the Tolsona springs and in the ground water, either by meteoric or connate water.

A third source of mud volcano water may be meteoric and/or other water circulated with a saline ground water formed by evaporation from glacial lakes and later reconcentrated by the growth of permafrost. This explanation is supported by the geological data. With the advent of each glaciation, the Copper River Basin drainage must have been dammed and a large lake formed. Water then rose to the level of the lowest outlets or until evaporation exceeded inflow, a condition that must have prevailed at various times during the history of the lake. This resulted in a higher concentration of salts in the lake and consequently was reflected in the ground water. After melting of the ice and draining of the lake, the thermal régime of the ground was controlled by the much lower mean annual temperature of the air. Permafrost then formed by progressive downward freezing from the surface of the ground and may be analogous in some respects to the selective freezing of sea water. In both sea water and soil, the freezing temperature is controlled largely by the salinity. The first ice formed is almost free of dissolved solids. According to Sverdrup (1956, pp. 6-9), the salinity of frozen sea water varies from 0 to 1.5 per cent and depends mainly upon the rapidity of freezing and temperature changes after freezing. With slow freezing, trapped brine gradually moves downward and results in a freshening of the top layers which produce potable water upon melting. This process of gravity flow of high-density saline water, convection, and selective freezing operating on ground water, would result in further concentration of salts. The freezing point of the present saline ground water lies between 28.0° F and 29.0° F. It is reasonable to assume that this temperature has never been reached below the zone of seasonal change in the ground. Hence, only the less saline water is selectively frozen, although brine-rich water may occur in taliks scattered throughout the permafrost as well as below it. It is proposed that the ground water became more saline after each glaciation when the salt content was increased by evaporation of lake water and was reconcentrated in interglacial stages by the progressive downward development of permafrost. Eventually, the ground water could reach a composition comparable, in many respects, to connate water or oil-field brine.

Discussion of Analytical Data

In Tables III and IV the results of analyses of mud volcanoes and ground waters in the Copper River Basin are compared with chemical concentrations and ratios found in oil-field brines and in sodium chloride volcanic springs as given by White (1957b, p. 1666). Representative analyses (Table III) indicate that ground water from a seep and a deep well near Glennallen consist largely of calcium, sodium, and magnesium chloride, that Drum springs contain sodium bicarbonate and chloride water, and that Tolsona mud volcanoes discharge water mainly of calcium and sodium chloride. The two mud volcano groups exhibit so many differences that they probably are composed, at least in part, of waters of different origin. According to White (1957b, p. 1672), the *silica* content of oil-field brines commonly ranges from 10 to 50 ppm, and is higher than that of ocean water but is significantly lower than that of volcanic hot springs. Only water samples from the Drum cones have a silica content greater than 50 ppm; Tolsona spring and ground waters range from 7 to 19 ppm silica.

White states (1957b, p. 1673) that in most volcanic hot springs *calcium* and *magnesium* values commonly are very low relative to the alkalies. This also is true in neutral and slightly acid waters which have abundant carbon dioxide, probably because these waters have been at a much higher temperature at depth. Thus relatively low calcium and magnesium content of the Drum springs may be an indication of a high-temperature volcanic source, while the high calcium of the Tolsona springs may be due to a low-temperature source. The high proportion of magnesium and calcium relative to sodium in analyses of ground water is more suggestive of salt concentration by evaporation.

Sodium and *potassium* concentrations in natural waters seldom are indicative of a specific origin. Hem (1959, p. 86) states that water saturated with sodium chloride or in contact with evaporite materials may contain up to 100,000 ppm sodium, that "sea water contains about 10,000 ppm of sodium" (p. 87), and that

TABLE IV

COMPARISON OF CHEMICAL CONSTITUENT RATIOS OF WATERS FROM THE COPPER RIVER BASIN (TABLE III) WITH A RANGE OF VALUES FOR OIL FIELD BRINES AND SODIUM CHLORIDE VOLCANIC SPRINGS

Ratios	Drum Group		Tolsona Group			Ground Water		Approximate Range*	
	Shrub	Lower Klawasi	Nickel Creek	Tolsona No. 1	Tolsona No. 2	354-foot well	Copper River seep	Oil-field brines	NaCl volcanic springs
HCO_3/Cl	0.6125	0.5832	0.00989	0.01612	0.00508	0.0082	0.0400	0.0001 –1.0	0.01 –3.0
SO_4/Cl	0.00	0.05328	0.02527	0.00	0.000582	0.00	0.0106	0.0 –1.0	0.01 –0.5
F/Cl	0.000033	—	0.000044	0.000034	—	—	—	0.00001–0.001	0.0005 –0.1
Br/Cl	—	—	0.001868	0.001916	—	—	—	0.0001 –0.01	0.0001 –0.001
I/Cl	—	—	0.000242	0.000417	—	—	—	0.00003–0.02	0.00001–0.005
B/Cl	0.01	—	—	0.003945	—	—	—	0.00001–0.02	0.01 –0.1
K/Na	0.0293	0.0416	0.00923	0.01288	0.0065	0.038	—	0.001 –0.03	0.03 –0.3
Li/Na	—	—	0.00308	—	—	—	—	0.0001 –0.003	0.003 –0.03
$\dfrac{Ca+Mg}{Na+K}$	0.0617	0.02299	1.0766	0.19025	0.4158	2.027	3.70	0.01 –5.0	0.001 –0.2

*After D. E. White, 1957b, p. 1666.

"In brines . . . potassium concentrations of several thousand ppm can occur and some hot springs yield water containing 100 ppm or more" (p. 91).

The combined carbon dioxide or *carbonate-bicarbonate* content may be roughly the same in water from entirely different sources. White (1957b, p. 1670) concludes that most brines of low to moderate salinity contain more than 140 ppm of combined carbon dioxide, the average of ocean water. In oil-field brines, bicarbonate is considered generally to be the product of decomposition of organic matter and the activity of sulphate-reducing bacteria (Hem, 1959, p. 97; White 1957b, p. 1670). If this were the cause for the exceptionally high bicarbonate content of the Drum springs, the gas, and possibly also the water, would be richer in sulphides and hydrocarbons. White states, however, that "volcanic hot springs of the NaCl type commonly contain much combined CO_2" (1957b, p. 1671).

The two analyses of Drum springs differ markedly in sulphate; one yields a value of 666 ppm and the other zero, as compared to a maximum of 230 ppm in a Tolsona spring and 60 ppm in the ground water. White states that "most oil-field brines are relatively low in sulfate as compared to sea water" (1957b, p. 1669) and that "the sulfate of volcanic NaCl springs commonly ranges from 50 to more than 100 ppm . . . " (p. 1670).

Of the halogens, only chloride has been determined in all samples. The other halogens, except for a fluoride analysis of Shrub mud volcano water, have been determined only from Nickel Creek and Tolsona No. 1 water. Because the halogen content of water is so variable and inconclusive of origin, no attempt has been made to compare these values with those of known water.

A comparison of the ratios of the chemical concentrations may be more significant than a direct comparison of constituents in determining the origin of a water (Table IV). The HCO_3/Cl ratio of brines (White, 1957b, p. 1666) ranges from 0.0001 to 1.0 and of volcanic hot springs from 0.01 to 3.0. Most of the samples fall in the overlap of these two ranges (Table IV). These ratios derived from Drum spring waters are midway between the end members for the sodium chloride volcanic springs, while ratios for Tolsona springs and ground waters lie midway in the range for oil-field brines. The area of overlap for SO_4/Cl ratios of brines and of volcanic springs is so great, and the analyses of Drum and Tolsona mud volcano waters cover such a wide range, that comparison is of little value. The available F/Cl ratios obtained from springs of both groups fall within the range of brines. The Br/Cl ratios for Tolsona No. 1 and Nickel Creek spring water fall in the range of brines, but I/Cl ratios fall in the overlap in ranges of brines and sodium chloride springs. The K/Na ratio is more distinctive, however. The average values of 0.035 for the Drum springs and of 0.038 for samples of ground water lie above the range for brines and within the range for volcanic hot springs (Table IV). In water from Tolsona springs, however, this ratio lies within the range for oil-field brines.

The Ca+Mg/Na+K ratios of water samples from the Drum cones fall within the overlap between brines and volcanic springs but they are closer to the mean of ratios of the latter (White, 1957b, p. 1666). The Ca+Mg/Na+K ratios of the ground water and samples from Tolsona springs are quite high and lie within the range for brines.

A study of analyses of Tolsona mud volcano water and ground water (Table IV), suggests that selected chemical constituents and their ratios fall within the range of values for brines as compiled by White (1957b, p. 1666). On the other hand, the same data from samples of Drum spring water lie solely within the range of values for sodium chloride volcanic water or appear to be more closely related to it than to the mean brine values when an overlap of the two ranges occurs. However, clear-cut support for either a connate or a volcanic source for the water of either group is lacking.

History of Cones

Most of the Copper River Basin mud volcanoes began forming prior to, and perhaps some during, the last major glaciation. Postglacial growth of these cones has been negligible. Probably only one cone has formed largely in Recent time.

The presence of a mantle of glacial drift on the Shrub and Upper Klawasi cones and their drumlinoid form indicate that these cones were formed sometime before the last major glaciation. The interpretation of the surficial deposits on the Shrub and Upper Klawasi mud volcanoes as a glacial drift mantle is based on two lines of evidence. Erratics up to five feet in diameter appear too large to have been brought up from buried drift, even by explosive eruption, although blocks up to one foot in diameter were thrown several hundred feet in the air during the 1951 eruption of the Lake City hot springs (White, 1955, p. 1113, 1119). The lack of a saline clayey silt component in the surficial gravelly sand is further evidence against a possible effusive origin for the mantle. On the other hand, the large size of the cones and the presence of vents which discharge gas, water, and mud at or near the crest suggest that the cones are composed of clayey silt and have only a thin glacial drift cover. The drumlinoid form of the cones suggests further that the drift mantle is thin on the steep, south margin and thicker on the long, northern slope.

The main growth of the Shrub and Upper Klawasi cones ceased prior to the last major glaciation, but intermittent activity on a minor scale has continued to the present. The prominent esker systems which appear to head in the Shrub and Upper Klawasi cones may be indicative of thermal activity in glacial time which caused large quantities of outwash to be released to subglacial drainage channels by thawing of ice covering the cones. The presence of erratics at the crest of the cones indicates a period of dormancy at the close of glacial time followed in postglacial time by minor activity which formed a small basin in the southwest side of the Shrub cone and a crater in the Upper Klawasi cone. Deposition of mud has been limited largely to drainageways leading from the cones. Since the basin was formed on the Shrub cone, activity has decreased to a point of near-dormancy. After development of the crater in the crest of the Upper Klawasi cone, postglacial activity decreased to the point where, in 1941, only two small pools were present on a crater floor sunken well below the crest. More recently, the pools have expanded into one and gas activity has increased.

The Nickel Creek mud volcano, because of the drift at its crest and the steep-sided nature of its cone, also probably formed before the close of the last major glaciation. The cone is not drumlinoid, however, and the erratics at the crest are small and could have been emplaced by explosive activity, although saline mud is not associated with them. The subsidiary cone on the south flank of the main cone may have been formed in postglacial time. It is now inactive and its crater has been almost filled by mud draining from the main crater which has areas of only minor activity. Construction of the cone was probably by quiet, gradual accretion of mud with perhaps a few explosive phases. Postglacial activity of the main cone is assumed to have been entirely of a quiet nature and deposition of mud has been limited to the crater and along two or three drainageways off the cone.

Because the Tolsona No. 1 and No. 2 and the Shepard mud volcanoes are covered by lacustrine deposits, they are assumed to have formed during, or prior to, draining of the last large glacial lake. The small size of the cones and the lack of observed angular rock fragments in their deposits suggest that they formed largely by quiet, gradual accretion of mud rather than by explosive action. Lacustrine deposition caused only slight modification of the original cone forms and postglacial activity has resulted in the deposition of minor amounts of clayey silt adjacent to vents and drainageways. Activity has ceased on the Shepard cone, is sharply reduced on the Tolsona No. 2 cone, and is only moderate on the Tolsona No. 1 cone.

The Lower Klawasi mud volcano, one of the largest of all the cones, was formed principally in postglacial time. Otherwise, it would have been overridden by ice during the last major glaciation because of its topographic position. Overriding would have steepened its sides and resulted in the typical crag-and-tail form of the Upper Klawasi and Shrub cones unless its preglacial size was so small that postglacial mud deposition completely buried the earlier form and masked its modification by glacial ice. The presence of buried mature spruce stumps in the modern crater wall of the Lower Klawasi cone suggests that quiescent periods, each probably lasting several hundred years, alternated with eruptive phases during which fairly coarse clastic particles and mud were rapidly deposited. The violence of these eruptions has been lessening as suggested by the decreasing number of pebbles upward in the crater wall. Destruction of vegetation on the northern slope followed a stable interval and probably was due to rapid but relatively quiet discharge of fluid mud. A recent period of lessened activity is indicated by the present low level of the pool, the incision of its outlet in the crater rim, and the apparent minor decrease in gas and water discharge during the last five years.

CONCLUSIONS

Definitive geologic and geochemical evidence is insufficient to accept any one hypothesis for the origin of the spring and/or ground water. Only by an over-all evaluation of the geology and by extensive analyses of water and gas can an attempt be made to conclude a connate, magmatic, meteoric, or compound origin for water of an area (White, 1957b). However, Allen and Day (1935, pp. 40-1) and Rankama and Sahama (1950, p. 280) believe that even in most volcanic systems, most of the water is meteoric and is only heated in the thermal system. White (1957a, p. 1643) also concludes that, "In the waters [of volcanic association] that have been studied, the volcanic component probably does not exceed 5-10 per cent of the total." Similarly, Lane (1908, pp. 502-7) recognizes the multiple sources of brines and believes that many of them may consist of mixtures of connate and meteoric waters and that some connate water may have been non-marine.

Water of the Drum and Tolsona mud volcano groups and ground water in the Copper River Basin are related chemically and all could be derived in part from volcanic, connate, or meteoric water in which salts have been concentrated in ground water by evaporation of lake water and invading permafrost. Water and gas from the Drum springs are the most distinctive and seem to include a definite volcanic component. Gas from the Tolsona springs probably originates from buried Cenozoic marsh or coal deposits, or from porous, non-petroliferous beds of pre-Tertiary (Cretaceous?) age.

REFERENCES

ALLEN, E. T., and DAY, A. L.   1935.   Hot springs of the Yellowstone National Park; Carnegie Inst. Washington Pub. no. 466, pp. 1-525.

ANDREASON, G. E., et al.   1958.   Aeromagnetic map of the Copper River Basin, Alaska; U.S. Geol. Surv., Geophys. Inv. Map GP-156.

CHAPIN, THEODORE.   1918.   The Nelchina-Susitna region, Alaska; U.S. Geol. Surv., Bull. 668, 67 p.

DEGOLYER, E.   1940.   Direct indications of the occurrence of oil and gas, in Elements of the petroleum industry; A I.M.E., Mudd Memorial vol., pp. 21-5.

DUTRO, J. T., JR., and PAYNE, T. G.   1957.   Geologic map of Alaska; U.S. Geol. Surv., geol. map, scale 1:2,500,000.

FERRIANS, O. J., JR., and SCHMOLL, H. R.   1957.   Extensive proglacial lake of Wisconsin age in the Copper River Basin, Alaska (abstract); Bull. Geol. Soc. Amer., vol. 68, no. 12, p. 1726.

FERRIANS, O. J., JR., NICHOLS. D. R., and SCHMOLL, H. R.   1958.   Pleistocene volcanic mud-flow in the Copper River Basin, Alaska (abstract); Bull. Geol. Soc. Amer., vol. 69, no. 12, p. 1563.

FROST, R. E.   1952.   Interpretation of permafrost features from airphotos; Natl. Research Council, Highway Research Board, Special Rept., no. 2, pp. 223-46.

GRANTZ, ARTHUR.   1953.   Preliminary report on the geology of the Nelchina area, Alaska; U.S. Geol. Surv., open file report.

HEM, J. D.   1959.   Study and interpretation of the chemical characteristics of natural water; U.S. Geol. Surv., Water-Supply Paper 1473.

HOWELL, J. V. (co-ordinating chairman)   1957.   Glossary of geology and related sciences; Amer. Geol. Inst., Natl. Acad. of Sci., Natl. Research Council, Pub. 501, Washington, D.C.

MENDENHALL, W. C.   1900.   A reconnaissance from Resurrection Bay to the Tanana River, Alaska in 1898; U.S. Geol. Surv., 20th Ann. Rept., pt. 7, pp. 265-340.

MILLER, D. J., PAYNE, T. G., and GRYC, GEORGE.   1959.   Geology of possible petroleum provinces in Alaska: Southern Alaska; U.S. Geol. Surv., Bull. 1094.

MOFFIT, F. H.   1938.   Geology of the Chitina valley and adjacent area, Alaska; U.S. Geol. Surv., Bull. 894.

———   1954.   Geology of the eastern part of the Alaska Range and adjacent area; U.S. Geol. Surv., Bull. 989-D, pp. 65-218.

NICHOLS, D. R.   1956.   Permafrost and ground-water conditions in the Glennallen area, Alaska; U.S. Geol. Surv., open file rept.

NICHOLS, D. R., and WATSON, J. R., JR.   1955.   Preliminary report on engineering perma-frost studies in the Glennallen area, Alaska (abstract); Bull. Geol. Soc. Amer., vol. 66, no. 12.

OVCHINNIKOV, A. M., IVANOV, V. V., and YAROTSKY, L. A.   1959.   On the origin of carbonic-acid gas in mineral waters (A criticism and discussion of A. A. Smirnov's views on the nature of CO_2); Internatl. Geol. Rev., vol. 1, no. 5, pp. 51-5. (Translated by Dean Miller from a paper published in Sovetskaya Geologiya, 1958, no. 1, pp. 145-9.)

PIIP, B. I.   1937.   The thermal springs of Kamchatka: Sovet po izucheni in proizvodi-tel'nejkli sil (SOPS; Seriia Kamchatskaia vyp. 2 (summary in English, pp. 247-51).

RANKAMA, K., and SAHAMA, T. G.   1950.   Geochemistry; University of Chicago Press.

SCHRADER, F. C.   1900.   A reconnaissance of a part of Prince William Sound and the Copper River district, Alaska in 1898; U.S. Geol. Surv., 20th Ann. Rept., pt. 7, pp. 341-424.

SVERDRUP, H. U.   1956.   Arctic sea ice, pt. 6 *in* Book 1, Dynamic North; U.S. Navy, Office, Chief Naval Operations Polar Proj. Op-O3A3.

UNITED STATES ARMY.   1955.   Certain engineering aspects of the geology along the Glenn and Richardson highways, Copper River Basin, Alaska (preliminary report); Dept. of the Army, Office of the Chief of Engineers, Engineer Intelligence Study 190, compiled by Military Geol. Branch, U.S. Geol. Surv.

————   1956.   Terrain and construction materials, Denali, Alaska; Dept. of the Army, Office of the Chief of Engineers, Engineer Intelligence Study 248, compiled by Military Geol. Branch, U.S. Geol. Surv.

UNITED STATES DEPARTMENT OF COMMERCE.   Undated.   Description of triangulation stations, Alaska; U.S. Coast and Geodetic Survey no. 66.

WASHBURN, BRADFORD.   1941.   A preliminary report on studies of the mountains and glaciers of Alaska; Geogr. J., vol. 98, pp. 219-27.

WHITE, D. E.   1955.   Violent mud-volcano eruption of Lake City hot springs, northeastern California; Bull. Geol. Soc. Amer., vol. 66, no. 9, pp. 1109-30.

————   1957a.   Thermal waters of volcanic origin; Bull. Geol. Soc. Amer., vol. 68, no. 12, pt. 1, pp. 1637-58.

————   1957b.   Magmatic, connate, and metamorphic waters; Bull. Geol. Soc. Amer., vol. 68, no. 12, pt. 1, pp. 1659-82.

Essay on Palaeoclimatology from Palaeozoic to Quaternary Eras in Northern Ellesmere Island[1]

MICHEL BROCHU

GEOLOGICAL, morphological, and palaeontological evidences permit a reconstitution of the great lines of the Palaeoclimatic evolution in Northern Ellesmere Island from the Palaeozoic era (probable time of the deposition of the sediments of the plateau of Lake Hazen) up to the Quaternary. For this last geological period, in particular, the suggested conclusions are likely to be of interest in many connex fields of geology, namely biogeography and especially in relation to the extension and the characteristics of the biotopes during the Quaternary era.

[1]Abstract only.

The Origin of
Cold High Salinity Water in
Foxe Basin and Foxe Channel[1]

N. J. CAMPBELL

ABNORMALLY LOW TEMPERATURES, —1.80 to —1.98° C, and relatively high salinities, 33.75 to 34.07 parts per thousand were observed in Foxe Basin and Channel in the summer of 1955. These conditions represented a significant deviation from the normal character of the waters found in Hudson Bay and Hudson Strait. The distribution of this water is considered in terms of possible source areas, oceanography, and mechanisms of formation. It is believed that this type of water attains its peculiar characteristics locally in Foxe Basin through freezing on tide flats. By this means both the temperature and the salinity are altered as fresh water is removed in the freezing process. Subsequent mixing and redistribution of the water brings it into the deep areas of Foxe Channel where it remains relatively undisturbed for long periods of time.

[1]Abstract only.

Oceanic Observations in the Canadian Arctic and the Adjacent Arctic Ocean[1]

A. E. COLLIN

THE FIRST OCEANOGRAPHIC OBSERVATIONS recorded in the Canadian Eastern Arctic were taken by Sir W. E. Parry in Lancaster Sound in 1821. Since that time oceanographic reconnaissance has been extended throughout the Eastern Arctic including Baffin Bay, Lancaster Sound, Foxe Basin, and Hudson Strait.

In 1915 the Canadian Fisheries Expedition carried out oceanographic measurements in the region of Hudson Bay and Hudson Strait and in 1928 the Danish Godthaab Expedition completed an extensive oceanographic survey in Baffin Bay and the connecting passages. Two surveys have been conducted in the Western Arctic in the region of the Beaufort Sea and in 1954 oceanographic stations were occupied at intervals through the Northwest Passage.

Since 1954 the oceanographic coverage in the Canadian Arctic has been greatly extended through the activities of the Canadian ice-breaker "Labrador." Over 700 stations have been conducted from this ship in Arctic waters. In the Western Arctic, oceanographic surveys in the southern Beaufort Sea have been under the direction of the Fisheries Research Board. In the region north of M'Clure Strait oceanographic observations have been taken from the IGY drift station T-3 and more recently by parties of the Polar Continental Shelf Project of the Department of Mines and Technical Surveys.

A comparison of Arctic oceanographic measurements reveals that during the summer, temperatures of 0.0° C to 0.5° C can be expected in the Eastern Arctic above 100 m. Within the same depth range temperatures of —1.0° C to —1.5° C are common for the Western Arctic. In the area northwest of Isachsen, Northwest Territories, water temperatures increase with depth below 200 m to a maximum of 0.4° C at 380 m. At greater depths there is a gradual decrease in temperature to 0.0° C at 1000 m. Below 1000 m water temperatures varies little from —0.3° C.

Salinity determinations in areas of continuous ice cover in the Western Arctic show that a layer of very low salinity develops under the ice during the summer and that, in general, salinities in these regions are lower than those in the Eastern Arctic within the depth interval zero to 150 m. Below 150 m salinity content is slightly higher in the Western Arctic, being approximately 34.9°/00 at 500 m.

The content of dissolved oxygen reaches a summer maximum of 10.0 ml/1 at the surface in the Beaufort Sea and a consistent minimum of 6.4 ml/1 occurs at 150 m. In Lancaster Sound and Hudson Strait slightly lower values were recorded in September 1957 and 1959.

[1]Abstract only.

Thrust Fracture Pattern in Young Sea Ice[1]

MOIRA DUNBAR

A CURIOUSLY REGULAR rectilinear thrust pattern in young sea ice has been described by field workers in the Beaufort Sea and in coastal areas of Labrador and Greenland. This paper seeks to show, with air photo illustrations, how widespread and characteristic this fracture pattern seems to be, and to point out similarities in deformation structures between sea ice and land masses.

[1]Abstract only.

The Ground Ice in the Deer Bay Area, Ellef Ringnes Island, Northwest Territories[1]

DENIS ST-ONGE

EXTENSIVE GROUND ICE mantled by 0.3 to 0.6 m of silts occurs in the Deer Bay area on Ellef Rignes Island in the Canadian Arctic Archipelago. Profiles from construction pits and observations made on river-bank exposures and mudflows suggest that this ground ice is continuous in areas below 75 m above sea-level. The possibility of its origin owing to burial of sea ice is discussed.

[1]Abstract only.

Section Three

LOGISTICS AND EXPLORATION

Field Methods and Logistics:

Arctic Canada[1]

R. G. BLACKADAR

ABSTRACT

The Geological Survey of Canada has pioneered in the use of light aircraft for reconnaissance surveys in the north and this approach has resulted in the mapping of more than 500,000 square miles during the past eight years. The organization and field operations of certain of these projects are described. Attention is also given to some of the older and more conventional techniques, such as coastal reconnaissance, which are of value in certain parts of the north.

PROBLEMS OF TRANSPORTATION, supply, and mapping technique have faced officers of the Geological Survey of Canada almost from its inception in 1842. Fortunately, the increased tempo of development in Canada has resulted in great improvements in transportation, but as mineral development is often the reason for opening up new territory, those who are responsible for mineral exploration must often pioneer in virgin country. In the early days, this meant taking canoes into country known only to Indians or making extensive trips by snowshoe and dog-sled; nowadays it may mean developing new techniques to extend the use of aircraft.

Prior to World War II, virtually no continuous geological mapping had been done in the far north and much of the information appearing on the then current maps dated back to the mid-nineteenth century when the British Naval expeditions, culminating in the feverish activity of the Franklin Search, led to much geographical and some geological mapping.

On the mainland, scattered reconnaissance mapping had been carried out, particularly in the western part of Mackenzie District, and in areas of economic interest some detailed mapping had been done; but in Keewatin District, except for a small amount of coastal work along Hudson Bay and the epic journeys of J. B. Tyrrell in the last decade of the nineteenth century, the geological map reflected little but surmise.

Indeed, in 1940 only 11 per cent of Canada had been mapped geologically, although the Geological Survey of Canada had been active for ninety-eight years. The situation is quite different today; published geological maps now cover 43 per cent of the country or 1,650,000 square miles, and much of this increase has been in the area north of the sixtieth parallel.

The developments which are resulting in the opening up of northern Canada have depended primarily on the use of aircraft, and, unlike older parts of Canada where development was a gradual movement out from established centres, we find, in the north, pockets of activity in widely separated places often accessible only by

[1]Published by permission of the Director, Geological Survey of Canada.

aircraft. Such a mode of development requires an entirely different approach in geological surveying if information is to keep pace with development (ideally it should precede development), and it is fortunate that aircraft suitable to the requirements of geologists became available when they were needed.

Much of the remarkable progress made since 1946 is the result of the introduction of helicopters and light fixed-wing aircraft for survey work as well as transportation. Table I shows the number of air operations since 1952, the areas mapped, and the types of aircraft used.

TABLE I

AIR OPERATIONS GEOLOGICAL SURVEY OF CANADA 1952-9

Year	Name	Area mapped	Type of aircraft
1952	Keewatin	57,000 sq mi	2—Hiller 360
1954	Baker	67,000 sq mi	2—Bell 47 D-1
1955	Thelon	61,000 sq mi	2—Bell 47 D-1
1955	Franklin	100,000 sq mi	2—Sikorsky S-55
1957	Mackenzie	100,000 sq mi	2—Bell 47 D-1
1958	Western Queen Elizabeth Islands	23,800 sq mi	1—Piper Super Cub PA-18A
1958	Pelly	incomplete	1—Bell 47 G-2
1959	Coppermine	65,000 sq mi	2—Bell 47 G-1
1959	Banks-Victoria	110,000 sq mi	2—Piper Super Cub PA-18A

Impressive as these airborne operations are, they have by no means completely replaced older methods, especially where detailed rather than regional information is desired. One can hardly refer to these older methods as "more conventional," for the use of aircraft has become so widespread in surveying techniques that such projects are no longer novel and now they too must be considered conventional. For example, a light Piper Super Cub aircraft, first used in the far north in 1956, was used by the Geological Survey in 1958 in the Western Queen Elizabeth Islands and by 1959 had become accepted as a standard aid in geological mapping by both government and private organizations.

OLDER METHODS

Prior to 1952, field work of the Geological Survey in northern continental areas was confined mainly to mapping standard 4 mile to 1 inch map-areas. Here the network of lakes and rivers provided access to most parts of a map-area. In this type of mapping, fixed-wing aircraft, usually Norseman or Beaver types, are used to move the traversing teams from one navigable area to another and to supply the field parties. The aircraft are only required intermittently and the responsibility for providing fuel is usually left with the operator.

Field work has been carried out in the Arctic Archipelago on a continuing basis since 1949, using both older methods, such as canoe, boat, or dog-team travel and, since 1955, a combination of both old and new, depending upon the region.

Canoes or small schooners have been used along the coasts of Baffin Island. The schooners, technically called peterheads, are manned and usually owned by a local Eskimo. In practice, a party operating in this manner consists of two geologists, two assistants, and a crew of three or four men. Traversing parties are placed on shore and walk from tidewater to tidewater. Because the base is mobile, there is no need to return over areas previously traversed and the use of a relatively large, though not excessively roomy boat for living quarters obviates the time-consuming task of pitching camp. The peterhead boat has a shallow draft and can reach most parts of a given coast. They are powered by 4-cylinder inboard gasoline engines and carry sails for emergency use. The cost varies from $25 per day for an open boat to $40 per day for the larger peterhead type. This includes the services of the crew, who are hired and paid by the owner of the boat.

In the Arctic islands, the snowfall is light and the land is snow-free and thus available for geological investigation long before the water routes are ice-free. To take advantage of this condition, reconnaissance survey teams have often combined spring travel by dog-team with water transportation in the summer. As in the case of boats, camp can be moved while the geologist is traversing. Although overland travel by sled becomes impossible in most regions during June, local conditions often permit travel by dog-team on the sea ice to within a week or ten days of the commencement of the open water period.

Dog teams have been used in widely separated parts of the north. For projects in the Queen Elizabeth Islands, teams have been flown from Resolute to the field area and where prior information indicated a paucity of game, dog food has also been taken in. Elsewhere teams have been hired locally and on-the-spot hunting has supplied food for them. Costs vary from place to place, but are usually between $10 and $12 a day for team and driver. Under reasonably good snow conditions, a dog can pull about a 100-pound load and, as the average team consists of ten to fifteen dogs, well over half a ton can be moved per team.

Because there are few navigable rivers in the Arctic Archipelago, inland geological exploration was a difficult and time-consuming process before the advent of light fixed-wing aircraft and helicopters.

AIRBORNE OPERATIONS

Introduction

About ten years ago, as field work was extended to more remote areas, it became apparent that the task of the Geological Survey, to complete at an early date an economical and otherwise acceptable geological reconnaissance of Canada, was a task not only increasingly laborious, but actually impossible to fulfil with methods then in use, without a vast increase in staff and expenditures. Yet it was obvious that if the then impending development of the north was to proceed in an orderly fashion, it was essential that a complete geological reconnaissance be made available to the public at as early a date as possible. If conventional methods were not acceptable, then new services would have to be adapted to the needs of the Survey.

Helicopter Surveys

Light helicopters had first been used by the Department of Mines and Technical Surveys in 1950, to assist topographical surveys, when the need for a rapid completion of exploratory work was as keenly appreciated as in the Geological Survey. The success of this venture led to the planning of a long-range programme of geological exploration centred on the helicopter. Common sense dictated that the first use of helicopters should be in areas presenting a minimum of operational and terrain problems, thus permitting the basic operational technique to be quickly appreciated and then to extend the technique, modified wherever necessary, to more exacting areas. The relatively accessible, low-lying, barren grounds of southern District of Keewatin were chosen for the first operation. The operation proved eminently successful and, as had been anticipated, the experience gained in this summer proved invaluable in the planning and execution of two further operations (Baker, 1954 and Thelon, 1955) in similar but more remote parts of the North-west Territories mainland. On these projects, an area of some 10,000 square miles was divided into a series of pie-shaped wedges each with a radius of about forty or fifty miles. This resulted in about eighteen traverse routes, involving a little more than 100 miles each. When a block was completed, the camp was moved by fixed-wing aircraft to the centre of another block and the process was repeated. In more recent years, similar projects have been carried out on the far northern northern islands, in the upper Mackenzie valley, and in central Yukon, as well as in regions south of lat. 60°.

Operations carried out in the Barrens where Precambrian rocks predominate have used the helicopter primarily as a traversing tool, thus providing a low-flying observation platform from which the geologist can make observations, rather than from canoes or from foot traverses. In contrast, those projects carried out in areas where younger strata predominate have utilized the helicopter to place field teams in critical areas where detailed work is indicated, as well as to make low-level observations and trace easily recognizable geological contacts.

During "Operation Mackenzie," which centred on the upper Mackenzie valley, helicopters were used to obtain stratigraphic data, to map bedrock and structures, to map surficial deposits, to set out a geologist and an assistant for one day, and to establish subcamps. In "Operation Franklin" in the Arctic Archipelago, the helicopters were also used for basic transportation, because the operating capabilities of fixed-wing aircraft would not be sufficiently flexible under prevailing physical conditions.

To give detailed operating figures or comparative data for recent operations would merely duplicate material already available in Bulletin 54 of the Geological Survey, and therefore only a generalized outline of such operations will be given.

Table II contains comparative costs for the major northern helicopter operations of the past eight seasons.

In order to avoid the drawbacks of a "crash programme," planning for all aircraft operations is begun twelve to eighteen months in advance of the actual field work. The area to be surveyed, which varies between 35,000 square miles, the optimum for one helicopter, and 100,000 square miles for more complex operations, is chosen and the objectives to be attained are formulated. These may vary from the presentation of a reconnaissance map of bedrock and surficial geology and the outline of favourable prospecting areas in Precambrian terrain, to the delineation of potentially valuable oil or gas areas and the extension of stratigraphic information to regions previously unknown in areas of younger rocks. Early in the planning of a project, an intensive study is made of the available air

TABLE II

Mapping Costs: Helicopter Operations*

Operation	Area mapped (sq mi)	Cost/sq mi mapped ($)
"Keewatin" (1952)	57,000	3.63
"Baker" (1954)	67,000	2.15
"Thelon" (1955)	61,000	2.21
"Franklin" (1955)	100,000	3.17
"Mackenzie" (1957)	100,000	1.68
"Coppermine" (1959)	65,000	2.15

*These costs are exclusive of the salaries of the geologists, the cost of equipment, and general overhead.

photographs and existing geological information. All parts of Canada are now covered by trimetrogen air photographs and within a few years vertical coverage, now available for a large part of the north, will also be completed. In "Operation Franklin," the consideration of earlier geological information devolved into a study of reports dating back 100 years and more.

Following this study and with the over-all objectives of the project in mind, sites are chosen for base camps and the over-all traversing programme is laid out. In general, it is found that from ten to twenty traverses can be run from each base if helicopters are used as traversing tools. The setting up of traverse routes permits the evaluation of the over-all gas consumption. In most operations, both fixed-wing aircraft and helicopters will be used and, if they can be chosen so that all use the same octane rating of gasoline, considerable simplification in caching arrangements results. For most Geological Survey operations, fuel is sent to the proposed area in the summer prior to the project, thus taking advantage of cheaper water routes. However, if gas cannot be sent in ahead of time, it can be flown from a railhead or other distribution point by fixed-wing aircraft. The fuel is distributed throughout the field area in the spring prior to break-up, using ski-equipped aircraft. Food may also be sent in advance, but if distributing centres are within range it is more satisfactory to service the camp from these centres, thus providing fresh food and mail service. To date, tents have been used almost exclusively for housing. In recent years, use has been made of propane gas for cooking, although naphtha is still favoured for heating in areas where wood is not available.

Although different areas require different approaches, most parties are weighted in favour of experienced geologists with fewer student assistants than are common on standard mapping parties. The reasons for this choice are varied, but of major importance are the facts that in this form of traversing there is little opportunity for checking work, and also that in some areas, notably among younger rocks, the observer must often grasp the significance of a local fact to the regional setting. Also, it has been calculated that the cost of one helicopter-geologist hour is about $350, which certainly emphasizes the need for experienced men on such projects.

"Operation Franklin," carried out in 1955 in the Queen Elizabeth Islands, required variations in the techniques developed for operations on the mainland. In the area to be surveyed, there was but one year-round airfield and a paucity of lakes, unpredictable sea-ice conditions, and long periods of low overcast that required low-level flights, all indicating that helicopters would have to be used exclusively. An assessment of the distances and loads involved indicated that helicopters heavier and larger than those previously used (Bell 47 D-1) would be required, and the Sikorsky S-55 was chosen. But, unless fitted with floats, this machine is not permitted to fly over extensive stretches of open water, and, because floats would have taxed the carrying capacity of the machines and have served no other useful purpose, they were not employed. Thus an operational plan which avoided crossing open water had to be evolved. This requirement introduced a limiting factor, required strict timing, and had the effect of reducing the density of field stations.

In the spring, gasoline deposited the previous year by sea transport at Resolute was airlifted by DC-3 to three main sites and several intermediate staging sites. Field work using the helicopters began on June 13 and terminated September 15. The survey technique primarily involved placing a geologist and assistant in a critical area and after three or four days moving them to another location. An average of eight or nine geological teams were involved, and, in the course of moving them, continuous observations were made from the air by a geologist who always accompanied each flight and also served as navigator.

With eight or nine parties in the field at one time, fairly rigorous control had to be exercised at the base camp and three geologists were always attached to the base. These men compiled the field data as it was accumulated, prepared flight plans and evaluated the fuel requirements for each flight, and made on the spot adjustments of the logistics worked out prior to the field season. They also prepared rations and were always available to accompany a flight. Because of the continuous daylight in the north in summer, operations were conducted on a twenty-four-hour basis. Original planning called for three pilots and two mechanics, but for part of the time only two pilots were available.

In summing up the helicopter operations the following points can be made. Operations "Keewatin," "Baker," and "Thelon" were carried out in country where older techniques would have called for canoe travel. It is estimated that in three seasons these projects completed what would have required forty party-years using the older methods for the same coverage, 185,000 square miles; had these projects been attempted in three years employing the older methods, then ninety to one hundred and twenty geologists and assistants would have been needed each season. The stratigraphic data obtained in the plains sections on "Operation Mackenzie" could have been obtained in three or four seasons using the same number of geologists, but employing conventional methods of travel. However, much of the information obtained from beyond the plains could not reasonably have been acquired using conventional methods.

LIGHT FIXED-WING AIRCRAFT

The Piper Super Cub has proved most successful for use in the Arctic, despite the dire predictions made when the first projects were undertaken. These aircraft are specially equipped with very large low pressure tires (5 psi pressure; 25 × 11 × 4 inch tire) which permit landings on quite rough terrain. The take-off run of the aircraft is about 200 feet and the landing roll is somewhat greater, both distances of course depending on the wind speed. In Canada, this aircraft is licensed to carry a disposable load of 700 pounds and when fitted with the large tires has a cruising speed of 95 mph. The aircraft has a fuel tank holding thirty gallons, and the weight of this gasoline plus that of the pilot and the emergency equipment must be subtracted from the 700-pound disposable load. Thus at maximum cruising range the aircraft can carry 250 pounds payload — adequate for the average geologist! The aircraft is well adapted to making landings for spot observations and for placing a geologist and camping equipment at sites where detailed information is required. An example of the versatility of this aircraft is shown by the 249 landings made at different points on Melville, Prince Patrick, Borden, and other adjacent islands during the 1958 operation. A Piper Super Cub supported operation is considerably less expensive than the comparable helicopter project would be. The gas consumption of the Cub is six-and-one-half gallons per hour or about half that of the helicopter now commonly used (Bell 47 G-1) and the basic charter costs are also considerably lower. The aircraft have been used for transportation in moving base camps from point to point, to carry a geologist to critical outcrops to be examined, and to serve as a vantage point from which formational contacts and other structural features could be recognized and located accurately on air photographs and maps.

No paper could hope to cover all the complexities which arise in planning operations in the north, but an effort has been made to give some idea of the detailed planning necessary to ensure success in carrying out field work in Arctic regions.

Modern Methods of Exploration in the Canadian Arctic Islands with Special Respect to Logistic Problems

T. A. HARWOOD

ABSTRACT

The author will examine, from 1946 onwards, the development of present methods of exploration and logistics in the Arctic islands and northern Canada; the types of ship involved, ice conditions generally encountered and methods of cargo handling. Some discussion will be directed to aircraft logistics and types of planes so far used in the Arctic islands, and relate this to the bearing strength of ice. Consideration will be given to the effect of climate on logistics and reference will be made to some specialized foods produced to reduce logistic loads. The paper as a whole will attempt to relate climate and weather, sea ice, navigation, permafrost conditions, and communications to logistics.

THE CANADIAN ARCTIC ISLANDS make up one of the last areas in the world to become accurately known. Exploration has been going on since 1576 and yet until very recently (1949) many of the coastlines and the major features were inaccurately sketched and detailed mapping was almost completely lacking. As late as 1948 two large islands in Foxe Basin were added to the map and since that time other major inaccuracies have been corrected. The situation has lately been changed by the R.C.A.F. programme of survey trimetrogon photography and the present programme of vertical photography by private companies, "Operation Franklin" by the Geological Survey, and many other major hydrographic, oceanographic, and geodetic surveys in the past decade. Despite all this, detailed exploration—geographical, geological, and oceanographical—will go on for years.

Broadly speaking, exploration in the Canadian Arctic has fallen into four phases, each associated with a specific motive which gave impetus to this exploration. They are: (1) the search for the Northwest Passage; (2) the search for the Franklin Expedition; (3) the lure of the North Pole; and (4) in recent years exploration for its own sake, mainly for geological or geographical information.

EARLY HISTORY OF EXPLORATION IN THE ARCHIPELAGO

Many people today underestimate the scale of effort in the early days in the search for the Northwest Passage and therefore through the Arctic Archipelago. One of the first voyages undertaken was by Martin Frobisher, who in 1576 discovered Frobisher Bay and found what he considered an exciting mineral deposit. As a result he organized a further expedition. When Martin Frobisher finally assembled his convoy for his mining ventures, he found he had fifteen ships under his command. As everyone knows, the operation was a failure. However, the interesting part is that it was not until 1954 that a larger convoy under a single commander entered the Canadian Arctic. This was Task Group 48.2 under the flag of a Canadian in H.M.C.S. "Labrador."

At the same time it is of interest to note that Frobisher, and in fact John Davis, Hudson, and Baffin, who by 1612-13 had delineated all the major entrances and channels on the European side of the Arctic Basin, made their voyages at the beginning of the period of 1550-1850 which has been called the "little ice age." The first attempts at the Northwest Passage and thus into the Arctic islands were therefore commenced at an unfortunate time. From 1616 (the dates of Baffin's last expedition) until the end of the Napoleonic Wars there was no serious seaward exploration along our northern coasts until the first British naval expedition of 1818 under Sir John Ross. Ross, who followed Baffin's routes and saw Lancaster Sound but mistook it for a closed bay, turned home. W. E. Parry, his second in command, was not defeated, however, and the next year with the Admiralty blessings sailed right through Lancaster Sound to Melville Island where he wintered at "Winter Harbour."

This was the first penetration into the Arctic Archipelago. The vexatious problem of discovering the Northwest Passage still persisted in the minds of the English, however, and from 1821-5 Parry made two further voyages, one of which ended in near disaster when his ship was wrecked on the east coast of Somerset Island. By this time the Admiralty was losing interest and private enterprise in the shape of a wealthy distiller, Booth, financed a second voyage for Ross in 1829. Again a ship was lost.

Between 1829 and 1840 the mainland coast was traced by Simpson, Rae, and Franklin. All this brings us to the most famous of all Arctic expeditions, and perhaps ultimately the most fruitful, for it led directly to the discovery and crude delineation of nearly all the coasts of those Arctic islands which interest us today. This was Sir John Franklin's expedition of 1845 which sailed into Lancaster Sound and was never seen again. After twelve years and some twenty separate expeditions, both public and private, it was established that Franklin had gone exactly where his orders sent him and possibly owing to this fact left very few records.

The Franklin Expedition more or less closed the seaward exploration of the Arctic islands for over forty years. In 1898, however, the famous and distinguished scientist and explorer, Otto Sverdrup, in the equally famous ship "Fram," began a four-year Norwegian expedition. Sverdrup, thwarted in his intention of exploring North Greenland from the Smith Sound area, wintered in his second year in Jones Sound. In four years of hard sledging Sverdrup more or less finished the exploration of the Arctic islands, leaving only Brock, Borden, Meighen, and Lougheed Islands to be discovered by Stefansson during the Canadian Arctic Expedition of 1913-18.

Prior to the Canadian Arctic Expedition two voyages were made by Canadian ships into the Arctic islands for purposes of sovereignty, one of which, commanded by Captain Bernier, wintered at Parry's old quarters at Winter Harbour on Melville Island. Since 1918 many small parties have worked in the Canadian Arctic and much exploration has been done, although little work was undertaken in the Arctic islands, as Ellesmere and Baffin islands attracted the most interest.

1946 ONWARDS

At the end of World War II it was realized that the Arctic Basin has considerable strategic significance; its importance to the great powers and possible great powers—North America, Western Europe, Japan—China—can easily be appreciated. This strategic importance, however, can really only be measured in terms of communications across it, that is, by aircraft and now by the ICBM. Any system of defense must include the capability to detect attacking aircraft. This can of course, as is well known, be done by radar lines; in future some similar techniques may have to be developed for ICBM's. The obvious and basic consideration for the major powers, therefore, was to attempt to push their aircraft warning and defensive systems northward.

In addition, about the same time in Canada, as a result of proposals put forward as early as 1938-9 by McPherson and Hubbard, it became apparent that a northerly extension of the meteorological network of North America should be considered. It was Hubbard who undertook the task of arousing interest in the United States and Canada. His plan involved the establishment of two main stations, one in Greenland and one in the Canadian Archipelago, which could be reached by sea supply. The main stations would then serve as advance bases from which a number of smaller stations on a 250-mile grid might be established. The plan required that surface ships proceed to the central base locations during the summer months and in the following winter stations would be built and preparations made for spring activities when flying conditions in the Arctic were most favourable. After the site was chosen and a suitable landing strip located on the ice, supplies would then be flown in by heavy transport aircraft.

Thus in 1946 the United States Weather Bureau jointly with the Canadian Meteorological Service installed the first of the joint Arctic weather stations at Resolute Bay. The network of satellite stations was extended to the Arctic Archipelago in the following years on a 250-mile grid to Mould Bay, Isachsen, Eureka, and Alert. This new method of development and exploration was dependent entirely on suitable ice-breakers and aircraft. Fortunately, as a secondary result of World War II, ice-breakers of considerable power and size were available and could be used for the initial supply and re-supply of the main bases. Similarly the C.54 (DC.4) also became available and could be used in the heavy airlift involved in setting up the satellite stations. Thus, although he was killed in its realization, Hubbard's vision of a high Arctic meteorological net came into being.

It has been the existence of these joint Arctic weather stations in the Arctic islands together with their airstrips and other facilities which have made it possible to carry out the more complex exploration of recent times in this area. Furthermore, it is from the lessons successfully learned and applied in the setting up and re-supply of these stations between 1946 and 1952 that allowed the development of the newer and larger centres like Foxe, Thule, Cambridge Bay, Cape Parry, and many others, which have in their turn vastly eased the problems of the modern developer and geological exploration companies.

For each operation under present-day conditions in the Arctic, there is always a fairly complex logistic problem to be solved. Obviously the nature and complexity of the problem depends on the size of the operation, but in general some ground rules on *modus operandi* can be laid down. Many of these rules have been developed over the years and have been incorporated into the standard operating procedure of the United States Navy for operations in the Arctic and Antarctic. They are the key to a successful operation and can be summarized for discussion under the following headings: (*a*) transport, (*b*) man, (*c*) climate, and (*d*) planning.

Transport

In all operations north of the mainland the crux of the logistics problem is the successful sea operation. There are a few simple but important rules for any sea-borne operation which has to unload over an open beach, and these have to be followed if success is to be achieved: (*a*) the operation must take place when ice conditions allow but not so far advanced in the season to disallow the efficient sorting and redistribution of cargo; (*b*) the ship should be combat-loaded and cargo-palletized; that is, first requirements must come ashore first; (*c*) the beach areas chosen must allow under the worst conditions stowage of all cargo, or at the best that percentage calculated to remain on the beach when the landing operation and the removal operation are running hand-in-hand; (*d*) the number of landing craft, cranes, bulldozers, stone boats, and trucks necessary for beach operations must be related properly to the ship's speed of discharge; (*e*) the cargo beach areas must be properly marked; (*f*) adequate food, shelter, and sleeping quarters must be provided on the beach for the beach crews; (*g*) fuel oil and if possible gasoline should be unloaded in bulk and stored in drums on the beach.

Of all these simple and fairly obvious rules, the one most violated in the past and which has caused the most trouble is that regarding adequate beach areas. A thousand tons of cargo covers approximately 20,000 square feet and additional space must be left for handling. Once the beach parties become cramped for space, disorganization begins.

It is apparent from the above that any ship chosen must have heavy booms, twenty-five tons minimum, and a major lift of fifty tons to handle the necessary heavy equipment.

Palletization deserves some mention, because it affects both the ship and the aircraft. The size or volume of palletization of course depends on a number of factors; the major considerations if an airlift is involved are: (*a*) the size of the aircraft door; (*b*) the floor loading; and (*c*) the fork lifts available. Naturally, the larger the pallet the more efficient the operation, provided of course that, once on board the aircraft, the pallet can be moved, a detail often forgotten. Palletization should also be carried out prior to loading the ship.

So far it has been assumed that the sea operation must be combined with an air operation in the high islands. This may not be necessarily so; in fact, heavy tractor-train operations are probably quite feasible over reasonable distances. Here again are some simple rules which should be kept in mind with regard to the efficiency and safety of a tractor train. (*a*) The break-even point between large aircraft using ice strips and tractor trains should be determined. At distances over 200 miles there is every probability that in the Arctic islands the aircraft may prove the cheaper, particularly in midwinter. (*b*) Ice thickness and therefore strength of the ice in the channels between the islands are at the maximum in early spring, but once insolation exceeds radiation the strength of sea ice quickly diminishes. Tidal currents also erode the undersurface of sea ice rapidly. Thus, if at all possible, the route should be reconnoitred in the summer for such danger points (tidal swirls). (c) The tide crack always constitutes a danger and an obstacle, particularly in the spring. Any approach from the shore to the beach must always be reconnoitred.

If, however, it is assumed that an aircraft operation is both essential and economical, then one must also assume, certainly in the preparatory stages of an exploration programme, that the landings will be made on an ice strip. Again one can give a few simple rules which if followed will provide maximum safety and economy in the operation of such a strip and thus of the airlift. (a) The behaviour of sea ice and freshwater ice when loaded differ radically and different bearing strengths must be assumed. Tables of ice strength are given in SIPRE Report no. 36 and should be consulted. (b) For any kind of ice over shallow water there are differing critical speeds for taxiing. If a plane taxis at these critical speeds, the ice will inevitably fail, irrespective of thickness and strength. (c) Runway discipline must be maintained. De-icer and hydraulic fluid and fuel oils are fatal to an ice strip at the most critical period, that is, in the spring. (d) Even should apparent failure take place under the undercarriage, final failure is not likely to take place and the aircraft can safely be moved.

Aircraft operations on the whole are very much the same as farther south except for pre-heating engines. It has been both the U.S.A.F. and R.C.A.F. experience that in midwinter a safe ratio of Herman Nelson heaters per engine at any fixed overnight base is not less than 5/4 or five per four-engine aircraft.

Man

In all such operations man must be involved, and in the end it is his heat loss which determines whether he is cold and inefficient or warm and efficient. In the barrens the most important source of heat loss to man and buildings is wind chill, that is, loss of heat in kilogram calories per square metre per hour from a surface over which turbulent air is flowing. To this there are two solutions: the first is proper clothing; the second is to choose operating areas where wind is light. The latter can usually be obtained if some worth-while preliminary reconnaissance is carried out. Clothing, however, is a problem which even to date is most controversial, but two points stand out: (a) the clothing must not be so bulky as to impede the movement of the man; (b) it must, to be adaptable to changing temperature and work output (of the man), be layered. At the moment there are a number of such articles of clothing available in the United States and Canada, the best of which is probably patterned on the Canadian or United States Army type winter clothing.

There is a low limit of temperature below which little normal activity can be undertaken. By experience this has been found to be about —30° F or 1600 wind-chill units.

Man, however, has to maintain his fuel supply and here is another area of great controversy—food. The Canadian Army lays down a ration which ranges from approximately 4500 to 5000 kg calories per day for the Arctic under combat conditions. The problem is not the reduction of bulk, which can easily be achieved by the provision of a high-calorie diet, such as fat or pemmican, but that such a diet is often inedible under conditions of extreme tension. Moreover, in winter the melting of snow and ice for water to rehydrate such foods becomes a truly wearisome business. The tendency therefore in the past ten years has been, where possible, to use pre-cooked frozen foods, although the susceptibility to solar radiation and other possibilities of thawing increase the bacteriological hazards from such food. This in many ways outweighs the disadvantages of dehydration and for this reason the Defence Research Medical Laboratories have gone back to dehydration. Most successful so far has been the freeze-drying process whereby the time required both to dehydrate and to rehydrate steaks, chops, and roasts has been greatly reduced (Hulse, 1956). But what is perhaps more important is that the product is stable, easily reconstituted, most palatable, and makes cooked meats hardly distinguishable from the fresh. Steaks, chops, and chickens in packages up to 50 lbs. in dry weight were successfully used on the D.R.B. Lake Hazen expedition in northern Ellesmere Island. It is the Defence Research Board's feeling that despite the disadvantages of rehydration, the weight and volume advantage of compressed dehydrated food is so great that it will be the basis for future food research for combat troops and perhaps field exploration in the Arctic.

Climate

The more obvious climatic conditions which affect the means of transport in the Arctic Islands can be summarized as sea ice conditions, weather, and permafrost conditions.

Sea ice conditions. In 1946 the sea ice conditions in the Canadian Arctic and particularly in the Arctic Archipelago were to all intents and purposes unknown, but commencing that year a United States Navy Task Force carried out a long-range air reconnaissance of ice conditions in Baffin Bay and the Arctic islands in connection with the establishment of the joint weather stations. Similar flights, with aircraft assigned by the U.S.N. Task Force employed in the yearly re-supply of the stations, were continued into 1949 when the R.C.A.F. took over the whole pro-gramme, but it was not until 1952 that any of the ice observations taken either by the Canadian or by the U.S.N. groups were made on a systematic basis with a view to both historical and synoptic records. Initially the records obtained were held by the U.S.N. Hydrographic Office with the information freely available to Canada. In 1956 the Joint Committee on Oceanography, now the Canadian Committee, set up a working group for "ice research in navigable waters" and at the same time the Meteorological Branch of the Department of Transport set up an ice reporting and forecasting organization for the eastern Arctic. This group's task has now been extended to cover the whole Canadian Arctic. Ice information, historical or synoptic, about any particular area or affecting all ship operations, can now be obtained in Canada or the United States from the following: (*a*) the Ice Central and Central Office of the Meteorological Branch, Department of Transport, Toronto; (*b*) the Geographical Branch, Department of Mines and Technical Surveys, Ottawa; (*c*) the secretary of the "Working Group on Ice in Navigable

Waters," Defence Research Board, Department of National Defence, Ottawa; and (d) the U.S.N. Hydrographic Office, Washington, D.C.

It is hoped that, within a year, through the research organization mentioned earlier, there will also be available to the public an "ice probability atlas of the Canadian Arctic" which will show the probability of encountering ice, together with its navigability, at 325 critical spots in the Canadian Arctic north of the mainland. In the meantime, good references are Dunbar's paper, "The pattern of ice distribution in the Canadian Arctic," and the Arctic pilot. The limits of the fast ice given in this paper for midsummer are representative of conditions as they exist today; that is, it is impossible to penetrate by any ship northwest of a line extending from Axel Heiberg Island–southern Ringnes Islands–central Melville Island–Prince Patrick Island.

This line of fast ice in midsummer is naturally a climatic phenomenon and follows very closely the average 35° isotherm for July and August. Ice conditions in the Arctic islands can be divided as follows: (a) land fast ice—in the waters northwest of a line from Prince Patrick Island through to Ringnes and Axel Heiberg islands; (b) close pack—in the waters of Viscount Melville Strait, Norwegian Bay, Barrow Strait, and south of the line mentioned in (a), and in northern Robeson Channel but excluding the northwestern fjords of Ellesmere Island; and (c) open pack and open water—Lancaster Sound, Prince of Wales Strait, Nansen Sound, Baffin Bay, and the channels leading south from Lancaster Sound, but not Committee Bay or the channels around King William Island.

Weather. It is natural that owing to the existence of the joint Arctic weather stations there is now a wealth of information concerning the high Arctic, but for ships and heavy aircraft the following details related to climate must be kept in mind in planning an operation. (a) Ice conditions can vary widely and extraordinarily from year to year. (b) The combination of wind and the darkness which implies total radiation makes the effects of extreme temperatures far more severe on men and machinery than a comparable temperature in lower latitudes where there is some insolation. (c) Without radio aids VFR flying is severely restricted after the air temperatures rise above 28°-32° F and is of course difficult in twilight and darkness. The only good VFR flying weather is between early March and mid-June. Thus major air activity by small aircraft should be planned for these periods. (d) The frequency of blowing snow and/or whiteout at any airstrip is more important than fog at any present or future strip in the Arctic islands. Underestimation of this factor leads to many unnecessary abortive flights. (e) For those working in the eastern islands, the so-called Baffin Bay Trough low has an influence out of proportion to its pressure gradient simply because migratory lows enter it from time to time. When this happens sea and air operations can be seriously affected, particularly in late summer and in the fall. (f) Throughout most of the archipelago, precipitation is extremely light and in some places the driest in the world. As a result, large camps, if poorly located in the spring, will inevitably suffer from a lack of water.

Permafrost. Permafrost conditions will seriously affect mobility of vehicles during the thaw period and if beaches or distribution areas and camp sites are chosen where the soils are cohesive, or the granular material can break down into cohesive materials, complete immobility may arise. Extreme care should also be taken in the choice of routes for vehicles and, where possible, roads should be built. It is fully realized that vehicles are now available which will in most cases traverse such terrain, but such vehicles add cost to the operation far beyond their real return since they increase what is known as the logistic "tail."

Planning

From these factors it can be seen that operating in the high Arctic must require a relatively higher support and logistic effort than farther south. The question is how much more. A good answer has been obtained from the experiences of the R.C.A.F. and the Geodetic Survey on the shoran survey operations carried out between 1946-7. The figures obtained from both these agencies are given here and have been put on a man/year basis. It was found that: (*a*) in the south or habitable areas four tons per man/year were required; (*b*) in the north but below the tree-line eight tons per man/year were required; (*c*) above the tree-line to the Arctic coast, northern Quebec, and southern Baffin Island twelve tons per man/year were required; (*d*) in the high Arctic and the Arctic islands sixteen tons per man/year were required.

The figures do not include aircraft fuels or aircraft support at the main stations when an airlift is involved. In confirmation of these figures, the supply and maintenance of the joint weather stations (satellites) require an airlift of about nineteen tons per man/year. The Russians claim their floating stations need approximately sixteen to nineteen tons per man/year, and the Defence Research Board "Operation Hazen" in northern Ellesmere Island ran about fourteen tons per man/year. It should be re-emphasized that these figures do not include aircraft fuels, or any more exotic piece of heavy equipment than a TD-14 type tractor to maintain the airstrips, but do include building material, stoves, power plants, etc. It is now possible to plan for such an operation but, because the sealift is a yearly event, much of the planning must be done far in advance. Experience has shown that such planning should follow roughly the sequence given below. (*a*) High-level planning conference for general details—housing, personnel, scale of programme, etc. D-12 months. (*b*) Staff planning of weights and cubes from port to beach and onto site, adding fuels required for lying from main base. D-11 months. (*c*) Arrangement of administrative details with government agencies, such as ice-breaker escort, licenses, etc. D-11 to D-10 months. (*d*) Arrangement of charter for ships and aircraft and plan airlift. D-10 months. (*e*) Visit to the area and completion of arrangements for storage at sites. D-10 months. (*f*) Purchasing and ordering and completion of arrangements for storage, arrangements for stevedoring in a southern port and on the beach. Continuing from D-9 to D-2 months. (*g*) Final planning conference, readjustment of cube, weights, fuel loads, charters, etc., and priorities for beach and unloading. Details for ice-breaker escort. D-2 months. (*h*) Final conference with stevedores and shipowners. D-1 month. (*i*) Combat load. D-½ month. (*j*) Sail ship, arrive site. D-Day ± 10 days. (*k*) Unload and rearrange material for onward shipment. D-Day. (D-Day will probably be in late August.)

Transfer of material from the beach to airstrip should in all probability take a further two weeks. This may be done in conjunction with an airlift in the fall, but not, of course, if the airlift is in the spring. Period of airlift will depend on size of aircraft available, number of crews available, speed of loading, and the minimum period with ramp-loading aircraft (twenty-four-hour flying in full daylight will be approximatey two weeks per 400 tons on a 250-mile leg).

It can be seen that at least twelve to eighteen months must be allowed for the logistics phase in any plan for an exploration programme in the Arctic Archipelago by any large party. Costs therefore, the least of which will be the ship charter, are high.

Extremely careful planning and organization is therefore obviously necessary. Furthermore, in the Arctic high morale is an essential; poor organization will inevitably lower morale with resulting confusion and a possible breakdown of the programme. Strangely enough the time for preparation for an Arctic operation has not been materially reduced since Frobisher's or Sir John Franklin's day.

CONCLUSIONS

Exceptionally successful expeditions based on these methods of operation have been carried out in the Arctic islands since 1946. These operations have been carried out by government agencies such as the R.C.A.F. and the Geodetic Survey 1950-7, "Operation Franklin" by the Geological Survey, "Operation Hazen" by the Defence Research Board, the site surveys of the DEW Line by Canadian Aero Surveys and Spartan Air Services, and numerous oceanographic and hydrographic operations by the ships of the U.S.N., Royal Canadian Navy, and the Canadian Hydrographic Service. Each has played an important part in the mapping, charting, and geological reconnaissance, so that the high Arctic and the Arctic Archipelago are now reasonably well known from a reconnaissance point of view. There is, however, a vast amount of exploration and research still to be undertaken. Perhaps the most important is further research into the role the over-all heat budget of the Arctic and the effect of darkness during the Arctic night have on hemispheric circulation, and consequently on the problem of long-range forecasting. This is obviously what Hubbard had in mind when he first proposed the joint Arctic weather stations. It appears that both meteorological and geological research in the area will go hand-in-hand during the next decade. Co-operation between the disciplines and agencies involved is, therefore, essential to the mutual solution of their many problems, not the least of which will be logistics.

REFERENCES

ANDERSON, L.   1958.   A theoretical analysis of sea ice strength; Trans. Amer. Geophys., vol. 39.

BUTKOVITCH, T. R.   1956.   Strength studies of sea ice; SIPRE Research Paper no. 20, Snow Ice and Permafrost Research Establishment, Wilmette.

CANADA, CANADIAN HYDROGRAPHIC SERVICE.   1959.   Department of Mines and Technical Surveys, Pilot of Arctic Canada; vol. I. Queen's Printer.

CANADA, DEFENCE RESEARCH BOARD.   1957.   Bathymetric Chart of the Arctic Ocean; Chief Cartographer's Office, Dept. of Mines and Technical Surveys, Ottawa.

CANADA, DEPT. OF TRANSPORT.   1952.   A review of the establishment and operations of the joint Arctic weather stations at Eureka, Resolute, Isachsen, Mould Bay and Alert and a summary of scientific activities at these stations 1946-51; Dept. of Transport, Ottawa.

CYRIAX, R.   1939.   Sir John Franklin's last expedition; London.

DUNBAR, M.   1954.   The pattern of ice distribution in Canadian Arctic seas; Trans. Roy. Soc., Canada, ser. III, sect. IV.

DUNBAR, M., and GREENAWAY, K. R.   1957.   Arctic Canada from the air; Queen's Printer.

HARE, F. K., and ORVIG, SVENN.   1958.   Arctic circulation; Arctic Meteorology Research Group, McGill University, Publication no. 12, Montreal.

HILDES, J. A.   1956.   Some physiological aspects of Arctic warfare; Canadian Services Medical Journal, Ottawa, October.

HULSE, J. H.   1956.   Food in the Arctic; Canadian Food Industries, Gardenvale, P.Q., November.

——— 1956.   Food for men in the Arctic; Canadian Food Industries, Gardenvale, P.Q., November.

LAMB, H. H.   1959.   Our changing climate past and present; Weather, Royal Meteorological Society, London, October.

MALMGREN, F.   1927.   On the properties of sea ice; Sci. Res. Norwegian North Pole Expedition with the Maud, 1918-25, 1 (5).

PETTERSSEN, S., JACOBS, W., and HAYNES, B.   1956.   Meteorology of the Arctic; Technical Assistant to Chief of Naval Operations for Polar Projects, U.S. Navy, Washington, D.C.

RAE, R. W.   1951.   Climate of the Canadian Arctic Archipelago; Meteorological Branch, Dept. of Transport, Toronto.

SIPLE, P. A., and PASSEL, C. F.   1945.   Dry atmospheric cooling in sub-freezing temperatures; J. Amer. Phil. Soc., vol. 89, p. 177.

SMITHIES, W.   1958.   Canadian experience with dehydrated foods; Activities Report, vol. 9, no. 4.

SWITHINBANK, C. W. M.   1958.   An ice atlas of the North American Arctic; Proceedings of the Conference on Sea Ice, National Academy of Sciences, National Research Council, Washington, D.C.

TAYLOR, ANDREW.   1955.   Geographical discovery and exploration in the Queen Elizabeth Islands; Geographical Branch, Department of Mines and Technical Surveys, Mem. 3.

UNITED STATES, NATIONAL ACADEMY OF SCIENCES.   1956.   Proceedings of the conference conducted by the Division of Earth Sciences, National Academy of Sciences on Arctic sea ice; National Academy of Sciences, National Research Council, Washington, D.C.

UNITED STATES NAVY HYDROGRAPHIC OFFICE.   1958.   Oceanographic atlas of the Polar seas, Part II; Washington.

WILSON, H. P., and MARKHAM, W. E.   1954.   Terminal weather conditions at Eureka, Isachsen, Mould Bay, Resolute; Department of Transport, Toronto.

Forecasting

Sea Ice Conditions

W. E. MARKHAM

ABSTRACT

The various types of ice are first discussed together with a summary of modern terminology. Seasonal ice conditions in the western hemisphere are described with the emphasis on navigable periods in the various waterways. Ice movement is then dealt with followed by a brief summary of the current Canadian ice observing and forecasting programme.

PRIOR TO THE LATE NINETEENTH CENTURY, ice was regarded merely as an obstruction to navigation and although the various explorers probably developed some rules of thumb regarding its behaviour, no concerted study was ever made. In 1881 the vessel "Jeanette" bearing the Delong Expedition was crushed by ice off northeastern Siberia, and some years later identifiable parts of the wreckage were found on the west coast of Greenland. This led a Norwegian, Fridjof Nansen, to two conclusions. First, that the general ice movement was northward from eastern Siberia, becoming southward in the region between Greenland and Spitsbergen, and second, that a ship constructed to withstand the ice could be used as a drifting platform to reach the Pole or at least a sufficiently high latitude from which the Pole could be reached by sledge. The revolutionary vessel "Fram" was the result and in 1895 Nansen reached 86°13′ N by sledge, while the ship itself later reached 85°57′ N before drifting south and free of the ice the following summer. During this voyage, if it can be called that, a broad scientific programme was carried out and one of the findings was that the average ice movement was 28 degrees to the right of the wind direction at about 2 per cent of the wind speed.

Activity in their Arctic regions was expanded in the 1930's by the U.S.S.R. and led to the Papanin North Pole expedition of 1937-8 and also the development of their Northern Sea Route from Murmansk to Vladivostok. This route is economically important to the Russians and provides ample reason why they have been, for the past twenty-five years, leaders in the development of the Arctic. Great north-flowing rivers like the Lena, the Ob, and the Yenisei provide cheap and convenient transport to the coastline, whereas east-west railway construction would be hampered by extensive marshes and mountain ranges. In the western hemisphere, on the other hand, Arctic development has had no such impetus and as a result our northland is chiefly restricted to isolated defense, meteorological, or mining outposts.

Following the expansion of the United States Air Force Base at Thule, Greenland, in 1951, during which millions of dollars damage to shipping was done, the United States Hydrographic Office was directed to undertake ice observing and forecasting. Since 1955, when the DEW line was established, Canada has become more and more active in this field.

One of the early requirements for a scientific approach to the problems associated with sea ice is a standardized terminology. The latest development in this field is an international nomenclature provided by the World Meteorological Organization, which is the basis for the terms used in North America. To describe a group of ice floes in a few words, we must include their size, their number, and their thickness. The thickness of ice is of course related to the duration and severity of the freezing weather which produced it. Relatively thin ice—up to about eight inches in thickness—is termed "new ice." Subcategories in this group include the following. Frazil crystals denote a suspension of small spicules of pure ice in the sea water, the first signs of freezing. The terms slush and sludge apply as the suspension thickens and begins to develop some hardness. Pancake and lily-pad ice are round floes with a raised rim resulting from further hardening. Wave or swell action in this case prevents the formation of a solid ice-sheet and also causes collisions between the floes, which produces the rim. If the water is still, ice rind in a thin but uniform ice layer will result, which can then grow into young ice as it thickens. All these forms of ice are black to greyish in colour when viewed from above because of their relative transparency, and are the unstable forms of ice which may form and disappear over relatively short periods of time.

Compressing motion of new ice floes with respect to each other produces "rafting" in which one floe overrides another with little fracturing. This process can occur repeatedly and build up considerable thicknesses over a short period of time even when the basic floes are only about six inches or so thick.

The next major category relating to age is termed "winter ice." This is defined as ice of one year's growth which has a thickness from eight inches to six feet. The distinction between young ice and winter ice is not sharp and is primarily one of colour. Sea ice is actually composed of a matrix of pure ice crystals interspersed with brine cells which expand or decrease in size with the temperature. At a thickness of about eight inches, sea ice has lost its transparency because of these pores and becomes white to greenish-white in colour. Because of greater thickness, compressions of winter floes cannot proceed as easily as in new ice and a greater proportion of fracturing occurs along the floe edges, building up what is known as a pressure ridge. When newly formed, a pressure ridge is a tent-shaped mass of broken pieces of ice piled haphazardly, which may reach twenty to thirty feet in height and extend for miles across a field of ice. The height of these ridges, although truly a measure of the amount of pressure exerted, does at the same time give an aerial ice observer clues as to the thickness of the underlying ice which may happen to be snow-covered. Only the broad terms medium or thick winter ice are used descriptively; these are distinguishable because the ridges themselves must retain positive buoyancy. A certain amount of control is thus imposed on the possible height of a ridge by the surrounding ice thickness.

The matter of buoyancy of ice is an important factor for submarines operating in the Arctic. Pressure ridges of ten to twenty feet are not uncommon in thick winter ice in that area and, even though the surrounding floes provide some support, a downward obstruction of fifty to sixty feet may be found beneath them. During her cruise across the Polar Basin, the "Nautilus" encountered some extending over 100 feet below the main ice.

The final category is "polar ice," defined as floes which have survived one or more summer melting seasons. This is the thickest form of sea ice and rather than the greenish-white of winter ice, these floes are powder blue in colour. In North America no distinction is attempted between one-year-old ice and that many years old, but it is possible that the intensity of the blue tint is a rough measure of this age. Polar floes, besides being heavier, are also considerably harder than winter ice. This can be explained by the process of freshening of sea ice. As mentioned previously, sea ice is a matrix of ice crystals and brine cells when it is first formed. During the ensuing summer, the brine seeps downward and by fall the upper portion of the heavy floes are effectively pure ice with a melting point of 32° F instead of 28.6° F. Crystal growth and flooding of the vacated pores and tubules by fresh water are probably additional factors affecting the strength and also the colour.

These then are the basic types of ice with which we have to deal. There remains amount of ice and floe size. Amount is usually indicated as tenths of the sea surface covered by ice in a specified area, but this is sometimes generalized to the terms very open, open, close, and very close pack ice. Further descriptive terms relating to age or topography can of course be added as appropriate. As far as size is concerned, the breakdown is by longest dimension from vast floes which are over five miles across, through big and medium to small floes which are from 30 to 600 feet, and finally to ice cakes and brash which are the wreckage forms of the larger floes.

There is only one other term which should perhaps be amplified a little and that is "puddle." As indicated above, the surface of most ice floes becomes irregular during the winter and spring months as a result of rafting and ridging. When the air temperatures rise during the long daylight hours of early summer, melting of the snow cover begins even before 32° F is reached. Melt-water gradually accumulates under the snow and before long collects in a depression, forming a puddle of fresh water. Once these puddles develop, they become the centres for further melting of the floes because of their greater absorption (compared to ice) of incident radiation. Eventually they melt their way through to the sea below, forming what is known as "burnt puddle."

The Arctic Ocean itself is of course always ice covered, but even in the winter months short leads and cracks may appear as a result of the stress of winds or currents. Data on currents in the Basin are rather scarce, but the general ice motion is approximated by a large anticyclonic eddy centred near 80° N 150° W and covering the area from the Arctic islands to the dateline and from Alaska to the Pole. It rotates about once every seven to eight years or about one mile per day in its outer sections. The rotation has sometimes been attributed to an underlying water current, but the mean wind flow in the area is a simpler and more tenable explanation as no driving force for the current has ever been deduced. The mean ice motion in the North American Arctic is thus on-shore from Greenland to western Ellesmere and roughly parallel to the general coastline from Axel Heiberg Island to Point Barrow.

In general, ice concentrations vary from almost 100 per cent in the winter months to about 90 per cent in midsummer, but this last figure can vary considerably over large areas. North of Greenland incidence of under 95 per cent ice is rare, while in the Beaufort Sea it may fall as low as 70 per cent over large areas. Proportions of polar to winter ice vary from one season to another, but average about 80 per cent in summer and 85 per cent in winter. It has been calculated that with a 10 per cent loss of ice as a result of the permanent south-flowing current along the East Greenland coast, the average age of polar ice in the Basin is eight to nine years although some floes—up to 10 per cent—will be more than twenty years old. Average ice thickness over the ocean is roughly seven to ten feet in summer and ten to thirteen feet in winter, but variations from this figure are great in any specific area.

Ship navigation within the polar pack is not feasible even with today's powerful ice-breakers, although this does not preclude scientific sorties along its edges. Talk of forcing passage to the Pole is probably only talk and would have only propaganda value now that submarines can cruise there under the pack. The general northwest coast of our islands is thus blocked to surface ships except in those special weather situations which produce leads between the shore and the moving pack ice.

The channels within the islands can conveniently be separated into two sections. The first is to the northwest of a line running from Banks Island to Victoria to Prince of Wales, Bathurst, Devon, and finally Ellesmere islands. Within this area, access to the Arctic Ocean is through broad and straightforward passages, but to the southeast the waterways are constricted and island-studded. Entry of polar ice from the ocean is thus unhindered but exit to the southeast is difficult, resulting in an area where a mixture of polar and winter ice collects. The broad channels along latitude 74° N from Lancaster Sound to McClure Strait are included within this group, because of the reduction of width from Melville Sound to Barrow Strait where a number of islands help to obstruct the flow. Through the entire area there is a gentle southeastward drift of water from the Arctic Ocean, which assists in the influx of polar ice but does not have the strength to carry the floes continuously through the narrower outlets. Access to these shores is thus eased a little by the appreciable proportions of winter ice, but at the same time the abundance of polar floes included prevents appreciable disintegration during the summer. Ice-breakers could probably penetrate this area regularly if there were need for it, but their operations would have to be concentrated in the month of August, when the ice is at its weakest.

Besides the expense of ice-breaker operation, the additional factor of persistent winds must also be considered. Although the summer is normally a quiet period in the Arctic, it is quite possible that a period of ten days or more with moderate winds in a particular direction may develop. If the winds happen to be on-shore in the landing area, operations may have to be delayed to the point where retreat southward by the ice-breaker is necessary. In other words, two alternate leading areas with different exposures are desirable if financially feasible. One of the DEW line sites in Foxe Basin is continually affected by this phenomenon and lack of this second beach will one day result in supplies being unloaded at an alternate site for later air delivery. This almost happened in 1959.

The other section of the island waterways are much more accessible and are related to two separate centres of ice disintegration: the Cape Bathurst area in the west and the north open water in the east.

In the western Arctic the general ice movement of the polar pack, as has already been described, is away from the area between Cape Bathurst and Cape Kellet on Banks Island. In addition, sea-bottom topography and tidal flow combine to produce vertical water currents in this same area. Eastward from Cape Bathurst, the water is quite deep—150-200 fathoms—while to the west it decreases abruptly to 20-50 fathoms at the same distance from shore. In flowing across this ridge, the tides produce vertical currents which transport heat upwards and reduce the rate of ice formation. These factors together result in an area where leads are very common and where open water becomes prevalent at the first sign of warming in the spring. Early in May 1957, when air temperatures were in the range 0-15° below, we found a lead over twenty miles wide in this area although there had been no storms or even fresh east winds to produce it.

A centre such as this has a tremendous effect in assisting ice break-up, since the sun's radiant heat is mostly absorbed by the exposed water instead of mainly reflected by the ice and snow. The warm waters of the Mackenzie are an added source of heat, so that by the first days of July a large area of open pack ice extends from the Mackenzie to Banks Island and Cape Bathurst. During July and August this disintegration spreads steadily to the east and west, aided in part through Dolphin and Union Strait by another area where tidal currents are of assistance. The remainder of the continental coast is characterized by relatively narrow channels where ice movement is hampered by shoals and numerous small islands. To a large extent the ice here melts where it is formed, and disappears rapidly in Coronation Gulf, Simpson Strait, and Rae Strait (followed shortly by the intervening Queen Maud Gulf area) during the first half of August. Navigation can be started early in July in the western Arctic, and by late August access to the whole coastline can be quite straightforward and remain so until late September. There are some possible difficulties however.

In the Beaufort Sea, intrusions of polar ice can cause delays following periods of protracted northwest winds, although the wide-open area involved makes the occurrence of high over-all ice concentrations unlikely. Ship progress is delayed in passing through it, rather than held up completely until it has dispersed. In the east, heavy polar ice can also be forced south from Victoria Straits under the same weather circumstances, but here, because of the narrow shipping lanes, high concentrations and a complete stop in traffic may result.

In normal circumstances, the close of shipping as it is carried out at present is dictated by the formation of slush ice along the shores, preventing small boat operations. In busy harbours, the construction of jetties would make these small boats unnecessary and would lengthen the season by several weeks, because at this time the off-shore area is still easily navigable.

In the eastern Arctic, the comparable centre of disintegration is known as the North Open Water. This phenomenon, known back in the whaling days, is of more complicated origin than the Cape Bathurst centre, but it is known that winds are a much more important factor and that subsurface currents, although present to a degree, have a much less important role. This area of unstable ice extends from the narrow portion of Smith Sound southward to about latitude 76° N and lies midway between Devon Island and Cape York. During the quiet periods in the winter it may be covered by high concentrations of young ice, but it clears quickly during storms and in early June provides the same water-warming influence as in the western Arctic. Its expansion in size is mainly on its southern boundary, because in this direction winter ice prevails while to the north proportions of polar ice are significantly greater. In this area there is the added feature of the outer ice boundary. At its maximum, ice cover extends 200 miles off the Labrador and Baffin Island coasts, then eastward to Greenland near latitude 60. A more normal extent, however, is 100-150 miles, and above the Arctic Circle the edge approaches the Greenland coast, reaching it more gradually in the vicinity of Disko Island. Retreat of the boundary early in the season is most evident in the north portion, as leads and reduced concentrations spread through Melville Bay to join with the North Open Water. This is commonly accomplished in late June and subsequently the ice pack contracts gradually to the central Baffin Island shore, finally disappearing in the period from mid-August to early September. At the same time this retreat is occurring, the disintegration also progresses through Jones and Lancaster sounds to the Hell Gate–Cornwallis Island line by early August and thence into the various inlets leading south and north.

Throughout this last area, water movement is principally into Lancaster Sound and Baffin Bay, although there is some motion into Prince Regent Inlet and thence through Fury and Hecla Strait. In Baffin Bay and Davis Strait there are two major currents—a cold one from Smith, Jones, and Lancaster sounds which flows southward along the Baffin coast and a warm one moving northward on the Greenland side with several variable and transitory eddies branching off to the west. This water movement explains in part the relatively ice-free conditions in southwest Greenland compared to Baffin Island at the same latitude.

The navigation season in the east thus begins any time from late June to mid-August and interference by drift ice can extend into September. Freeze-up, however, is slightly later than in the west and is not well under way until November. Ice-breaker escort is advantageous early in the season, but during September and October is rarely necessary. Relatively easy access to Baffin, Devon, Cornwallis, and Somerset islands and to the southern part of Ellesmere Island is possible with about a two- to three-month shipping season.

The Hudson Bay shipping route does not really lie within the Arctic, but for completeness let it be stated that its navigation season extends from mid-July to mid-October. Foxe Basin, on the other hand, has been used very little and until the installation of the DEW lines in 1955 was comparatively unknown and even unsounded. The shipping season there extends from late August to early October in most years, but lingering ice is much more of a threat and polar ice may also intrude from Fury and Hecla Strait. This lingering ice poses a threat to the Churchill route and strings of Foxe Basin ice sometimes move into Hudson Strait in mid-season, causing delays and interruptions in operations.

The Canadian Ice Reconnaissance and Forecasting programme developed from the initial Gulf of St. Lawrence ice patrols operated by the Marine Branch of the Department of Transport during the war years. The responsibility for this service has recently been transferred to the Meteorological Branch and has been expanded to include ice forecasting in addition to giving bulletins of existing conditions. More and more of the re-supply work connected with the DEW and Pine Tree lines is being taken over by Canada each year. Our ice reconnaissance and forecasting has expanded in a similar fashion, replacing the American service initially provided. Forty-eight-hour and five-day ice forecasts make up the main part of the programme, with thirty-day and seasonal forecasts issued to a few major operators. Forecasts and reconnaissance now cover almost all the Canadian coastal waters in which shipping operates. In the future freeze-up forecasts will be developed and operations extended into the St. Lawrence and Great Lakes areas.

Freeze-Up and Break-Up of the Lower Mackenzie River, Northwest Territories[1]

J. ROSS MACKAY

ABSTRACT

Freeze-up and break-up data of varying reliability have been collected for the Mackenzie River and studied with particular reference to the lower course between Fort Good Hope and the Beaufort Sea. Most attention is given to a discussion of Fort Good Hope conditions, but similar results have been obtained for other stations.

Variations in dates of freeze-up and break-up have been analysed by standard statistical procedures (t-test, Spearman coefficient of rank correlation, and Kendall coefficient of concordance), using a level of significance of $P = .01$, in order to determine long-term trends in open season and the relation between early and late freeze-ups and break-ups. The results show long-term fluctuations in open season from 1876-1955.

Cumulated degree days, summed for the site of freeze-up, have proved of no statistical significance in the study of freeze-up in the lower Mackenzie valley. Good correlations are obtained between freeze-up and weighted air temperatures (Rodhe, 1952) when air temperatures are based upon those of estimated flow-travel time of river water issuing from Great Slave Lake.

Break-up in the lower Mackenzie valley can be correlated with temperatures in the Fort Simpson area. By using four weighted mean ten-day Fort Simpson temperatures, from April 1 to May 10, estimates can be made of break-up at Fort Good Hope. Reindeer Station break-up can be correlated with Fort Good Hope break-up, which occurs ten to twenty days earlier. Computations involving regression analysis were carried out on a high-speed electronic computer.

THE NORTHWARD-FLOWING MACKENZIE RIVER has a length of 1071 miles from its head in Great Slave Lake to its mouth at the Beaufort Sea and a total length of 2635 miles when measured to the source of its remotest tributary, Finlay River. Great Slave Lake, Mackenzie River, and the Beaufort Sea freeze over in winter with ice whose thickness ranges from about five to eight feet.

In the absence of roads and railways, commercial water transportation on the Mackenzie River is of vital importance to the economy of the river settlements and to the more distant settlements of the Yukon and Northwest Territories which are supplied by Mackenzie River transportation facilities. Consequently, a study of the environmental conditions affecting freeze-up and break-up, and therefore the length of the open (navigation) season, is of both practical concern and theoretical interest.

GENERAL

Sources of Information

The freeze-up and break-up data used in this study have come from published sources, government officers, missionaries, the Royal Canadian Mounted Police,

[1]Published with permission of the Director, Geographical Branch, Department of Mines and Technical Surveys, Ottawa, Canada.

traders, and others. In the lower Mackenzie valley, fairly complete records go back to 1876 for Fort Good Hope, 1896 for Arctic Red River, 1935 for Aklavik, 1937 for Reindeer Station, and 1940 for Lang Trading Post. The reliability of the dates varies considerably from place to place and year to year, depending on the criteria used to define freeze-up and break-up, the interest of the observer, transcription errors, and so forth. At some settlements the dates recorded by two or more observers have differed by as much as two weeks for both freeze-up and break-up. However, statistical analyses of the dates show that the records kept by a single organization (e.g. a mission) or an individual tend to be internally consistent, because of adherence to specified definitions for freeze-up and break-up, even though the definitions do not conform to standards applied elsewhere.

Freeze-up is usually defined as the date when the river freezes over and the ice stops moving or jams, although open holes may persist for some time afterwards.

FIGURE 1.   Location and flow-travel time map. The numbers refer to one estimate of daily flow-travel time of Mackenzie River water from Great Slave Lake to Beaufort Sea.

The dates of first skim or drift ice may be reported at some settlements and the records often list "ice running in river" for several weeks prior to freeze-up.

Break-up is usually defined as either the time when the ice starts to move in the spring, or else when the main river ice clears out of the river at the place of observation. Both dates have been reported for some stations. In contrast to freeze-up, the date of break-up is, in a sense, immutable. If, for example, break-up is defined as the moment when the river ice commences to move — a definition commonly applied — no subsequent event can alter the date of break-up. However, freeze-up is not so easily defined. If it is based upon the formation of a given quantity or type of ice, a warm spell or a change in river régime may cause its destruction. Therefore, a freeze-up date may be provisional until confirmed by subsequent events.

There are only five weather stations along the Mackenzie River with climatic records of long enough duration to use: Fort Providence, Fort Simpson, Fort Norman (and nearby Norman Wells), Fort Good Hope, and Aklavik. These weather stations are spaced from 100 to 300 miles apart, a long distance when day-to-day weather conditions are considered in relation to those of the river. The climatic records are of varying length, but none are as long as the longest freeze-up and break-up records. The published daily weather data give only temperature (maximum and minimum) and precipitation. Other weather information would prove of value in a study of freeze-up and break-up, but none is available in published form.

Adequate river discharge and water temperature data are also unavailable. The gauge records for open water seasons at Fort Simpson (since 1938) and Norman Wells (since 1943) cannot, at present, be converted into discharge. The first discharge measurements of the Mackenzie River were made at infrequent intervals in 1958 and 1959 near Fort Providence and Fort Simpson. No data are available for discharges of the principal tributaries of the Mackenzie River, such as the Liard, Great Bear, and Arctic Red Rivers. The very few water temperature records (Thomas, 1957) are not of sufficient detail or continuity to help in relating climatic conditions to water temperature.

Scope of the Work

The study of environmental factors affecting freeze-up and break-up has been based, of necessity, upon very limited data: (1) freeze-up and break-up records of varying length, continuity, and reliability; and (2) daily temperatures and precipitation for five widely separated weather stations with records of relatively short duration. Early in the study, it was found that precipitation data were of little importance in the study of freeze-up. The extent and depth of snow are important, through run-off at break-up, but snow survey data are not available. Therefore, the study has been confined by the nature of the data primarily to a statistical analysis of air temperatures in relation to freeze-up and break-up dates.

When freeze-up and break-up in a standing body of water, such as a lake, are related to air temperature, the prevailing air temperature may be measured at all times at one station. However, the air temperatures affecting a given mass of river water are those in constant contact with the water as it moves downstream. Therefore, in order to analyse the air-water relation, an estimate was required of the flow-travel time of the Mackenzie River so that prevailing air temperatures could be obtained or interpolated from nearby weather stations. This was done by estimating "average" daily flow-travel times using hydrographic data (Great Slave Lake and Mackenzie River pilot, 1958), hydrographic charts, and other information. The surface velocity which is recorded in the navigational aids is normally that measured along well-defined navigation channels or routes and so does not represent either a mean surface velocity or a mean velocity of the river. Moreover, the flow varies seasonally and is affected by input of tributary rivers. For these reasons, different flow-travel times have been analysed in relation to air temperatures, although only one flow-travel time (i.e. that of fourteen days from Great Slave Lake to the mouth) is reported here. This flow-travel time probably represents a minimum value.

Regression analysis was used extensively in this study. Where there were two or more independent variables in the regression analysis, a high-speed electronic computer was used. In interpreting the results of the regression analyses, it is emphasized that there is non-independence in successive climatic observations; that is, there is autocorrelation among the climatic variables, such as temperature at Fort Simpson on three successive days. Therefore the rigorous application of significance tests is questionable (Mills, 1956, p. 629) and so they have not been stressed. In some calculations, because the climatic records are not complete, a limited amount of interpolation has been necessary.

Studies have been made of freeze-up and break-up for all the lower Mackenzie River settlements, but the present report deals primarily with Fort Good Hope. Fort Good Hope is more typical of the lower Mackenzie River than those stations in the Mackenzie delta and its records are the longest.

Notation

The notation used in this paper for regression analysis is based upon that of Mills (1956); that for air-water temperature relations upon Rodhe (1952); and that for rank correlation upon Siegel (1956).

X_1 Dependent variable

$X_2, X_3 - X_n$ Independent variables

S_1 Standard deviation of X_1

$S_{1.23-n}$ Standard error of estimate of X_1 when estimates are based on X_2, $X_3 - X_n$

r Coefficient of correlation

$R_{1.23-n}$ Coefficient of multiple correlation between X_1 and a combination of other variables including X_2, $X_3 - X_n$

a Constant term in regression equation

$b_{12.34-n}$ Coefficient of net or partial regression; the coefficient of X_2 in an equation in which X_1 is the dependent variable and X_2, X_3, $X_4 - X_n$ are independent variables

t Student's t or Fisher's t used to test significance of a mean

F_{95} F_{99} The 95th and 99th percentile values of the F-distribution used in comparison of variances

Yi Cumulated percentile deviation for year i

Z_i Mean length of open season or temperature for n years of record

Z Length of open season or mean annual temperature for year i

T_v Mean daily air temperature at time v in ° C

t_v Time at period v; here in days

Δt Period of time; here one day

τ Water temperature at time v in ° C

k Constant with inverse dimension of time; k^{-1} may be expressed in days

r_s Spearman coefficient of rank correlation

W Kendall coefficient of concordance.

Previous Work

Although most of the rivers and lakes in Canada are frozen over each winter and freeze-up and break-up are important events in the lives of many Canadians, technical literature on freeze-up and break-up is surprisingly sparse. This is also true for Alaska (cf. Williams, 1955) and continental United States (cf. Shipman, 1938; Swenson, 1942). Descriptions and some mean dates of freeze-up and break-up for the Mackenzie River are given by Kindle (1920, 1921) and Lloyd (1943). Publications of the Canadian Department of Transport summarize freeze-up and break-up dates (Meteorological Branch CIR-3156, ICE-2, 30 January, 1959) and maximum winter ice thicknesses in rivers and lakes in Canada (Meteorological Branch CIR-3195, ICE-4, May 1959).

The relation of environmental factors, such as air temperature, wind velocity, cloudiness, and humidity, to water temperatures and ice formation in standing bodies of fresh and salt water has been studied intensively (e.g. Callaway, 1954; U.S.G.S. Prof. Paper 269, 1954). By comparison, however, there have been relatively few studies dealing with the relation of environmental factors, water temperatures, and ice formation in large rivers. Currie (1954) discusses the physical and environmental factors governing freeze-up and break-up in northern Canada, but he does not relate them to specific water bodies and local conditions. The relation between cumulated degree days (summed temperatures above or below freezing) with freeze-up and break-up has been investigated for specific locations (Burbidge and Lauder, 1957; Shipman, 1938; Stankiewicz, 1947; Swenson, 1942), but not cumulated degree days as summed for flow-travel time along a river.

European literature on freeze-up and break-up is more voluminous than that in North America. Cumulated degree-days, snow melt run-off, and other data recorded at specific locations have been correlated with freeze-up and break-up. (NOTE: Abstracts in the *Arctic Bibliography* — prepared under the direction of the Arctic Institute of North America—and those in the Snow, Ice, and Permafrost Research Establishment abstract volumes of the United States Army include abstracts of many articles written on freeze-up and break-up of Russian rivers. However, the author unfortunately has been unable to obtain any relevant quantitative information from the abstracts or from the few original articles which he has been able to consult.)

LONG-TERM VARIATIONS IN FREEZE-UP
AND BREAK-UP

An examination of the freeze-up and break-up dates for Fort Good Hope and Arctic Red River, whose records go back much longer than the climatic records, shows that the dates have undergone long-term variations, just as have those of climatic elements elsewhere. The mean freeze-up and break-up dates for Fort Good Hope for ten-year periods are given in Table I. Two "break-ups" are shown, one for Fort Good Hope and the other for the Ramparts, a seven-mile stretch of fast water which terminates a mile upstream from Fort Good Hope.

As shown in Table I, the ten-year mean dates of first-ice, freeze-up, and break-up have varied considerably. The maximum range in first-ice at Fort Good Hope is 6.7 days; freeze-up, 9.2 days; break-up (Fort Good Hope), 6.6 days; break-up (Ramparts), 4.4 days; and open season, 14.1 days. In order to test for the significance of long-term variations of mean dates of freeze-up and break-up, the twenty-year period of 1876-95 was compared with that of 1936-55. Twenty-year periods were selected, rather than ten-year periods, in an endeavour to reduce the bias of observational and other errors. The t test (Mills, 1955, pp. 240-2) was used to test the significance of the differences between the means of 1876-95 and 1936-55. The mean dates of freeze-up were November 5.0 and 11.6 with standard deviations of 5.3 and 7.9 days. The mean break-up dates were May 19.1 and May 14.6 with standard deviations of 4.6 and 4.7 days. The mean dates of freeze-up and break-up differ significantly at a probability of $P = .01$. The conclusion is, therefore, that freeze-up was significantly earlier and break-up significantly later in the period 1876-95 as compared with 1936-55. Consequently, the length of open season also changed significantly, by an average of 10.6 days, in the same period — a 6 per cent change.

The standard deviations for freeze-up — for Fort Good Hope as well as for other stations — are usually greater than those for break-up. This may reflect larger variations in date of freeze-up or, in part at least, the criteria used for recording the dates, because break-up dates are often more precisely defined than freeze-up dates (see above).   .

TABLE I

MEAN VALUES OF DATA SHOWN

Date	First-ice (Fort Good Hope)	Freeze-up (Fort Good Hope)	Break-up (Fort Good Hope)	Break-up (Ramparts)	Open season (Fort Good Hope) days
1876–85	Oct. 15.9	Nov. 4.5	May 17.2	May 25.1	171.3
1886–95	" 20.7	" 5.7	" 20.9	" 26.5	168.8
1896–1905	" 17.0	" 3.6	" 19.9	" 26.8*	167.7
1906–15	" 22.5	" 7.6	" 17.5	" 23.9	174.1
1916–25	" 19.6	" 6.8	" 19.9	" 27.3	170.9
1926–35	" 17.6	" 9.3	" 15.8	" 23.8	177.5
1936–45	" 22.6	" 9.6	" 14.3	" 23.2	179.3
1946–55	" 22.1	" 12.8	" 15.0	" 22.9	181.8

*This figure is for nine years.

To determine whether early or late freeze-up was associated significantly with an early or late break-up and first-ice over the period of record, the Spearman rank correlation coefficient r_s (Siegel, 1956, pp. 202-13) was calculated for the mean dates of Table I, the results appearing in Table II. The value of r_s at $P = .05$ and $P = .01$ are .64 and .83.

TABLE II

SPEARMAN RANK CORRELATION COEFFICIENT FOR MEAN VALUES GIVEN IN TABLE I

	First-ice	Freeze-up	Break-up (Fort Good Hope)	Break-up (Ramparts)
First-ice	1.00	0.71	0.36	0.52
Freeze-up	0.71	1.00	0.75	0.83
Break-up (Fort Good Hope)	0.36	0.75	1.00	0.86
Break-up (Ramparts)	0.52	0.83	0.86	1.00

Spearman's method of rank correlation is non-parametric, involving no assumption about the parameter of the population sampled. The degree of correlation is a measure of the concordance between two rankings, which in the present analysis are pairs of eight ten-year means for the period 1876-1955. At the level of $P = .01$, freeze-up and break-up at Fort Good Hope are significantly correlated with break-up at the Ramparts. At $P = .05$, first-ice and freeze-up at Fort Good Hope, and also freeze-up and break-up at Fort Good Hope, are significantly correlated. Only the correlation between first-ice at Fort Good Hope with break-up at the Ramparts and Fort Good Hope have a level of significance below $P = .05$.

Three conclusions may be drawn from the results of the t-test and the Spearman rank correlation tests: (1) the records of first-ice, freeze-up, and break-up for Fort Good Hope and the Ramparts from 1876 to 1955 are internally consistent; (2) there have been significant long-term variations in freeze-up and break-up dates, and hence for the open season and the navigation period; and (3) in the decades when mean freeze-up dates were early, mean break-up dates were late.

The Kendall coefficient of concordance, W (Siegel, 1956, pp. 229-39), may be used to obtain the best estimate of the "true" rankings of first-ice, freeze-up, and break-up variations, by decades, at Fort Good Hope and the Ramparts. Whereas the Spearman rank correlation coefficient (r_s) shows the correlation between two ranks, Kendall's coefficient of concordance (W) may be used for the four columns of first-ice, freeze-up, and break-ups in Table I. The value of W is .76 and it is significant at $P = .01$. By applying Kendall's criteria, the best estimate of the "true" rankings is the sum of the rankings. The lowest summed rank is 6.0 for the decade 1936-45 and the highest is 28.5 for 1896-1905. The decade 1946-55 has a rank of 7, one greater than that of 1936-45. As the difference in ranks is not significant, the two decades may be combined to form a twenty-year period. Similarly, the three decades of 1876-85, 1886-95, and 1916-25 may be combined into one uniform, but broken, period. The results, shown in Table III, give "best" estimates of the open seasons, with the days rounded off.

TABLE III

ESTIMATES OF OPEN SEASONS

Decade	Open Season (days)
1876–85	171
1886–95	171
1896–1905	168
1906–15	176
1916–25	171
1926–35	176
1936–45	181
1946–55	181

The rankings of decades according to length of open season in Table III are good estimates of the actual differences among them. Inasmuch as a long open season reflects warmer air temperatures than a short one, the length of open season indicates trends of temperature fluctuations in the lower Mackenzie valley since 1876.

Long-term trends of freeze-up and break-up have also been studied by comparing five-year running averages of open season and mean annual temperature, and by cumulated percentual deviations (cf. Kraus, 1955; Crowe, 1958). The trends of five-year running averages (plotted at year three) of open season for the Mackenzie River stations resemble those of five-year running averages of mean annual temperature at centrally located climatic stations, such as Fort Simpson. In Figure 2 the cumulated percentual deviations are derived from

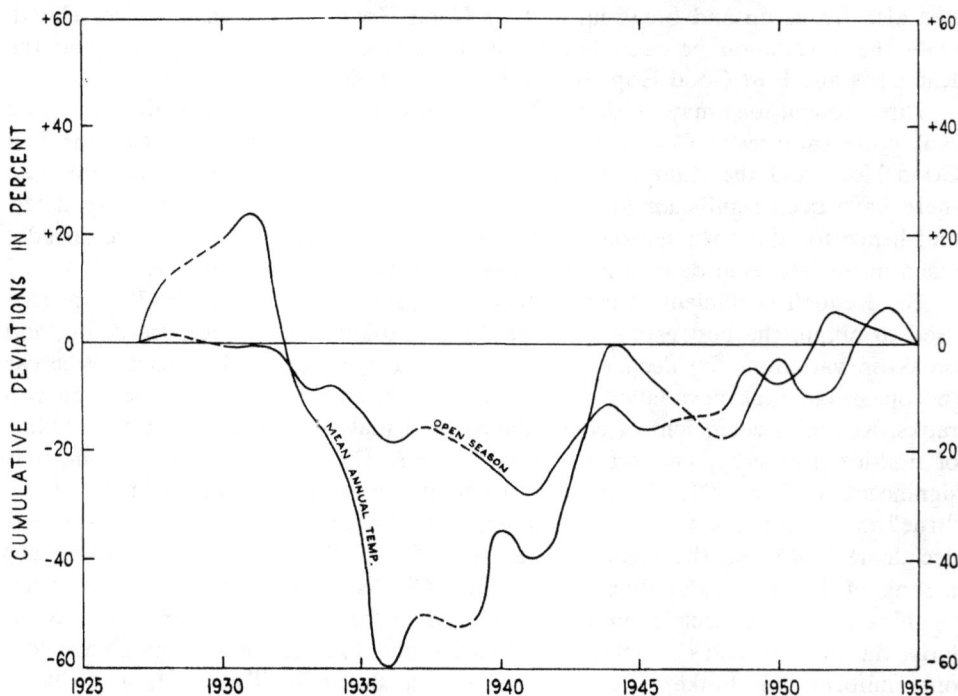

FIGURE 2.   Cumulative percentual deviations, Fort Good Hope.

(1)
$$Y_i = \frac{100}{\bar{Z}} \sum_{i=1}^{n} (Z_i - \bar{Z}).$$

Although the two percentual curves are not identical — and they should differ because open season at Fort Good Hope is affected also by conditions upstream — the similarity of the curves demonstrates the dependence of open season upon air temperature over long periods of time.

FREEZE-UP

Air Temperature and Ice Formation

The relation between air temperature (omitting the effect of wind, cloud cover, radiation, etc.) on ice formation in a standing body of water, such as a lake or a sea, has been discussed at length by Rodhe (1952). According to Rodhe, the cooling of the water surface in response to past temperatures is

(2)
$$\tau_n = \tau_o e^{-k(t_n - t_v)} + (1 - e^{-k\Delta t}) \sum_{v=1}^{n} T_v e^{-k(t_n - t_v)}.$$

In applying equation (2) to ice formation in the Baltic, Rodhe equated τ_n to zero, and by starting the series sufficiently long ago for the lower boundary condition (τ_o) to be neglected he obtained an equation of the form:

(3)
$$0 = \sum_{v=1}^{n} T_v e^{-k(t_n - t_v)}.$$

The unknown parameter, k, may be solved for in at least two ways. Rodhe, in his detailed study of k values in the cooling of Baltic water, took a number of τ series based upon constants of k^{-1} about five to ten days apart. He then selected the best k^{-1} value, through interpolation between series, by comparing their standard errors. Another estimate of k may be obtained by multiple regression analysis of net (for example, $b_{12.34}$) regression coefficients. Equation (3), when expanded, is a multiple regression equation in n variables. The net regression coefficients will be, within the limits of agreement between theory and observation, powers of e^{-k}, so that the solution of equation (3) will give an estimate of k. The regression constant (a) may be related to τ_o, the lower boundary condition.

However, when the above approach of Rodhe's is applied to a river, conditions are much more complicated than for a standing body of water. Specifically, as the Mackenzie River flows from Great Slave Lake, the τ series cannot be carried back indefinitely so that the lower boundary condition may be neglected. The temperatures (τ_o) of Mackenzie water leaving Great Slave Lake are unknown and the Mackenzie River receives large tributaries, such as Liard River, with unknown discharges and water temperatures.

Great Slave Lake contains an area of about 11,170 square miles — an area greater than Lake Ontario or Lake Erie — and, being deep, a large volume of water. The annual inflow, from all sources, comprises about 0.8 per cent of its total volume (Rawson, 1950) so that it would take over 100 years for the Mackenzie River outflow to "empty" the lake. Such a large body of water changes its mean temperature very slowly. In the summer, the surface waters in the shallow — less than sixty feet deep — western end of Great Slave Lake, west of Hay River, may exceed 13° C (Allen, 1958). By the end of September, the waters west of Hay River are thoroughly mixed and complete circulation may occur by September 30 (Rawson, 1947, 1950). That is, the temperature of water leaving Great Slave Lake after early October should have a mean temperature of less than 4° C. When allowance is made for flow-travel time, the water which freezes in the Mackenzie River, with few exceptions, should have left Great Slave Lake with its temperature (τ_o) between 0° C and 4° C.

The downstream changes in Mackenzie River temperature (τ_v) are also affected by the inflow from tributary rivers, such as the Liard (entering at Fort Simpson), Great Bear (at Fort Norman), Hare Indian (at Fort Good Hope), and Arctic Red River (at Arctic Red River). During the autumn, water temperatures of these tributary rivers are probably lower at their mouths than that of Mackenzie water because: (1) first-ice usually begins to run up to three weeks earlier on the tributary rivers before the corresponding portions of the Mackenzie River have running ice; and (2) the tributaries usually freeze over one to two weeks earlier. However, the large size and rapid flow of the Mackenzie River undoubtedly contributes to its later freeze-up.

Progress of Freeze-Up

Mean dates of freeze-up and break-up for the period of 1946-55 are listed in Table IV. Too much reliance should not be placed upon differences of several days between stations, because some freeze-up dates, for example those of Fort Simpson and Fort Providence, are highly variable from year to year. In the Mackenzie delta area, the smaller channels, such as those on which Lang Trading Post and Aklavik are situated, freeze before large swift channels, such as the main branch of the Mackenzie River. It is interesting to note the early freeze-up of Arctic Red River compared to that of the Mackenzie River at the settlement of Arctic Red River. It normally takes five to six weeks for freeze-up to progress upstream from Reindeer Station to Fort Providence. The mean monthly temperature for the month (i.e. thirty days) preceding the mean freeze-up date at Lang Trading Post and Aklavik is about —0.5° C; Reindeer Station —0.4° C; Fort Good Hope —9° C; Fort Norman —11° C; and Fort Simpson —14° C. Therefore an increasingly long period of subfreezing temperatures is required for freeze-up with distance upstream on the Mackenzie River. Fort Simpson, for example, normally has two months with mean temperatures below 0° C before freeze-up occurs.

Freeze-up and break-up records for the Mackenzie delta locations of Aklavik, Lang Trading Post, and Reindeer Station are short, going back only to 1937-40. The freeze-up dates show no significant correlation (at $P = .05$) between either frost-sums (degree-days below 0° C) or frost-sums less cumulated temperatures above 0° C, when Aklavik temperatures are used. The results are as expected, because the water which freezes has come from upstream and so has been subjected to temperatures quite different from those prevailing at Aklavik. This shows the futility of applying cumulated degree-days to freeze-up at one place on the river when the water has been chilled earlier by low temperatures prevailing hundreds of miles upstream. Consequently, flow-travel time temperatures should be used.

Freeze-Up at Fort Good Hope

A number of different types of multiple regression studies were made of freeze-up in relation to air temperature for Fort Good Hope and stations downstream from the settlement. As Fort Good Hope has the longest record of freeze-up and break-up dates, and is more typical of the main Mackenzie River than the Mackenzie delta settlements, only Fort Good Hope data will be given here. The results of two multiple regression studies are reported below.

In the first study, equation (3) was used with $n = 8$ and flow-travel times as given in Figure 1. The mean daily temperatures at Fort Providence for flow-travel days 1 and 2 were averaged to give T_1; daily Fort Simpson means were used for T_2, T_3, and T_4; daily Fort Norman means for T_5 and T_6; and Fort Good Hope means for T_7 and T_8, the latter being freeze-up day mean temperature. The regression equation was computed for the period 1933-55. The standard error of estimate was 7° C, a high figure. However, if the year 1949 is omitted, the value is reduced to 6° C, and if 1938 is omitted, it drops to 5.2° C. The coefficient of multiple correlation (R) is .87 and it is significant at the 1 per cent level. Although the regression equation gives seemingly good results, the net regression coefficients are both positive and negative, and no consistent value of k can be derived from them. The probable reason is autocorrelation among the temperatures (for example, T_2, T_3, and T_4) at the various stations. Interpolation between stations might result in better k values.

TABLE IV

MEAN FREEZE-UP AND BREAK-UP DATES, 1946–55

Location	Distance from Great Slave Lake	Freeze-Up	Break-Up
Fort Providence	50	Nov. 24.9	May 18.6
Fort Simpson			
Mackenzie above Fort Simpson	208	Nov. 27.5	May 15.4
Mackenzie below Fort Simpson	218		May 11.5
Liard River			May 6.1
Fort Norman	513	Nov. 15.3	May 14.2
Norman Wells	565	Nov. 10.9	May 15.1
Fort Good Hope			
Ramparts	680	Nov. 5.6*	May 22.9
Settlement	684	Nov. 12.8	May 15.0
Arctic Red River settlement			
Arctic Red River	898	Oct. 8.7	May 25.1
Mackenzie	898	Nov. 1.5	May 26.8
Lang Trading Post	970	Oct. 9.3**	May 29.1
Aklavik	999	Oct. 9.4**	June 0.1
Reindeer Station	1004	Oct. 18.1	June 2.5

*Seven years of record.
**Nine years of record.

In order to estimate long-term temperature effects on freeze-up, the date of freeze-up at Good Hope was studied in relation to mean ten-day temperatures at Fort Good Hope and upstream weather stations. In general, the highest correlation was between freeze-up at Fort Good Hope and mean temperatures at Fort Simpson. The date of freeze-up (X_1) at Fort Good Hope has been calculated as a linear function of mean ten-day temperatures at Fort Simpson where X_2 is the mean ten-day temperature for October 11-20; X_3 for October 21-31; X_4 for November 1-10; X_5 for November 11-20; and X_6 for November 21-30. The period was 1928-55 with an average freeze-up date of November 9.3.

TABLE V

FORT GOOD HOPE FREEZE-UP REGRESSION ANALYSIS

Variables	S_1	$S_{1.2-n}$	$R_{1.2-n}$	$F_{.99}$
X_2 X_3 X_4	7.8	5.7	.68	sig.
X_2 X_3 X_4 X_5	7.8	3.3	.91	sig.
X_2 X_3 X_4 X_5 X_6	7.8	3.2	.92	sig.

The results are not of great prognostic value because mean freeze-up date is at the end of X_4 period, but the decreases in standard errors are of value in interpreting temperature effects.

An attempt was also made to estimate the value of k by interpolation from τ series with k values of 0, .01, .20, .30, and .40 calculated for the period 1946-55 with $n = 8$ and flow-travel times as given in Figure 1. In order to compare the τ series, mean temperatures (T_v) were substituted in equation (3) and a constant added so that each τ series was equal to zero when the different k values were used. Then the standard error of each τ series, in days, was calculated.

TABLE VI

FORT GOOD HOPE k VALUES, $n = 8$

k	0	.01	.10	.20	.30	.40
Standard error (in days)	2.45	2.45	2.88	2.84	3.19	3.26

The mean freeze-up date for the period was November 13 with a standard deviation of 8.25 days. The τ series with $k = 0$ and .01 had the lowest standard errors. It is interesting to note the standard error of $k = .20$ is the lowest in the range of $k = .10$ to .40, an observation which is also true for Reindeer Station (cf. Rodhe's break in curve, 1952, pp. 188-9). On theoretical grounds it may be shown by using mean temperatures for stations along the Mackenzie, in equation (2), that k must have a value of less than about .05 if the equation holds true. This is consistent with the small k values obtained above.

BREAK-UP

Break-up is usually associated with melting of ice to form side channels along the river banks and a rise in water level to lift, crack, spread, and float the ice downstream (Brown, 1957; Burbidge and Lauder, 1957, pp. 4-7; Currie, 1954, pp. 8-9; Williams, 1955). Although river ice is thinned and weakened over many weeks, break-up is usually initiated by a rapid rise in river level resulting from snow melt and run-off.

Progress of Break-Up

In Table IV, mean break-up dates are given for the decade 1946-55. The Fort Providence date is not reliable, as two definitions for break-up appear to have been used. The Liard River usually breaks up a week to ten days before the Mackenzie River just above the Liard River mouth. By all accounts, break-up is a spectacular sight (McConnell, 1888-9, pp. 86D and 87D) with Liard water cutting a swath across the still-frozen Mackenzie River. The upper part of the Mackenzie may not break up until two weeks after the river has broken below the mouth of the Liard (Kindle, 1920, p. 391). In the river stretch between Great Slave Lake and Fort Simpson, the Mackenzie River receives no tributary of a size sufficient to raise its water level appreciably. As the level of Great Slave Lake fluctuates only three or four feet in a year, and the rise from May to June (i.e., at break-up time) averages only about one foot, river levels upstream from Fort Simpson do not show the rapid rise due to influx of run-off waters that the portion downstream does. This is a very important factor in the delay of break-up above Fort Simpson.

As shown in Table IV, break-up between Fort Simpson and Fort Good Hope is nearly synchronous. Break-up at the Ramparts is delayed by ice jams in the narrow canyon. From Arctic Red River to Reindeer Station the process of break-up moves gradually downstream.

In the analysis of Fort Good Hope break-up dates, good correlations were usually obtained between date of break-up and temperatures at Fort Simpson, 477 miles upstream. This is probably because Fort Simpson climatic conditions are reasonably representative of the lower Liard and Mackenzie rivers. Table VII shows the results of a regression study where X_1 is date of break-up in May at Fort Good Hope; X_2, X_3, and X_4 are mean temperatures at Fort Simpson for the periods of April 1-10, April 11-20, April 21-30, and May 1-10. The period is 1928-55.

In Figure 3, the actual break-ups for Fort Good Hope are compared with computed break-ups, based upon the mean ten-day temperatures at Fort Simpson from April 1 to May 10. The results, as computed on May 10, give a reasonable measure of agreement between actual and computed break-ups, even when break-up occurs a week to ten days after the end of the computation period.

TABLE VII

FORT GOOD HOPE BREAK-UP REGRESSION ANALYSIS

Variables	S_1	$S_{1.23\text{---}n}$	$R_{1.2\text{---}n}$	$F_{.99}$
X_2	4.95	4.67	.33	not sig.
$X_2\ X_3$	4.95	4.56	.38	not sig.
$X_2\ X_3\ X_4$	4.95	4.33	.46	not sig.
$X_2\ X_3\ X_4\ X_5$	4.95	3.48	.71	sig.

In view of the large heat storage capacity of Great Slave Lake in comparison to annual inflow and outflow, an attempt was made to determine if long-term climatic records would aid in the study. Break-up and length of open season at Fort Good Hope were correlated with mean temperatures at Fort Simpson for the current year and the four preceding years, the five preceding years, the ten preceding years, the ten preceding years grouped in pairs, and the four seasons. Only the open season in terms of the current and preceding four yearly Fort Simpson means was significant at the 95 per cent level.

Because break-up at Reindeer Station takes place, on the average, from two to three weeks later than break-up upstream from Fort Good Hope, a comparison was made of Reindeer Station break-ups with those of Fort Good Hope, the Ramparts, and Fort Simpson above the Liard mouth on the Mackenzie River, the period being 1937-58 inclusive. The coefficients of correlation (r) are .72, .48, and .30 respectively, with that of Fort Good Hope significant at $P = .01$, the Ramparts at $P = .05$, and Fort Simpson not significant at $P = .05$. The standard error of Reindeer Station break-up, as a linear function of Fort Good Hope break-up which occurs about two weeks earlier, is 2.7 days.

CONCLUSIONS

The present report is based largely upon a statistical analysis of variations in freeze-up and break-up dates through time and upon a statistical correlation of the dates with estimated flow-travel time–mean air temperatures. Although most attention has been given to a discussion of Fort Good Hope conditions, similar results have been obtained for the other stations studied. The principal conclusions are as follows. (1) Freeze-up and break-up dates can be correlated significantly with air temperatures without direct reference to other factors such as snow cover. (2) Cumulated degree-days, when summed for a location in the lower Mackenzie

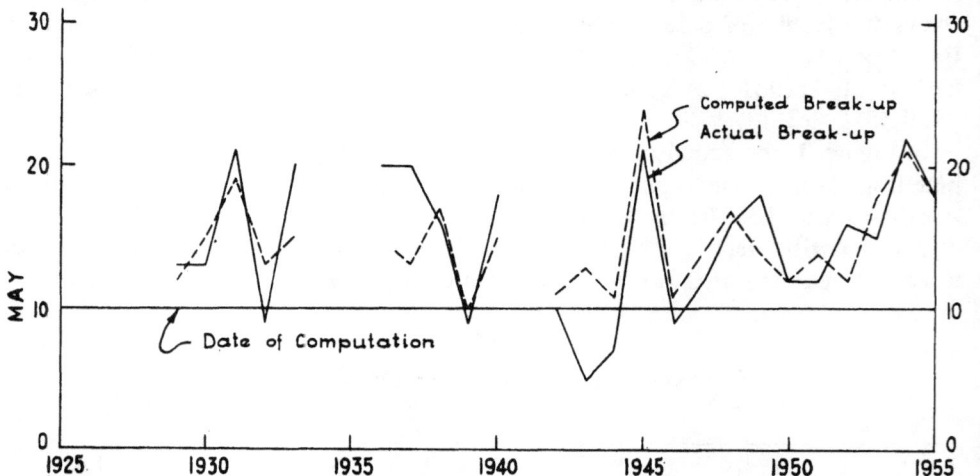

FIGURE 3.   Regression analysis of break-up, Fort Good Hope. Actual break-up and computed break-up, based upon variables X_2, X_3, X_4, X_5 in Table VII, are plotted for comparison. Break-up in 1943 is the earliest on record in the period 1876–1958.

valley, are not significantly related to freeze-up or break-up at the station, because upstream conditions are not considered. (3) There have been long-term variations in freeze-up, break-up, and length of open season. Since 1876, the shortest open season at Fort Good Hope has been about 168 days in the period 1896-1905; the longest open season, averaging 181 days, occurred 1936-55. (4) The length of open season at Fort Good Hope reflects current mean annual temperature at Fort Simpson. (5) The Mackenzie River tributaries contribute in general to earlier freeze-up and break-up of the Mackenzie River. (6) About two weeks of subzero flow-travel time temperatures will produce a freeze-up at Reindeer Station, but two months exposure of water to subzero temperatures may be required at Fort Providence. (7) Standard deviations of freeze-up usually exceed those of break-up. (8) Equation (3), with small k values, for example, .01, may be used in Fort Good Hope freeze-up studies. (9) Freeze-up at Fort Good Hope may be estimated From October and early November temperatures at Fort Simpson, 477 miles upstream. (10) Break-ups at Fort Simpson and Fort Good Hope are nearly synchronous. (11) Fort Good Hope break-up may be correlated significantly with April and early May temperatures at Fort Simpson. (12) Break-up at Reindeer Station for the period of study may be estimated from that at Fort Good Hope with a standard error of three days or less.

It seems likely that when water temperatures and discharge data are available for the Mackenzie River and its principal tributaries, freeze-up and break-up correlations with environmental factors will allow fairly accurate predictions to be made.

ACKNOWLEDGMENTS

The work in this article was supported by the Geographical Branch, Department of Mines and Technical Surveys, Ottawa. The writer would like to thank T. G. Douglas, H. Figgures, J. W. Goodall, L. P. Holman, M. Houldcroft, A. Huget, G. Kraus, K. Lang, A. Sherwood, J. H. Webster, the Roman Catholic Missions at Arctic Red River, Fort Good Hope, and Fort Norman, and the Royal Canadian Mounted Police, for their help in obtaining freeze-up and break-up data. Miss B. J. Morrison was most careful and efficient in undertaking a large number of seemingly endless calculations. Dr. W. H. Mathews, Department of Geology, University of British Columbia, has offered helpful suggestions on the manuscript.

REFERENCES

ALLEN, W. T. R.   1958.   Surface water temperature of northern Canadian lakes; Meteorological Branch, Canada Department of Transport (ms.).

BROWN, R. J. E.   1957.   Observations on break-up in the Mackenzie River and its delta in 1954; J. Glaciol., vol. 3, pp. 133-41.

BURBIDGE, F. E., and LAUDER, J. R.   1957.   A preliminary investigation into break-up and freeze-up conditions in Canada; Meteorological Branch, Canada Department of Transport, CIR—2939 TEC—252.

CALLAWAY, E. B.   1954.   An analysis of environmental factors affecting ice growth; Hydrographic Office, U.S. Navy, Tech. Rept. TR-7.

CROWE, R. B.   1958.   Recent temperature fluctuations and trends for the British Columbia coast; Meteorological Branch, Canada Department of Transport, CIR—3137 TEC—288.

CURRIE, B. W.   1954 (?)   Prairie Provinces and Northwest Territories, ice, soil temperatures; Physics Department, University of Saskatchewan.

HYDROGRAPHIC SERVICE.   1958.   Great Slave Lake and Mackenzie River pilot; Surveys and Mapping Branch, Ottawa.

KINDLE, E. M.   1920.   Arrival and departure of winter conditions in the Mackenzie River basin; Geog. Rev., vol. 10, pp. 388-99.

———   1921.   Mackenzie River driftwood; Geog. Rev., vol. 11, pp. 50-3.

KRAUS, E. B.   1955.   Secular changes of tropical rainfall regimes; Quart. J. Roy. Met. Soc., vol. 81, pp. 198-210.

LLOYD, T.   1943.   The Mackenzie waterway: a northern supply route; Geog. Rev., vol. 33, pp. 415-34.

McCONNELL, R. G.   1888-9.   Report on an exploration in the Yukon and Mackenzie basins, N.W.T.; Geol. Surv., Canada, Ann. Rept., vol. 4 (N.S.), pp. 1D-163D.

METEOROLOGICAL BRANCH.   1959a.   Break-up and freeze-up dates of rivers and lakes in Canada; Canada Department of Transport, CIR—3156, ICE—2.

———   1959b.   Maximum winter ice thicknesses in rivers and lakes in Canada; Canada Department of Transport, CIR—3195, ICE—4.

MILLS, F. C.   1955.   Statistical Methods; Henry Holt and Co.

RAWSON, D. S.   1947.   Great Slave Lake; Fish. Res. Bd. Canada Bull., vol. 72, pp. 45-68.

———   1950.   The physical limnology of Great Slave Lake; J. Fish. Res. Bd. Canada, vol. 8, pp. 3-66.

RODHE, B.   1952.   On the relation between air temperature and ice formation in the Baltic; Geografiska Annaler, vol. 34, pp. 175-202.

SHIPMAN, T. C.   1938.   Ice conditions on the Mississippi River at Davenport, Iowa; Trans. Amer. Geophys. Un., vol. 19, pp. 590-4.

SIEGEL, S.   1956.   Nonparametric statistics for the behavioral sciences; McGraw-Hill Book Co., Inc.

STANKIEWICZ, M. J.   1947.   Break-up can be foretold; Pulp and Paper Mag. Canada, vol. 48, pp. 118-20.

SWENSON, B.   1942.   Opening and closing dates of river navigation in the United States; Monthly Weather Review, vol. 70, pp. 280.

THOMAS, J. F. J.   1957.   Mackenzie River and Yukon drainage basins in Canada, 1952-53; Industrial Water Resources of Canada, Ottawa, Water Survey Rept. no. 8.

UNITED STATES GEOLOGICAL SURVEY.   1954.   Water-loss investigations; Prof. Paper 269, Lake Hefner studies, Technical Rept.

WILLIAMS, J. R.   1955.   Observation of freeze-up and break-up of the Yukon River at Beaver, Alaska; J. Glaciol., vol. 11, pp. 488-95.

Transportation in the Canadian North

R. A. HEMSTOCK

ABSTRACT

Petroleum reservations have recently been taken on some eighty million acres in the continental Canadian north and another one hundred million acres in the Arctic islands. There seems little doubt that large oil reserves will be located in these areas. Industry, however, must first face the problem of getting into the remote locations, for reconnaissance and exploration and then of developing the resources and moving crude oil to market.

The problems of distances and climate, together with lack of usual facilities make transportation a key in any northern programme. The best solution can only be arrived at by careful study of field conditions and the needs of various phases of the oil industry.

It would appear that rapid development of normal means of travel will be needed together with the use of special vehicles designed for terrain conditions typical of the north.

The area is large and methods must be found whereby people and products can move freely and cheaply from one area to another. There are virtually no railroads and very few trails and roads and unlike Russia there is only one navigable river on which transportation is seasonal. There are a few all-weather airports but they are widely scattered. The Arctic Sea is open for only a matter of weeks in August and lies like a great wall effectively sealing the north coast from the shipping lanes of the world.

Economic exploitation of the northern resources will depend on a carefully planned programme which will often require several integrated means of transportation. There is a great scope for some of the newer vehicles which are being developed and it may be that several of these will be required for the economic development of resources.

EXPLORATION AND RESEARCH in the Canadian North will lead to the discovery of natural resources of great potential value. One of the barriers, however, to efficient exploration for, and development of, these resources is the lack of transportation into remote areas. It is the purpose of this paper to discuss the present means of transport and to attempt to evaluate the future needs in this field. The area discussed will be that part of Canada lying north of the provinces. Development of the Arctic by industry will necessitate travel through the sub-Arctic. Moreover, the development of various sectors of the north will be closely linked. Success in one place undoubtedly will bring more interest and effort in other areas.

It has often been said that transportation is the greatest single problem in the development of the North. This is still true. Not only has cost of transportation been a limiting factor (Mathews and Dalton, 1959), but in addition it is usually impossible to reach the destination desired. If products from the resources of the Arctic are to compete in world markets, then we must find ways and means to lower the cost of finding and developing these resources and of moving the products to markets. Another important factor will be the ability to provide the amenities of modern living in the North and thus remove from northern development the stigma of privation and isolation. The National Research Council has pointed out that at present it costs as much to transport a house to Aklavik as it does to buy it in

Edmonton (Legget and Dickens, 1959). The present-day Canadian is not a pioneer and has little desire to "rough it." Adequate transportation is essential to insure competent and happy personnel.

PRESENT FACILITIES

Air travel has made possible much of the present northern development and the aeroplane will surely be a key to future progress. There are at present, however, few all-weather airports and extensive facilities are not available for servicing aircraft. It is anticipated that rapid development of ground facilities will take place in the next few years. This, together with further development of turbo-prop freighters with their low cost operation, and the established contribution of small bush planes, will put the aeroplane at the forefront of Arctic transport.

Fixed-wing aircraft available at present will freight for from $.40 to $2.00 per ton mile, depending upon the type of aircraft, length of haul, and local conditions. This is exclusive of ground facilities and makes no provision for wear on the planes which may be serious if they are taken into inadequate airstrips. The new turbo-prop freighters in which Canada has such a great interest show promise of very low cost operation (reported at about $.15 per ton mile) under suitable conditions. Capital costs are extremely high, however, and only larger developments will support this sort of craft. The smaller "bush" planes, so familiar in the north, have remarkably good operational properties but are limited in the weight and bulk they can carry.

The helicopter has also made a valuable contribution, being important for reconnaissance and for movement of personnel or light emergency freight. Capital costs are high for helicopters and loads are limited to about 4000 pounds with the models available at present. Freight costs have been estimated at about $4.00 per ton mile but costs vary with conditions of the area and the service required. Larger helicopters will be developed in the future but there will probably be little change in their relative use in the over-all transportation picture.

Further contributions of science in the form of revolutionary aircraft may also be of great importance. Some of these are: the Fairey-Rotodyne, half-airplane, half-helicopter, a craft which lands and takes off as a helicopter but switches to normal fixed-wing flight when in the air, thus having good speed and range; the aircar, or hovercraft, may also have possibilities because it can fly over ice, water, or tundra, provided the surface is not too uneven. It remains to be seen just how economically these vehicles will operate. Certainly, if Arctic territories are to be developed, then continued research on aircraft and extensive development of ground facilities are necessary.

The Mackenzie is the most important navigable river in the Canadian Arctic, despite the fact that it is closed by ice for about eight months of the year. Water transportation on this artery has been and will continue to be an important link, especially in the development of petroleum resources in the Mackenzie Basin area. Large tonnages can be transported at reasonable rates in spite of the need for long-term planning and the inflexibility imposed by short seasonal operation. It is not likely that there will be any significant change in this form of transport. The

Arctic sea is also frozen for all but a few weeks of the year, thus effectively sealing the whole northern coast of Canada. A few of the islands have slightly better access to the Atlantic. Improvement in shipping may be expected with the development of newer and better ice-breakers, greater knowledge of the movements of sea ice, and more experience in predicting Arctic weather. The commercial development of submarine tankers and freighters also holds exceptional promise for the Canadian Arctic. If this form of vessel can be made economical, the northern coast and Arctic islands will become accessible to the markets of the world. Petroleum products, in particular, could easily be moved by this form of craft.

It should be pointed out, however, that there are formidable obstacles to be overcome before such tankers become a reality. The Electric Boat Division of General Dynamics Corporation has undertaken a feasibility study of a 40,000 DWT– knot nuclear submarine. Costs of this vessel are estimated to be four to five times that of a conventional surface vessel of similar capacity. The speed of the submarine would be about two-and-a-half times that of the surface vessel, but shaft horsepower would be about ten times as great. As a result costs — at least at the present stage of development — might well be excessive.

There is a possibility that, with greater knowledge of sea ice and the use of nuclear-powered ice-breakers, surface vessels can be used over a long enough season to be attractive. From the study, *Economics of Nuclear and Conventional Merchant Ships*, by the United States Atomic Energy Commission (Conklin *et al.*, 1958), it has been concluded that: (*a*) nuclear ships can now compete economically with conventional ships on long trade routes at high speeds and the competitive position of nuclear ships will improve with time; (*b*) nuclear fuel costs are generally lower than conventional fuel costs but capital costs are 10 to 50 per cent higher; and (*c*) ore carriers and tankers have the best potential for economic application of nuclear power to ship propulsion.

With the advances in technology that may be expected in the next decade, it may well be possible to build ocean-going vessels to serve the Arctic on a routine economic basis. Research in these fields is going on in both Britain and the United States. Surely Canada should be taking a much more active part in these developments, for they appear necessary for access to much of the Canadian Arctic.

There has been very little use of overland routes until recently. The Canadian Government "roads to resources" programme, together with the extension of rail facilities into the southern region of the Territories, will do much to help, but at best these are only thin and limited lines into a vast country. Development will require either a much more extensive road network or vehicles that can traverse virgin terrain with little or no road preparation. The latter seem to provide the greater possibility of success. Vehicles capable of travelling across typical northern terrain have become practicable in the past two years. Mechanical improvement is still required, but the principle and economics appear sound. A significant improvement in the over-all transportation problem may be expected from this source.

The Musk-Ox is one example of a successful cross-country vehicle (Hemstock, 1959). It has an over-all length of forty-eight feet, a width of ten feet, and is powered by a 335 hp diesel engine. It is capable of carrying twenty tons cross country. The load is distributed on tracks 52 inches wide which run nearly the full length of the vehicle. This machine has logged over 1,600 miles in its initial trials during the past season. As would be expected, some minor mechanical difficulties have shown up but all major components are performing well. Another large cross-country transporter which also travels on wide track areas is the Nodwell Transporter. This vehicle has completed its second season in the north and has proven its ability to carry ten tons cross country.

It is likely that overland transportation in future development will involve some rail traffic, considerable trucking over main highways and winter roads, and movement from the ends of the road or rail or water transport by cross-country vehicles. Costs will dictate that as much of the freighting as possible will be by conventional means. Rail costs will be 4 to 5 cents per ton mile; trucking costs from 5 to 20 cents per ton mile on highways, and 20 to 50 cents per ton mile on winter roads, depending on the conditions involved.

Cross-country vehicles now operate in the $1.00 to $2.00 per ton mile range. This may be reduced, however, with improvements in the vehicles. The apparent higher costs of travel with these vehicles can be justified where road construction costs can be saved or where access during a particular season is worth a premium (Thomson, 1959). An access trail will often cost $3,000 to $4,000 per mile, exclusive of maintenance, and in the case of freighting for exploration work, where perhaps only a few hundred tons of equipment are moved, road investment alone

FIGURE 1.   Imperial Musk-Ox.

will amount to several dollars per ton mile. It is for this reason that cross-country vehicles are deemed important in the development of the North.

Future Development

A review of the transportation problems of the Canadian Arctic shows that there has been rapid improvement in the past year or two and that these improvements will have a significant effect on the development of the North. It is also obvious, however, that in many cases only a start has been made and that further research is needed. New aircraft will fill a most important role and will continue their contribution to development. Water transportation facilities should be expanded where feasible along the rivers and every effort should be made by Canada to speed up the development of ocean-going vessels that can operate all year around in the Canadian Arctic. Land transportation will be greatly improved by the "roads to resources" programme and the main arteries provided will give access into remote areas heretofore inaccessible. Development of natural resources will, however, require much additional overland transportation from these main arteries. For this phase of the programme the continued development of overland vehicles is deemed most important.

Government support to improve transportation will be most helpful if it recognizes the varying needs of a new country. Exploration features wide probing of large areas and must be accomplished with a minimum capital outlay, work best performed by aircraft and cross-country vehicles. Development of resources requires permanent facilities and if tonnages are sufficient will justify road construction. Exploration will therefore best be helped by installation of air facilities including airstrips, navigational aids, and weather stations. Development of the located resources will be aided by construction of permanent airfields and overland access by road or rail, or in the Arctic islands by regular boat service.

Another field of investigation that is most important for proper and continued development in the North is the design and construction of permanent facilities. It is tied closely to the transportation problem because initially at least, equipment, housing, food, and fuel all have to be moved into the North. Efforts should be extended to design housing that is lighter, more comfortable, and requires less fuel to heat, and to design equipment that is light, flexible, and well adapted to use in remote areas. Every improvement in this field results indirectly in an improvement in transportation facility.

From an engineering standpoint, there is a great challenge in northern development. The engineer who works in the north must not be afraid to try new engineering methods and he must above all not be afraid of problems which do not exist. Success will result from proper use of fundamental information and thoroughly co-ordinated application of engineering principles.

In conclusion, it should be pointed out that the aeroplane or the atomic submarine will not alone provide the answer to Arctic development. Rather, such development will depend on a co-ordinated transportation system designed to use vehicles, aircraft, and ships as advantageously as possible. Each has its own place in the economy of a new country.

REFERENCES

CONKLIN, D. L. *et al.*   1958.   Economics of nuclear and conventional merchant ships; United States Atomic Energy Commission.

HEMSTOCK, R. A.   1959.   The musk-ox—Canada's answer to tough terrain; Oil & Gas Journal, August 17.

LEGGET, R. F., and DICKENS, H. B.   1959.   Building in northern Canada; National Research Council of Canada, DBR 62.

MATHEWS, TED C., and DALTON, JAMES W.   1959.   Supply and transportation can mean tough sledding in Alaska; Oil & Gas Journal, March 16.

THOMSON, J. G.   1959.   Development of North hampered by muskegs; Oilweek, December 11.

Regional Geology and Petroleum Prospects: Lower Mackenzie Basin and Arctic Coastal Area, Northwest Territories

J. C. SPROULE

ABSTRACT

The historical background and present status of geological exploration in the lower Macken-zie Basin area are described. Special attention is given in this description to local and regional structure in relation to oil occurrence. Stratigraphy and structural relationships to the better-known portions of the Western Canadian Basin to the south and to the Arctic Islands Basin areas to the north are pointed out. The potential oil reserves of this large area are assessed in general terms and reference made to the possible future internal and external market situation.

IN THIS SYMPOSIUM the stress is almost entirely on technical aspects of the Arctic, the geology, geophysics, climatology, transportation techniques, and related sub-jects. Many of the conclusions arrived at, however, are of no use to humanity if they do not lead to an economic evelution of the Arctic as a human habitation and/or as an area from which the human race can derive benefits in the form of the gainful production of natural resources. Although the United States and Scan-dinavian countries hold considerable land within the Arctic Circle, the two countries most concerned are Canada and the Soviet Union. The Soviet Union has done much to understand and to exploit their Arctic areas. Indeed, they have for many years had cities established there, and railways serving them. Canada has for many years carried on scientific explorations of the Arctic, but has not seriously con-sidered exploitation and related settlement of the Far North, including the Arctic, in the foreseeable future. That, of course, is quite understandable. Canada is a large country with a relatively small population. We have sufficient lands and natural resources in climatically favourable areas to satisfy the existing population. It is only recently that our government has come to realize the exploitation of the Far North appears capable of bearing fruit in the near future, using the means at our disposal. My objective in this paper should be made clear: it is to lay only a sufficient foundation to permit an assessment of the present and future of the oil and gas business in the lower Mackenzie Basin area. The blocks in that foundation are the component subjects referred to; they are, of necessity, limited in volume. The best way in which to show that a general coverage of the subject is necessary to an understanding of the petroleum possibilities is to suggest that, just as ore is only a worthless country rock until such time as circumstance raises it to economic status, so oil is a worthless mineral until such time as the related economic conditions elevate it to the category of a proven oil reserve. I am going to by-pass much fine technical detail, dear to many geologists, in favour of a minimum summary of the related factors, both technical and economic. Much of the technical detail is available from the sources referred to in the references.

ACCESS AND GENERAL GEOLOGIC AND GEOGRAPHIC SETTING

The general setting can be most easily described by use of illustrative material (Figure 1*). The subject area is, as illustrated, a part of the lower Mackenzie River drainage area, referred to loosely as the lower Mackenzie sedimentary basin. This particular index map has been selected to show the location of the project area with respect to the following.

(a) The Western Canadian sedimentary basin area of British Columbia and the Prairie Provinces, where a considerable knowledge of the geology and oil and gas prospects has already been acquired. Criteria established for the southern provinces can, therefore, be used in evaluating the prospects of the area under study.

(b) The northern extension of the Western Canadian sedimentary basin across the Arctic islands. Because of the early stage of our knowledge, we are taking a certain amount of licence in making this interpretation but the evidence available to date leaves us on fairly safe ground, at least so far as much of the stratigraphic section is concerned. We do not mean to imply that exact identity in the geological formations continues across the islands. If it did, it would not speak well for the petroleum prospects. On the other hand, we do say that the geologic and structural assemblages and combinations are similar, circumstances that speak well for the prospects.

(c) The flanking position of the cordilleran area to the west and the Precambrian Shield to the east. The Shield is the source of one of the great mineral-bearing areas of the North American continent and potentially one of the greatest on earth.

The geological relationship of the lower Mackenzie Basin to the partly developed areas to the south and the undeveloped areas to the north is most interesting for what it means in terms of a strict geological evaluation. The geographic relation of this same area to the areas of vast metallic mineral resources of the Shield and to the igneous areas of the Cordillera is of utmost importance to an understanding of the future potential of the oil resources, the metallic mineral resources, and of the entire north country. That geographic relation in our opinion holds the key to the future of the Northwest Territories and, to a considerable extent, of Canada as a whole.

In the matter of access to the area and transportation routes within the area, please refer to Figure 2*. Passenger access to the Northwest Territories is principally by air. There is a commercial line from Edmonton, Alberta, down the Mackenzie to its confluence with the Arctic Ocean, via Hay River and Norman Wells, and another by way of British Columbia to Mayo and Dawson, in the

*See separate container of figures.

Yukon. There is a graded highway to the head of the Mackenzie River water navigation system at the outlet of Great Slave Lake. The Mackenzie River is the principal freight traffic artery into the heart of the Northwest Territories. Another motor road to the Yukon passes by way of the Alaska Highway, with a branch to Mayo and beyond, to Inuvik. The northern portion of this road is still under construction, although we understand it has been temporarily suspended. A limited amount of freight is transported into the Northwest Territories by way of the Athabasca and the Slave rivers. A small amount of traffic is sea-borne for a short season each year along the Arctic coast. Within the area, transportation is by barge, canoe, aircraft, and by winter road and trail.

These transportation facilities are adequate to support the present culture and the present industry which comprises: (*a*) fur trading throughout the area. (*b*) fishing, which centres at Hay River, Great Slave Lake; (c) mining at Yellowknife, Pine Point, Radium, Mayo, Rankin Inlet, and elsewhere; (*d*) oil production at Norman Wells. The limitations to population and industrial growth in this area are natural obstructions and have been due mainly to the climate and terrain in relation to agriculture. These two principal hindrances to progress in the Territories are very real but they are not, in my opinion, as serious as they are commonly believed to be. Our ignorance of conditions in the Far North is nearly as profound as the ignorance of those people from the Southern States who believe Southern Canada to be a land of continuous ice and snow. It is true that the Arctic only gets a maximum of three or four months of fine weather in the summer, but how much more than that do we get here? To be more specific, records from the Meteorological Division, Department of Transport, show that over twenty-two years the average temperature at Aklavik was 18° below zero in January, 56° above zero in July, and averaged 12° below zero during the winter months (November to March) and 52° above zero during the summer months (June to August). Comfort in that area in the winter months is dependent principally on proper dress and space heating.

The subject of the terrain in relation to agriculture is a very big one and I do not intend to go deeply into it, other than to point out that a certain amount of grain can and has been grown in the southern part of the Northwest Territories; that garden vegetables are commonly grown in the vicinity of the Arctic Circle and farther north; and that timberland is present in the upper Mackenzie area. The science of agriculture, like other sciences, is making great progress. It has been found that once the insulation of moss and other vegetation is removed from the permafrost, the latter recedes and growth of crops is encouraged; muskeg areas can be drained and the acid soil treated; culture chemicals can be added to the soil to promote the rate of growth. All these factors and others can and will permit the soil in the Territories to support a large local population.

As I see it, the keynote to the future of the subject area, part of a vast undeveloped area of over five million square miles, lies in its geological setting, inasmuch as geological factors control the metallic mineral resources, the oil and gas resources, and agriculture.

The geological setting is in many ways as ideal as one could wish (excluding the natural barriers mentioned above). We have a centrally located sedimentary basin which will unquestionably produce great reserves of oil and gas, with relatively little expenditure of effort. The combination of a large undeveloped mineral wealth and a cheap source of hydrocarbon energy is a most desirable situation. As a measure of the importance of hydrocarbon energy, it has been estimated that about 60 per cent of all Canada's energy consumption is from that source.

EXPLORATION AND DEVELOPMENT

Alexander Mackenzie was the first white man to explore the lower Mackenzie area. He followed this route to the Arctic in 1789. Nearly forty years later, in 1826, John Richardson explored the Arctic coastline from the south of the Mackenzie River to the Coppermine River. Sir John Franklin was followed by Collinson, McFarlane, Petitot, and numerous others, mostly fur traders and, along the coast, whalers. Most of the geological work done prior to the explorations of the Canol Project, a United States government war measure, was very sketchy and purely reconnaissance. The Canol reports prepared during the war under the supervision of Dr. T. A. Link were compiled and published by Dr. G. S. Hume in 1954. Very little of significance has been published since that time, although more geological work, most of it by oil companies, has been done over the past seven years than was done during all previous exploration.

The earliest of the white man's industries in the lower Mackenzie Basin was trapping. The oil industry dates back to about 1920, with the discovery of Norman Wells. Pitchblende was discovered at Port Radium in 1930 and gold at Yellowknife in 1933. These earlier discoveries were followed by the discovery of lead and zinc at Pine Point and silver, lead, and zinc near Mayo. A more recent discovery, reputed to be the largest potentially commercial body of tungsten known, has been made in the southern Yukon near the British Columbia border. All these isolated discoveries have taken place in what appears to be an inaccessible wasteland. All have been developed or are being developed without benefit of assistance in the form of a ready and cheap source of energy, such as could be provided by oil or gas. If these widely scattered mineral deposits in a relatively barren wasteland can be developed and produced without a convenient source of energy, it behooves us to consider what might be done in the case of other mineral deposits in the same general area if such sources of energy are available.

To give only one example of a situation known at present, I would refer to the very large deposit of iron on Baffin Island. A cheap source of fuel oil, possibly available from the nearby island archipelago, might enable this mineral to be pelletized and made ready for the European market, reputed to be hungry for iron. The same applies to other mineral deposits in the same general area. In my opinion some organized collaboration between the mining and the oil industries could work wonders.

GEOLOGY AND STRUCTURE

The glacial drift and alluvium mantling the bedrock of the lower Mackenzie Basin is much thinner than in basin areas farther south. One result is that large areas are relatively well exposed and the stratigraphy and structure can be much better known with less effort. Another result is that oil and gas seepages are abundant and are of such a nature as to speak well for the oil and gas possibilities.

TABLE I
A SIMPLIFIED TABLE OF GEOLOGICAL FORMATIONS

Era	Period	Formation–Group		Lithology
Cenozoic	Tertiary	(Eocene)		clay, shale, sandstone, continental
Mesozoic	Cretaceous			sandstone, shale
	Jurassic			shale, sandstone
Palaeozoic	Permo-Pennsylvanian			conglomerate, sandstone, shale, limestone
	Mississippian			limestone, chert, shale sandstone
	Devonian	Imperial	largely marine	sandstone, shale
		Fort Creek		black bituminous shale
		Ramparts		reefoid limestone, shale
		Bear Rock		dolomite, limestone, breccia, evaporites
	Ordovician–Silurian	Ronning		limestone, dolomite, graptolitic shales
	Cambrian	Macdougal		shale, sandstone, limestone, dolomite, evaporites
				sandstone, argillite
Proterozoic	———————— ?	Katherine		

It would serve no good purpose in a paper of this sort to describe in detail the characteristics of the geological formations summarized above. It should suffice to refer only to the several known prospective source and reservoir horizons and to the fact that numerous structural and stratigraphic traps of the type that are commonly known to contain oil can be expected to occur in the area. The principal prospective horizons include the basal sands of the Cretaceous, Imperial sands, local Fort Creek reefoid developments, reefoid developments of the Upper and Lower Ramparts, and of Silurian and Ordovician formations. In the western basin (Eagle Plains) area, we should add to these the sands of the Permo-Pennsylvanian and Mississippian.

The structure of the area is dominated by several strong tectonic trends, four of which are indicated by numerals (1), (2), (3), and (4), on the block diagram (Figure 3*) of the stratigraphic and other regional relations of the lower Mackenzie Basin. One of these, involving the Franklin Mountains, appears to bend sharply westward a few miles north of Norman Wells, but we have reason to believe that faults and possibly older folds which may be related to this trend, and possibly related to an older line of weakness, carry on in a northwesterly direction across the basin area. A second trend, represented, in direction at least, by the Mackenzie Mountains, extends in an east-west direction; we have reason to believe that this trend also is part of an old and dominant feature, as witness the Fort Good Hope High, which is certainly pre-Devonian (Ramparts) in age. This trend is now known to continue eastward to Coronation Gulf and the Coppermine area. It is such a strong feature as to extend the sedimentary basin to Coronation Gulf through an area north of Great Bear Lake formerly mapped as Precambrian. A third prominent tectonic feature, the Richardson Mountains, extends north-south, from the west end of the Mackenzie Mountains to the vicinity of the Mackenzie delta. A fourth folded and faulted trend, truncated and buried for the most part beneath later Mesozoic and Tertiary flat-lying rocks, is represented by a sharp bend in the northern end of the Richardson Mountains and by the Cambro-Silurian rocks of the Inuvik area. It is believed that this trend extends through the Liverpool Bay area, at least across Bathurst and Parry peninsulas, although it may branch north to cross the Arctic islands. Quite aside from the several local trends in the Eagle Plains area, there are one or more additional older trends which can be detected in the framework of the basement rocks themselves, on a basis of photo interpretations. It is these latter features that are likely to prove of greatest economic significance, inasmuch as they are likely to have controlled reefing and the development of potential clastic reservoirs, at least through Palaeozoic time.

Oil and Gas Seepages

Numerous seepages of oil and gas are known in the lower Mackenzie Basin. These seepages are in all known cases closely related to folding and/or faulting. For example, a gas seepage reported by Glacier Explorers crosses the Ontaratue River along the extension of one of the faults that appears to continue northeastward from the Franklin Mountains. Another, a light gravity (17° A.P.I.) oil seepage located at Rond Lake, about fifty-five miles almost due north of Fort Good Hope, appears to issue from near a north-south fault and joint system. This seepage area is currently being tested by drilling by Western Decalta Petroleum Ltd. Other large seepages, in the form of oil sands, occur about twenty miles east of Fort Good Hope, at Belot Lake and at several other localities in the general area; most of this group of seepages are in Lower Cretaceous sands that apparently occur as convenient "sponges" in which to soak up migrating oil. East of Fort Good Hope, the Cretaceous section is very similar in character and thickness to the McMurray Basal Cretaceous oil sands. Elsewhere in the general area, in the Eagle Plains area of the Yukon, and beyond into Alaska, seepages are fairly common and widespread. The only indicated commercial discovery to date in the entire area is one of the two Eagle Plains wells, reported to be productive of light gravity oil from the Permo-Pennsylvanian. It is of further significance that the two Eagle Plains wells are the only two deep tests that have been drilled in the area, excluding those drilled during the Canol Project in the Norman Wells area. The deep stratigraphic test currently being drilled by Richfield and Associates in the central portion of the basin should yield interesting results.

MARKET PROSPECTS

Most of those who are acquainted with the oil potential of the lower Mackenzie Basin will agree that it has substantial promise, given a market incentive. When mention of markets for that oil is made, however, we immediately think in terms of the excessive cost of the pipeline that will transport oil that must move ultimately to some foreign country. Such thinking is logical, considering the fact that we do not now have markets within our own country, but overlooks the potential for future markets within our borders and the mutual benefit that the mining and oil industries could be to one another in their joint struggle to make otherwise worthless minerals economic.

The principal markets for the lower Mackenzie Basin and other Arctic oil on the island archipelago, as we see them, are as follows.

Japan. Japan has already shown a preference for dealing with Canada for our oil.

Europe. Geographically the English Channel area is less than half the distance (*via* the Arctic Ocean) to the same area from the Middle East, using present transportation facilities.

Atlantic and Pacific seaboard of North America. It is not at all impossible that the Montreal market problem might be solved by Arctic oil. The west coast of the United States is definitely a potential future market.

The Interior of Canada. Canada is a great and growing country. We have every reason to believe that, although we are anxious now, and possibly will be for some time to come, to export as much surplus oil and gas as we can produce, our own consuming needs will ultimately far exceed anything we can at present visualize. Even so, when that time comes, we will still have vast reserves of surplus energy in the form of oil, gas, oil sands, coal, and uranium.

CONCLUSIONS

The lower Mackenzie Basin contains a thick geological section that is satisfactory from the standpoint of oil source and reservoir rocks. Suitable geological structures and potential reservoir trends appear to be abundant and widespread; they should not be difficult to find. As a result of the preceding, the incidence of discovery of oil per drilled well there should be high and the finding cost of the oil relatively low.

Granted that the above is correct, we are still confronted with the very difficult problem of how to dispose of the oil found. Studies of the nearby Precambrian Shield to the east and the mineral-producing areas to the west, in the Yukon and in Alaska, indicate that the potential in terms of metallic mineral resources is also great. Studies of the climate and of the agricultural possibilities lead us to believe that limited agriculture is not impossible and that it could go far towards supporting a fair population, an important factor in the development of such a terrain. Advances made by Soviet Russia in similar areas have already shown that the area can support a substantial population and be developed to the good of the whole nation.

Finally, we would not conclude that the oil industry should plunge headlong into a development programme in Canadian Arctic areas without giving considerable thought to markets, both at home and abroad. On the other hand, it is reasonable to assume that those concerned with the development of both the metallic and the non-metallic mineral resources would do well to co-operate with one another in their planning in order to lighten the burden for both groups. Authorities agree that the area is rich in natural resources and that it is merely a question of time and technical know-how before these resources are developed. Why wait fifty years if we can do it in twenty-five? Why wait twenty-five years if we can do it in ten?

Operational Report on a Gravity Meter Survey Conducted on the Arctic Coastal Plain

L. N. INGALL AND
R. J. COPELAND

ABSTRACT

This report describes the operational aspects of a gravity meter survey conducted during June and July 1959 on the Arctic Coastal Plain. Aspects discussed include personnel, logistics, equipment, mode of operation, and operational statistics.

IN JUNE AND JULY 1959, the British American Oil Company conducted a gravity meter survey on oil and gas permits held on the Arctic Coastal Plain. A helicopter was used for transportation and a float-equipped Beaver aircraft for support. The area covered by the survey is centred near Parsons Lake and is situated forty-five miles north of Inuvik in the Northwest Territories.

The topography in the area ranges from low relief plain and "knob and kettle" topography, to undulating hills up to 900 feet in height. Lakes abound in the area; they are of varying sizes and many have precipitous banks.

Ground transportation is extremely slow because of the rough tundra surface and the circuitous routes necessary to cover relatively short distances. This is a serious handicap to the conduct of a successful gravity meter survey because it is necessary to travel considerable distances within limited times. The area is, however, suited to efficient helicopter operation as it is near sea level and has little vegetation over three feet in height.

The gravity crew commenced operations on June 21, 1959, and suspended operations five weeks later on July 25. Weather conditions varied considerably during this period. Warm, sunny periods were interspersed with cold, wet periods. Ground fog disrupted operations on occasion. The fog became more frequent as the season progressed.

PERSONNEL

The crew of eleven men was composed of gravity and aircraft personnel as follows:

Gravity Personnel	*Aircraft Personnel*
Party chief	Two helicopter pilots
Assistant party chief	Beaver pilot
Operator	Helicopter engineer
Surveyor	Beaver engineer
Computer	
Cook	

LOGISTICS

Before beginning operations, all gravity personnel and one helicopter pilot were flown to Inuvik by regular commercial airline. The remainder of the aircraft personnel flew to Inuvik by helicopter and Beaver aircraft.

The supply base for the survey operation was Inuvik, where fuel and camp provisions were obtained. Crew equipment and other supplies were sent from Edmonton, Alberta, by truck to Hay River and by barge to Inuvik.

Emphasis was on mobility in planning the operation. All camp equipment used was portable, including tents for the personnel. The Beaver aircraft was used to move personnel and equipment to each camp location, as well as to transport helicopter fuel and camp supplies.

EQUIPMENT

Both aircraft used on the survey are well known. The Beaver is a popular work-horse in the Canadian north and needs no description here. The helicopter was a Bell 47 G-1, capable of carrying two passengers and the survey equipment.

One gravity meter was used throughout the survey. It was a standard Worden meter with a scale constant of 0.0828 milligals per dial division. Range of this instrument without resetting is 66 milligals and the total range of the instrument is 4,100 milligals. Drift characteristics of the meter are excellent: no significant calibration change was observed during the survey.

A Photo-Transit developed by Mr. W. O. Bazhaw was used to obtain the vertical and horizontal control necessary for the location and reduction of the gravity stations. Basic components of the Photo-Transit are (1) a Graflex Tele-optar lens of focal length twenty-five cm used as the transit telescope, and (2) a thirty-five mm Leica camera adapted for use with the telephoto lens and fitted with an engraved scale. This scale provides the means by which distance and elevation change can be calculated. Reduction of results is carried out in a special film-reader by projecting each film frame onto a calibrated ground-glass screen. An optical device fitted to the transit allows the magnetic bearing of each shot to be photo-graphed at the same time as the rod. The rod itself consists of three distinctively-painted Lucite targets equally spaced on an aluminum pipe.

There are several advantages to using the Photo-Transit: a permanent record is obtained of each shot taken by the surveyor; chances of human error are reduced considerably, for the surveyor merely levels and orients the transit and then exposes a film frame—a sequence of operations which provides information for obtaining both horizontal and vertical control; the eight-degree field of view of the transit objective allows a considerable change in elevation to be recorded for a level setup of the instrument. For example, an elevation change of ninety-two feet can be recorded at a distance of a quarter of a mile.

MODE OF OPERATION

The operations team comprised the surveyor, a helicopter pilot, and the meter operator, who also acted as the surveyor's rodman.

Gravity readings were made at half-mile intervals on three-mile loops so that the average backsight or foresight for the surveyor was a quarter of a mile. Problems of terrain necessitated a change of this station interval at times. The helicopter carried both the operator and the surveyor to and from work from the base; however, the helicopter carried only one passenger at a time during actual operations because the operator and surveyor continually "leapfrogged" each other.

A typical operational day consisted of four "runs," each run of from two to two-and-one-half hours' duration. This period of time was adopted as being most practical considering both the drift control of the gravity meter and the refuelling of the helicopter. A base station observation was made with the meter at the end of each run, making it unnecessary to establish sub-bases. There were twenty-four hours of daylight; however, the crew worked from 7:00 A.M. to 6:00 P.M. This work period was selected to prevent disruption of crew members' routine habits and because the lighting was best for photography.

Two morning runs were separated from two afternoon "runs" by a midday meal. The helicopter pilots changed shift each noon so that both men shared each day's flying.

An area of 300 square miles, representing 250 miles of linear traverse, was efficiently covered from a centrally located base. No point in such an area is more than twelve minutes helicopter flying time from this base.

A gravity base was established at each camp and special base tie runs were made during the survey by Beaver aircraft or helicopter to obtain accurate base gravity differences. Gravity meter calibration checks were made in Calgary, Alberta, before and after the survey, and calibration stations established in the area were used to check the scale constant of the meter during the survey.

Vertical loop closures obtained were consistently better than twelve inches in twelve miles; however, the horizontal control provided by the magnetic traverse method used was not adequate. The horizontal component of the earth's magnetic field, or compass directive force, is small in the high geomagnetic latitudes prevailing in the Canadian Arctic. It is influenced appreciably by local anomalies, as well as by diurnal and storm variations in the earth's magnetic field. The method adopted to obtain reliable horizontal control was to fit the magnetic traverses to mapped identifiable geographic features, such as lakes and creeks. The surveyor was able to relate each traverse to several features during the course of each survey run.

Accurate maps became available for the first time when an airborne magnetometer survey of the area was initiated in 1958. These maps were constructed from Shoran-controlled Royal Canadian Air Force air photos and were found to be adequate for reduction of the gravity stations.

To supplement control obtained in the survey network, the Beaver aircraft and helicopter flew several long traverses to obtain regional gravity information. Horizontal control for the second-class stations observed on these traverses was obtained by locating each station at an identifiable map feature. Vertical control was obtained by observing the stations at, or near, sea-level. In the few cases where stations were not observed near sea-level station elevations were established using an aneroid barometer.

REDUCTION OF RESULTS

All gravity observations were reduced to mean sea-level using an elevation correction based on density information obtained during the survey. Latitude corrections were made using Lambert and Darling's tables with second differences, based on the 1930 International Gravity Formula. At the conclusion of the survey a special tie run was made to the proposed site of the Dominion Government absolute gravity station at the Inuvik airport.

OPERATIONAL STATISTICS

Days spent on project: 34.

Lost time: 9½ days, comprising: helicopter out of action, 5½ days; ground fog (6 half-days), 3 days; excessive wind, ½ day; earthquake activity, ½ day.

Area covered by survey network: 600 square miles (equivalent approximately to 17 townships 6 mi. by 6 mi.).

Miles of linear traverse: 565 miles, comprising 470 miles in the survey network, and 95 miles of traverse for regional control.

Total number of stations observed: 1089.

Average daily production: 44.5 stations (excluding lost time due to the causes listed above).

CONCLUSIONS

Mobility and speed of operation are essential to the conduct of a successful gravity meter survey. These essentials cannot be achieved by using existing methods of ground transportation on the tundra but an efficient gravity meter survey can be conducted using a helicopter and Beaver aircraft as described in this paper.

Marine Seismograph and Sparker Survey in the Mackenzie River, Northwest Territories

JIMMIE G. MEADOR

ABSTRACT

The Mackenzie River, Northwest Territories, flows from the Great Slave Lake northward 1000 miles to the Arctic Ocean with an average gradient of half a foot per mile. The sediments range in age from Pleistocene drift to Palaeozoic.

A Sparker traverse of the Mackenzie River starting at Hay River was carried out down to the Sans Sault Rapids, a distance of 650 miles. The Sparker instrument recorded seismic reflections from the river bottom, and from one to three sub-bottom horizons to a depth of 400–1,600 feet in a continuous profile. At a rate of forty miles per day the Sparker recorded the geologic configurations every three feet of the near-surface formations, giving an accurate reconnaissance of the area.

In addition, a marine seismic survey was conducted in selected areas at a rate of seventy shot locations per day, mapping the geologic horizons down to basement.

Sparker and marine seismic were used to detail the Norman Wells reef oilfield beneath the Mackenzie River. These data may be used as references to correlate reflections with other data recorded.

DURING JULY AND AUGUST 1959, Accurate Exploration Ltd. conducted a Sparker survey and a marine seismic programme in the Mackenzie River, Northwest Territories, from Hay River to mile 610, plus approximately thirty-five miles up the Liard River from Fort Simpson. The object of the survey was to perfect a technique of mapping the near-surface geology with the Sparker instrument and to detail selected anomalies with marine seismic (Figure 1). The Norman Wells field was one of the locations selected for detailing, for use as a case reference, and for evaluation of the results.

The Sparker system is a continuous seismic profiler utilizing a high voltage spark discharge as an energy source and a hydrophone as receiver. The marine seismic method incorporated the use of a floating seismic cable with multiple pressure-type geophones, the output being recorded on magnetic tape. The dynamite was suspended and detonated in the water. Survey location was established by photographing a short-range radar screen.

In most of the area the Palaeozoic outcrops are near the surface. River elevation along the traverse ranges from 510 feet at Great Slave Lake to approximately 130 feet at Sans Sault Rapids (mile 610), giving a gradient of a half a foot per mile. The temperature ranges from a mean annual maximum of 85° F to a mean annual minimum of —50° to —60° F.

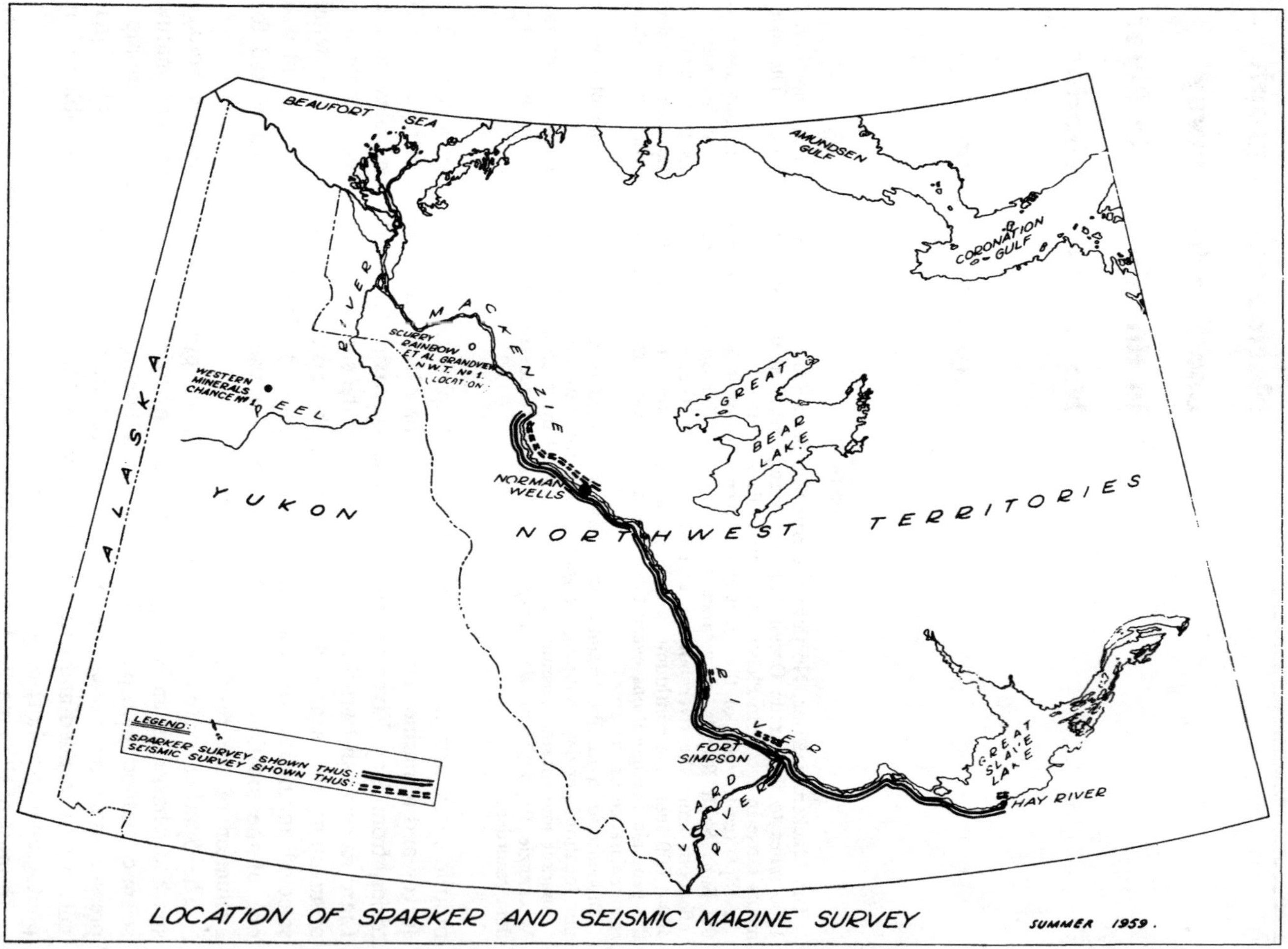

FIGURE 1.   Location of Sparker and seismic marine survey, summer 1959.

EVALUATION

New exploration tools or techniques are usually tested in an area of known geology. For this reason, both instruments were used to detail the Norman Wells oilfield.

The Sparker profile, although limited in depth of penetration to 1,500 feet, is capable of mapping horizons beneath the top of the Palaeozoic erosional surface. Thus, in a conformable section, structure may be predicted at greater depth. In a large portion of the Mackenzie River traverse, the Slave Point or equivalent and other economically potential formations can be correlated by the continuous seismic profiler.

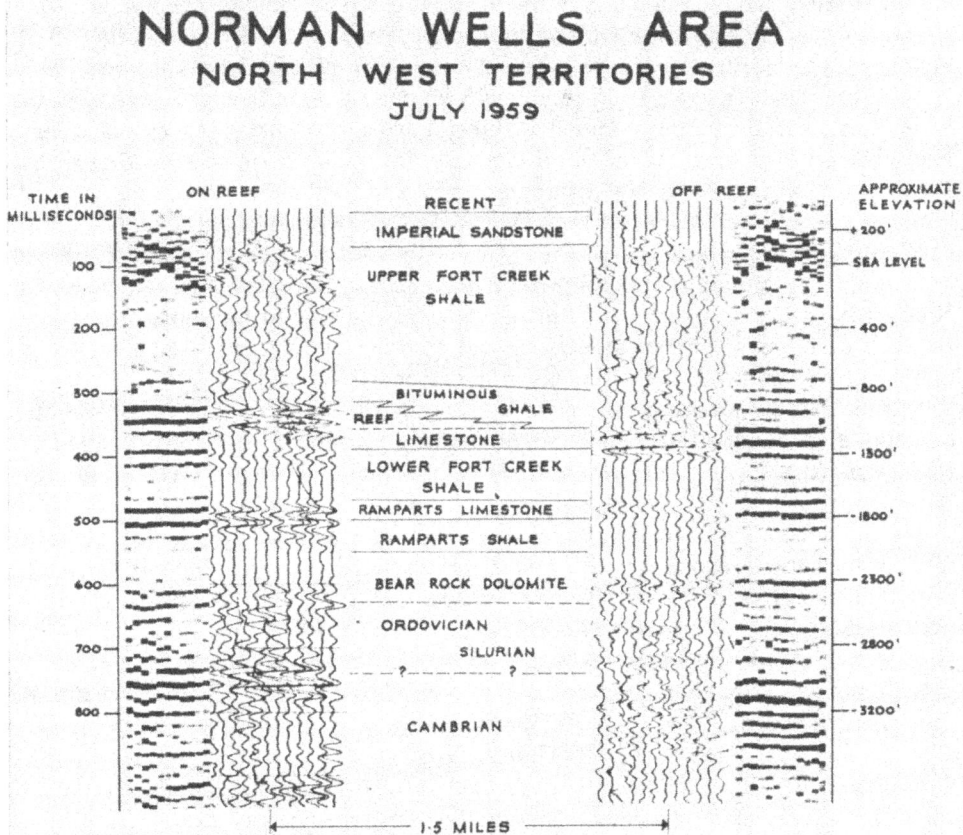

FIGURE 2. Norman Wells area, Northwest Territories, July 1959. Marine seismogram and corresponding variable density comparison (with estimated geological correlation).

The Sparker Instrument

Using interval velocities obtained from the Reflection Report Seismograph Survey, Norman Wells area, No. 32, depths were computed for the Sparker reflections and compared to the structures obtained from well control.

The Sparker record made approaching the Norman Wells field from the southeast showed a strong reflection at approximately 200 feet below sea-level which dipped down to approximately 500 feet below sea-level and continued at that depth to the end of the record shown in this figure. The strong reflection correlates with the top of the Devonian Imperial sandstone.

Marine Seismic

The marine seismic shooting resulted in good record quality (Figure 2) which simplified the interpretation of the subsurface structure. Seismic profiling in the vicinity of the Norman Wells reef yielded data which clearly defines the reef flanks. A reflection from the reef itself along with a continuous reflection from the underlying Ramparts formation, yielded a direct means of establishing the amount of reef present. The presence of approximately 700 feet of upper Fort Creek shale above the Norman Wells reef compared with approximately 1200 feet above the "off-reef" limestone equivalent may be noted. This is a situation ideally suited for exploration by reflection seismograph, as no reflections directly precede the reef reflection. The situation encountered in the central plains of Alberta, where the Nisku–Ireton–Leduc sequence is difficult to resolve owing to the short time interval between component energy reflections and complicated by the instrumental automatic volume control, is fortunately not present.

Seismic record quality is enhanced by shooting in the river rather than shooting on land as a result of the good coupling with "ground." Poor record quality is mainly associated with areas where there is an aerated surface with glacial drift and river deposit.

Marine seismic is a means of evaluating a sizable land spread for structure, section thickness, record quality, and shooting technique. A detailed exploration programme can then be strategically located taking full advantage of this initial high-grading. Marine profiling can be carried out at less than one-half the per-mile cost of land work, owing to the speed of marine progress (ten miles per day or more) and the elimination of the need for bulldozers, drills, and large camp and transportation costs.

CONCLUSIONS

The use of the water bodies in the Northwest Territories appears to be the solution to the need for a rapid, flexible, low cost preliminary examination of the subsurface, allowing the operator to locate his follow-up programme most rewardingly.

Use of marine seismic in major rivers and lakes, and off-shore with a portable continuous seismic profiler of the Sparker or gas exploder type in minor lakes and streams, is a combination naturally suited to the early stages of Arctic and sub-Arctic exploration.

REFERENCES

ATLAS OF CANADA.   1955.   Department of Mines and Technical Surveys, Section 22.
PALLISTER, A. E.   1959.   Oilweek, November 6.

Geochemical Prospecting for Base Metals in Schuchert Dal, Northeast Greenland[1]

ALAN REECE AND

A. L. MATHER

ABSTRACT

The bringing into production of a lead-zinc mine at Mesters Vig, northeast Greenland in 1955 stimulated the search for minerals in that area by Nordisk Mineselskab A/S, Copenhagen. The general exploratory techniques used in 1956-7 are described. Simple geochemical techniques successfully revealed the presence of minor galena veins below the permafrost level and covered by 2 m of glacial drift.

IN 1948 A GEOLOGICAL PARTY, led by Lauge Koch, discovered lead-zinc mineralization in the vicinity of Mesters Vig, northeast Greenland (Figure 1). As the discovery appeared to have economic possibilities, Nordisk Mineselskab A/S (Northern Mining Company) was formed in 1952 and an exploration and drilling programme started. An orebody was found at Blyklippen, some 8 km from the coast and 12 km northwest of Mesters Vig. The deposit totalled more than 400,000 tons of ore, the average grade being 20 per cent lead and zinc over 9.1 m. Production began in 1956 and the first shipments of concentrates were made that summer.

The success of this operation encouraged the Company to continue exploration in other parts of their concession (70°00′ N to 74°30′ N, between the ice-cap and the sea) in 1955. A large part of the concession was reconnoitred and limited geochemical work was done by R. C. Pargeter. This work was continued in 1956 by one of the present authors and followed up by more detailed work in specific areas by both authors in 1957.

GENERAL ACCOUNT

1955 and 1956 Seasons

During the first two seasons field parties covered a large part of the concession, which measures 500 km from north to south and 250 km from east to west. Helicopters based on the mine at Blyklippen were invaluable for this work. Reconnaissance flights and spot landings were made in the most interesting areas and followed up by the establishment of light-weight camps allowing detailed examination on the ground. Bell G-2 helicopters were used in 1955 but longer range aircraft were required in 1956 and Sikorsky S-51 machines were chartered.

[1]This paper is published with the permission of Nordisk Mineselskab A/S, Copenhagen.

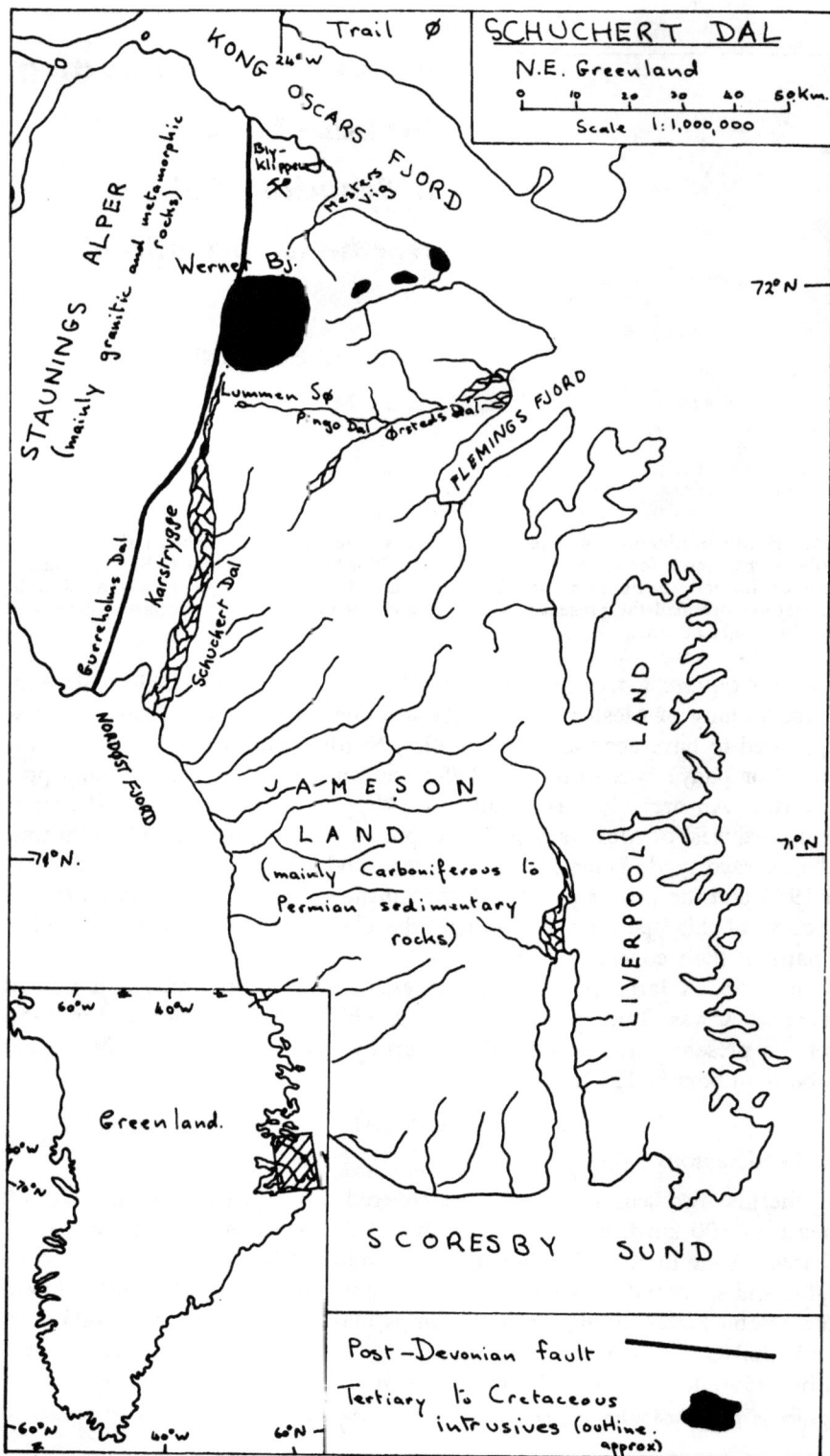

FIGURE 1. Scoresby Sund-Kong Oscars Fjord area, northeast Greenland.

Much of the work consisted of searching for mineralized float and tracing it back to its source. In some areas mineralized float was abundant, but the source was not apparent; these areas were surveyed geochemically. Some stream sediments were also sampled.

1957 Season

On the results of work of the previous two seasons, it was decided that in 1957 the main effort would be concentrated in the area around Schuchert Dal, a large glaciated valley 70 km south-southwest of Blyklippen (Figure 1). The presence of at least one potentially economic lead-zinc deposit was known on the east side of Schuchert Dal, near a small lake called Lummen Sø. To develop this prospect, a drilling camp was established at Lummen Sø and most of the exploration was based at this camp. The shortest route from the mine at Blyklippen to Lummen Sø is about 80 km, but it involves travelling over glaciers in the Werner Mountains, a difficult if not hazardous undertaking for a tractor train. The route chosen went southeast along the coast of Kong Oscars Fjord to Flemings Fjord, and then west up Øersted Dal and Pingo Dal to Lummen Sø, a total distance of about 120 km (Figure 1). The camp was established by two D-6 caterpillar tractors, which made three to four journeys during April and early May, before the winter snow and sea ice began to thaw. A dog team was used to lay out caches for summer work in Schuchert Dal. A small party of men with one tractor, fitted for bulldozing, was left at Lummen Sø to prepare drill sites and an air strip, before the main party arrived in Greenland in early July.

Helicopters were not used in 1957, but instead two fixed-winged aircraft. One of these, a twin-engine de Havilland Rapide, provided transport between the airfield at the mine and the temporary air strip at Lummen Sø. It was also used for free-dropping supplies to small field parties working out from the Lummen Sø drilling camp. The other aircraft, a small single-engine, four-seater monoplane, KZ-VII (Skandinavisk Aero Industri A/S), was used for reconnaissance work. In the hands of an experienced pilot this plane could land and take off from unprepared ground which was reasonably level and contained no boulders larger than about 20 cm in diameter, or other obstacles. A minimum distance of 150 m was needed for take-off and landing. Although this plane was not able to land in as many places as a helicopter, it was perfectly adequate for the type of work involved and was, of course, more economical and reliable.

The geochemical work consisted of a reconnaissance survey for about 15 km on each side of northern Schuchert Dal (Figure 2), and a detailed survey in the Lummen Sø area (Figure 3*). A laboratory was set up in a hut at the Lummen Sø camp, where all the samples were received from the field parties. The hut had a floor area of nine square yards, which was quite adequate. The sampling teams were composed of British and Danish undergraduates. Students of geology and geography were found to be most satisfactory for this work. Two undergraduate chemistry students were employed in the laboratory where they were able to carry out the analytical work with a high degree of accuracy and productivity.

*See separate container of figures.

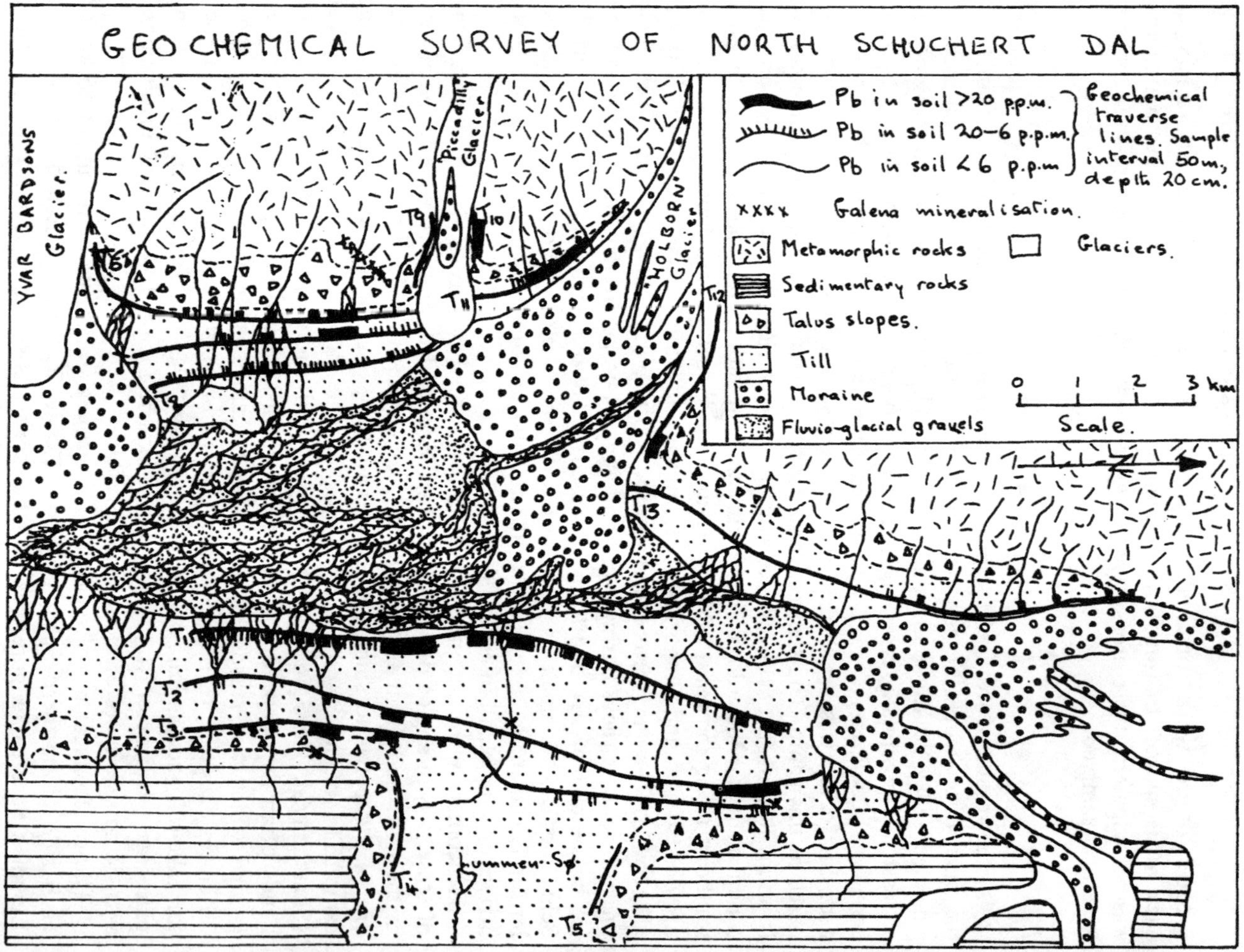

FIGURE 2. Reconnaissance geochemical survey of northern Schuchert Dal.

GEOLOGICAL BACKGROUND

Topography and Bedrock Geology

A large irregular tongue of land trends approximately southeast between Scoresby Sound and Kong Oscars Fjord (Figure 1). Topographically, as well as geologically, it is divided into two distinct regions, approximately by long. 24.5° W. To the west lie Staunings Alper, a glaciated mountainous area of alpine peaks and ridges rising 2000 to 2500 m above sea-level, and composed of a metamorphic complex of granitic rocks which were folded during the Caledonian (Palaeozoic) orogeny. To the east undulating plateaux, rarely exceeding 1000 m in height, are separated by wide U-shaped valleys. The rocks consist mainly of continental sediments of Carboniferous to Permian age which are only moderately folded, and sometimes faulted. A major post-Devonian fault, trending just east of north, separates these sedimentary rocks from the older metamorphic rocks in the Staunings Alper region. A complex intrusion, probably of Tertiary age, and composed of various igneous rocks, lies just east of the fault and southwest of Mesters Vig (Figure 1). This intrusion and the adjacent sedimentary rocks form the glaciated Werner Mountains. Schuchert Dal is a large, straight, U-shaped valley, 70 km long and 5 km wide, trending south-southwest from the Werner Mountains to Nordøst Fjord in Scoresby Sound. The northern part of this valley is occupied by a glacier, fed partly from the Werner Mountains and partly from Staunings Alper. The terminal moraine is situated about 75 km from Nordøst Fjord. South of the moraine a braided melt-water stream fills the wide flat valley floor in summer. Three other glaciers just reach the northern half of Schuchert Dal from the west. Various streams enter the valley from the west and east.

The west wall of the northern part of Schuchert Dal is composed of the Caledonian metamorphic complex mentioned above, but exposures of folded sedimentary rocks are found on the floor of the valley close to the west side. Consequently, the post-Devonian fault must lie very close to the west wall. About half-way down Schuchert Dal this fault trends southwest for about 15 km and then continues south–southwest down Gurreholms Dal, a minor valley parallel to Schuchert Dal and farther to the west. Between Schuchert Dal and Gurreholms Dal the sedimentary rocks appear to be block-faulted and slightly tilted.

Mineralization

The mineralized zone is confined to the northern part of Schuchert Dal, within a few kilometres of Lummen Sø. A number of small quartz-galena veins with southwest and south-southeast trends occur in this area. Minor sphalerite and copper minerals are associated with the galena. It is tentatively suggested that the mineralization is directly associated with the Werner Mountains intrusion, and that its location is determined by the fracture pattern near the post-Devonian fault. In this connection it is interesting to note that molybdenite has been found in the centre of the Werner Mountains intrusion. The distances from this locality to the mineralizations at Lummen Sø and Blyklippen are approximately the same, that is, 20 km. This suggests mineral zoning around the intrusive.

Superficial Deposits

Four main types of superficial deposit are found in Schuchert Dal: (1) talus at the foot of the steep valley walls, (2) ground moraine forming a gently sloping pediment below the talus slopes, (3) fluvioglacial gravels which form the wide, flat floor of the valley, and (4) terminal moraines in front of glaciers. There are, of course, no sharp divisions between these deposits. Fragments of talus become mixed with the till of the ground moraine, and boulders from the terminal moraines are included in the gravels, but generally the four types of superficial deposits form easily recognizable, distinct units (Figure 2).

The valley profile is characterized by high walls with steep, bare rock surfaces which rise above the talus slopes. Below the talus lies the extensive, gently sloping pediment. In northern Schuchert Dal this pediment is more than 1.5 km wide between the talus and the fluvioglacial gravels at the valley bottom.

The geochemical prospecting described in this paper was carried out on the till deposits of the pediment. These tills were never less than 2 m thick, and consisted of ill-sorted material ranging in size from clay to boulders with a diameter of 1 m. In summer the permafrost level was 1 to 2 m below the surface, and the overlying waterlogged material became mobile and subject to mass movement.

RECONNAISSANCE GEOCHEMICAL SURVEY

Sampling Methods

In mountainous areas of the Arctic, rocks disintegrate rapidly as a result of frost action, and oxidation is slight. These two factors suggest that the dispersion of lead from galena veins is a physical rather than a chemical process. Undoubtedly solifluction plays a major role in the dispersion of fragments of galena derived from veins located at, or near, the surface. Mud flows commonly occur on the pediments, and one would expect that the fine grains of mineralized rock might become intimately mixed with the till and get carried to the base of the slope.

The dispersion of lead downslope was investigated by laying out three traverse lines on the east side of the valley below known mineralized areas. The position of the lines coincided approximately with the 450, 300, and 150 m contours (Figure 2). The upper traverse line, T_3, was situated on the pediment slope just below the talus edge, T_1 was at the base of pediment, and T_2 was approximately midway between these lines. Similar lines were laid out on the west side of the valley (traverses T_6, T_7, T_8).

The till was sampled at a depth of 20 cm to avoid contamination from possible organically-enriched surface material. The traverse lines were sampled at 50 m intervals.

Results

Preliminary analyses showed that the background value for zinc was in the order of 60 to 80 ppm, while the lead content was less than 6 ppm. High zinc values of 200 to 500 ppm were observed in the till close to the galena veins, and showed a marked decline to background values (60 to 80 ppm) within a distance of only

100 to 200 m. The contrast between anomalous and background values was much greater in the case of lead.

Lead analyses of the till showed a fan-shaped pattern of anomalous values from the Lummen Sø veins to the foot of the pediment slope, a distance of more than 1.5 km. The lead values decrease gradually from around 2000 ppm near the veins to between 10 and 50 ppm. The dilution is caused by fine detritus from the higher rock and talus slopes. Similar fan-shaped dispersion patterns were observed below other known galena veins, some of which occurred above the talus line. In all of these examples anomalous lead values were found at the foot of the pediment slope at a distance of more than 1.5 km from the veins. Lead anomalies on the lower traverse lines led, in one instance, to the discovery of an outcropping vein. In several other places the anomalies provided targets for future investigation.

DETAILED GEOCHEMICAL SURVEY

Sampling Methods

The Lummen Sø vein lies near the surface on a steeply dipping slope where the overlying till is subject to considerable solifluction. In order to ascertain the extent of the mineralization, the area surrounding the vein was examined in detail by geochemical methods. Till samples were taken at 20 m intervals along a number of traverse lines spaced 50 m apart and oriented at right-angles to the general strike of the vein.

The samples were taken at a depth of approximately 2 m by driving pinch bars, 1.5 m and 2.5 m long, consecutively into the till until the top of the permafrost was reached. Samples were taken from the bottom of the hole by means of an auger. These samples were taken at the lowest level to which the ground thaws during the summer. It seems likely that the large particles of galena derived from the near-surface veins will sink to this level and, in contrast with the overlying mobile material, not move very far.

Results

Analyses of the samples taken in the vicinity of Lummen Sø showed a distinct area of high lead values. This area of anomalous till we trenched down to bedrock by means of a bulldozer. As a result, a number of minor quartz-galena veins and galena boulders were exposed in the bedrock and on the surface of the permafrost "table" (Figure 3*).

This detailed geochemical survey is, of course, only applicable where the overburden is thin, say less than 5 m. On the lower pediment slope where the overburden is thick, it is unlikely that any underlying mineralization would be apparent as anomalous metal values at the surface.

CONCLUSIONS

The reconnaissance geochemical survey indicates that analyses of the lead content in the overburden at the foot of the pediment slope can detect galena-bearing veins, which lie at, or near, the surface, more than 1.5 km up the valley slopes. The fan-shaped pattern of the lead anomalies over the pediment makes it possible to take samples as widely apart as 150 m at the foot of the slope. Zinc and copper may be used also as indicator elements, provided that the background content of the host and associated rocks is low.

It is evident, therefore, that considerable areas similar to the Arctic mountainous terrain described above may rapidly be reconnoitred by geochemical methods. The number of personnel is considerably less than that required for the usual, less efficient, visual reconnaissance.

The follow-up procedure of a detailed geochemical survey provides a means of pin-pointing the mineralized bedrock below the relatively large area of overburden containing anomalous lead values. The same procedure may also be used rapidly to scan areas where quartz float has been noted at the surface during reconnaissance.

Appendix: Methods of Analysis

The instructions outlined below are those currently in use at the Geochemical Prospecting Research Centre, Imperial College, London (Stanton, 1960). These methods differ from those published by Stanton and Gilbert (1956) and Gilbert (1960) in that benzene has been substituted for carbon tetrachloride and certain changes have been made in the composition of the buffer. Detailed lists of reagents and equipment for these methods are given in both the recent publications.

Abbreviated Instructions for the Determination of Lead

Reagents Required for 1,000 Determinations
  (1) 0.15 g dithizone (analytical reagent)
        0.01 per cent solution in benzene
        0.001 per cent solution in benzene
  (2) 6 litres benzene (recrystallized)
  (3) 2 litres carbon tetrachloride
  (4) 1.5 kg tri-ammonium citrate (analytical reagent)
  (5) 300 g hydroxylamine hydrochloride (analytical reagent)
  (6) 1 g thymol blue
  (7) 1.2 litres ammonium hydroxide, S.G. 0.88 (analytical reagent)
  (8) 200 g potassium cyanide (analytical reagent)
  (9) 1.2 litres nitric acid (analytical reagent)
 (10) 1 g lead nitrate (analytical reagent)
 (11) Metal-free water from resin column used throughout.

Preparation of Buffer Solution. Dissolve 120 g of tri-ammonium citrate and 20 g of hydroxylamine hydrochloride in 1 litre of water, and add 5 ml of 0.04 per cent thymol blue. Add ammonia (S.G. 0.880) until the indicator changes to a distinct blue, and then add 18 g of potassium cyanide. Purify with 0.01 per cent dithizone in carbon tetrachloride, followed by carbon tetrachloride extraction. The buffer is liable to some deterioration if not prepared daily.

Preparation of Standards. To 10 test tubes containing 10 ml of buffer solution add respectively 0, 0.5, 1.0, 1.5, 2.0, 2.5, 3.0, 3.5, 4.0, and 4.5 μg of lead and 1 ml of 7.5 per cent nitric acid. Add 5 ml of 0.001 per cent dithizone in benzene and shake vigorously for 30 seconds.

Procedure

(1) Weigh 0.1 g of sieved sample, usually minus-80-mesh (British Standards) fraction, into a 16 × 150 mm test-tube, previously calibrated at 3 ml and 10 ml.

(2) Add 3 ml of 25 per cent nitric acid and digest on a hot plate for 1 hour.

(3) Dilute to 10 ml with water and mix by shaking.

(4) Pipette a 2 ml aliquot of the above extract into 10 ml of buffer solution, contained in a 19 × 150 mm test-tube previously calibrated at 10 ml.

(5) Add 5 ml of 0.001 per cent dithizone in benzene and shake vigorously for 30 seconds.

(6) Compare with standards (colour matching may be facilitated by the use of a colour matching tube obtainable from the General Electric Co., London).

(7) Repeat on a smaller aliquot if colour exceeds top standard. For an aliquot of 0.5 ml or less, add also 1 ml of 7.5 per cent nitric acid.

Abbreviated Instructions for the Determination of Zinc

Reagents Required for 1,000 Determinations

(1) 0.15 g dithizone (analytical reagent)
      0.01 per cent solution in benzene
      0.001 per cent solution in benzene

(2) 6 litres benzene (recrystallized)

(3) 2 litres carbon tetrachloride (analytical reagent)

(4) 600 g potassium bisulphate (fused and crushed)

(5) 400 g sodium thiosulphate (pentahydrate)

(6) 1.5 kg sodium acetate, trihydrate (analytical reagent)

(7) 50 ml glacial acetic acid

(8) 600 ml concentrated hydrochloric acid (analytical reagent)

(9) 1.0 g zinc (analytical reagent)

(10) Metal-free water from resin column used throughout.

Preparation of Buffer Solution. Dissolve 250 g sodium thiosulphate in 1500 ml water. Add 1 kg sodium acetate, dissolve, then add 30 ml glacial acetic acid. Make up to 2 litres. Extract 1 litre at a time with 50 ml portions of 0.01 per cent dithizone until dithizone remains green. Remove excess dithizone by extraction with carbon tetrachloride until organic layer is colourless. Dilute to 4 litres with water.

Preparation of Standards. 100 μg zinc per ml: dissolve 0.1 g of zinc in 10 ml concentrated hydrochloric acid and dilute to 1 litre with water (store in a polythene reagent bottle; this solution is stable).

5 μg zinc per ml: dilute 5 ml of the above zinc solution to 100 ml with water (store in a polythene reagent bottle; this solution must be prepared freshly each week).

Pipette 0, 0.05, 0.1, 0.2, 0.3, 0.4, 0.5, 0.6 and 0.7 ml of the 5 μg per ml zinc solution into test tubes containing 5 ml of buffer solution. Add 5 ml of 0.001 per cent dithizone and shake vigorously for 1 minute. The standard colours range from green through blue to purple. They must be prepared each day and stored in the dark.

(1) Place 0.1 g sample, usually minus-80-mesh (British Standards) fraction, in a 16 × 150 mm Pyrex test-tube, add 0.5 g potassium bisulphate, and fuse to a quiet melt. Add 5 ml N hydrochloric acid and warm to complete solution of the melt. Add 5 ml water and mix well.

(2) Add 5 ml of buffer solution to a test-tube (use a polythene wash bottle) and pipette an aliquot of 1 ml of the extract into the tube.

(3) Add 5 ml of 0.001 per cent dithizone.

(4) Cork the tube and shake vigorously for 1 minute.

(5) Compare the colour of the organic phase with extracts obtained from standard solutions.

(6) Repeat the test on a smaller aliquot if the colour is denser than the top standard.

REFERENCES

GILBERT, M. A. 1959. Field and laboratory methods used by the Geological Survey of Canada in geochemical surveys, no. 1, Laboratory methods for determining copper, zinc and lead; Geol. Surv., Canada, paper 59-3.

STANTON, R. E. 1960. Geochemical Prospecting Research Centre, London; Technical Communication no. 00.

STANTON, R. E., and GILBERT, M. A. 1956. Analytical procedures employed at the Imperial College; Geochemical Prospecting Centre, London, Technical Communication no. 4.

Activities of

Norsk Polarinstitutt

(Norwegian Polar Institute)

in Svalbard

THORE S. WINSNES

ABSTRACT

Since 1906 Norwegian expeditions have gone to Svalbard every summer (except during World War II), carrying out hydrographic, topographic, and geological surveys. After the establishment of Norwegian sovereignty in Svalbard through the international treaty of 1925, the Norges Svalbard-og Ishavs-undersøkelser (Norwegian exploration in Svalbard and the Polar Seas) was founded as a continuation of the state-supported Norwegian Spitsbergen Expedition. The results are published in a series of *Skrifter* (papers) and *Meddelelser* (shorter notes). In 1948, the Norsk Polarinstitutt was founded as an expansion of the former institution; it has now a staff of twenty-one.

In the course of the years the western waters of Svalbard have been charted and 11 charts issued. During the summers of 1936 and 1938, the whole of Svalbard was photographed from the air, and the new colour maps at a scale of 1:100,000 are based on these photographs. Eight of the standard maps have been published. The geological exploration of Svalbard has made possible collections of fossils and rock specimens, resulting in important palaeontological and stratigraphic contributions. Stress has been laid on an investigation of the Tertiary coal, which is being mined in several places. The areas surveyed geologically will be included in maps based on the standard 1:100,000 topographic maps. Special glaciological investigations are also being undertaken. In addition, Norsk Polarinstitutt acts as an adviser to foreign expeditions, furnishing necessary maps and information.

Norsk Polarinstitutt will continue to operate along these well-established lines. Charting of the northern areas of the Norwegian Sea will be started, and a pilot sailing of Svalbard waters will be completed. Co-operation with other institutions in similar fields of activity will continue.

THE SCIENTIFIC EXPLORATION of Svalbard started in 1827 when the geologist B. M. Keilhau, Professor at the University in Oslo, visited Bjørnøya (Bear Island) and Spitsbergen. Later followed a period when predominantly Swedish expeditions were doing fundamental scientific exploration, although Norwegian sealing skippers also made essential contributions to the exploration of Svalbard in those years, followed by oceanographic expeditions in 1872, 1876, and 1878. In 1906, a new era in the Norwegian exploration of Svalbard began with an expedition financed by Prince Albert of Monaco, who had a team carrying out topographic and geological exploration in the northwestern parts of Svalbard. The next year this project continued, and as a result of work which showed the rich Downtonian and Devonian fauna in this area, an extensive Norwegian exploration of these formations was initiated. In the following years, 1909–15 and 1917–25, the Norwegian government supported the expeditions with grants. Members of the expeditions were also recruited partly from government institutions. Mapping, charting, and geological survey work continued in these years; botanical, zoological, and other investigations were also undertaken. The leader of the majority of these expeditions was Adolf Hoel. The results of the expeditions were published in a series: "Resultater av de norske statsunderstøttede Spitsbergenekspeditioner" (Results of the state-supported Norwegian Spitsbergen expeditions).

In a treaty signed in Paris in 1920, the sovereignty of Svalbard was assigned to Norway. On August 14, 1925, Norway took over the sovereignty which was proclaimed in Longyearbyen that year. According to the treaty the subjects of powers who signed the treaty enjoy the same rights in Svalbard as do Norwegians.

In 1928, the Norges Svalbard- og Ishavs-undersøkelser (Norwegian exploration in Svalbard and the Polar Seas) was founded, under the jurisdiction of the Ministry of Commerce, and the publication series changed its name to "Skrifter om Svalbard og Ishavet" (Publications on Svalbard and the Polar Sea). In 1929, another series, "Meddelelser" (Communications), appeared containing shorter papers of more popular character. In the years 1906–26 a total of twenty-one Norwegian expeditions visited Svalbard.

The staff of Norges Svalbard- og Ishavs-undersøkelser (N.S.I.U.) consisted of Adolf Hoel, leader and geologist, Anders K. Orvin, and Gunnar Horn, geologists, Wilhelm Solheim, Bernhard Luncke, Alfred Koller, and J. Sartorius, topographers, and office personnel, ten to twelve persons in all. In the summer of 1927, expeditions again went to Svalbard, and before World War II eighteen expeditions were sent out. Since 1929 a hydrographic surveyor has been attached to the staff. In addition, twenty-four expeditions went to northeast or southeast Greenland, one to Davis Strait, and two to Franz Josefs Land. As it is, of course, impossible for the institution to employ experts in every branch of its work, it has always co-operated with institutions engaged in similar work in Norway, such as the Geographical Survey of Norway and the Geological Institute of the University in Oslo.

In the years 1906–28 the financial contributions to the Norwegian Svalbard expeditions were 1.7 million kroner from the government, 55,000 kroner from scientific funds and institutions, and 365,000 kroner from private sources, the total amounting to 2.12 million kroner. From 1927 to 1944, the amount given to N.S.I.U. totaled 2.2 million kroner, plus 160,000 kroner from funds and subscribers. The greater part of this money was used for the scientific expeditions.

In the years immediately following World War II, expeditions again went north, and Dr. Anders K. Orvin was then acting director of the institution. In 1948, the institution was expanded under the name of Norsk Polarinstitutt. Professor Dr. H. U. Sverdrup was appointed director and Dr. Anders K. Orvin, associate director; the other personnel are one head clerk, three geologists, one glaciologist, two hydrographic surveyors, three topographic surveyors, one geodetic surveyor, one meteorologist, one librarian, two draftsmen, and five other office personnel, the staff totalling twenty-one persons. Professor Sverdrup died in 1957, and Dr. Orvin was appointed his successor, a position he retained until his retirement in 1960 at which time Dr. Tore Gjelsvik was appointed director. The office of Norsk Polarinstitutt has been the old university observatory, Observatoriegaten 1, Oslo, a building more than 100 years old, but plans for new, modern premises are well advanced. The library contains about 15,000 volumes of polar literature and 7,000 to 8,000 papers, reprints, and periodicals.

The activities of Norsk Polarinstitutt in the polar regions are restricted to certain fields of work because of the historical background, but apart from these, special funds can be used in other fields of investigation, such as ornithology, botany, and Pleistocene geology, and a number of expeditions in these fields have been financed by Norsk Polarinstitutt.

The work carried out by the permanent personnel involves the mapping of territories, sounding and charting of the ocean and coastal areas, geological mapping and investigations, including special examination of coal deposits, palae-ontological and geomorphological studies, glaciological investigations in Svalbard and Norway, and also special meteorological work. The maps and charts are being printed in Norges Geografiske Oppmåling (Geographical Survey of Norway), but are prepared entirely by Norsk Polarinstitutt. The ordinary budget of Norsk Polar-institutt now averages about 900,000 kroner annually, but since much of the work in Svalbard is useful to the coal companies, a part of the expenditure is covered by a grant of 200,000 kroner from the Svalbard budget.

During the years 1949–52, when the Norwegian-British-Swedish Antarctic Expedition led by the head clerk of Norsk Polarinstitutt, Capt. John Giæver, was maintained, the budget was considerably higher. This was also the case during the Norwegian I.G.Y. expedition to the Antarctic, led by the geodetic surveyor of Norsk Polarinstitutt, Mr. Sigurd Helle, in 1956–60 and the mapping expedition of 1958-9, in co-operation with the Royal Norwegian Air Force, led by the chief topographic surveyor of N.P.I., Mr. Bernhard Luncke. The Svalbard expeditions from 1946 to 1959 totalled 79 field parties.

FIGURE 1.   Office of Norsk Polarinstitutt, Oslo.

FIGURE 2.   Expedition ship used after the war, a former Nova Scotia schooner of 135 gross tons, 115 feet in length.

ORGANIZATION OF THE EXPEDITIONS

Each expedition is equipped as an independent field party with one leader and two or three assistants. They are fitted out with tents, food for the summer in standard crates, and a 17-foot boat, equipped with an outboard motor of 5 horse-power. The boat is a dory which, because of its steadiness and flat bottom can be beached anywhere and pulled onto shore with a pulley. In this way the expedi-

FIGURE 3.   A 17-foot dory equipped with outboard motor and used by field parties.

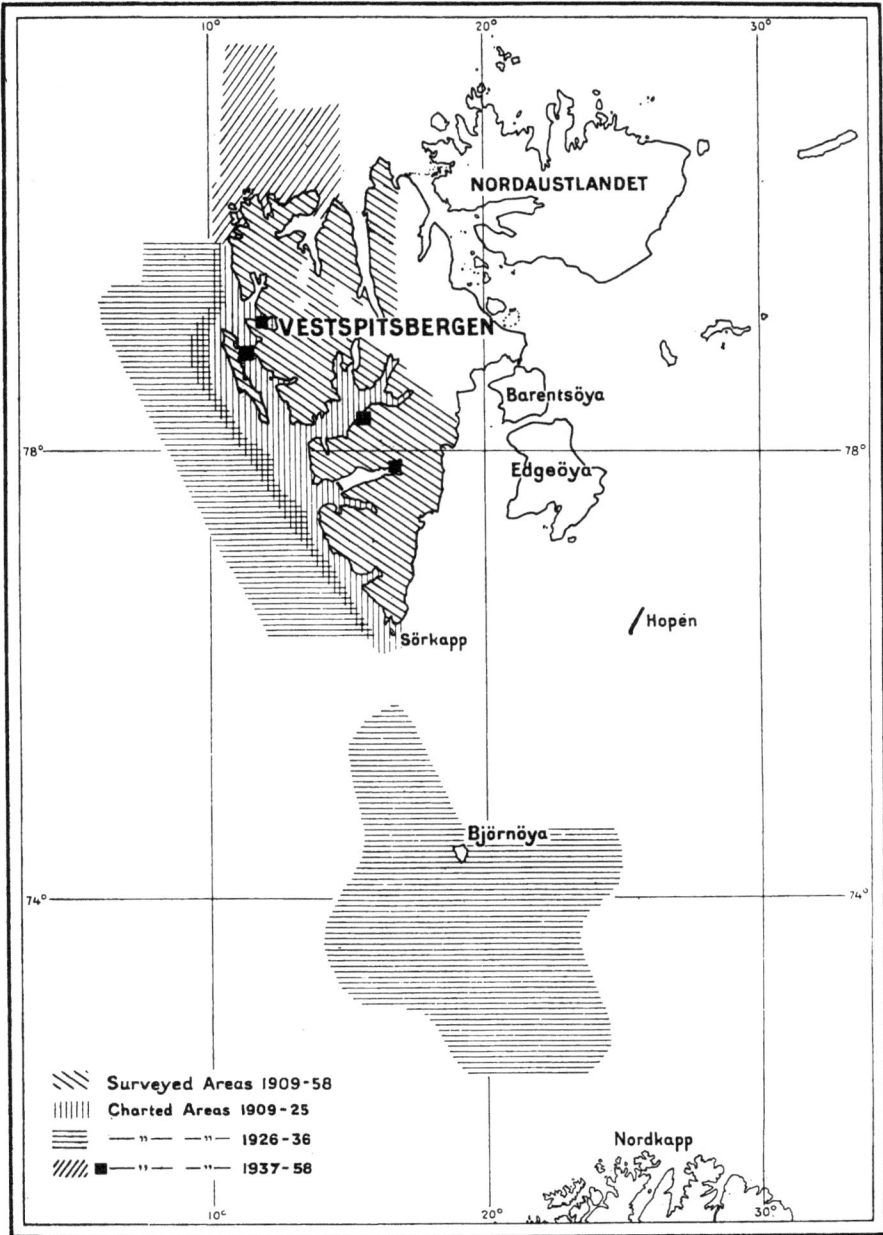

FIGURE 5. General map showing the areas in Svalbard that have been hydrographically and topographically surveyed.

FIGURE 4.   3 Base camp tents used by field parties.

tion is able to move along the coast and cover a considerable area. For inland journeys smaller tents, man-hauled sledges, skis, and special food are being used.

A vessel of the necessary size is chartered for the transportation of people and and equipment from Norway. From this vessel the field parties are landed at their working places and are moved long distances. During the stay in Svalbard the ship is used for the hydrographic survey. The expeditions leave from Ålesund in western Norway in mid-June and return at the beginning of September.

The work carried out is outlined as follows.

HYDROGRAPHIC SURVEY

The hydrographic survey is carried out with a boat party taking soundings in coastal waters, fjords, and harbours. The men live on the shore in tents. In early years they were stationed on board the expedition vessel, but this vessel is needed for many other tasks, and the present practice seems more valuable. The boat is 30 feet long and is equipped with a diesel engine of 20 horsepower. This party also undertakes tidal observations. The larger vessel operates off the coast as far as reliable positions are obtainable by angle measurements. The work is much hampered by fog and low clouds. During the years between the two world wars the Bjørnøya waters and the west coast of Vestspitsbergen were charted, and a pilot map produced for the Bjørnøya waters. Also the Greenland waters have been hydrographically surveyed to a considerable extent. In post-war years a number of harbours and the coastal waters in northwestern Spitsbergen have been charted. Outside Svalbard the waters round the island Jan Mayen have been surveyed in the course of four summers. The results of this work have been published in several charts. In 1957 and 1958 geomagnetic observations were carried out at 85 stations at Spitsbergen.

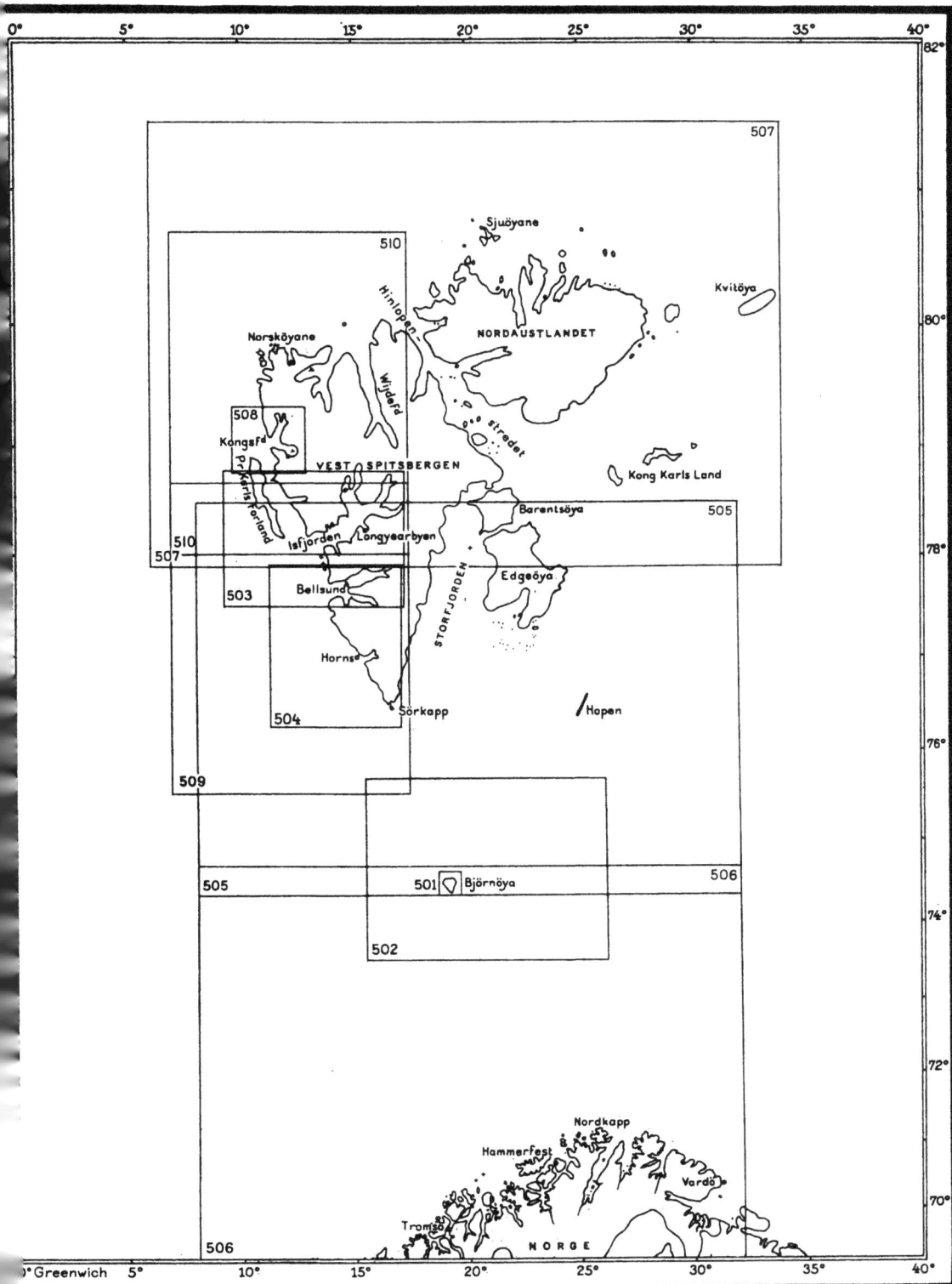

FIGURE 6.   Charts in the Svalbard area.

In addition to this work K. Z. Lundquist, hydrographer, has also been responsible for supervising the lighthouses and radio beacons in Svalbard.

TOPOGRAPHIC AND GEODETIC SURVEY

The older maps of Svalbard were based mainly on terrestrial photogrammetric methods of measurement and triangulation. In 1936 and 1938, however, the whole of Svalbard was photographed from the air, and 5,500 18 × 18 cm photographs were taken with a Zeiss R.M.K. camera with a focal length of 21 cm. The photographs were taken obliquely in order to have a considerably larger area covered by a stereoscopic pair. In later years additional photography (including some vertical) has been done to fill in gaps in the earlier work and to get a more up-to-date record of the glacier fronts. The utilization of photographs for map construction with the stereoplanigraph has to be preceded by continued triangulation to obtain points of control.

The maps are constructed on aluminium plates at a scale of 1:50,000 with contour lines at 50 m intervals. They are published at 1:100,000 in five colours. There are approximately thirty sheets covering Vestspitsbergen, the largest island.

FIGURE 7.　Stubendorff glacier (78°50′N–16°30′E), one of about 6,000 18 × 18 cm oblique air photographs of Svalbard.

Eight of these sheets have been published so far, one is being drafted, and two are under construction. During 1959 two sheets covering the island of Jan Mayen also appeared. In order to provide a greatly needed general survey map a simple construction covering the whole of Spitsbergen in four sheets, at 1:200,000 and 1:400,000, is under way; here maps issued by other nations, mostly English, are also being incorporated.

The geodetic survey consists of base measurements and triangulations. Tidal measurements are carried out over a period of 30 days and more, to determine the

Topographical maps   1 : 100 000

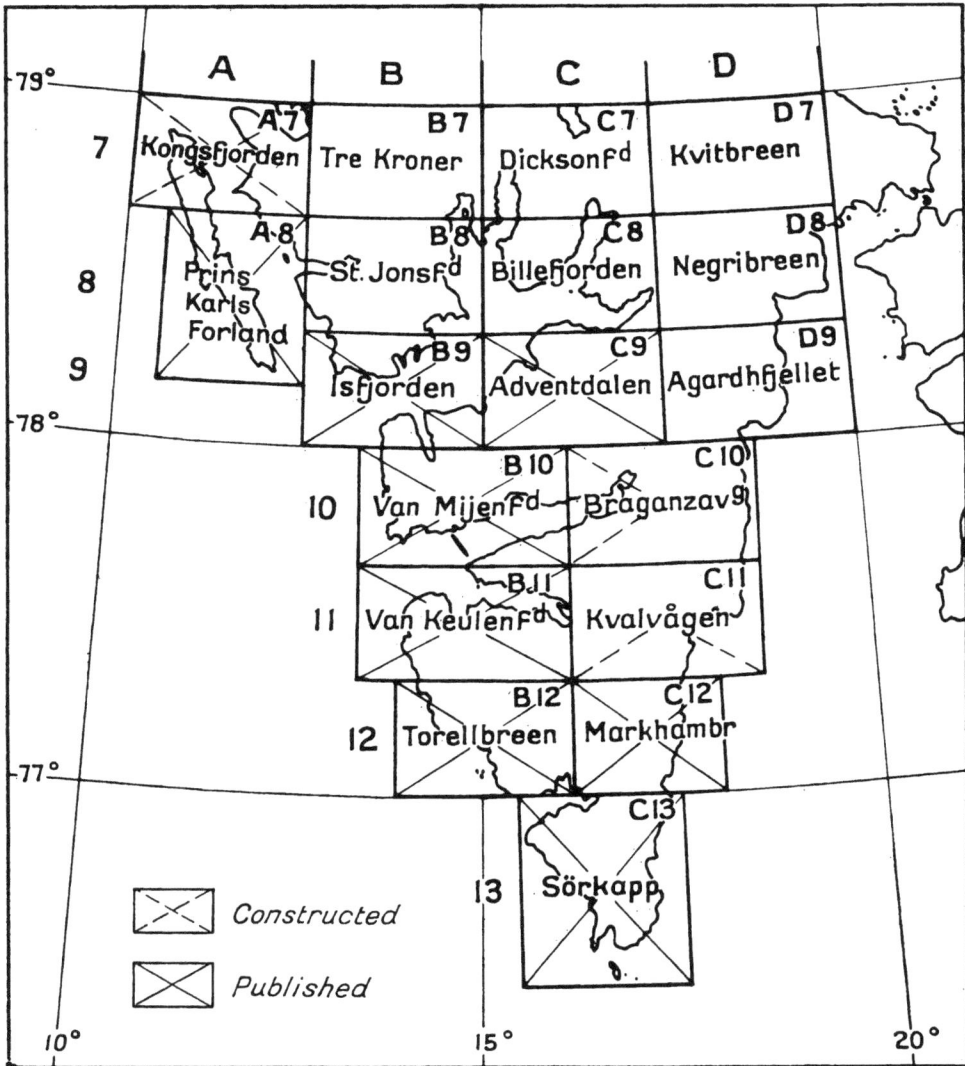

FIGURE 8.   Standard maps at the scale of 1:100,000, published or in work.

mean half-tide, thus giving a zero-plane for determination of spot heights on land. Since the summer of 1959 a Telurometer set is being used to obtain larger triangulation bases.

GEOLOGICAL SURVEY

The geology of Svalbard is most interesting because of the existence of rocks from nearly all geological periods. The sedimentary sequence from the Upper Silurian and up to the Tertiary contains a vast number of fossils, relating the geological history of Svalbard. In these sediments, especially from the earlier part of the Tertiary period, are found the coal seams which are worked today.

Norsk Polarinstitutt is charged with the task of making geological and palaeontological investigations and geological mapping. Special stress has been laid on the examination of the coal deposits, both in Bjørnøya and in Vestspitsbergen. There is still extensive and protracted geological work to be carried out in Svalbard. With new maps and aerial photographs, conditions for detailed geological investigation are much more favourable than they were in the pioneer period when few maps existed. Extensive work has been done, however, on stratigraphy and palaeontology, and rocks have been collected from all over the archipelago. The collections have been studied at the institute, or by specialists in Norway or other countries.

With regard to the geological mapping, extensive material is available in manuscript, either redrawn on the new topographical maps or drawn directly on them, and it is planned soon to start publication of geological maps at a scale of 1:100,000.

Since the war coal investigations have been carried out in the area between Isfjorden and Van Mijenfjorden. From these we have obtained a fairly good knowledge of the various coal strata and their relations. Some drilling has been done by the coal companies as well. A good deal remains to be done, however, to give a more intimate knowledge of the coal reserves of the region. A more particular branch of study has been the investigation of spores and pollen. In addition, investigations on the older metamorphic rocks (the Hecla Hoek sequence along the west coast of Vestspitsbergen) were started in 1952 in the southernmost part and will be continued north of Isfjorden. The finding in 1952, and later, of Cambrian and Ordovician fossils now permits a clearer understanding of the stratigraphy. Special investigations of the microfauna of the Carboniferous and Permian sediments are also being carried out. A third geologist, working in the field of petrography and mineralogy, was engaged by the institute in 1960.

GLACIOLOGICAL INVESTIGATIONS

The glaciological work going on in Svalbard includes a continuous record of the various glacier fronts by comparison of older maps and photographs and the later series of aerial photographs. Thus the peculiar movement of cold (meaning below freezing point) glaciers is studied.

In the southern part of Vestspitsbergen a glacier with a well-defined accumulation area, Finsterwalderbreen, has been selected for a special study. Every second year accumulation and ablation is measured during the summer, and the movement is recorded. Meteorological observations are taken, and by means of thermistors at

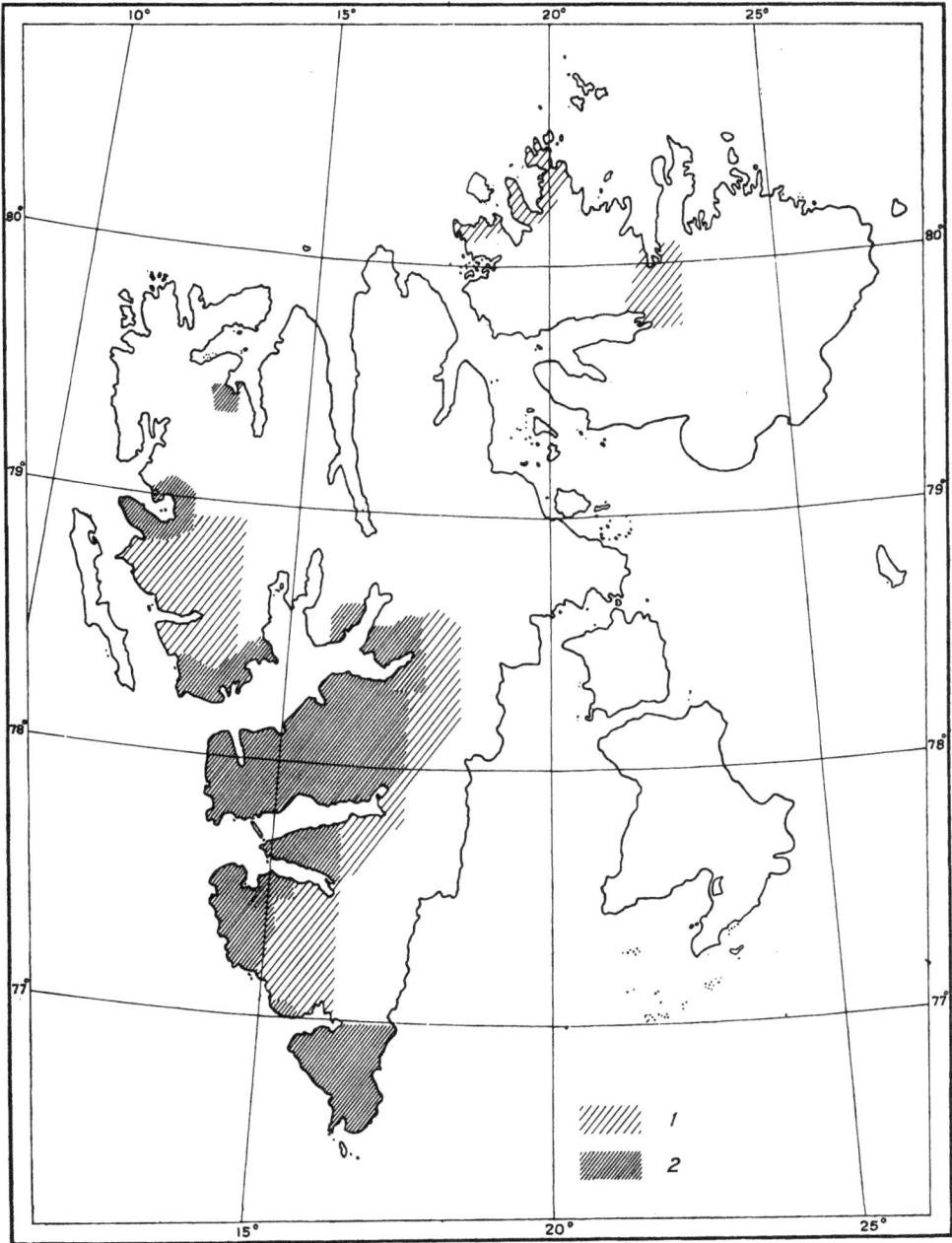

FIGURE 9. Geologically mapped areas in Spitsbergen. (1) Based on older maps on a scale of 1:100,000 or less. (2) Based on recent maps on a scale of 1:50,000, field work completed.

different depths, the inner temperature is examined. Important contributions to the knowledge of glaciers are being gained.

OTHER WORK UNDERTAKEN BY NORSK POLARINSTITUTT

Among the important tasks of Norsk Polarinstitutt in post-war years has been the building of navigational aids in Svalbard. As early as 1932 a lighthouse and two lanterns were erected at Isfjorden under the direction of Norges Svalbard- og Ishavs-undersøkelser. Since 1946 eleven lighthouses, four radio beacons and one land radar station have been erected and partly maintained by Norsk Polarinstitutt.

Assistance is regularly given for transportation of other Norwegian and foreign expeditions.

The expedition members of Norsk Polarinstitutt also make many observations outside their special working fields, on the wild life, flora, etc.

The geographical names in Svalbard have been given by expeditions from a number of countries, and thus are far from uniform. To bring order to this chaos about 360 maps and 500 books have been examined. More than 10,000 names were investigated, about 3,300 of which were officially recognized. On more recent maps many new names have been added. Dr. A. K. Orvin has been in charge of this important task.

Considerable work is done by the institute in its advisory capacity by supplying adequate maps and local knowledge to other Svalbard expeditions.

FUTURE TASKS

The charting of Svalbard waters will continue on the northern coast but, in addition, extensive charting of the northern area of the Norwegian Sea is planned. By using electronic instruments for determination of position, this work will not be so hampered by weather conditions. "Sailing Directions" for Svalbard waters are in the process of being completed, and the topographic survey will issue the standard maps from time to time. The geologists will continue their survey work, with emphasis on coal investigation and mapping.

During the International Geological Congress, 1960, and the International Geographical Congress in the same year, two different excursions to Svalbard were organized by Norsk Polarinstitutt.

A list of published maps and charts and a list of papers issued in the series "Skrifter" and "Meddelelser" follows.

MAPS AND CHARTS

The following topographical maps and charts have been published separately:

Maps

Bjørnøya. 1:25 000. 1925. New edition 1944.
Bjørnøya. 1:10 000. [In six sheets.] 1925. Out of print.
Adventfjorden—Braganzavågen. 1:100 000. 1941.
Svalbard. 1:2 000 000. 1958.
Topografisk kart over Svalbard. Blad C 13. Sørkapp. 1:100 000. 1947.
Topografisk kart over Svalbard. Blad B 10. Van Mijenfjorden. 1:100 000. 1948.
Topografisk kart over Svalbard. Blad C 9. Adventdalen. 1:100 000. 1950.
Topografisk kart over Svalbard. Blad B 11. Van Keulenfjorden. 1:100 000. 1952
Topografisk kart over Svalbard. Blad B 12. Torellbreen. 1:100 000. 1953.
Topografisk kart over Svalbard. Blad B 9. Isfjorden. 1:100 000. 1955.

Topografisk kart over Svalbard. Blad C 12. Markhambreen. 1:100 000. 1957.
Topografisk kart over Svalbard. Blad A 8. Prins Karls Forland. 1959.
Austgrønland. Eirik Raudes Land frå Sofiasund til Youngsund. 1:200 000. 1932.
Jan Mayen. 1:100 000. 1955. Førebels utgåve.
Dronning Maud Land. Sør-Rondane. 1:250 000. 1957.
Topografisk kart over Jan Mayen. Blad 1. Sør-Jan. 1:50 000. 1959.
Topografisk kart over Jan Mayen. Blad 2. Nord-Jan. 1:50 000. 1959.
   Preliminary topographical maps ⌐1:50 000⌐ covering claims to land in Svalbard and a preliminary map of Hopen 1:100 000 may be obtained separately.
   In addition, Norsk Polarinstitutt has prepared a wall map: Norden og Norskehavet. Revised edition, 1957. This map is to be obtained through H. Aschehoug & Co. (W. Nygaard), Oslo. 1:2 500 000.

Charts

501. Bjørnøya. 1:40 000. 1932.
502. Bjørnøyfarvatnet. 1:350 000. 1937.
503. Frå Bellsund til Forlandsrevet med Isfjorden. 1:200 000. 1932.
504. Frå Sørkapp til Bellsund. 1:200 000. 1938.
505. Norge—Svalbard, nordre blad. 1:750 000. 1933.
506. Norge—Svalbard, søre blad. 1:750 000. 1933.
507. Nordsvalbard. 1:600 000. 1934.
508. Kongsfjorden og Krossfjorden. 1:100 000. 1934.
509. Frå Storfjordrenna til Forlandsrevet med Isfjorden. 1:350 000. 1936.
510. Frå Kapp Linné med Isfjorden til Sorgfjorden. 1:350 000. 1936.
511. Austgrønland, frå Liverpoolkysten til Store Koldeweyøya. 1:600 000. 1937.
512. Jan Mayen. 1:100 000. 1955.
513. Svalbard-Havner. Adventfjorden, 1:25 000; Sveagruva, 1:15 000; Ny-Ålesund, 1:25 000; Forlandsrevet, 1:60 000. 1959.

PAPERS

Resultater av De Norske statsunderstøttede Spitsbergenekspeditioner

1. HOEL, A. The Norwegian Svalbard Expeditions, 1906-1926. 1929.
2. RAVN, J. P. J. On the mollusca of the Tertiary of Spitsbergen. 1922.
3. WERENSKIOLD, W., and OFTEDAL, I. A burning coal seam at Mt. Pyramide, Spitsbergen 1922.
4. WOLLEBÆK, A. The Spitsbergen reindeer. 1926.
5. LYNGE, B. Lichens from Spitsbergen. 1924.
6. HOEL, A. The coal deposits and coal mining of Svalbard. 1925. Out of print.
7. DAHL, K. Contributions to the biology of the Spitsbergen char. 1926.
8. HOLTEDAHL, O. Notes on the geology of northwestern Spitsbergen. 1926.
9. LYNGE, B. Lichens from Bear Islands (Bjørnøya). 1926.
10. IVERSEN, T. Hopen (Hope Island), Svalbard. 1926.
11. QUENSTEDT, W. Mollusken a. d. Redbay- u. Greyhookschichten Spitzb. 1926.
1—11: Vol. I. From no. 12 the papers are only numbered consecutively.

Skrifter om Svalbard og Nordishavet

12. STENSIÖ, E. A:SON. The Downtonian and Devonian vertebrates of Spitsbergen. Part I. Cephalaspidae. A, Text, and B, Plates. 1927.

Skrifter om Svalbard og Ishavet

13. LIND, J. The Micromycetes of Svalbard. 1928.
14. KLER, R., and FJELDSTAD, J. E. Tidal observations in the Arctic. 1934.
15. HORN, G., and ORVIN, A. K. Geology of Bear Island. 1928.
16. JELSTRUP, H. S. Determinations astronomiques. 1928.
17. HORN, G. Beiträge zur Kenntnis der Kohle von Svalbard. 1928.
18. HOEL, A., und ORVIN, A. K. Das Festungsprofil auf Spitzbergen. Karbon-Kreide I. Vermessungsresultate. 1937.
19. FREBOLD, H. Das Festungsprofi auf Spitzbergen. Jura und Kreide. II. Die Stratigraphie 1928.
20. —————— Oberer Lias und unteres Callovien in Spitzbergen. 1929.
21. —————— Ammoniten aus dem Valanginien von Spitzbergen. 1929.

22. HEINTZ, A.   Die Downtonischen und Devonischen Vertebraten von Spitzbergen.   II Acanthaspida. 1929.
23. ——— Die Downtonischen und Devonischen Vertebraten von Spitzbergen. III. Acanthaspida. Nachtrag. 1929.
24. HERITSCH, F.   Eine *Caninia* aus dem Karbon des De Geer-Berges. 1929.
25. ABS, O.   Untersuchungen ueber die Ernährung der Bewohner von Barentsburg, Svalbard. 1929.
26. FREBOLD, H.   Untersuchungen ueber die Fauna, die Stratigraphie und Paläogeographie der Trias Spitzbergens. 1929.
27. THOR, S.   Beiträge zur Kenntnis der invertebraten Fauna von Svalbard. 1930.
28. FREBOLD, H.   Die Altersstellung des Fischhorizontes, des Grippianiveaus und des unteren Saurierhorizontes in Spitzbergen. 1930.
29. HORN, G.   Franz Josef Land. Nat. Hist., Discovery, Expl., and Hunting. 1930.
30. ORVIN, A. K.   Beiträge zur Kenntnis des Oberdevons Ost-Grönlands.   HEINTZ, A. Oberdevonische Fischreste aus Ost-Grönland. 1930.
31. FREBOLD, H.   Verbr. und Ausb. des Mesozoikums in Spitzbergen. 1930.
32. ABS,, O.   Ueber Epidemien von unspezifischen Katarrhen der Luftwege auf Svalbard. 1930.
33. KLÆR, J.   *Ctenaspis*, a new genus of Cyathaspidian fishes. 1930.
34. TOLMATCHEW, A.   Die Gattung *Cerastium* in der Flora von Spitzbergen. 1930.
35. SOKOLOV, D., und BODYLEVSKY, W.   Jura- und Kreidefaunen von Spitzb. 1931.
36. SMEDAL, G.   Acquisition of sovereignty over polar areas. 1931.
37. FREBOLD, H.   Fazielle Verh. des Mesozoikums im Eisfjordgebiet Spitzb. 1931.
38. LYNGE, B.   Lichens from Franz Josef Land. 1931.
39. HANSSEN, O., and LID, J.   Flowering plants of Franz Josef Land. 1932.
40. KLÆR, J., and HEINTZ, A.   The Downtonian and Devonian vertebrates of Spitsbergen. V. Suborder Cyathaspida. 1935.
41. LYNGE, B., and SCHOLANDER, P. F.   Lichens from N. E. Greenland. 1932.
42. HEINTZ, A.   Beitr. zur Kenntnis d. devonischen Fischfauna O.-Grönlands. 1931.
43—46. BJØRLYKKE, B.   Some vascular plants from southeast Greenland, Collected on the "Heimen" Expedition in 1931: Preliminary report. LID, J.   Vascular plants. LYNGE, B. Lichens. OMANG, S. O. F.   Beiträge zur Hieraciumflora. 1932.
47. LYNGE, B.   A revision of the genus *Rhizocarpon* in Greenland. 1932.
48. VAAGE, J.   Vascular plants from Eirik Raude's Land. 1932.
49. SCHAANNING, H. THO. L.   1. A contribution to the bird fauna of East-Greenland. 2. A contribution to the Bird Fauna of Jan Mayen.   Zool. Res. Norw. Sc. Exp. to East-Greenland. I. 1933.
50. JELSTRUP, H. S.   Détermination astronomique de Mygg-Bukta au Groenland oriental. 1932.
51. BIRKELAND, B. J., et SCHOU, GEORG.   Le climat de l'Eirik-Raudes-Land. 1932.
52. KLÆR, J.   The Downtonian and Devonian vertebrates of Spitsbergen. IV. Suborder Cyathaspida. 1932.
53. 1. MALAISE, R.   Eine neue Blattwespe. 2. ROMAN, A.   Schlufwespen. 3. RINGDAHL, O. Tachiniden und Musciden. 4. GOETGHEBUER, M.   Chironomides du Groenland oriental, du Svalbard et de la Terre de François Joseph.   Zool. Res. Norw. Sc. Exp. to East-Greenland. II. 1933.
54. VARTDAL, H.   Bibliographie des ouvrages norvégiens relatifs au Grœnland (Y compris les ouvrages islandais antérieurs à l'an 1814). 1935.
55. OMANG, S. O. F.   Uebersicht ueber die Hieraciumflora Ost-Grönlands. 1933.
56. DEVOLD, J., and SCHOLANDER, P. F.   Flowering plants and ferns of southeast Greenland. 1933.
57. ORVIN, A. K.   Geology of the Kings Bay Region, Spitsbergen. 1934.
58. JELSTRUP, H. S.   Détermination astronomique à Sabine-Øya. 1933.
59. LYNGE, B.   On *Dufourea* and *Dactylina*: Three Arctic lichens. 1933.
60. VOGT, TH.   Late-Quaternary oscillations of level in S. E. Greenland. 1933.
61. 1. BURTON, M.   Report on the sponges. 2. ZIMMER, C.   Die Cumaceen.   Zool. Res. Norw. Sc. Exp. to East-Greenland. III. 1934.
62. SCHOLANDER, P. F.   Vascular plants from northern Svalbard. 1934.
63. RICHTER, S.   A contribution to the archaeology of north-east Greenland. 1934.
64. SOLLE, G.   Die devonischen Ostracoden Spitzbergens. 1935.
65. 1. FRIESE, H.   Apiden. 2. LINDBERG, H.   Hemiptera. 3. LINNANIEMI, W. M.   Collembolen.   Zool. Res. Norw. Sc. Exp. to East-Greenland. IV. 1935.

66. 1. NORDENSTAM, Å.   The Isopoda.   2. SCHELLENBERG, A.   Die Amphipoden.   3. SIVERTSEN, E.   Crustacea Decapoda, Auphausidacea, and Mysidacea.   Zool. Res. Norw. Sc. Exp. to East-Greenland. V.   1935.
67. JAKHELLN, A.   Oceanographic investigations in East Greenland waters in the summers of 1930-1932.   1936.
68. FREBOLD, H., und STOLL, E.   Das Festungsprofil auf Spitzbergen.   III. Stratigraphie und Fauna des Jura und der Unterkreide.   1937.
69. FREBOLD, HANS.   Das Festungsprofil auf Spitzbergen.   IV. Die Brachiopoden- und Lamellibranchiatenfauna des Oberkarbons und Unterperms.   1937.
70. DAHL, EILIF, LYNGE, B., and SCHOLANDER, P. F.   Lichens from southeast Greenland. 1937.
71. 1. KNABEN, NILS.   Makrolepidopteren aus Nordostgrönland.   2. BARCA, EMIL.   Mikrolepidopteren aus Nordostgrönland.   Zool. Res. Norw. Sc. Exp. to East-Greenland. VI. 1937.
72. HEINTZ, A.   Die Downtonischen und Devonischen Vertebraten von Spitzbergen.   VI. Lunaspis-Arten aus dem Devon Spitzbergens.   1937.
73. Report on the Activities of Norges Svalbard- og Ishavs-undersøkelser 1927-1936.   1937.
74. HØYGAARD, ARNE.   Some investigations into the physiology and nosology of Eskimos from Angmagssalik in Greenland.   1937.
75. DAHL, EILIF.   On the vascular plants of eastern Svalbard.   1937.
76. LYNGE, B.   Lichens from Jan Mayen.   1939.
77. FREBOLD, HANS.   Das Festungsprofil auf Spitzbergen.   V. Stratigraphie und Invertebratenfauna der älteren Eotrias.   1939.
78. ORVIN, ANDERS K.   Outline of the geological history of Spitsbergen.   1940.
79. LYNGE, B.   Et bidrag til Spitsbergens lavflora.   1940.
80. The place-names of Svalbard.   1942.
81. LYNGE, B.   Lichens from north east Greenland.   1940.

Norges Svalbard- og Ishavs-undersøkelser—Skrifter

82. NILSSON, TAGE.   The Downtonian and Devonian vertebrates of Spitsbergen.   VII. Order Antiarchi.   1941.
83. HØEG, OVE ARBO.   The Downtonian and Devonian flora of Spitsbergen.   1942.
84. FREBOLD, HANS.   Ueber die Productiden des Brachiopodenkalkes.   1942.
85. FØYN, SVEN, and HEINTZ, ANATOL.   The Downtonian and Devonian vertebrates of Spitsbergen. VIII.   1943.
86. The Survey of Bjørnøya (Bear Island) 1922-1931.   1944.
87. HADAČ, EMIL.   Die Gefässpflanzen des "Sassengebietes" Vestspitsbergen.   1944.
88. Report on the activities of Norges Svalbard- og Ishavs-undersøkelser 1936-1944.   1945.
89. ORVIN, ANDERS K.   Bibliography of literature about the geology, physical geography, useful minerals, and mining of Svalbard.   1947.

Norsk Polarinstitutt—Skrifter

90. HENIE, HANS.   Astronomical observations on Hopen.   1948.
91. RODAHL, KÅRE.   Vitamin sources in Arctic regions.   1949.
92. —— The toxic effect of polar bear liver.   1949.
93. HAGEN, ASBJØRN.   Notes on Arctic fungi.   I. Fungi from Jan Mayen.   II. Fungi collected by Dr. P. F. Scholander on the Swedish-Norwegian Arctic Expedition, 1931. 1950.
94. FEYLING-HANSSEN, ROLF W., and JØRSTAD, FINN A.   Quaternary fossils.   1950.
95. RODAHL, KARE.   Hypervitaminosis A.   1950.
96. BUTLER, J. R.   Geochemical affinities of some coals from Svalbard.   1953.
97. WÅNGSJÖ, GUSTAV.   The Downtonian and Devonian vertebrates of Spitsbergen.   Part IX. Morphologic and systematic studies of the Spitsbergen cephalaspids.   A, Text, and B, Plates.   1952.
98. FEYLING-HANSSEN, ROLF W.   The barnacle Balanus Balanoides (Linné, 1766) in Spitsbergen.   1953.
99. RODAHL, KÅRE.   Eskimo metabolism.   1954.
100. PADGET, PETER.   Notes on some corals from Late Paleozoic rocks of inner Isfjorden, Spitsbergen.   1954.
101. MATHISEN, TRYGVE.   Svalbard in international politics, 1871-1925.   1954.

102. RODAHL, KÅRE.   Studies on the blood and blood pressure in Eskimo and the significance of ketosis under Arctic conditions.   1954.
103. LØVENSKIOLD, H. L.   Studies on the avifauna of Spitsbergen.   1954.
104. HORNBÆK, HELGE.   Tidal observations in the Arctic, 1946-52.
105. ABS, OTTO, und SCHMIDT, HANS WALTER.   Die arktische Trichinose und ihr Verbreitungsweg.   1954.
106. MAJOR, HARALD, and WINSNES, THORE S.   Cambrian and Ordovician fossils from Sørkapp Land, Spitsbergen.   1955.
107. FEYLING-HANSSEN, ROLF W.   Stratigraphy of the marine Late-Pleistocene of Billefjorden, Vestspitsbergen.   1955.
108. —————   Late-Pleistocene deposits at Kapp Wijk, Vestspitsbergen.   1955.
109. DONNER, J. J., and WEST, R. G.   The Quaternary geology of Brageneset, Nordaustlandet, Spitsbergen.   1957.
110. LUNDQUIST, KAARE Z.   Magnetic observations in Svalbard, 1596-1953.   1957.
111. SVERDRUP, H. U.   The stress of the wind on the ice of the polar sea.   1957.
112. ORVIN, ANDERS K.   Supplement I to The place-names of Svalbard, dealing with new names, 1935-55.   1958.
113. SOOT-RYEN, TRON.   Pelecypods from East-Greenland.   1958.
115. GROOM, G. E., and SWEETING, M. M.   Valleys and raised beaches in Bünsow Land Central Vestspitsbergen.   1958.
116. SVENDSEN, PER.   The algal vegetation of Spitsbergen.   1959.

MEDDELELSER

1. PETTERSEN, K.   Isforholdene i Nordishavet i 1881 og 1882.   Optrykk av avisartikler. Med en innledn. av A. Hoel.   Særtr. av Norsk Geogr. Tidsskr., b. 1, h. 4.   1926.   Out of print.
2. HOEL, A.   Om ordningen av de territoriale krav på Svalbard.   Særtr. av Norsk Geogr Tidsskr., b. 2, h. 1.   1928.   Out of print.
3. —————   Suverenitetsspørsmålene i polartraktene.   Særtr. av Nordmands-Forbundet, årg. 21, h. 4 & 5.   1928.   Out of print.
4. BROCH, O. J., FJELD, E., og HØYGAARD, A.   På ski over den sydlige del av Spitsbergen. Særtr. av Norsk Geogr. Tidsskr., b. 2, h. 3-4.   1928.
5. TANDBERG, ROLF S.   Med hundespann på eftersøkning efter "Italia"-folkene.   Særtr. av Norsk Geogr. Tidsskr., b. 2, h. 3-4.   1928.
6. KJÆR, R.   Farvannsbeskrivelse over kysten av Bjørnøya.   1929
7. NORGES SVALBARD- OG ISHAVS-UNDERSØKELSER, Jan Mayen.   En oversikt over øens natur, historie og bygning.   Særtr. av Norsk Geogr. Tidsskr., b. 2, h. 7.   1929.   Out of print.
8. I. LID, JOHANNES.   Mariskardet på Svalbard.   II. ISACHEN, FRIDTJOV.   Tidligere utforskning av området mellem Isfjorden og Wijdebay på Svalbard.   Særtr. av Norsk Geogr. Tidsskr., b. 2, h. 7.   1929.
9. LYNGE, B.   Moskusoksen i Øst-Grønland.   Særtr. av Norsk Geogr. Tidsskr., b. 3, h. 1. 1930.   Out of print.
10. NORGES SVALBARD- OG ISHAVS-UNDERSØKELSER, Dagbok ført av Adolf Brandal under en overvintring på Øst-Grønland 1908-1909.   1930.   Out of print.
11. ORVIN, A. K.   Ekspedisjonen til Øst-Grønland med "Veslekari" sommeren 1929.   Særtr. av Norsk Geogr. Tidsskr., b. 3, h. 2-3.   1930.
12. ISACHSEN, G.   I. Norske Undersøkelser ved Sydpollandet 1929-31.   II. "Norvegia"-ekspedisjonen 1930-31.   Særtr. av Norsk Geogr. Tidsskr., b. 3, h. 5-8.   1931.
13. Norges Svalbard- og Ishavs-undersøkelsers ekspedisjoner sommeren 1930. I. ORVIN, A. K. Ekspedisjonen til Jan Mayen og Øst-Grønland.   II. KJÆR, R.   Ekspedisjonen til Svalbardfarvannene.   III. FREBOLD, H.   Ekspedisjonen til Spitsbergen.   IV. HORN, G.   Ekspedisjonen til Frans Josefs Land.   Særtr. av Norsk Geogr. Tidsskr., b. 3, h. 5-8.   1931.
14. I. HØEG, O. A.   The fossil wood from the Tertiary at Myggbukta, East Greenland.   II. ORVIN, A. K.   A fossil river bed in East Greenland.   Særtr. av Norsk Geol. Tidsskr., b. 12.   1931.
15. VOGT, T.   Landets senkning i nutiden på Spitsbergen og Øst-Grønland.   Særtr. av Norsk Geol. Tidsskr., b. 12.   1931.
16. HØEG, O. A.   Blütenbiologische Beobachtungen aus Spitzbergen.   1932.
17. —————   Notes on some Arctic fossil wood, with a redescription of Cupressinoxylon Polyommatum, Cramer.   1932.

18. ISACHSEN, G. OG ISACHSEN, F.   Norske fangstmenns og fiskeres ferder til Grønland 1922-1931. Særtr. av Norsk Geogr. Tidsskr., b. 4, h. 1-3. 1932.
19. ——— Hvor langt mot nord kom de norrøne grønlendinger på sine fangstferder i ubygdene. Særtr. av Norsk Geogr. Tidsskr., b. 4, h. 1-3. 1932.
20. VOGT, TH.   Norges Svalbard- og Ishavs-undersøkelsers ekspedisjon til Sydøstgrønland med "Heimen" sommeren 1931. Særtr. av Norsk Geogr. Tidsskr., b. 4, h. 5. 1933.
21. BRISTOWE, W. S.   The spiders of Bear Island. Reprinted from Norsk Entomol. Tidsskr., b. 3, h. 3. 1933.
22. ISACHSEN, F.   Verdien av den norske klappmyssfangst langs Sydøst-Grønland. 1933.
23. LUNCKE, B.   Norges Svalbard- og Ishavs-undersøkelsers luftkartlegning i Eirik Raudes Land 1932. Særtr. av Norsk Geogr. Tidsskr., b. 4, h. 6. 1933.
24. HORN, G.   Norges Svalbard- og Ishavs-undersøkelsers ekspedisjon til Sydøstgrønland med "Veslemari" sommeren 1932. Særtr. av Norsk Geogr. Tiddskr., b. 4, h. 7. 1933.
25. ORVIN, A. K.   Norges Svalbard- og Ishavs-undersøkelsers ekspedisjoner til Nordøst-Grønland i årene 1931-1933. Isfjord fyr og radiostasjon, Svalbard. Særtr. av Norsk Geogr. Tidsskr., b. 5, h. 2. 1934.
26. GRIEG, J. A.   Some echinoderms from Franz Josef Land, Victoriaøya and Hopen. Collected on the Norwegian Scientific Expedition 1930. 1935.
27. MAGNUSSON, A. H.   The Lichen-Genus Acarospora in Greenland and Spitsbergen. Reprinted from Nyt Magazin for Naturvidensk. B. 75. 1935.
28. BAASHUUS-JESSEN, J.   Arctic nervous diseases. Reprinted from Skandinavisk Veterinär-Tidskrift, no. 6, 1935.
29. I. KOLSRUD, O.   Til Østgrønlands historie. II. OSTERMANN, H.   De første efterretninger om østgrønlændingerne 1752. Særtr. av Norsk Geogr. Tidsskr., b. 5, h. 7. 1935.
30. TORNØE, J. KR.   Hvitserk og Blåserk. Særtr. av Norsk Geogr. Tidsskr., b. 5, h. 7. 1935.
31. HEINTZ, A.   Holonema-Reste aus Devon Spitzbergens. Sonderabdr. aus Norsk Geol. Tidsskr., b. 15. 1935.
32. ORVIN, A. K.   Norges Svalbard- og Ishavs-undersøkelsers ekspedisjoner i årene 1934 og 1935. Særtr. av Norsk Geogr. Tidsskr., b. 5. 1935.
33. OSTERMANN, H.   Dagbøker av nordmenn på Grønland før 1814. 1935
34. LUNCKE, B.   Luftkartlegningen på Svalbard 1936. Særtr. av Norsk Geogr. Tidsskr., b. 6. 1936.
35. HOLTEDAHL, O.   On fault lines indicated by the submarine relief in the shelf area west of Spitsbergen. Særtr. av Norsk Geogr. Tidsskr., b. 6, h. 4. 1936.
36. BAASHUUS-JESSEN, J.   Periodiske vekslinger i småviltbestanden. Særtr. av Norges Jeger- & Fiskerforb. Tidsskr. h. 2 og 3. 1937.
37. ORVIN, A. K.   Norges Svalbard- og Ishavs-undersøkelsers ekspedisjoner til Øst-Grønland og Svalbard i året 1936. Særtr. av Norsk Geogr. Tidsskr., b. 6, h. 7. 1937.
38. GIÆVER, JOHN.   Kaptein Ragnvald Knudsens ishavsferder. Sammen-arbeidet efter hans dagbøker, rapporter m.v. 1937.
39. OSTERMANN, H.   Grønlandske distriktsbeskrivelser forfattet av nordmenn før 1814. 1937.
40. OMANG, S. O. F.   Ueber einige Hieracium-Arten aus Grönland. 1937.
41. GIÆVER, JOHN.   Norges Svalbard- og Ishavs-undersøkelsers ekspedisjoner til Øst-Grønland sommeren 1937. Særtr. av Norsk Geogr. Tidsskr., b. 6, h. 7. 1937.
42. SIEDLECKI, SANISLAW.   Crossing West Spitsbergen from south to north. Særtr. av Norsk Geogr. Tidsskr., b. 7, h. 2. 1938.
43. SOOT-RYEN, T.   Some pelecypods from Franz-Josef Land, Victoriaøya and Hopen. Collected on the Norwegian Scientific Expedition 1930. 1939.
44. LYNGE, B.   A small contribution to the lichen flora of the Eastern Svalbard Islands. Lichens collected by Mr. Olaf Hanssen in 1930. 1939.
45. HORN, GUNNAR.   Recent Norwegian Expeditions to south-east Greenland. Særtr. av Norsk Geogr. Tidsskr., b. 7, h. 5-8. 1939.
46. ORVIN, ANDERS K.   The settlements and huts of Svalbard. Særtr. av Norsk Geogr. Tidsskr., b. 7, h. 5-8. 1939.
47. STØRMER, PER.   Bryophytes from Franz Josef Land and eastern Svalbard. Collected by Mr. Olaf Hanssen on the Norwegian Expedition in 1930. 1940.
48. LID, JOHANNES. Bryophytes of Jan Mayen. 1941.
49. I. HAGE, ASBJØRN.   Micromycetes from Vestspitsbergen. Collected by dr. Emil Hadač in 1939. H. HADAČ, EMIL.   The introduced flora of Spitsbergen. 1941.
50. VOGT, THOROLF.   Geology of a Middle Devonian cannel coal from Spitsbergen. HORN, GUNNAR.   Petrology of a Middle Devonian cannel coal from Spitsbergen. 1941.
51. OSTERMANN, H.   Bidrag til Grønlands beskrivelse, forfattet av nordmenn før 1814. 1942.

52. ——— Avhandlinger om Grønland 1799-1801. 1942.
53. ORVIN, ANDERS K. Hvordan opstår jordbunnsis?.   Særtr. av Norsk Geogr. Tidsskr., b. 8, h. 8. 1941.
54. STRAND, ANDR. Die Käferfauna von Svalbard. Særtr. av Norsk Entomol. Tidsskr., b. 6, h. 2-3. 1942.
55. ORVIN, ANDERS K. Om dannelse av strukturmark. Særtr. av Norsk Geogr. Tidsskr., b. 9, h. 3, 1942.
56. TORNØE, J. KR. Lysstreif over Noregsveldets historie. I. 1944.
57. ORVIN, ANDERS K. Litt om kilder på Svalbard. Særtr. av Norsk Geogr. Tidsskr., b. 10, h. 1. 1944.
58. OSTERMANN, H. Dagbøker av nordmenn på Grønland før 1814. 2. 1944.
59. ——— Dagbøker av nordmenn på Grønland før 1814. 3. 1944.
60. AAGAARD, BJARNE. Antarktis 1502-1944. 1944.
61. ——— Den gamle hvalfangst. 1944.
62. ——— Oppdagelser i Sydishavet fra middelalderen til Sydpolens erobring.   1946.
63. DAHL, EILIF, og HADAC, EMIL. Et bidrag til Spitsbergens flora. 1946.
64. OSTERMANN, H. Skrivelser angaaende Mathis Iochimsens Grønlands-Ekspedition.   1946.
65. AASGAARD, GUNNAR. Svalbard under og etter verdenskrigen. 1946.
66. RICHTER, SØREN. Jan Mayen i krigsårene. 1946.
67. LYNGAAS, REIDAR. Oppføringen av Isfjord radio, automatiske radiofyr og fyrbelysning på Svalbard 1946. Særtr av Norsk Geogr. Tidsskr., b. 11, h. 5-6. 1947.
68. LUNCKE, BERNHARD. Norges Svalbard- og Ishavs-undersøkelsers kartarbeider og anvendelsen av skrå-fotogrammer tatt fra fly. Særtrykk av Tidsskrift for Det norske Utskiftningsvesen Nr. 4, 1949, 19, binds 7. hefte.
69. HOEL, ADOLF. Norsk ishavsfangst. En fortegnelse over litteratur. 1952.
70. HAGEN, ASBJØRN. Plants collected in Vestspitsbergen in the summer of 1933.   1952.
71. FEYLING-HANSSEN, ROLF W. Conglomerates formed in situ on the Gipshuk coastal plain, Vestspitsbergen. 1952.
72. OMDAL, KIRSTEN. Drivisen ved Svalbard 1924-1939. 1952.
73. HEINTZ, A. Noen iakttagelser over isbreenes tilbakegang i Hornsund, V. Spitsbergen. 1953.
74. ROOTS, E. F. Preliminary note on the geology of western Dronning Maud Land.   1953
75. SVERDRUP, H. U. The currents off the coast of Queen Maud Land. 1953.
76. HOEL, A. Flateinnholdet av breer og snøfonner i Norge. 1953.
77. FEYLING-HANSSEN, ROLF W. De gamle trankokerier på Vestspitsbeergens nordvesthjørne og den formodede senkning av landet i ny tid. Særtr. av Norsk Geografisk Tidsskrift, b. XIV, Heft 5-6. 1954.
78. ROER, NILS. Landmålerliv i Dronning Maud Land. Særtr. av Norsk Tidsskrift for Jordskifte og Landmåling nr. 3. 1953.
79. MANUM, SVEIN. Pollen og sporer i tertiære kull fra Vestspitsbergen.   Særtr. av »Blyttia«, bind XII. 1954.
80. THORSHAUG, K., and ROSTED, A. FR. Researches into the prevalence of trichinosis in animals in Arctic and Antarctic waters. Særtrykk av Nord. Vet.-Med. 1956, B. 8, Nr. 2.
81. LIESTØL, OLAV. Glacier dammed lakes in Norway. 1956.
82. FJELDSTAD, J. E. Harald Ulrik Svedrup, 15. nov. 1888-21 aug. 1957. S. Richter, H. U. Sverdrups forfatterskap.
83. LØNØ, ODD. Reinen på Svalbard. 1959.

Geological Investigation of Greenland: Past and Present

K. ELLITSGAARD-RASMUSSEN

GREENLAND IS ABOUT fifty times the size of its parent country, Denmark. It has an area of 2,175,600 square kilometres, the greater part of which is covered by the ice-cap. The narrow, ice-free belt along the coast, which holds the greater interest for geologists, is itself greater than Denmark. As the coastline is very intricate and extended, the normal means of communication is by boat; locally, access may be difficult or impossible for extended periods, because of sea-ice. In the past, voyages from Denmark to Greenland have always been time-consuming and inconvenient.

The native population, known as Greenlanders, today totals about 25,000. For several generations they have been increasingly dependent upon support from Denmark and very few native communities are now self supporting.

The climate of Greenland is strongly influenced by both the inland and the sea-borne ice. The field season in which geological investigations can be usefully pursued is limited to a few summer months.

In view of all these factors it is apparent why the geological exploration of this country has to be carried out by successive expeditions. Supplies and equipment must be brought to Greenland at the start of each campaign.

Prior to the nineteenth century nothing was known of the geology and mineralogy of Greenland. In the early part of that century mineral collectors and mineralogists first began to visit the country. From the very beginning the exploration of Greenland has often been controlled by peculiar circumstances. In 1806 the German mineralogist K. L. Giesecke visited Greenland. Owing to the war between England and Denmark, he had to remain there until 1813. During his seven years of enforced exile, he assembled an excellent collection of mineral specimens which still constitute a fundamental part of the collection in the Mineralogical Museum in Copenhagen. After Giesecke's visit, other expeditions followed at varying intervals throughout the nineteenth century.

By the end of that century, geological surveys were established in most European countries. At this time in Denmark the initiative was taken to start a Greenland Survey. A committee was formed and a leader appointed; however, the enterprise was unsuccessful and the activity of the new Survey petered out.

Several classical expeditions with varied scientific interests went to Greenland in the following years and, although geology was not the dominant interest, taking its place among surveying, botany, archaeology, etc., geological knowledge increased considerably.

In the decade prior to World War II, Dr. Lauge Koch directed several expeditions to East Greenland. His expeditions contributed in a conspicuous way to the elucidation of this area which is both remote and difficult of access. The end of this pre-war decade still saw a Greenland Survey as an unfulfilled dream, despite the exhortations of some prominent men. Its formation was imminent when war came in 1940. In 1946, however, the new Greenland Geological Survey sent its first expedition to Greenland. The men who took the initiative were two professors at the University of Copenhagen, Dr. Arne Noe-Nygaard (professor of mineralogy), and Mr. A. Rosenkrantz (professor of geology), and Dr. H. Ødum, director of the Geological Survey of Denmark. Since then, its work has continued without interruption.

Besides the regular Survey expeditions, several other expeditions have visited Greenland since the war. In the significant Danish East Greenland expeditions, Dr. Lauge Koch has continued the exploration work which he commenced in pre-war days.

It is interesting to note that when the Survey was formed in 1946 West Greenland was the least explored part of Greenland (from a geologist's viewpoint). It would seem that the less accessible regions have had a stronger attraction for Danish explorers and geologists. North and East Greenland had received more attention than West Greenland. This concentration on the more difficult terrain might be too easily explained as a result of romantic emotions; more realistically it can be attributed to Danish geological tradition and education. Danish geologists have always been more familiar with sedimentology, palaeontology, stratigraphy, and quaternary geology than with the geology of the Precambrian. This seems logical enough when it is remembered that Denmark is almost lacking in rocks of Precambrian age, in strong contrast to Greenland. These factors help to explain why the vast gneiss areas of West Greenland were still virtually untouched when geological surveying commenced in 1946.

The Survey has had a vigorous and increasing activity since its formation thirteen years ago. The Danish government has provided the necessary finances and the Mineralogical Museum of the University of Copenhagen has acted as host and guardian to the young Survey in a generous fashion. The geologists of the Mineralogical Museum and the Survey are housed in the same building and share various institutional facilities; for example, the library, laboratories, and map-drawing room. Laboratory work is divided between personnel of both organizations and problems in research receive attention from both teams. Besides the overlapping interests of office and laboratory, the Museum personnel have their lecturing commitments for the University and the Survey geologists their annual expedition to Greenland.

This team-work by the two organizations is due mainly to the pressure of circumstances; so far it has worked perfectly. In the future it will be necessary to conduct some of the enterprises more independently, but for the time being it is impossible to make any significant change, owing to the inadequate size of the building.

The Survey's expeditions to Greenland commence each year at the beginning of June and continue until the middle of September. An average of about 100 days are available for field work, but this number is usually reduced because of rain and bad weather. Attempts have been made to continue mapping while wintering on Greenland but have been given up as impracticable.

The expeditions are self-contained. Provisions, camping equipment, instruments, etc., are brought from Copenhagen. The Survey has its own vehicles, motor cruisers, and aircraft. When the Survey was first formed some fifteen geologists travelled to Greenland each year. Since then, activity has steadily increased and today a party of about seventy participants, of which twenty-five are geologists, goes annually to Greenland.

Each summer since the war, groups of geologists have been active in different localities, predominantly in West Greenland. One group has mapped Disko Island and the Nugssuaq peninsula in northwest Greenland and another has mapped between Disko Bay and Cape Farewell in southwest Greenland. Almost the whole of the Nugssuaq peninsula has been surveyed and mapped in detail. The Mesozoic and Tertiary sediments were carefully studied and a valuable collection of fossils taken home to Copenhagen. The Tertiary basalts of Disko Island were mapped at the same time. Geological maps of these areas will be published.

The second major area chosen for mapping was the ice-free part of West Greenland from Disko Bay to Cape Farewell. From 1946 to 1954 a systematic reconnaissance was carried out within the Precambrian orogenic belt between Disko Bay and Frederikshaabs Isblink. During these years it was impossible to cover the inland areas fully and most of the work was done in coastal areas.

It has been possible to set up a chronology based on degree of metamorphism and structural analysis. The results of the work done during these years have already been published and a preliminary map on a scale of 1:500.000 will be published in the near future.

After 1954 techniques and mapping methods were changed and investigation of a further area commenced. Inland areas as well as coastal areas have been subjected to more thorough investigation. Work, of course, progresses more slowly, but this has been partly counteracted by the employment of more geologists and the use of helicopters. Thus it has been possible to map an area extending from Ivigtut cryolite mine to the Ilimaussaq alkaline intrusion in South Greenland. In all, about 8,000 square kilometres have been mapped using field maps on a scale of 1:20.000. The final map will be published on a scale of 1:100.000 in the future. Some results from this area have already been given by Asger Berthelsen under the title: "On the Chronology of the Precambrian of Western Greenland."

In addition to general geological mapping, the Survey has undertaken several other tasks. Exploration for radioactive raw materials in Greenland has been delegated to the Survey by the Danish Atomic Energy Commission and the geological supervising of a tentative uranium-mining project has been conducted by GGU officials. Geophysical and geochemical prospecting have recently commenced.

From a geological viewpoint, the southermost tip of Greenland has received little attention as yet. As yet a topographical map is not available for the whole area. Despite this, some geological surveying will be carried on there in the next few years. The rocks are of Precambrian age and consist of metamorphosed sediments, migmatites and younger granites of various types.

An area of Precambrian agpatides occurs north of the Nugssuaq peninsula in North Greenland. Surveying of this district will commence in 1961, initially on a small scale. When the reconnaissance results are obtained it will be possible to plan future work in the region.

Jacobsen–McGill University Expedition Axel Heiberg Island 1959–61

GEORGE JACOBSEN

ABSTRACT

The Jacobsen–McGill Arctic Research Expedition to Axel Heiberg Island is the joint effort of McGill University and George Jacobsen. Its main objective is to gain knowledge of the Pleistocene and Recent physiographic evolution of central Axel Heiberg Island, including the geology and permafrost of the area.

The expedition will run for three years and will include surveying, geology, glaciology, hydro-glaciology, glacial meteorology, seismic and gravity surveys, geomorphology, permafrost, and botany. At its completion the base camp will be used as a permanent arctic research station for McGill University.

IN MARCH 1959, several scientists at McGill University who were interested in expanding arctic research formed a committee under the chairmanship of the author. At an early stage in their discussions, the group agreed that they should organize a major arctic research expedition which would have three main objectives: firstly, to develop integrated studies across the whole field of earth sciences to describe accurately the physical environment of a circumscribed area in the Canadian High Arctic; secondly, to show that although government direction of scientific research in Arctic Canada was necessarily and inevitably increasing, it was also most necessary and desirable for Canadian universities and private corporations and individuals to organize and direct major arctic research, as government departments, bound by their terms of reference, had less time for pure research and the training of post-graduates in arctic research; thirdly, to plan to make the base camp of the expedition into a permanent high arctic research station for McGill University, which would be open to scholars from all over Canada.

Axel Heiberg Island was chosen as the site of the expedition for several reasons. Its interior had been visited by the author in 1953, but remained scientifically unknown except for some recent geological investigations by R. Thorsteinsson and E. T. Tozer during "Operation Franklin" in 1955. The island is in many senses a link between the eastern group of northern islands (Ellesmere and Devon islands) and the western group (Ellef Ringnes, etc.). It supports the most westerly of the large northern ice-caps and reflects in the geology and geomorphology both eastern and western characteristics. It has the further advantage of lying between northern Ellesmere Island where there have been several scientific exploring parties in the last few years — including the Canadian IGY party at Lake Hazen — and Ellef Ringnes Island, the present headquarters of the Department of Mines and Technical Surveys' Polar Continental Shelf Project.

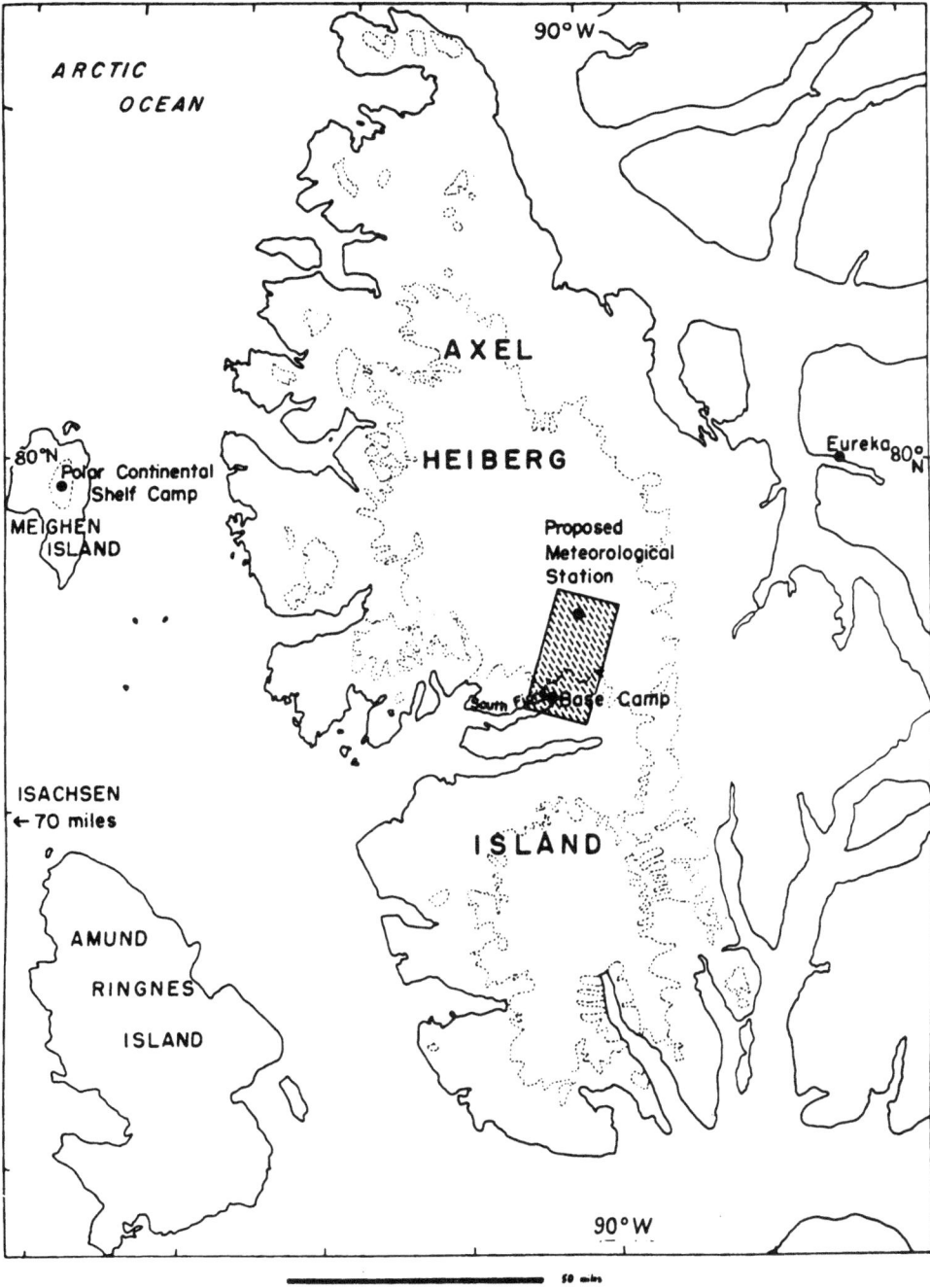

FIGURE 1.

Once the site of the expedition was decided, the committee went on to make three additional decisions by the end of April. They believed little would be accomplished scientifically in less than three years and the first year's field work would be largely of a reconnaissance nature and should start immediately. In the second and third years when large parties would be on the island, a permanent building would be required for a base camp; a building that would probably be left on the island at the end of 1962 to act as a continuing centre for research. Thirdly, Dr. Fritz Müller-Battle was offered the full-time position of field leader. Dr. Müller-Battle had been at McGill University and the Arctic Institute of North America from 1954 to 1956 and had already done considerable field work in Greenland and northwest Canada. In 1957 he had returned to Switzerland from a Himalayan expedition. He accepted the committee's offer and arrived in Canada at the end of June; he has since been appointed an associate professor at McGill University.

RECONNAISSANCE PARTY, 1959

A reconnaissance party, consisting of F. Müller-Battle, leader and glaciologist, G. Jacobsen, permafrost research, E. H. Kranck, geologist, and W. P. Adams, geographer, left Montreal for Axel Heiberg Island in mid-July. The party was flown to Eureka, Ellesmere Island, by the Royal Canadian Air Force and Nordair Ltd. From Eureka a chartered Piper Cub, equipped with large low-pressure balloon tires to enable it to land on unprepared terrain, ferried the scientists and their equipment to a preliminary camp on the west of Axel Heiberg Island. After many reconnaissance flights a permanent base was established at the head of South Fiord. From this base there is access by a major outlet glacier to the largest ice-cap of the island. There are several other types of glacier and also gypsum domes in the vicinity.

The principal achievements of the first year were the choice of the base site where a small airstrip could be built if necessary, and the definition of the main scientific objectives. Scientific work during reconnaissance included a theodolite survey to prepare a preliminary map, glacier and meteorological measurements, a geological examination of South Fiord and adjacent areas, and the choice of a site for the deep rock drilling in 1960. The party left the island at the end of August.

Meanwhile, in preparation for 1960 and a party of twenty to twenty-five members, two fibreglass houses including laboratory facilities, a plywood hut, aviation gasoline, food, and much of the equipment was shipped from Montreal by the Canadian Government ice-breaker "d'Iberville" to Eureka in July and is now in storage there.

MAIN OPERATION, 1960 AND 1961

Approximately half of the 1960 party left Montreal in April and flew to the base camp site via Eureka. The building was erected at the site and the food, equipment and gasoline ferried over from Eureka. Shortly afterwards the party was joined by two Piper Supercub aircraft for transportation throughout the summer. The remainder of the party arrived in May and June.

The main objective of the expedition in 1960 was to gain knowledge of the Pleistocene and Recent physiographic evolution of central Axel Heiberg Island,

FIGURE 2. Sketch map of expedition area.

including the geology and the permafrost of the area. Research was in the following eight categories.

Surveying. A general map, 1 to 50,000, of the expedition area was prepared based on vertical air photos taken by Spartan Air Services in 1959 from a height of 30,000 feet. Detailed maps of selected areas on a scale of 1 to 5,000 were made based on new aerial photography or terrestrial photogrammetry. The Axel Heiberg triangulation was connected with the Polar Continental Shelf survey by tellurometre. This programme was organized by Dr. Blachut of the National Research Council.

Geology. The tectonics of the area were examined and a study made of the gypsum domes. A rich collection of fossils from Upper Triassic to Upper Cretaceous was gathered.

Glaciology (including hydro-glaciology). A study of the mass balance of the three main types of glaciers in the area was conducted, based on studies of accumulation, ablation, and ice-movement through main profiles. Special studies were made of the ice-temperature régime.

Glacial meteorology. Determination of the heat balance of the glaciers was made by the establishment of a meteorological station in the accumulation area and another one in the ablation area.

Seismic and gravity surveys. Longitudinal and cross-profiles of the ice-cap and the main outlet glacier were made by geophysical methods.

Geomorphology. A study of the former extent and age of the glaciation, an analysis of the development of glacial landforms under "high arctic" conditions, and an examination of the distribution and age of the elevated strandlines were made.

Permafrost. During 1961 a reference hole to the base of permafrost will be drilled and instrumented with temperature measuring apparatus, to determine the depth of permafrost, the temperature profile, and the short- and long-term irregularities. The cores will be examined for stratigraphy and conductivity.

Botany. A study of lichens and pollen analysis of certain soils for age determination and palaeoclimatological study was made.

The most important fact of the research programme outlined above is that it was planned to achieve an integrated and comprehensive symposium of knowledge of a circumscribed area in the Canadian Arctic. Each scientist directed the main effort in his field towards results which would complement and interlock with the work of scientists in other disciplines. This co-ordinated team effort produced an organic whole of knowledge in earth sciences of central Axel Heiberg Island and should serve as guide for future arctic research projects. Detailed reports are being published by the McGill University Press.

The Arctic Institute

of North America

A. T. BELCHER

ABSTRACT

The Arctic Institute of North America is a non-profit research organization incorporated in Canada and the U.S.A. with offices in Montreal and Washington. It has pioneered basic Arctic studies in North America and has financed over 300 field studies in all subjects since 1946, and also provides such basic research tools as the *Arctic Bibliography* which cross-refers 50,000 English abstracts from papers in over twenty languages and every subject. In recent years it has also undertaken contract research on a cost and overhead basis, so that it played a significant role in both polar regions during the IGY.

Such preparations as the *Arctic Bibliography*, the Glacial Map of Canada, and the basic revision for Canada's new "Arctic pilot," are typical of the work to which the Institute is well suited, and there have also been compilations of subject and site bibliographies.

The Library of the Institute in Montreal stands well in the top four of the world's Arctic libraries; is the largest single collection in Canada; and serves as a mainstay of the Carnegie–McGill Arctic scholarships which are unique in the western world.

In 1960 the Institute financed nineteen private field parties; undertook research on the Ellesmere ice-shelf for the United States Air Force; and installed research stations on Devon Island, Northwest Territories, Canada. It hopes to raise funds in order to provide Arctic data for the new continuing project for charting North Atlantic marine biology.

THE ARCTIC INSTITUTE of North America is an international non-profit research organization which was incorporated in Canada and the United States in 1946, with financial support from the National Research Councils of Canada and the United States. In spite of an annual grant from the Canadian Government *via* the Department of Northern Affairs and the fact that the budget in recent years has been expanded by undertaking contracted research work on a cost plus overhead basis, its existence depends upon private donations. The Institute's interests soon spread to cover the Antarctic regions, but a number of movements towards a change in name were rejected because of the very real tradition behind it, involving the unique quality of the Institute, and the early start in North America polar sciences.

The Institute is concerned only with basic research. It does not pretend to be a technical survey or an engineering research laboratory but, as will be shown, its facilities and abilities have many currently practical applications, not the least being a large fund of original data, basic materials, and various research tools. Its Board of twenty-four Governors includes a great number of scientific leaders in Arctic studies from universities in both countries, and such eminent geologists as Dr. C. S. Lord, Chief Geologist of the Geological Survey of Canada, and Dr. J. C. Reed, at present Staff Coordinator of the United States Geological Survey, who became Executive Director of the Institute in April 1960. The programmes of this symposium have presented work by twelve Fellows of the Arctic Institute and four Governors.

The association with universities has always been very strong, beginning with the McGill group, whose advanced workers provided a great deal of the impetus to the Institute in its early days and with whom there is still a close collaboration. Other associations have grown with the universities of British Columbia, Toronto, Dartmouth, Harvard, Yale, California, and others, but we are far from realizing the full possibilities of a properly funded working co-operation with all the university potential of the continent. For example, there is a serious shortage of Arctic teaching staff, and an appalling shortage of funds for university students and research projects. In 1958 we could finance only 50 per cent of the desirable field projects, and this year we are $187,000 short of the money needed to finance those applications which have been given first-class rating by our Research Committee which is drawn from the continent's universities. There are many other projects and people that should also be helped at this time, to increase their potential value to Arctic workers, and future generations of workers need to be recruited and trained.

Our projects for 1960 show that there is a need for help. That year we installed research stations in Devon Island, one of the least known parts of the Canadian Archipelago. There is a coastal station with facilities for bathymetric, oceanographic, and geological studies, and a satellite ice-cap station helps us understand the relation between the sea and the land-ice. This programme is to continue over at least three years and is one of the most intensive programmes ever privately undertaken in the Canadian islands. Already we have guaranteed $40,000, but we must find another $100,000 if this station is to run at even minimum efficiency. After the initial installation, we plan to invite highly qualified workers from all sources to share these much-needed facilities with our own team. McGill University runs a similar station on Axel Heiberg, but with different objectives. This station is $50,000 short of funds and despite our unanimous and enthusiastic support of its plans, we could do no more than make a token grant of $1,000 to them in 1960. All this is in addition to the fact that we could not raise funds for about $37,000 for highly rated individual applications.

In addition there is a new and valuable project of world significance to which the Institute should contribute. A panel on North Atlantic Biogeography has been set up to prepare a series of charts aimed towards discovering the relation between biological productivity and distribution of marine animals and plants, and the various elements of the marine environment. These elements include bathymetry, bottom sediments, properties of water masses, courses of currents, seasonal changes in temperature, movement of water, direction and force of winds, air temperatures, and solar radiation. This vast enterprise will be produced over many years as a continuing atlas of constantly revised material. The Arctic Institute of North America possesses most of the known material to prepare the Arctic sections of this series, but has no funds for personnel and chart drafting. Until such funds become available this knowledge cannot be presented to a scientific audience.

Among the research tools provided by the Institute, first mention must go to the *Arctic Bibliography*. We have now produced eight volumes of this basic work, providing cross-indexed abstracts from 50,000 works on every Arctic subject without exception, translated, moreover, from seventeen languages. Volume 9 is now being prepared and will take us up to 54,000 abstracts, all cross-referenced and indexed by author. Each volume takes about one year to produce and costs over $120,000.

Next on the list I must rate our polar library. As an Arctic library it rates in the world's top four, of which the Russian Institute's collection in Leningrad is, of course, the outstanding first. The Arctic Institute's collection of material is intensively used by McGill and the Institute itself, but is accessible throughout the continent *via* the Interlibrary Loan system. Because of the Carnegie-McGill Arctic scholarship system this material is used by a larger number of students working for senior degrees than is any other polar collection. Accession lists are sent around the world and are available regularly to anyone simply by a request to the librarian.

The Institute has directly financed over 300 field parties since 1948 and the many manuscript reports are held in our offices, some as yet unpublished. In addition we have given advice and logistics aid and loaned equipment to many other parties. We have also undertaken contract research on a cost and overhead basis for private companies and government. Of course a condition of such contracts is that we achieve the required objectives in our own way, and that results eventually become available to the general public. We maintained an International Geophysical Year station in the Brooks Range of Alaska and participated in Antarctic IGY programmes, including biological studies and seismic traverses of the ice-cap. Studies of the Ellesmere Island ice shelf have been started for the Geophysics Research Directorate of the United States Air Force, in full collaboration with the Defence Research Board of Canada, and we shall be in the field again this spring with the promise of significant results this year.

Although we are little concerned with development, our uniquely long history of basic research in North America's Arctic has seen us at the outset of many significant developments. The first winter oceanographic station in North America was made by a young Institute grantee, and we were ten years and more ahead of other bodies in initiating and promoting oceanographic research. One of our first grantees investigated the possibilities of predetermining ground conditions for engineering purposes by the use of aerial photographs. This process is now so commonly used that it is hard to realize that the pioneer days were so recent. The Institute is still outstanding in this field, as the production of the Glacial Map of Canada in 1958 has shown, and aerial photographic interpretation was also a tool in our basic preparation for the new "Arctic pilot" which we undertook for the Canadian Hydrographic Office in 1956-7.

We suffer many handicaps. At the outset we were regarded as bearded explorers in a Boy Scout phase and of late we have been regarded as long-haired university dreamers of no practical purpose. This is far from the truth. In 1959 our budget came very close to the million dollar mark and we were actively associated with more practical research ventures than at any other time in our history. The burden on our limited funds has increased significantly and there is an awakening to the many urgent needs in basic tools, training, and facilities. The fact is that nowhere in the continent can Arctic environmental data be assembled more rapidly than at the Institute, and nowhere at such low cost, the total real cost being lower than that of any other organization. Academically and administratively the Arctic Institute has achieved an astonishing efficiency. Information about an area, person, or subject can be assembled rapidly in any required detail. Information as to what is *not* known is equally available and is certainly just as important. Bibliographies, agency listings like "Institutions of the USSR active in Arctic Research and Development," and personal contacts can all be readily arranged by the present organization. Not all the information is our own, of course, but our coverage is unique in the western world.

Not all the information is our own simply because we do not, and never will, provide competition for companies, surveys, and government agencies. Our aim is to fill the gaps; to encourage co-operation and mutual knowledge; to prevent random duplication of effort; to publish materials that have no other forum; to provide the tools that no other group has offered; and, above all, to provide adequate education for future generations. We depend on public support because it gives us independence; we are international because we believe science knows no boundaries; we are interdisciplinary because we believe that scientific disciplines also have no clear boundaries. We have a very clear vision of our role in helping provide opportunities and co-operation between all men regardless of race, employment, or training, in order to understand and use one of the world's last great unknown areas.